Advances in Poultry Science

Advances in Poultry Science

Edited by Benjamin MacClare

SYRAWOOD
PUBLISHING HOUSE

New York

Published by Syrawood Publishing House,
750 Third Avenue, 9th Floor,
New York, NY 10017, USA
www.syrawoodpublishinghouse.com

Advances in Poultry Science
Edited by Benjamin MacClare

International Standard Book Number: 978-1-68286-555-2 (Hardback)

Cataloging-in-Publication Data

Advances in poultry science / edited by Benjamin MacClare.
    p. cm.
Includes bibliographical references and index.
ISBN 978-1-68286-555-2
1. Poultry. 2. Animal culture. 3. Poultry industry. I. MacClare, Benjamin.
SF487 .A38 2018
636.5--dc23

TABLE OF CONTENTS

Preface.. IX

Chapter 1   **Reference Genes for Quantitative Gene Expression Studies in Multiple
             Avian Species**...1
             Philipp Olias, Iris Adam, Anne Meyer, Constance Scharff, Achim D. Gruber

Chapter 2   **Sulfonamide-Resistant Bacteria and their Resistance Genes in Soils Fertilized
             with Manures from Jiangsu Province, Southeastern China**................................13
             Na Wang, Xiaohong Yang, Shaojun Jiao, Jun Zhang, Boping Ye, Shixiang Gao

Chapter 3   **Selectively Adsorptive Extraction of Phenylarsonic Acids in Chicken Tissue
             by Carboxymethyl α-Cyclodextrin Immobilized Fe₃O₄ Magnetic Nanoparticles
             Followed Ultra Performance Liquid Chromatography Coupled Tandem Mass
             Spectrometry Detetion**...24
             Jing Jia, Wei Zhang, Jing Wang, Peilong Wang, Ruohua Zhu

Chapter 4   **A Draft *De Novo* Genome Assembly for the Northern Bobwhite
             (*Colinus virginianus*) Reveals Evidence for a Rapid Decline in Effective
             Population Size Beginning in the Late Pleistocene**...34
             Yvette A. Halley, Scot E. Dowd, Jared E. Decker, Paul M. Seabury, Eric Bhattarai,
             Charles D. Johnson, Dale Rollins, Ian R. Tizard, Donald J. Brightsmith,
             Markus J. Peterson, Jeremy F. Taylor, Christopher M. Seabury

Chapter 5   **Parents and Early Life Environment Affect Behavioral Development of Laying
             Hen Chickens**..52
             Elske N. de Haas, J. Elizabeth Bolhuis, Bas Kemp, Ton G. G. Groothuis,
             T. Bas Rodenburg

Chapter 6   **Activation of Duck RIG-I by TRIM25 is Independent of Anchored Ubiquitin**...........64
             Domingo Miranzo-Navarro, Katharine E. Magor

Chapter 7   **A Mixture of Cod and Scallop Protein Reduces Adiposity and Improves
             Glucose Tolerance in High-Fat Fed Male C57BL/6J Mice**................................76
             Hanne Sørup Tastesen, Alexander Krokedal Rønnevik, Kamil Borkowski,
             Lise Madsen, Karsten Kristiansen, Bjørn Liaset

Chapter 8   **Searching for Novel Cdk5 Substrates in Brain by Comparative
             Phosphoproteomics of Wild Type and Cdk5$^{-/-}$ Mice**................................88
             Erick Contreras-Vallejos, Elías Utreras, Daniel A. Bórquez,
             Michaela Prochazkova, Anita Terse, Howard Jaffe, Andrea Toledo,
             Cristina Arruti, Harish C. Pant, Ashok B. Kulkarni, Christian González-Billault

Chapter 9   **Multiple Exposure and Effects Assessment of Heavy Metals in the Population
             near Mining Area in South China**...101
             Ping Zhuang, Huanping Lu, Zhian Li, Bi Zou, Murray B. McBride

Chapter 10   **Getting More Out of Less – A Quantitative Serological Screening Tool for Simultaneous Detection of Multiple Influenza A Hemagglutinin-Types in Chickens**..**112**
Gudrun S. Freidl, Erwin de Bruin, Janko van Beek, Johan Reimerink,
Sjaak de Wit, Guus Koch, Lonneke Vervelde, Henk-Jan van den Ham,
Marion P. G. Koopmans

Chapter 11   **Experimentally Infected Domestic Ducks Show Efficient Transmission of Indonesian H5N1 Highly Pathogenic Avian Influenza Virus, but Lack Persistent Viral Shedding**..**122**
Hendra Wibawa, John Bingham, Harimurti Nuradji, Sue Lowther, Jean Payne,
Jenni Harper, Akhmad Junaidi, Deborah Middleton, Joanne Meers

Chapter 12   **A 20-Amino-Acid Deletion in the Neuraminidase Stalk and a Five-Amino-Acid Deletion in the NS1 Protein both Contribute to the Pathogenicity of H5N1 Avian Influenza Viruses in Mallard Ducks**..............................**133**
Yanfang Li, Sujuan Chen, Xiaojian Zhang, Qiang Fu, Zhiye Zhang, Shaohua Shi,
Yinbiao Zhu, Min Gu, Daxin Peng, Xiufan Liu

Chapter 13   **A Comparison of Next Generation Sequencing Technologies for Transcriptome Assembly and Utility for RNA-Seq in a Non-Model Bird**..................**145**
Findley R. Finseth, Richard G. Harrison

Chapter 14   **Origin and Loss of Nested LRRTM/α-Catenin Genes during Vertebrate Evolution**..**156**
Pavel Uvarov, Tommi Kajander, Matti S. Airaksinen

Chapter 15   **High-Resolution Melting-Curve Analysis of *obg* Gene to Differentiate the Temperature-Sensitive *Mycoplasma synoviae* Vaccine Strain MS-H from Non-Temperature-Sensitive Strains**..**167**
Muhammad A. Shahid, Philip F. Markham, Marc S. Marenda,
Rebecca Agnew-Crumpton, Amir H. Noormohammadi

Chapter 16   **Thermal Manipulation during Embryogenesis has Long-Term Effects on Muscle and Liver Metabolism in Fast-Growing Chickens**..**178**
Thomas Loyau, Sonia Métayer-Coustard, Cécile Berri, Sabine Crochet,
Estelle Cailleau-Audouin, Mélanie Sannier, Pascal Chartrin, Christophe Praud,
Christelle Hennequet-Antier, Nicole Rideau, Nathalie Couroussé,
Sandrine Mignon-Grasteau, Nadia Everaert, Michel Jacques Duclos,
Shlomo Yahav, Sophie Tesseraud, Anne Collin

Chapter 17   **A New Method to Monitor the Contribution of Fast Food Restaurants to the Diets of US Children**..**191**
Colin D. Rehm, Adam Drewnowski

Chapter 18   **Ecological, Social and Biological Risk Factors for Continued *Trypanosoma cruzi* Transmission by *Triatoma dimidiata* in Guatemala**........................**201**
Dulce M. Bustamante, Sandra M. De Urioste-Stone, José G. Juárez,
Pamela M. Pennington

Chapter 19   **Identification of Genes Related to Beak Deformity of Chickens using Digital Gene Expression Profiling**..**213**
Hao Bai, Jing Zhu, Yanyan Sun, Ranran Liu, Nian Liu, Dongli Li, Jie Wen, Jilan Chen

Permissions

List of Contributors

Index

PREFACE

Poultry is an important agricultural product in the world and is also involved in basic biological research on genetics biochemistry and diseases. It includes raising of various types of domestic birds for the production of meat and eggs. This book discusses theories and concepts related to poultry science. It is a valuable compilation of topics, ranging from the basic to the most complex advancements in this field. For someone with an interest and eye for detail, this book covers the most significant topics in the field of poultry science.

The researches compiled throughout the book are authentic and of high quality, combining several disciplines and from very diverse regions from around the world. Drawing on the contributions of many researchers from diverse countries, the book's objective is to provide the readers with the latest achievements in the area of research. This book will surely be a source of knowledge to all interested and researching the field.

In the end, I would like to express my deep sense of gratitude to all the authors for meeting the set deadlines in completing and submitting their research chapters. I would also like to thank the publisher for the support offered to us throughout the course of the book. Finally, I extend my sincere thanks to my family for being a constant source of inspiration and encouragement.

Editor

Reference Genes for Quantitative Gene Expression Studies in Multiple Avian Species

Philipp Olias[1][⑨]*, Iris Adam[2][⑨]*, Anne Meyer[1], Constance Scharff[2], Achim D. Gruber[1]

1 Institute of Veterinary Pathology, Freie Universität Berlin, Berlin, Germany, 2 Institute of Biology, Department of Animal Behavior, Freie Universität Berlin, Berlin, Germany

Abstract

Quantitative real-time PCR (qPCR) rapidly and reliably quantifies gene expression levels across different experimental conditions. Selection of suitable reference genes is essential for meaningful normalization and thus correct interpretation of data. In recent years, an increasing number of avian species other than the chicken has been investigated molecularly, highlighting the need for an experimentally validated pan-avian primer set for reference genes. Here we report testing a set for 14 candidate reference genes (*18S, ABL, GAPDH, GUSB, HMBS, HPRT, PGK1, RPL13, RPL19, RPS7, SDHA, TFRC, VIM, YWHAZ*) on different tissues of the mallard (*Anas platyrhynchos*), domestic chicken (*Gallus gallus domesticus*), common crane (*Grus grus*), white-tailed eagle (*Haliaeetus albicilla*), domestic turkey (*Meleagris gallopavo* f. *domestica*), cockatiel (*Nymphicus hollandicus*), Humboldt penguin (*Sphenicus humboldti*), ostrich (*Struthio camelus*) and zebra finch (*Taeniopygia guttata*), spanning a broad range of the phylogenetic tree of birds. Primer pairs for six to 11 genes were successfully established for each of the nine species. As a proof of principle, we analyzed expression levels of 10 candidate reference genes as well as *FOXP2* and the immediate early genes, *EGR1* and *CFOS*, known to be rapidly induced by singing in the avian basal ganglia. We extracted RNA from microbiopsies of the striatal song nucleus Area X of adult male zebra finches after they had sang or remained silent. Using three different statistical algorithms, we identified five genes (*18S, PGK1, RPS7, TFRC, YWHAZ*) that were stably expressed within each group and also between the singing and silent conditions, establishing them as suitable reference genes. In conclusion, the newly developed pan-avian primer set allows accurate normalization and quantification of gene expression levels in multiple avian species.

Editor: Johan J. Bolhuis, Utrecht University, Netherlands

Funding: The authors have no support or funding to report.

Competing Interests: The authors have declared that no competing interests exist.

* E-mail: olias@borcim.wustl.edu (PO); iris.adam@fu-berlin.de (IA)

⑨ These authors contributed equally to this work.

Introduction

The zebra finch (*Taeniopygia guttata*) and the domestic chicken (*Gallus gallus* f. *dom.*) have become widely used model organisms for biologists studying neurobiology and behavior, ecology as well as diseases and their transmission [1–5]. With both genomes sequenced and technologies adapted from traditional genetic model systems, the finch and the chicken are becoming increasingly amenable to genetic investigation [6–9]. The more than 10,000 avian species are known for their extraordinary differences in behavior and physiology. This offers opportunities to find a suitable species to address particular questions about the genetic background of migration, mating systems, parental care, flight, niche partition, vocal behavior or cognition, to name just a few.

Quantitative real-time polymerase chain reaction (qPCR) is currently the most economic, efficient and reliable method to measure gene expression levels in low to medium throughput approaches. It is sensitive, specific and reproducible even with limited mRNA copy numbers [10]. As in any quantitative study, it is necessary to correct for sample to sample variations in order to obtain reliable results. The most commonly used method is normalization to the expression of an internal control gene, also called reference gene. Hypothetically, the ideal reference gene is expressed at stable levels irrespective of tissue type, species, treatment, metabolism or sampling conditions. To date, no such ideal gene has been found and most likely does not exist, thus it is becoming clear that reference genes need to be established for each new experimental design, controlling for the numerous variables which might bias the results, i.e. number of cells and their transcriptional activity, RNA quality and reverse transcriptase efficiency [11]. In recent avian gene expression experiments, genes such as *18S* (18S ribosomal RNA), *ACTB* (beta actin) and *GAPDH* (glyceraldehyde 3-phosphate dehydrogenase) have been used most frequently [12–17] but several studies have shown that using single or inappropriate reference genes for normalization may dramatically bias the results of mRNA copy number quantification [10,18–20].

For mammals, multiple sets of reference genes have been published [21–25] but are of limited use for non-mammalian vertebrates including birds due to the phylogenetic distance. Certainly, the advent of Next Generation Sequencing and the now fully sequenced genomes of the domestic chicken, domestic turkey (*Meleagris gallopavo* f. *domestica*), zebra finch, duck (*Anas platyrhynchos*) and the collared flycatcher (*Ficedula albicollis*) has made it easier and faster to conduct gene expression studies in these species [7,9,26–28]. However, currently more than 10,000 avian species are known, more and more of which are also investigated at the

Table 1. Gene names and primer pairs tested on all nine avian species.

Gene symbol	GenBank accession number *G. gallus*	Primer sequences 5'-3' (forward/reverse)	Amplicon length range (bp)	Size on genomic level chicken (bp)
18S	AF173612	CGAAAGCATTTGCCAAGAAT	98–99	98
		GGCATCGTTTATGGTCGG		
ABL	XM_001233811	GCTGCTCGCTGGAACTCC	218	940
		GTGATGTAATTGCTGGGGACC		
GAPDH	NM_204305	GTGGTGCTAAGCGTGTTATCATC	269–270	534
		GGCAGCACCTCTGCCATC		
GUSB	NM_001039316	GGCAAACTCCTTCCGCAC	222–224	858
		TCATTGGCTACTGACCACATCA		
HMBS	XM_417846	CTGAAGAGAATGGGCTGGGA	113–115	1791
		TCTTGGTCTTTGGCACGAAC		
HPRT	NM_204848	GATGAACAAGGTTACGACCTGGA	181	1575
		TATAGCCACCCTTGAGTACACAGAG		
PGK1	NM_204985	AAAGTTCAGGATAAGATCCAGCTG	167	450
		GCCATCAGGTCCTTGACAAT		
RPL13	NM_204999	CCACAAGGACTGGCAGCG	135	434
		ACGATGGGCCGGATGG		
RPL19	NM_001030929	CCAACGAGACCAACGAGATC	152–153	629
		CATGTGCCGGCCCTTCC		
RPS7	XM_419936	TAGGTGGTGGCAGGAAAGC	156	1773
		TTGGCTTGGGCAGAATCC		
SDHA	XM_419054	TTGGTGGACAGAGTCTTCAGTT	238	1821
		GTGTTCTTTGCTCTAAAACGATG		
TFRC	NM_205256	GGAACTTGCCCGTGTGATC	111–113	723
		GTAGCACCCACAGCTCCGT		
VIM	NM_001048076	GGAACAATGATGCCCTGC	145	761
		GCAAAATTCTCCTCCATTTCAC		
YWHAZ	NM_001031343	GTGGAGCAATCACAACAGGC	222–224	326
		GCGTGCGTCTTTGTATGACTC		

molecular level [29–32]. The tools available to compare those less genetically amenable bird species are highly limited, one problem being the lack of sequence information to design PCR-primers. To date, avian reference genes have been published only for the domestic chicken [33,34], great tit (*Parus major*) [32], Japanese quail (*Coturnix c. japonica*) [35] and domestic pigeon (*Columba livia* f. *domestica*) [36].

This prompted us to develop a set of pan-avian PCR primers for the amplification of reference genes that can be tested for their suitability to normalize gene expression levels in avian studies involving qPCR analyses. We designed primer pairs for the amplification of 14 candidate reference genes classically used in mammalian research, including *18S*, *ABL* (Abelson murine leukemia viral oncogene homolog), *GAPDH*, *GUSB* (beta glucuronidase), *HMBS* (hydroxymethyl-bilane synthase), *HPRT* (hypoxanthine-guanine phosphoribosyl-transferase 1), *PGK1* (phosphoglycerate kinase 1), *RPL13* (60S ribosomal protein L13), *RPL19* (60S ribosomal protein L19), *RPS7* (40S ribosomal protein S7), *SDHA* (succinate dehydrogenase complex, subunit A), *TFRC* (transferrin receptor protein 1), *VIM* (vimentin) and *YWHAZ* (tyrosine 3-monooxygenase/tryptophan 5-monooxygenase activation protein, zeta polypeptide). For each of the nine avian species

investigated in this study we successfully established between six and 11 primer pairs.

To demonstrate the suitability of these genes for normalization purposes, we conducted an experiment investigating singing-induced gene expression changes in a striatal song control region, Area X of zebra finches. We measured the expression of three genes of interest, among them two immediate early genes known to be upregulated by singing [37–39], as well as 10 potential reference genes, on samples of birds that sang undirected song (not for a female conspecific) or remained silent for the same amount of time. Using the statistical algorithms geNorm [11], NormFinder [40] and BestKeeper [41] we determined the most stably expressed genes and employed them to normalize the expression of our genes of interest.

Material and Methods

Animal experiments were carried out in accordance with the guidelines provided and approved by the governmental institutions (LAGeSo, Berlin, Permit Number: T 0298/01). All zebra finch samples originated from a captive breeding colony approved by local authorities (Permit Number: ZH147). The brain of the

Table 2. Summary of primer test and efficiency for each primer pair and species.

Species	18S	ABL	GAPDH	GUSB	HMBS	HPRT	PGK1	RPL13	RPL19	RPS7	SDHA	TRFC	VIM	YWHAZ	No. of established primer pairs
Zebra finch	102.5	93.9	70.8	100.0	96.3	93.5	94.0	nd	nd	100.9	103.3	103.6	95.8	96.6	11
Cockatiel	nd	89.9	nd	91.3	nd	84.4	96.8	nt	nt	92.8	95.8	92.3	95.5	84.6	6
White tailed eagle	nd	94.0	nd	96.4	94.4	89.2	94.2	nt	nt	93.0	97.9	100.4	98.1	93.0	9
Humboldt penguin	nd	94.3	nd	93.0	84.7	87.5	93.6	nt	nt	95.9	94.5	97.7	98.8	94.8	8
Common crane	nd	nw	nd	90.8	91.0	89.0	94.2	nt	nt	94.8	86.1	94.6	98.8	95.8	7
Chicken	107.6	93.3	72.7	90.3	94.3	81.2	88.3	nd	nd	89.8	96.2	87.8	100.8	99.7	7
Turkey	nd	92.4	nd	96.0	98.3	88.0	95.9	nt	nt	91.3	96.8	99.4	101.0	98.8	9
Mallard	nd	nw	nd	83.4	nw	86.7	94.4	nt	nt	97.9	91.1	95.7	99.5	94.7	6
Ostrich	nd	93.1	nd	99.3	96.9	85.8	96.3	nd	nd	85.3	98.5	85.9	101.8	94.9	7

Numbers refer to the efficiency determined (bold: efficiency ok); nt; primer not tested; nw, primer tested but not working; nd: efficiency not tested.

ostrich was obtained from a slaughterhouse (Winkler, Neuloewenburg, Germany, registration number: DE-BB65011EG). All other birds had been euthanized for animal welfare reasons for causes unrelated to the present study in strict accordance with the German National Animal Protection law (Tierschutzgesetz in der Fassung der Bekanntmachung vom 18. Mai 2006 (BGBl. I S. 1206, ber. S. 1313), last amendment: Artikel 20 G vom 9. Dezember 2010 (BGBl. I S. 1934, 1940 f.)).

Sample collection

Three sets of tissues were collected. (1) We collected samples from nine phylogenetically distant avian species: The brains from a mallard (*Anas platyrhynchos*), a domestic chicken (*Gallus gallus domesticus*), a white-tailed eagle (*Haliaeetus albicilla*), a domestic turkey (*Meleagris gallopavo* f. *domestica*), a cockatiel (*Nymphicus hollandicus*) and an ostrich (*Struthio camelus*); from a common crane (*Grus grus*) we took blood and from a Humboldt penguin (*Sphenicus humboldti*) the lung with fungal pneumonia. (2) From adult zebra finches we collected brains and gonads (10 females, 10 males, below called 'tissue dataset'). (3) We collected microbiopsies of AreaX from another 12 adult zebra finches (below called 'song data set'). All samples were immediately snap frozen after dissection and stored at −80°C until further use.

Song data set. Adult male zebra finches were housed in free flight aviaries in our breeding colony and kept alone in sound attenuated chambers overnight. After lights were switched on in the morning at 8 a.m., song was monitored and recorded. Birds that did not sing during the first 30 min were sacrificed ('silent group'). Birds that started singing (while being alone, called 'undirected song') within the first two hours were sacrificed 30 min after the onset of song ('song group'). Birds were sacrificed by an isoflurane overdose and brains were immediately dissected, cut into hemispheres, embedded in Tissue-Tek O.C.T. compound (Sakura Finetek) and stored at −80°C. To obtain microbiopsies of Area X, hemispheres were mounted on a cryostat (Cryo-Star HM 560 Cryostat, MICROM) and 20 µm sagittal sections were cut from medial to lateral until Area X became visible. Then a 1 mm diameter coring tool (Harris Unicore) was used to mark Area X while still on the block. Subsequently, a 200 µm section was cut and Area X microbiopsies were taken with the coring tool. The frozen biopsy was immediately transferred to dry ice, while the remaining section was transferred into a 4% paraformaldehyde/ 0.1 M PBS solution for microscopic inspection of targeting. The procedure was repeated until Area X was no longer visible. Microbiopsies were stored at -80°C until further use.

Selection of candidate reference genes and primer design

Candidate reference genes were selected from previous reports for mammalian species [21–25] (Table 1). All primers except *RPL13* [36] were designed *de novo* on homologous gene segments of the chicken and zebra finch derived from GenBank (http://www.ncbi.nlm.nih.gov) and Ensembl (http://www.ensembl.org/index.html) databases using NetPrimer (http://www.premierbiosoft.com/netprimer/netprimer.html). Primers were designed to reside in the open reading frame (ORF) of genes to grant maximal conservation across all birds. ORFs were identified by Biowire Jellyfish 1.5. Criteria for primer design were a predicted melting temperature of 58°C, primer length of 15–25 nucleotides, a guanine-cytosine content of 40–70% and amplicon lengths smaller than 250 base pairs [10,42]. All primer pairs except those for 18S were designed to span exon-exon boundaries as identified by MEGA4 [43].

Table 3. Sequence similarity of all amplified products to the chicken (ICGSC Gallus gallus 4.0) and zebra finch (WUGSC 3.2.4/taeGut1) genomes.

Gene symbol	Sequence identity (%) to *G. gallus*/to *T. guttata*								
	A. platyrhynchos	*G. gallus*	*G. grus*	*H. albicilla*	*M. gallopavo*	*N. hollandicus*	*S. humboldti*	*S. camelus*	*T. guttata*
18S	100/100	100/100	100/100	98/98	100/100	98/98	97/97	100/100	100/100
ABL	nw	100/89	nw	95/92	98/88	97/93	97/93	94/89	89/100
GAPDH	95/94	100/92	93/95	94/95	98/93	93/96	93/94	94/92	92/100
GUSB	91/89	97/86	88/89	91/89	96/87	90/89	90/90	87/90	87/100
HMBS	93/89	100/89	88/93	92/96	97/98	92/90	88/92	88/92	89/100
HPRT	95/98	100/94	96/98	95/99	97/95	95/98	95/99	95/99	94/100
PGK1	92/95	100/92	93/96	93/97	98/92	93/97	93/95	93/96	92/100
*RPL13**	nd	100/85	nd	nd	nd	nd	nd	86/-	90/95
RPL19	nd	100/86	nd	nd	nd	nd	nd	89/84	85/97
RPS7	98/93	100/93	96/93	97/95	100/93	97/93	97/92	97/92	93/100
SDHA	93/89	100/90	94/93	95/93	98/91	92/91	93/91	90/90	90/100
TFRC	99/87	100/86	86/95	87/90	99/87	89/86	90/93	94/89	86/100
VIM	97/90	100/88	93/90	96/91	98/90	95/92	96/90	93/90	88/100
YWHAZ	97/95	100/96	96/95	96/97	99/96	97/97	97/97	97/96	96/99

*[36]; nd, not done; nw, not working.

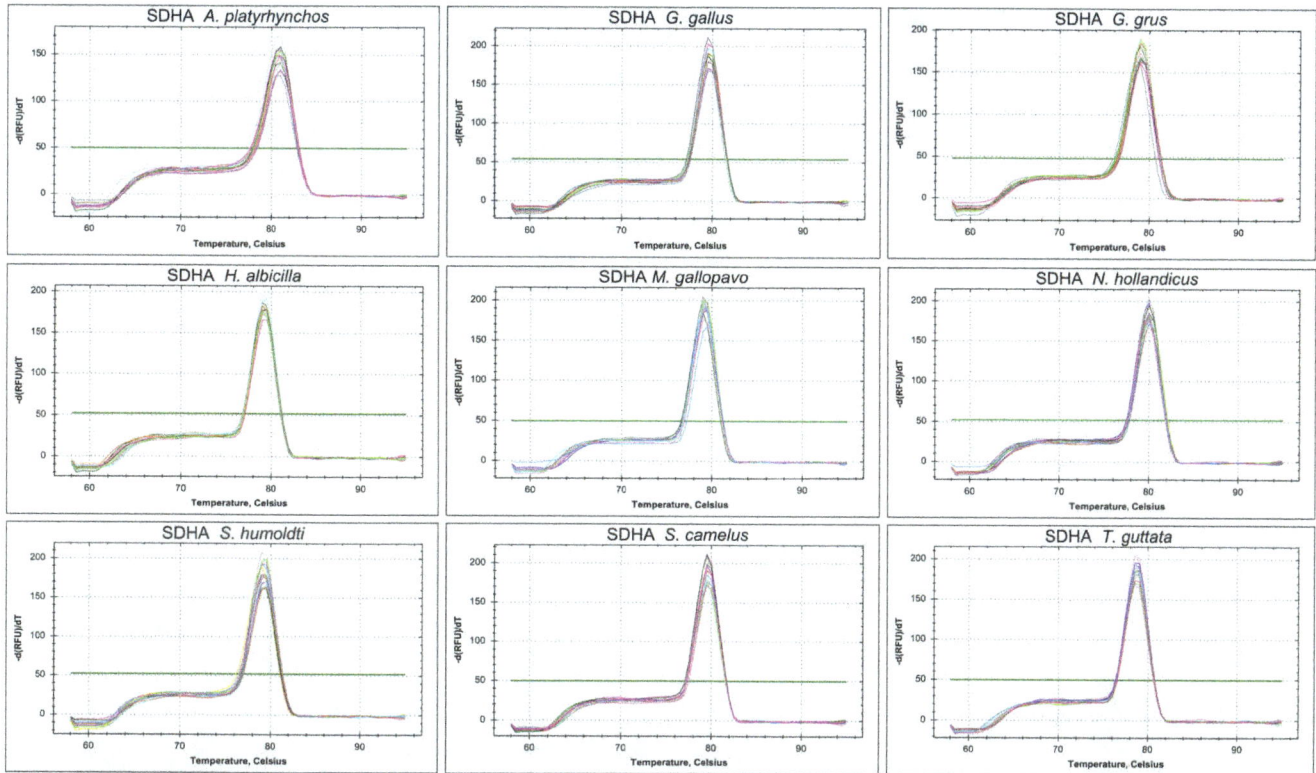

Figure 1. Specificity of primers for *SDHA* in qRT-PCR of nine avian species. Melting curves of five dilutions run in triplicates are shown with one panel per species. Species names are indicated in the title of each panel.

RNA extraction, cDNA synthesis and qPCR

All tissues except for Area X tissue samples ('song data set'). RNA was extracted and purified from approximately 100 mg of each sample using Trizol (Invitrogen, Carlsbad, CA) or the NucleoSpin RNA II kit (Macherey-Nagel, Düren, Germany) following the manufacturer's recommendations. Concentration and purity of total RNA were determined at 260/280 nm absorbance ratio to be above 1.8 in all cases (NanoDrop 1000 Spectrophotometer, Thermo Fisher Scientific, Wilmington, DE). RNA integrity numbers (RIN) of all tissue samples were measured using the RNA Nano kit on an Agilent 2100 Bioanalyzer (RNA 6000 Nano Kit, Agilent, Santa Clara, CA) and used only if above 8.0. For cDNA synthesis, 100 ng RNA was reverse transcribed at 25°C for 5 min followed by 30 min at 42°C in 20 µl containing 200 U iScript (Bio-Rad, Hercules, CA), 100 nmol MgCl₂, 50 ng random hexamers, 0.2 mmol DTT, 40 U RNase Out (Promega, Madison, USA) and 10 nmol of each dNTP. The initial tests for successful product amplification were run on the MX 3000P or the CFX96 qPCR detection system using the MX Pro 3.0 (Agilent) or CFX Manager 2.0 (Bio-Rad) software packages, respectively. All subsequent PCR experiments were performed on the CFX96 system. PCR reactions were carried out in 96-well polypropylene plates (Sarstedt, Nümbrecht, Germany) in a volume of 20 µl. 5 µl cDNA was added to 15 µl reaction mix containing 10 µl Brilliant SYBR Green QPCR Master Mix (Applied Biosystems, Carlsbad, CA) and 4.5 pmol of each primer. Cycling conditions were as follows: 10 min at 95°C followed by 40 cycles of 30 sec at 95°C, 1 min at 58°C, and 30 sec at 72°C and a subsequent melting curve analysis. All reactions were run in triplicate and the mean C_t-values were used for further analysis. If one of the three C_t-values deviated from the other two by more than 0.5 it was

excluded from further analysis or the experiment was repeated. Nuclease-free water was used for the no-template controls. PCR efficiencies (E) were calculated using the equation $E = 10^{-1/slope}$ by measuring a ten-fold dilution series over five orders of magnitude for each primer pair and for each avian species as indicated in Table 2. Specificity of the primers was confirmed by sequencing and sequence identity was evaluated with BLASTN [44] against coding sequences of the chicken and zebra finch derived from GenBank (Table 3).

Song data set. After microscopic verification of proper targeting, we extracted the RNA from the Area X microbiopsies of each animal individually, resulting in 12 separate RNA extractions. RNA was extracted using the NucleoSpin RNA XS kit (Macherey-Nagel) and quantified using the Qubit RNA assay (life technologies). For cDNA synthesis 70 ng RNA was reverse transcribed in 20 µL using 200 ng random hexamer primers, 10 nmol of each dNTP, 50 mM DTT, 200 U SuperScript III (Invitrogen) and 40 U RNasinPlus (Promega). The temperature program was chosen as follows: 5 min at 65°C, cool down on ice, 5 min 25°C, 45 min at 50°C, 15 min at 72°C. We included reverse transcriptase-free reactions to control for DNA contaminations. All cDNAs were diluted 10-fold with nuclease free water and 5 µL were used in each PCR reaction.

QPCR reactions were carried out in 96-well polypropylene plates (Sarstedt, Nümbrecht, Germany) in a volume of 20 µl and in triplicates. 5 µl cDNA was added to 15 µl reaction mix containing 10 µl KAPA SYBR FAST Universal QPCR mix (Peqlab) and 10 pmol (18 pmol in the case of FoxP2) of each primer. PCRs were run on a MX3005P system (Agilent) with the following cycling conditions: 10 min at 95°C followed by 40 cycles of 30 sec at 95°C, 30 sec at 65°C (64°C for FoxP2) and a

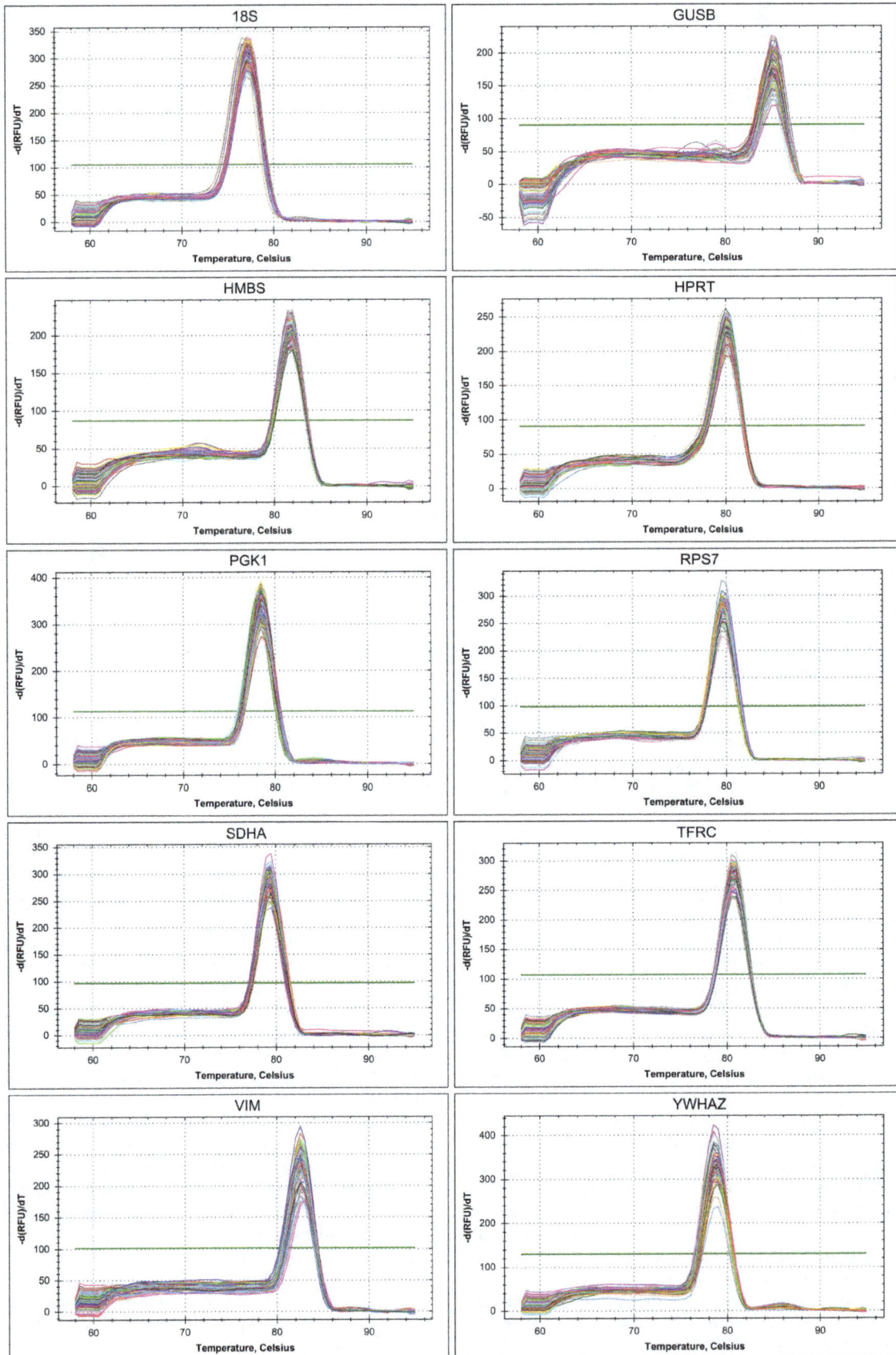

Figure 2. Specificity of 10 reference genes of the zebra finch with single peaks in melting curves. All amplifications were run in triplicate on 32 tissue samples.

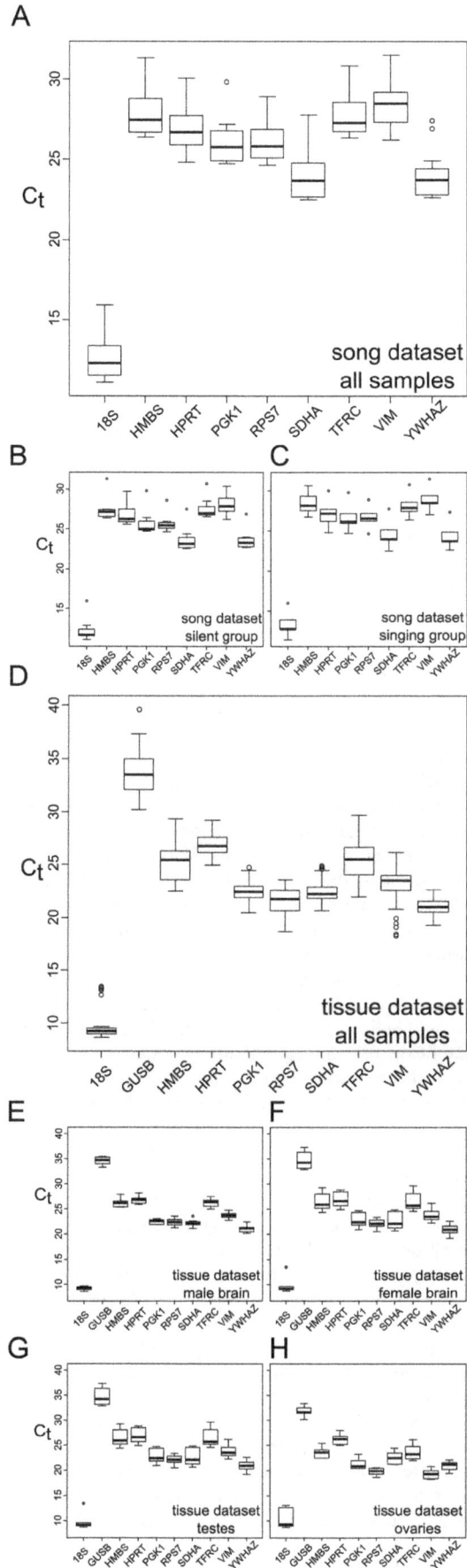

Figure 3. Expression levels of candidate reference genes for the zebra finch song (A–C) and tissue dataset (D–H). Values are given as cycle threshold numbers (C_t values) in (A) all Area X samples combined (n = 12), (B) silent group only (n = 7), (C) singing group only (n = 5), (D) tissue dataset all samples (n = 32), (E) male brain only (n = 10), (F) female brain only (n = 10), (G) testes only (n = 6) and (H) ovaries only (n = 6).

subsequent melting curve analysis. Primer pairs for *FOXP2* (CCTGGCTGTGAAAGCGTTTG/ATTTGCACCCGACACT-GAGC [8]), *EGR1* (ACTTCATCATCGCCATCCTC/TGGA-ATTGGGAAATGTTGGT) and *CFOS* (AGCTGGAGGAGGA-GAAGTCC/CTCCTCGGAGAAGCACAACT) were designed for the zebra finch only.

Analysis of expression stability

The stability of expression was evaluated for 10 zebra finch candidate reference genes (*18S, GUSB, HMBS, HPRT, PGK1, RPS7, SDHA, TFRC, VIM* and *YWHAZ*) on samples from 10 female and 10 male brains and six gonads of each sex, and samples from Area X of male zebra finches that sang before sacrifice (n = 5) or were silent (n = 7). Mean C_t values from qPCR runs were exported to Excel (Microsoft Excel 2010) and analyzed by the three commonly used statistical algorithms geNorm (version 3.5), NormFinder (version 0.953) and BestKeeper (version 1.0) [11,40,41]. For the geNorm analysis, we employed the comparative C_t method, taking into account the efficiency of each assay as outlined in the manual. The resulting values were used as input into the geNorm analysis. For the NormFinder analysis we followed the input procedure outlined in [45]. For the BestKeeper analysis the raw C_t values were entered into the Excel mask.

The song data set was analyzed for differences in gene expression using the Mann-Whitney-U-test.

Results

Normalization to reference genes is a crucial step to obtain meaningful and reliable results, when gene expression levels are quantified using qPCR. In this study we aimed to establish and test a set of primer pairs that could ideally be used on the entire avian class. We selected 14 candidate genes (*18S, ABL, GAPDH, GUSB, HMBS, HPRT, PGK1, RPL13, RPL19, RPS7, SDHA, TFRC, VIM, YWHAZ*) and designed non-degenerate primer pairs.

Establishing and testing primer pairs on the avian class

To test our primers, we collected samples from different branches of the avian tree in an attempt to represent phylogenetic diversity. We chose the zebra finch, cockatiel and white tailed eagle to represent land birds, the Humboldt penguin for water birds, the common crane for the gruiformes, the chicken, turkey and mallard for gallanserae and the ostrich for paleognaths. We tested all primer pairs except *RPL13* and *RPL19* on all species. The performance of each primer pair was evaluated for each species separately and was considered successful if the following criteria were met:

1. Amplification of a single product indicated by a single peak in the melting curve analysis

2. Sequence of the PCR product confirming amplification from the proper gene

3. Efficiency of amplification between 90 and 110%.

Concerning the first criterion, a single product was amplified by most primer pairs in most species (Figure 1, 2; Table 2), except for

Table 4. Stability ranking of expression analysis for the song data set with three different statistical methods.

Gene	Ranking		
	geNorm	Normfinder	BestKeeper
18s	**1**	**2**	5
HMBS	6	7	5
HPRT	8	9	5
PGK1	**2**	**3**	5
RPS7a	7	5	1
SDHA	**3**	6	5
TFRC	5	**4**	2*
VIM	9	8	2
YWHAZ	**4**	**1**	2

The first four ranked genes are shown in bold. Asterisks indicate which gene was chosen for normalization in case of multiple genes with one rank.

the following cases. The *ABL* primers did not yield any product on the mallard and crane cDNA but worked on the other species. *RPL13* and *RPL19* were only tested on zebra finch, chicken and ostrich samples because the melting curve analysis suggested amplification of more than one product. However, when we sequenced those samples we obtained clean traces with low background and no sign of more than one template in the sequencing reaction. Likewise, the melting curve of the *HMBS* amplicon on the mallard and cockatiel cDNA also suggested unspecific amplification but sequencing yielded the expected sequence.

With regard to the second criterion, sequencing of the PCR products confirmed the specificity of the amplification in all cases (Table 3; sequences are provided in Document S1). Sequence similarity to the zebra finch and chicken was higher than 84% and 85% in all cases, respectively.

Regarding the third criterion, we determined the amplification efficiency of each primer pair for each species (Table 2). For each avian species the amplification efficiencies of eight to 12 reference genes were determined and ranged between 70.8% and 107.6% (for details see Table 2). We did not determine the efficiency on all species for the *GAPDH* primers as the efficiency on the zebra finch and chicken samples was too low to be considered for the use in gene expression studies. Taken together we established between six and 11 primer pairs for each species.

Expression stability analysis of candidate reference genes for normalization in the zebra finch

To test if the established primers could be used for normalization in a proof of principle experiment, we conducted two independent screens with 10 candidate genes (*18S, GUSB, HMBS, HPRT, PGK1, RPS7, SDHA, TFRC, VIM* and *YWHAZ*) on zebra finch tissues (Figure 3).

1) Microbiopsies of the striatal song nucleus Area X, either taken 30 min after the onset of undirected singing or of a non-singing control (henceforth called "song data set").
2) Zebra finch entire brains (female and male separately), testes and ovaries.

We determined the expression stability of the 10 genes in both screens using three different algorithms: geNorm [11], NormFinder [40] and BestKeeper [41]. Table 4 and Table 5 display the rankings derived from each of the algorithms for the two data sets.

All raw data were prepared for input into the three software applications as described in the methods section. We excluded *GUSB* from the analysis in the song data set, as it was expressed below the detection limit of our assay.

GeNorm. The geNorm program is a Visual Basic application (VBA) tool for Microsoft Excel [11]. The algorithm assumes that the ratio of two reference genes is unaffected by any experimental treatment or condition. It calculates the pairwise variation of each gene with all other control genes. The stability value M for a gene is the average of all pairwise variations of this particular gene. By stepwise exclusion of the gene with the highest M values, the two most stable reference genes are determined. One critical assumption of the algorithm is that the tested genes are not co-regulated [46]. If they are, they will falsely be considered as suitable reference genes, because of their low co-variation.

In the case of the song dataset, geNorm analysis recommended *18S* and *PGK1* as the most stably expressed genes (Table 4). In contrast, the *18S* gene was one of the least stably expressed genes in all four tissues tested (Figure 3, Table 5).

Besides determining which genes are most suited as reference genes, the geNorm algorithm also gives a recommendation on how many genes should be used for normalization. A Vn/n+1 pairwise variation value below 0.15 indicates that n genes are sufficient for normalization and the gene n+1 should not be included. For the song dataset the V2/3 value (0.079) was already below 0.15 indicating that two genes are sufficient for normalization. For the other tissues the same was true if the tissues were analyzed separately. When all four tissues were analyzed together, three genes were recommended.

NormFinder. The NormFinder program, also a VBA based Microsoft Excel tool, is regarded as the most reliable of the three algorithms we employed [40]. In contrast to the other two approaches, NormFinder takes the experimental groups from which samples are drawn into account [46]. The algorithm distinguishes between variation across samples irrespective of the experimental condition, and variation across experimental groups due to experimental condition. To determine the most suited reference genes, NormFinder analyzes the inter- and intra-group variability and calculates a so called "stability value" which takes both measures into account. Consequently, the genes which are most suited as reference genes have the lowest stability values.

For the song dataset *YWHAZ* and *18S* were recommended as reference genes, though *18S* alone did not have the second lowest stability value, but the two genes together had a very low stability

Table 5. Results of testing the stability of expression analysis for the four different zebra finch tissues and three different statistical methods.

Gene	All			Brain			Gonads			Brain male			Brain female			Testes			Ovaries		
	gN	Nf	BK	gN	Nf	BK	gN	Nf	BK	gN	Nf	BK	gN	Nf	BK	gN	Nf	BK	gN	Nf	BK
18s	7	8	9	9	9	9	8	8	9	9	9	8	9	9	10	9	8	8	9	8	9
GUSB	10	9	7	10	10	10	10	9	8	10	10	10	10	10	9	10	10	10	10	10	10
HMBS	6	7	8	5	6	7	6	7	7	6	6	6	5	5	7	1	1	5	1	2	4
HPRT	2	3	2	1	1	3	3	3	4	1	1	1	2	2	4	3	5	2	2	1	1
PGK1	1	1	1	4	4	3	4	4	6	8	8	2	1	1	3	2	4	4	5	4	6
RPS7a	3	2	6	7	7	2	5	5	2	4	3	5	8	8	2	5	6	6	7	7	3
SDHA	4	4	4	2	2	5	1	2	1	3	3	4	3	3	6	6	3	9	4	3	6
TFRC	8	6	5	8	8	7	2	1	5	7	7	9	7	6	8	7	2	7	6	6	8
VIM	8	10	9	3	3	1	9	10	10	5	4	3	4	4	1	8	9	2	3	5	2
YWHAZ	5	5	2	6	5	6	7	6	2	2	2	6	6	7	4	4	7	1	8	9	5

Samples were either analyzed together (all), by tissue type (brains, gonads) or separately. gN, geNorm; Nf, Normfinder; BK, BestKeeper.
The first four ranked genes are shown in bold.

value. For the different tissues the results again differed (Table 5). One interesting aspect here was that *YWHAZ* was the second most stable gene in the male brain but only ranked as the seventh most stable gene in the female brain, hinting towards a sexually dimorphic expression in the brain.

BestKeeper. The BestKeeper Excel template offers several measures to detect the most stable genes [41]:

1) The standard deviation (SD) of the C_ts of all samples for one gene.
2) The correlation coefficient with the BestKeeper Index. The BestKeeper Index for each sample is the geometric mean of the C_ts of all reference genes.
3) The coefficient of variation of a potential reference gene.

The BestKeeper does not indicate how to weigh the importance of these three measures, leaving it up to the experimenter to choose which values to consider most relevant. As a guideline, the authors recommend to exclude all genes with an SD>1.5. However, in the song dataset all candidate genes displayed SDs higher than 1.5, consistent with a previous report finding that the BestKeeper algorithm may be better suited for analysis of gene expression in homogeneous cell populations rather than complex tissues [46]. Given that previous studies used various selection criteria, for example SD only [46], R^2 only [47], an unspecified combination of SD+CV [48] and in one case a self-designed decision criterion [49], we decided to rank all potential reference genes based on each of the values separately and then calculate the mean of the rankings to determine the final rank of a gene. Employing this method, *RPS7*, *TFRC*, *VIM* and *YWHAZ* were the four most stably expressed genes. We decided to use *RPS7* and *TFRC* for normalization as *TRFC*, unlike *VIM* and *YWHAZ*, was equally stable in all three values contributing to the final ranking (Table 4). For the other tissues the BestKeeper ranking largely agreed with the ranking of the other two programs (Table 5).

Reference gene validation

To evaluate the usefulness of the reference genes suggested by our data analysis, we measured the expression of *EGR1*, *CFOS* and *FOXP2* genes in the song dataset. It is known that *EGR1* and *CFOS* mRNA is upregulated in a sexually dimorphic subregion of the avian striatum, Area X, after the bird sings undirected song for at least 30 minutes [38,39]. *FOXP2* is downregulated by undirected singing that lasts for two hours. Whether *FOXP2* is also regulated as early as 30 minutes after the onset of song is currently unknown [50].

We normalized the expression of each of our genes of interest to the geometric mean of the two best genes recommended by each algorithm. Regardless of which genes we chose for normalization, we detected a significant upregulation of *CFOS* and *EGR1* by more than 7-fold in the singing animals compared to the silent controls (Figure 4). We failed to detect such a significant result for *FOXP2*. The upregulation of *CFOS* was most significant if normalized to *18S* and *YWHAZ* as suggested by NormFinder. The results for *EGR1* and *FOXP2* did not differ much between the different normalizations. These results support the suitability of our pan-avian primer set to identify appropriate reference genes for gene expression studies on avian cells and tissues.

Discussion

Gene expression studies in diverse bird species are used to explore the molecular mechanisms underlying physiology and behavior. Due to the richness of species, the comparative

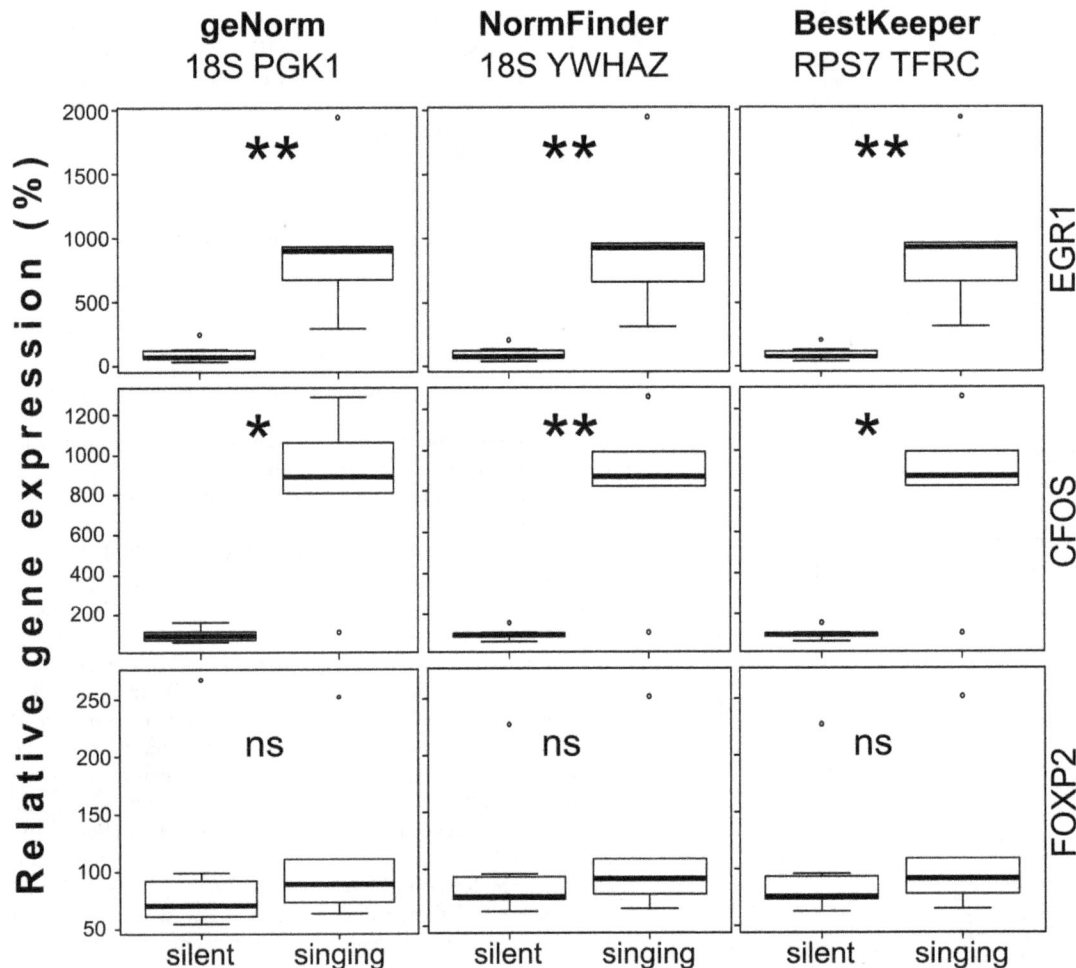

Figure 4. Singing induced gene regulation in Area X of adult male zebra finches. Relative gene expression levels for EGR1, CFOS and FOXP2 were calculated by normalizing to the geometric mean of the two genes recommended by each of the three different algorithms. Relative expression values of the singing group are compared to the silent group. Statistical differences were calculated using the Mann Whitney U test and are indicated in the diagram (p<0.05 *, p<0.01 **, no significant difference ns).

approach to understand the evolution of particular traits is especially promising in the avian lineage. As more genomes are being sequenced, this endeavor becomes increasingly feasible. Frequently quantitative gene expression data derived from qPCR experiments are used as a first step to probe functional consequences of gene activity [8,51,52]. For these data to be meaningful it is essential to choose an appropriate method to normalize the expression level of the gene of interest.

Because the genomes of only very few avian species have been fully explored to date [7,9,26,27], it can be difficult and time consuming to establish primers for qPCR which work on all species under study. Systematic evaluations of reference genes for avian species other than the chicken [33,53,54] are scarce. The present study aimed to establish primers for a set of 14 candidate reference genes and to test whether they can be used to normalize gene expression in qPCR experiments on any given avian species. First we tested the adequacy of the primer pairs for each of our candidate genes on nine different avian species spanning the phylogenetic tree from the ostrich to the zebra finch. Subsequently, we evaluated the suitability of the tested genes as reference genes on samples from zebra finch brain microbiopsies. Moreover,

we examined the expression levels of the candidate reference genes in gonads and brains from male and female zebra finches.

Of the 14 primer pairs tested, six to 11 turned out to match all our quality criteria on each species. Even though the *GAPDH* primers amplified properly on all species tested, we excluded the primers because of the low amplification efficiency for the zebra finch and chicken. Although we have not yet tested the primers *RPL13* and *RPL19* on all nine avian species of this study, our initial results from the chicken, ostrich and zebra finch indicate that both could serve as additional sets to the established ones. Additionally, we previously used *RPL13* and *RPL19* for normalization in the domestic pigeon (*Columba livia* f. *dom.*) [36] and the cockatiel (A. Meyer, personal communication).

The overall evaluation of all our analyses revealed that *PGK1* was consistently among the most stably expressed genes. Notably, the results calculated by the three algorithms employed (Best-Keeper, geNorm, NormFinder) varied considerably between different tissues and sexes. No single gene ranked as the most stable one in all tissues tested. This emphasizes the importance of determining the most reliable reference genes for each new experimental design [11]. This point gains even more emphasis when comparing the results of the stability analysis in the song dataset on the one hand and the total brain samples from male zebra finches on the other hand. *HPRT1* was the most stable gene when analyzing the entire male brain but the least stably expressed gene in Area X under the conditions tested (even though Area X is part of the brain and thus contributes to the first result). Another example of how differently genes can be expressed in similar tissues is *VIM* in the gonad samples and *PGK1* in the male and female brain samples. While *VIM* is stably expressed in ovaries, it is one of the least stably expressed genes in testes. Likewise, *PGK1* is one of the most stably expressed genes in the female brain, while it is among the least stable genes in the male brain. The sexually dimorphic variation of *PGK1 and YWHAZ* in the zebra finch brain might point towards hormonal regulation of gene expression or even a functional role in, e.g. the song system, which is fully developed only in male zebra finches.

As expected, 18S was the most abundant mRNA and *GUSB* the least abundant mRNA in all samples tested (Figure 3). In fact, in the song data set *GUSB* expression was below the detection limit of the qPCR-assay. The use of 18S to normalize gene expression in a qPCR experiment has been heavily debated in the past [20,55,56]. A control gene should be expressed at roughly the same level as the gene of interest to minimize the influence of the technical error [57]. It thus might be advisable to avoid using *18S* as reference gene for all but the most abundantly expressed genes. Additionally, *18S* might not be a suitable reference gene as it is produced by RNA Polymerase I whereas synthesis of mRNA is performed by RNA Polymerase II [20]. *18S* is not polyadenylated and one thus needs to be cautious to only use it if the experimental conditions allow it. For example, if the RNA for the experiment was extracted via oligo(dT) purification or reverse transcribed using oligo(dT)s, *18S* should clearly not be used as reference gene. In spite of this criticism, *18S* has repeatedly been used to normalize gene expression in qPCR experiments and we thus decided not to exclude it from our results [56,58–60].

In conclusion, the practical and easy-to-use pan-avian reference gene primer panel will greatly facilitate molecular research in multiple avian species.

Supporting Information

Document S1 FASTA file of all sequencing results yielded from the PCR products during establishing the primer pairs on nine different species.

Acknowledgments

We are grateful to R. Boussarhane, R. Hauck, M. Lierz and K. Müller for kindly contributing avian specimens. We also thank M. Schaerig and J. Enders for excellent technical assistance.

Author Contributions

Conceived and designed the experiments: PO IA AM ADG. Performed the experiments: PO AM IA. Analyzed the data: PO AM IA. Contributed reagents/materials/analysis tools: PO IA CS ADG. Wrote the paper: PO IA CS ADG.

References

1. Naurin S, Hasselquist D, Bensch S, Hansson B (2012) Sex-biased gene expression on the avian Z chromosome: highly expressed genes show higher male-biased expression. PloS One 7: e46854.
2. Holmes DJ, Ottinger MA (2003) Birds as long-lived animal models for the study of aging. Exp Geront 38: 1365–1375.
3. Dugas-Ford J, Rowell JJ, Ragsdale CW (2012) Cell-type homologies and the origins of the neocortex. PNAS 109: 16974–16979.
4. Bateson M, Feenders G (2010) The use of passerine bird species in laboratory research: implications of basic biology for husbandry and welfare. ILAR J 51: 394–408.
5. Ball GF, Balthazart J (2010) Japanese quail as a model system for studying the neuroendocrine control of reproductive and social behaviors. ILAR journal/ National Research Council, Institute of Laboratory Animal Resources 51: 310–325.
6. Scharff C, Adam I (2012) Neurogenetics of birdsong. Current opinion in neurobiology 23:29–36.
7. Warren WC, Clayton DF, Ellegren H, Arnold AP, Hillier LW, et al. (2010) The genome of a songbird. Nature 464: 757–762.
8. Haesler S, Rochefort C, Georgi B, Licznerski P, Osten P, et al. (2007) Incomplete and inaccurate vocal imitation after knockdown of FoxP2 in songbird basal ganglia nucleus Area X. PLoS Biol 5: e321.
9. Hillier LW, Miller W, Birney E, Warren W, Hardison RC, et al. (2004) Sequence and comparative analysis of the chicken genome provide unique perspectives on vertebrate evolution. Nature 432: 695–716.
10. Bustin SA (2000) Absolute quantification of mRNA using real-time reverse transcription polymerase chain reaction assays. J Mol Endocrinol 25: 169–193.
11. Vandesompele J, De Preter K, Pattyn F, Poppe B, Van Roy N, et al. (2002) Accurate normalization of real-time quantitative RT-PCR data by geometric averaging of multiple internal control genes. Genome Biol 3: RESEARCH0034.
12. Khurshid N, Agarwal V, Iyengar S (2009) Expression of mu- and delta-opioid receptors in song control regions of adult male zebra finches (*Taenopygia guttata*). J Chem Neuroanat 37: 158–169.
13. Thompson CK, Meitzen J, Replogle K, Drnevich J, Lent KL, et al. (2012) Seasonal changes in patterns of gene expression in avian song control brain regions. PloS One 7: e35119.
14. Mukai M, Replogle K, Drnevich J, Wang G, Wacker D, et al. (2009) Seasonal differences of gene expression profiles in song sparrow (*Melospiza melodia*) hypothalamus in relation to territorial aggression. PloS One 4: e8182.
15. Banerjee SB, Arterbery AS, Fergus DJ, Adkins-Regan E (2012) Deprivation of maternal care has long-lasting consequences for the hypothalamic-pituitary-adrenal axis of zebra finches. Proc R Soc London, B 279: 759–766.
16. Barber MR, Aldridge JR Jr, Webster RG, Magor KE (2010) Association of RIG-I with innate immunity of ducks to influenza. PNAS 107: 5913–5918.

17. Nerren JR, He H, Genovese K, Kogut MH (2010) Expression of the avian-specific toll-like receptor 15 in chicken heterophils is mediated by gram-negative and gram-positive bacteria, but not TLR agonists. Vet Immunol Immunopathol 136: 151–156.
18. Dheda K, Huggett JF, Chang JS, Kim LU, Bustin SA, et al. (2005) The implications of using an inappropriate reference gene for real-time reverse transcription PCR data normalization. Anal Biochem 344: 141–143.
19. Tricarico C, Pinzani P, Bianchi S, Paglierani M, Distante V, et al. (2002) Quantitative real-time reverse transcription polymerase chain reaction: normalization to rRNA or single housekeeping genes is inappropriate for human tissue biopsies. Anal Biochem 309: 293–300.
20. Radonic A, Thulke S, Mackay IM, Landt O, Siegert W, et al. (2004) Guideline to reference gene selection for quantitative real-time PCR. Biochem Biophys Res Commun 313: 856–862.
21. Chechi K, Gelinas Y, Mathieu P, Deshaies Y, Richard D (2012) Validation of reference genes for the relative quantification of gene expression in human epicardial adipose tissue. PloS One 7: e32265.
22. Goossens K, Van Poucke M, Van Soom A, Vandesompele J, Van Zeveren A, et al. (2005) Selection of reference genes for quantitative real-time PCR in bovine preimplantation embryos. BMC Develop Biol 5: 27.
23. Nygard AB, Jorgensen CB, Cirera S, Fredholm M (2007) Selection of reference genes for gene expression studies in pig tissues using SYBR green qPCR. BMC Mol Biol 8: 67.
24. Rubie C, Kempf K, Hans J, Su T, Tilton B, et al. (2005) Housekeeping gene variability in normal and cancerous colorectal, pancreatic, esophageal, gastric and hepatic tissues. Mol Cell Prob 19: 101–109.
25. Veazey KJ, Golding MC (2011) Selection of stable reference genes for quantitative rt-PCR comparisons of mouse embryonic and extra-embryonic stem cells. PloS One 6: e27592.
26. Dalloul RA, Long JA, Zimin AV, Aslam L, Beal K, et al. (2010) Multi-platform next-generation sequencing of the domestic turkey (*Meleagris gallopavo*): genome assembly and analysis. PLoS Biol 8: e1000475.
27. Ellegren H, Smeds L, Burri R, Olason PI, Backstrom N, et al. (2012) The genomic landscape of species divergence in *Ficedula* flycatchers. Nature 491: 756–760.
28. Huang Y, Li Y, Burt DW, Chen H, Zhang Y, et al. (2013) The duck genome and transcriptome provide insight into an avian influenza virus reservoir species. Nat Genet 45: 776–783.
29. Van Bers NE, Santure AW, Van Oers K, De Cauwer I, Dibbits BW, et al. (2012) The design and cross-population application of a genome-wide SNP chip for the great tit *Parus major*. Mol Ecol Res 12: 753–770.
30. Castoe TA, Poole AW, de Koning AP, Jones KL, Tomback DF, et al. (2012) Rapid microsatellite identification from Illumina paired-end genomic sequencing in two birds and a snake. PloS One 7: e30953.

31. Wolf JB, Bayer T, Haubold B, Schilhabel M, Rosenstiel P, et al. (2010) Nucleotide divergence vs. gene expression differentiation: comparative transcriptome sequencing in natural isolates from the carrion crow and its hybrid zone with the hooded crow. Mol Ecol 19 Suppl 1: 162–175.

32. Perfito N, Jeong SY, Silverin B, Calisi RM, Bentley GE, et al. (2012) Anticipating spring: wild populations of great tits (*Parus major*) differ in expression of key genes for photoperiodic time measurement. PloS One 7: e34997.

33. De Boever S, Vangestel C, De Backer P, Croubels S, Sys SU (2008) Identification and validation of housekeeping genes as internal control for gene expression in an intravenous LPS inflammation model in chickens. Vet Immunol Immunopathol 122: 312–317.

34. Yin R, Liu X, Liu C, Ding Z, Zhang X, et al. (2011) Systematic selection of housekeeping genes for gene expression normalization in chicken embryo fibroblasts infected with Newcastle disease virus. Biochem Biophys Res Comm 413: 537–540.

35. De Winter P, Sugden D, Baggott GK (2008) Effect of egg turning and incubation time on carbonic anhydrase gene expression in the blastoderm of the Japanese quail (*Coturnix c. japonica*). Brit Poult Sci 49: 566–573.

36. Olias P, Meyer A, Klopfleisch R, Lierz M, Kaspers B, et al. (2013) Modulation of the host Th1 immune response in pigeon protozoal encephalitis caused by *Sarcocystis calchasi*. Vet Res 44: 10.

37. Kimpo RR, Doupe AJ (1997) FOS is induced by singing in distinct neuronal populations in a motor network. Neuron 18: 315–325.

38. Jarvis ED, Scharff C, Grossman MR, Ramos JA, Nottebohm F (1998) For whom the bird sings: context-dependent gene expression. Neuron 21: 775–788.

39. Wada K, Howard JT, McConnell P, Whitney O, Lints T, et al. (2006) A molecular neuroethological approach for identifying and characterizing a cascade of behaviorally regulated genes. PNAS U S A 103: 15212–15217.

40. Andersen CL, Jensen JL, Orntoft TF (2004) Normalization of real-time quantitative reverse transcription-PCR data: A model-based variance estimation approach to identify genes suited for normalization, applied to bladder and colon cancer data sets. Cancer Res 64: 5245–5250.

41. Pfaffl MW, Tichopad A, Prgomet C, Neuvians TP (2004) Determination of stable housekeeping genes, differentially regulated target genes and sample integrity: BestKeeper - Excel-based tool using pair-wise correlations. Biotechnol Lett 26: 509–515.

42. Nolan T, Hands RE, Bustin SA (2006) Quantification of mRNA using real-time RT-PCR. Nat Protoc 1: 1559–1582.

43. Tamura K, Dudley J, Nei M, Kumar S (2007) MEGA4: Molecular Evolutionary Genetics Analysis (MEGA) software version 4.0. Mol Biol and Evol 24: 1596–1599.

44. Altschul SF, Madden TL, Schaffer AA, Zhang J, Zhang Z, et al. (1997) Gapped BLAST and PSI-BLAST: a new generation of protein database search programs. Nucl Acids Res 25: 3389–3402.

45. Latham GJ (2010) Normalization of microRNA quantitative RT-PCR data in reduced scale experimental designs. Methods Mol Biol 667: 19–31.

46. Hibbeler S, Scharsack JP, Becker S (2008) Housekeeping genes for quantitative expression studies in the three-spined stickleback *Gasterosteus aculeatus*. BMC Mol Biol 9: 18.

47. Maroufi A, Van Bockstaele E, De Loose M (2010) Validation of reference genes for gene expression analysis in chicory (*Cichorium intybus*) using quantitative real-time PCR. BMC Mol Biol 11: 15.

48. Chang E, Shi S, Liu J, Cheng T, Xue L, et al. (2012) Selection of reference genes for quantitative gene expression studies in *Platycladus orientalis* (Cupressaceae) using real-time PCR. PLoS One 7: e33278.

49. Ahi EP, Guethbrandsson J, Kapralova KH, Franzdottir SR, Snorrason SS, et al. (2013) Validation of reference genes for expression studies during craniofacial development in arctic charr. PLoS One 8: e66389.

50. Teramitsu I, White SA (2006) FoxP2 regulation during undirected singing in adult songbirds. J Neurosci 26: 7390–7394.

51. Koshiba-Takeuchi K, Mori AD, Kaynak BL, Cebra-Thomas J, Sukonnik T, et al. (2009) Reptilian heart development and the molecular basis of cardiac chamber evolution. Nature 461: 95–98.

52. O'Connell LA, Hofmann HA (2012) Evolution of a vertebrate social decision-making network. Science 336: 1154–1157.

53. Yue H, Lei XW, Yang FL, Li MY, Tang C (2010) Reference gene selection for normalization of PCR analysis in chicken embryo fibroblast infected with H5N1 AIV. Virol Sin 25: 425–431.

54. Tomaszycki ML, Peabody C, Replogle K, Clayton DF, Tempelman RJ, et al. (2009) Sexual differentiation of the zebra finch song system: potential roles for sex chromosome genes. BMC Neurosci 10: 24.

55. Arsenijevic T, Gregoire F, Delforge V, Delporte C, Perret J (2012) Murine 3T3-L1 adipocyte cell differentiation model: validated reference genes for qPCR gene expression analysis. PloS One 7: e37517.

56. Kuchipudi SV, Tellabati M, Nelli RK, White GA, Perez BB, et al. (2012) 18S rRNA is a reliable normalisation gene for real time PCR based on influenza virus infected cells. Virol J 9: 230.

57. Bustin SA, Nolan T (2009) Analysis of mRNA expression by real-time PCR. Real-Time PCR: Current Technology and Applications: 111–135.

58. Dridi S, Buyse J, Decuypere E, Taouis M (2005) Potential role of leptin in increase of fatty acid synthase gene expression in chicken liver. Domest Anim Endocrinol 29: 646–660.

59. Ding ST, Yen CF, Wang PH, Lin HW, Hsu JC, et al. (2007) The differential expression of hepatic genes between prelaying and laying geese. Poult Sci 86: 1206–1212.

60. Abruzzo LV, Lee KY, Fuller A, Silverman A, Keating MJ, et al. (2005) Validation of oligonucleotide microarray data using microfluidic low-density arrays: a new statistical method to normalize real-time RT-PCR data. Biotechniques 38: 785–792.

Sulfonamide-Resistant Bacteria and Their Resistance Genes in Soils Fertilized with Manures from Jiangsu Province, Southeastern China

Na Wang[1,2◐], Xiaohong Yang[3◐], Shaojun Jiao[2], Jun Zhang[3], Boping Ye[3*¶], Shixiang Gao[1*¶]

1 State Key Laboratory of Pollution Control and Resource Reuse, School of the Environment, Nanjing University, Nanjing, 210093, China, 2 Nanjing Institute of Environmental Science, Ministry of Environmental Protection of China, Nanjing, 210042, China, 3 School of Life Science and Technology, China Pharmaceutical University, Nanjing, 210009, China

Abstract

Antibiotic-resistant bacteria and genes are recognized as new environmental pollutants that warrant special concern. There were few reports on veterinary antibiotic-resistant bacteria and genes in China. This work systematically analyzed the prevalence and distribution of sulfonamide resistance genes in soils from the environments around poultry and livestock farms in Jiangsu Province, Southeastern China. The results showed that the animal manure application made the spread and abundance of antibiotic resistance genes (ARGs) increasingly in the soil. The frequency of sulfonamide resistance genes was *sul*1 > *sul*2 > *sul*3 in pig-manured soil DNA and *sul*2 > *sul*1 > *sul*3 in chicken-manured soil DNA. Further analysis suggested that the frequency distribution of the *sul* genes in the genomic DNA and plasmids of the SR isolates from manured soil was *sul*2 > *sul*1 > *sul*3 overall ($p < 0.05$). The combination of *sul*1 and *sul*2 was the most frequent, and the co-existence of *sul*1 and *sul*3 was not found either in the genomic DNA or plasmids. The sample type, animal type and sampling time can influence the prevalence and distribution pattern of sulfonamide resistance genes. The present study also indicated that *Bacillus, Pseudomonas* and *Shigella* were the most prevalent *sul*-positive genera in the soil, suggesting a potential human health risk. The above results could be important in the evaluation of antibiotic-resistant bacteria and genes from manure as sources of agricultural soil pollution; the results also demonstrate the necessity and urgency of the regulation and supervision of veterinary antibiotics in China.

Editor: Jose Luis Balcazar, Catalan Institute for Water Research (ICRA), Spain

Funding: The funding source of this study are the Commonwealth and Environmental Protection project granted by the MEP "The health risk assessment and management technology of veterinary medicine" (201109038), "Study of determination method, pollution level and pollution control strategy of antibiotic-resistant gene in China" (201309031). The funders had no role in study design, data collection and analysis, decision to publish, or preparation of the manuscript.

Competing Interests: The authors have declared that no competing interests exist.

* Email: yebp@cpu.edu.cn (BY); ecsxg@nju.edu.cn (SG)

◐ These authors contributed equally to this work.

¶ These authors also contributed equally to this work.

Introduction

In the past few decades, veterinary antibiotics have been widely used in many countries to treat disease and promote animal growth. However, this release together with antibiotic-resistant bacteria (ARB) is a great concern recently [1], primarily because the land application of antibiotic-polluted manure in agricultural practice not only introduced bacteria carrying antibiotic resistance genes (ARGs) into the soil but also had a significant effect on the ARB promotion and selection. In the soil, antibiotics provide a positive selective pressure for these bacteria [2]. The horizontal transfer of ARGs between bacteria is an important factor in resistance dissemination [3]. It is worth noting that some ARB in soil and manure are phylogenetically close to human pathogens, making genetic exchange more likely [3]. Evidence from the last 35 years demonstrates that there was consistent correlation between the use of antibiotic-contaminated manure on farms and the transfer of ARGs in human pathogens, as well as the direct shift of ARB from animals to humans [4]. Therefore, ARGs are recognized as new environmental pollutants, and special concern is warranted due to their potential environmental and human health risks.

The used amount of veterinary medicines in China is more than that of other countries. According to a 2007 survey, the usage of antibiotics in livestock was almost half of the total antibiotics produced in China, which was 210,000 tons [5]. It was approximately 10-fold higher than in the USA and approximately 300-fold higher than in the UK [6]. It would be a good chance to analyze the impact of livestock practices on ARGs in the environment in China, where the animal farm was large-scale and the antibiotics usage was great [7]. However, there are few reports on veterinary ARGs in China.

Sulfonamides are synthetic veterinary antibiotics that are the most widely used veterinary antibiotics in China, the European Union and some developing countries due to their low costs [8,9].

However, sulfonamides were ranged as "High priority" of veterinary medicines, due to the high potential to reach the environment [10]. Sulfonamide resistance is primarily mediated by the *sul*1, *sul*2 and *sul*3 genes encoding dihydropteroate synthetase (DHPS) with a low affinity for sulfonamides [11–13]. A wide range of bacterial species harbor these genes, which are located in transposons and in self-transferable or mobilizable plasmids with a broad host range; these genes manifest multiple antibiotic resistance that is co-selected by sulfonamides [14–16].

Numerous recent studies have focused solely on the prevalence of sulfonamide resistance genes in bacterial isolates from manured agricultural soils or on the quantification of the total ARGs from environmental soil media to reflect the resistance reservoir. Few studies have systematically covered the identity of sulfonamide-resistant (SR) bacteria and the distribution patterns of sulfonamide ARGs in the total soil DNA and in sulfonamide-resistant bacteria.

The objectives of this study were (i) to determine the influence of the fertilization with antibiotic-polluted manure on the selection of sulfonamide ARB and ARGs and (ii) to investigate the distribution pattern of the *sul*1, *sul*2 and *sul*3 genes in the total soil DNA and the identified SR bacteria. Furthermore, (iii) the identification of the SR bacteria genera and description of the genotypes in each genus were also conducted to identify resistant opportunistic pathogens that increased the risk of ARGs affecting public health. To the best of our knowledge, this is the first comprehensive study of sulfonamide ARB and ARGs in livestock and poultry farms in China. The present study could be important in the evaluation of the pollution of soils used for agriculture by ARB and ARGs from manure; this study also demonstrates the necessity and urgency for the regulation and supervision of veterinary antibiotics in China.

Materials and Methods

Sampling

Soil samples from 10 sites were studied, including four pig farms, four chicken farms, one non-arable agricultural area and one mountain forest. The animal feeding farms of different sizes and scales were selected (detailed information about the sampling sites and the person in charge of sampling are given in Table S1 in File S1). The study was permitted and approved by the Ministry of Environmental Protection, China. The land accessed was not privately owned or protected. No protected species were sampled. There were vegetable cultivation area and grain planting area, which were all fertilized with animal manure, in each animal feeding farm. Therefore, two replicates of 1 kg soil samples for each type in every animal feeding farm were collected from depth of 10 to 15 cm, loaded into sterile glass flasks. The soil samples of the same type in different animal feeding farms were mixed (50 g from each source) to processed within 1 to 2 days after collection. The following description was the name rule of samples: (i) samples from the vegetable region of pig farms collected in the winter, the mixture of which was marked as PVW; (ii) samples from the agricultural region of pig farms collected in winter, the mixture of which was marked as PAW; (iii) samples from the vegetable region of pig farms collected in the summer, the mixture of which was marked as PVS; (iv) samples from the agricultural region of pig farms collected in the summer, the mixture of which was marked as PAS; (v) samples from the vegetable region of chicken farms, the mixture of which was marked as CV; (vi) samples from the agricultural region of chicken farms, the mixture of which was marked as CA; (vii) non-arable soils (marked as NA) where manure was not used for a few years near a Nanjing chicken farm; and (viii) forest soil collected from the Fangshan mountain in the Jiangning district of Nanjing (manure and/or antibiotics were not used),

which was marked as F. Soil P represents the mixture of soil samples from a pig farm in winter, and soil C is the mixture of soil samples from a chicken farm. The manure (M) was obtained from chickens that were treated with sulfonamides.

For each sample, 100 g was taken for the isolation of SR bacteria and the measurement of sulfonamide residues, and the remainder was stored at 4°C for DNA extraction. Meanwhile, the concentration of sulfonamides in the samples was analyzed in this study using a previously published method [17].

Viable plate counts

The isolation of SR bacteria from the soil or manure was performed by cultivating bacteria on nutrient broth agar plates containing 60 μg/ml sulfadiazine (SDZ) [15] followed by the spread plate technique [17]. Total bacteria from samples M, F, NA, P and C were cultivated on nutrient broth agar plates without SDZ. In brief, 1.0 ml of each soil sample solution, which was prepared by dissolving 5 g of soil in 45 ml of sterile physiological saline (0.9% NaCl), was mixed with 9 ml of sterile physiological saline. The process was repeated to make additional serial 10-fold dilutions, i.e., 10^{-3}, 10^{-4}, 10^{-5} and 10^{-6}. After 2–5 days of incubation at 37°C, the number of resistant bacteria on the agar plates were counted to calculate the colony-forming units (CFUs) per gram of soil with the following formula: CFU/g soil $= 45 \times$ average colony number $\times$ dilution factor. For subsequent analyses, SR isolates were randomly picked from the plates of each soil sample, with a total of 237 SR bacterial isolates, including 6 isolates from M; 1 isolate from F; 2 isolates from NA; 65, 57, 25 and 25 isolates from PVW, PAW, PVS, and PAS, respectively; and 20 and 36 isolates from CV and CA, respectively. All bacterial strains were stored at −80°C in nutrient broth medium containing 15% glycerol.

DNA extraction

Total soil DNA was extracted from 0.5 g of soil using a PowerSoil® DNA Isolation Kit (MoBio Laboratories, Carlsbad, California, USA) following the manufacturer's instructions. SR isolates were cultured at 37°C overnight with constant shaking at 200 rpm/min in 5 ml of LB supplemented with 60 μg/ml SDZ. DNA extraction was performed with 3.0 ml of cultured SR isolates using the TIANamp bacteria DNA kit (Tiangen, Beijing, China). The plasmids were extracted with the Biomiga EZgene™ Plasmid Miniprep kit (Biomiga, USA) following the manufacturer's protocol. The genomic DNA and plasmids were examined by 1% and 1.5% agarose gel electrophoresis, respectively. Moreover, the λDNA and DNA5000 were used as the marker of genomic DNA and pasmid, respectively. Usually, the molecular weight of genomic DNA was greater than that of the plasmid.

The detection of the *sul*1, *sul*2, and *sul*3 genes in the SR isolates

The prevalence of the *sul*1, *sul*2, and *sul*3 genes in the genomic DNA and plasmids of the isolates was examined via PCR with gene-specific primers (Table S2 in File S1). The amplification conditions for the *sul*1 and *sul*2 genes were as follows: 94°C for 5 min; 30 cycles of 94°C for 30 s, 69°C for 30 s and 72°C for 45 s; and one cycle of 72°C for 7 min. The amplification conditions for the *sul*3 gene were 94°C for 5 min, 30 cycles of 94°C for 30 s, 52°C for 30 s and 72°C for 60 s, and one cycle of 72°C for 7 min. Gel electrophoresis was performed on 1.5% agarose gels. The CA01 (a bacteria from soil CA) plasmid containing the *sul*1 gene was used as the positive control for the detection of the *sul*1 gene; the M01 (bacteria from chicken manure) plasmid containing the sul2 and sul3 genes was used as the positive control for the detection of the sul2 or sul3 genes. *E. coli* DH5α cells were used as the negative control. When the PCR product appeared as a single clear band with the same migration profile as the corresponding gene control, the isolate was counted as positive for that gene.

Quantitative PCR

The relative abundances of the sul1, sul2, and sul3 genes in the soil DNA were determined in triplicate via SYBR Green-based real-time PCR on a CFX96 Touch Real-Time PCR Detection System. The primer sequences are listed in Table S3 in File S1. Each 10-μl reaction mixture contained 5 μl of SYBR Premix (Cwbio, China), 1 μl of 2 μM forward and reverse primer mix, 1 μl of template, and 3 μl of ddH$_2$O. The PCR conditions were 95°C for 10 min, followed by 39 cycles of 95°C for 15 s and 60°C for 60 s. The samples were assessed via $2^{-\Delta\Delta Ct}$ relative quantitative analysis to compare the relative abundance of the sul genes among samples. All samples were analyzed in triplicate. The CA01 (a bacteria from soil CA) plasmid containing the sul1 gene was used as the positive control for the detection of the sul1 gene; the M01 (bacteria from chicken manure) plasmid containing the sul2 and sul3 genes was used as the positive control for the detection of the sul2 or sul3 genes. *E. coli* DH5α cells were used as the negative control.

16S rRNA sequencing of SR isolates

The complete 16S rRNA gene was used to identify the genera present in the bacterial isolates. Genomic DNA was used as the template for the PCR amplification of the 16S rRNA gene using the universal bacterial 16S rRNA primers 27F and 1492R (Table S2 in File S1). Each 50-μl reaction mixture consisted of 1 to 4 μl of genomic DNA, Taq plus polymerase buffer containing 1.5 mM MgCl$_2$, 0.2 mM each of the 4 deoxynucleoside triphosphates (dNTPs), 1 mM each of the 27F and 1492R primers, and 1 U of Taq plus polymerase (Tiangen). PCR was performed using a Bio-Rad thermal cycler under the following conditions: 94°C for 5 min, followed by 30 cycles of 94°C for 30 s, 58°C for 30 s, and 72°C for 1.5 min, and 1 cycle of 72°C for 10 min. The PCR products were separated via electrophoresis on 1.0% agarose gels. The PCR amplicons were sequenced by Sangon (Shanghai, China). A pair-wise 16S rRNA gene sequence similarity was performed using the EzTaxon server (http://www.eztaxon.org/) [18] and NCBI BLAST (http://blast.ncbi.nlm.nih.gov/blast.cgi). A bacterial genus was considered present when a sample 16S rRNA gene sequence was ≥97% identical to the reference sequence of the bacteria in that genus.

Statistical analysis

The statistical analysis was performed using SAS 9.1. The group mean levels were analyzed via a one-way Analysis of Variance (ANOVA). Statistical significance was defined as a p-value≤0.05. This p-value was chosen because the standard error associated with CFU plating and qPCR results are generally approximately 5% of the mean. The mean and standard error (SE) displayed in the figures were generated using the means procedure without transformation.

Results and Discussion

Enumeration of the total culturable microbial populations and SR Bacteria in the soil

The number of total culturable microbial populations on the nutrient agar ranged from 1.96×10^7 to 9.75×10^7 CFU/g soil and that of the SR isolates on the nutrient agar ranged from 4.5×10^5

to 9.0×10^7 CFU/g soil (Figure 1), which were higher than those of the reported aquaculture-agriculture ponds (3.0×10^4 to 1.6×10^6 and 3.0×10^2 to 4.1×10^4, respectively) [19]. The higher numbers of total bacteria and SR isolates were found in chicken manure (9.75×10^7 and 9.00×10^7, respectively), which was most likely due to the amount of easily accessible nutrients in the manure that stimulated the growth of bacteria [20]. The number of SR bacteria from the soils affected by pig or chicken manure (3.02×10^6 to 9.40×10^6 CFU/g soil) was higher than that from non-arable soil (1.96×10^6 CFU/g soil) or forest soil (4.5×10^5 CFU/g soil). This difference was most likely due to the application of manure to the soil. Previous studies reported that manure from treated pigs was rich in antibiotics and bacteria carrying ARGs, which were both transferred to the soil via fertilization [3,10]. Furthermore, the number of SR isolates from the vegetable soils was significantly higher than that from the agricultural soils (5.96×10^6 and 3.02×10^6 CFU/g soil for PVW and PAW, respectively; 9.40×10^6 and 4.98×10^6 CFU/g soil for PVS and PAS, respectively; 7.50×10^6 and 4.11×10^6 CFU/g soil for CV and CA, respectively). Because liquid manure or wastewater was frequently used to irrigate the vegetable region, manure was more frequently applied to the vegetable soils than to the agricultural soils, and the repeated application of manure to the vegetable soils may have increased bacterial resistance. Additionally, the mean number of SR isolates from the winter soils (4.49×10^6 CFU/g soil for PW) was lower than that from the summer soils (7.19×10^6 CFU/g soil for PS). This difference most likely occurred because the temperature in the summer is more suitable for the growth of bacteria than that in the winter.

The concentration sums of sulfadiazine, sulfamerazine, sulfathiazole, sulfamethazine, sulfadimethazine and sulfamethoxazole were 4503, 0, 0.536, 35.6, 25.9, 15.8, 12.6, 239 and 193 μg/kg in the mixed samples of M, F, NA, PVW, PAW, PVS, PAS, CV and CA, respectively. The number of cultivable bacteria was not consistent with the concentration of antibiotic sulfonamides in the soil. The pollution level of sulfonamides was found to be significantly higher in chicken farms than in pig farms, but there was no significant difference among the numbers of cultivable bacteria.

Characterization of SR bacteria

All 237 SR isolates that were identified via 16S rRNA belonged to 26 typical soil bacteria genera, including *Achromobacter*, *Arthrobacter*, *Bacillus*, *Brevibacterium*, *Chryseobacterium*, *Citrobacter*, *Cupriavidus*, *Escherichia*, *Flavobacterium*, *Hydrogenophaga*, *Klebsiella*, *Lysinibacillus*, *Massilia*, *Microbacterium*, *Microvirga*, *Pseudomonas*, *Pseudoxanthomonas*, *Rhizobium*, *Rhodococcus*, *Shigella*, *Sphingobacterium*, *Sphingopyxis*, *Staphylococcus*, *Stenotrophomonas*, *Streptococcus*, and *Streptomyces*. *Bacillus* was the most prevalent genus in all 9 environmental samples with a frequency of 43.88%, followed by *Pseudomonas* and *Shigella* (11.39% and 8.02%, respectively; Figure 2). However, it is reported that *Acinetobacter* was abundant in pig wastewater in Vietnam [21]. Both pig- and chicken-manured soil samples were rich in bacteria species; for example, 12 genera were found in PVW and CA (see Figure S1).

Relative abundance of the sul genes in the soils

A qPCR analysis of sulfonamide resistance genes was performed on the total DNA extracted directly from the soil. There was significant variation in the relative quantities of the sul1, sul2, and sul3 genes in the DNA extracted from the eight types of soils (see Figure 3). The DNA from the pig-manured soils (PVW, PAW, PVS and PAS) contained relatively higher copy numbers of sul1

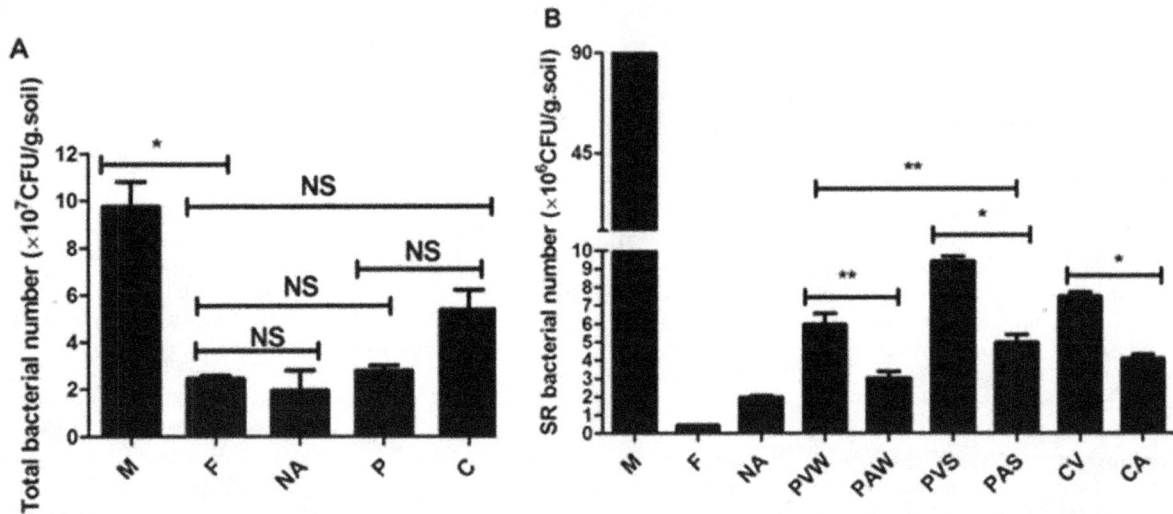

Figure 1. Numbers of cultivable bacteria. (M = Manure, F = Forest, NA = non-arable fied, P = Pig, C = Chicken, W = winter, V = vegetable garden soil, A = agricultural soil; *p≤0.05, **p≤0.01, n = 3; NS, not significant).

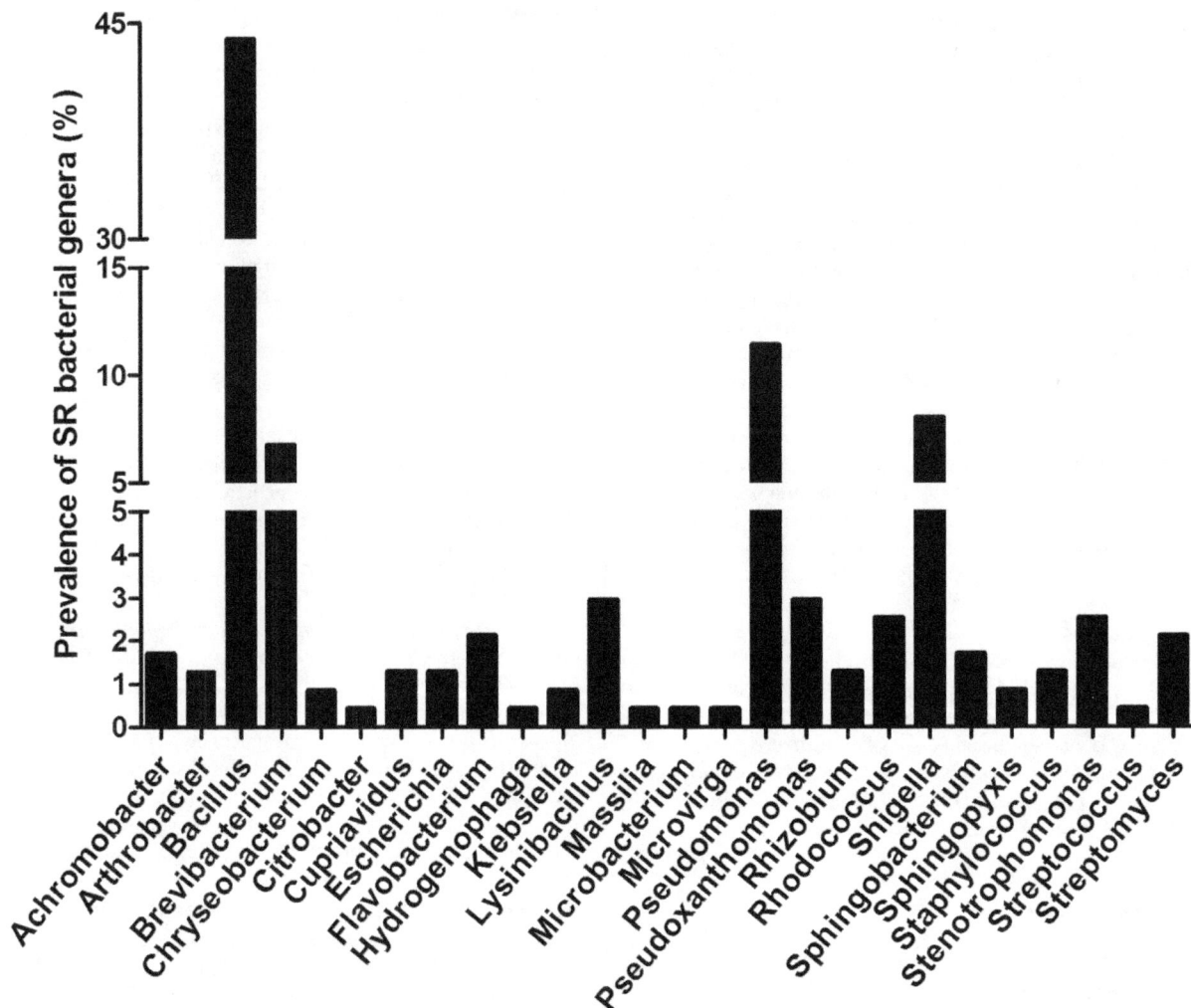

Figure 2. The genera of SR bacteria and their detected frequency in all sampling sites.

than *sul2*. Comparatively, the relative quantity of the *sul1* gene in the chicken-manured soils was lower than that of the *sul2* gene. Additionally, the *sul3* genes were detected at low relative quantities in the DNA extracted from the eight soils but were not detected via PCR in bacteria isolated from forest and pig-manured agricultural (summer) soils. The results of our study were consistent with other reports that demonstrated that the repeated application of manure from pigs or chickens treated with SDZ increased the transfer and abundance of ARGs in the soil [3,10,20]. Furthermore, good positive linear correlations were observed between the relative abundance of the *sul2* genes and the number of culturable SR isolates in the soil. For the *sul2* gene and sum of the three *sul* genes, the correlation coefficients (R^2) were 0.95 and 0.65, respectively ($p<0.05$). However, the abundance of *sul1* and *sul3* showed no significant correlation with the numbers of culturable SR isolates in the soil ($R^2 = 0.44$, $p>0.05$ for *sul1* and $R^2 = 0.39$, $p>0.05$ for *sul3*). This lack of a correlation could be attributed to the fact that the viable plate counts method only sampled microbes that were culturable and expressed their ARGs under those conditions, so most of the microbes carrying *sul1* and *sul3* genes may not be culturable. The other probable reason was that some "silent" or unexpressed *sul1* and *sul3* genes may be existed in the isolates of soils, which could be horizontally transferred or expressed under other conditions.

In brief, the number of culturable SR isolates in the soil can reflect the total relative abundance of the three *sul* genes, showing that the plate count method was effective in assessing the antibiotic resistance risk of the soil. Therefore, the diversity of ARGs enriched at the farm level should be the focus of more attention.

Distribution of *sul* genes in SR isolates

The number and percentage of isolates carrying the *sul* genes in their genomic DNA and plasmids are summarized in Table 1 and Table 2. The distribution and spread of SR genes in the soil microbes are sufficiently frequent to warrant special concern. The *sul1*, *sul2*, and *sul3* genes were all detected at a frequency of 100% in the genomic DNA and plasmids of the SR isolates from the manure sample, indicating that ARGs were extensively harbored in the chromosome and mobile genetic elements of the bacteria in manure, leading to the high potential of horizontal gene transfer of ARGs in soil. Interestingly, the *sul2* genes were only present in the genomic DNA of the isolates collected from forest soil and non-arable soil, which had no history of manure application. This finding may be attributed to the notion that the *sul1* and *sul3* genotype in genomic DNA maybe associate with the amended manure. However, the *sul1*, *sul2* and *sul3* genes were all located in the plasmids of the isolates from non-arable soil but were absent from the plasmids of the isolates from the forest soil; a potential explanation for this difference could be that the bacteria carrying *sul* genes in the manured soil may transfer to the nearby region by aerosolization or runoff, then horizontal transfer occurred in close bacteria via plasmids.

For the manured soil, the frequency distribution of the *sul* genes in the genomic DNA and plasmids of the SR isolates investigated overall followed a trend of *sul2* > *sul1* > *sul3* ($p<0.05$). This result was in contrast to several previous studies showing that the *sul1* gene was more prevalent than the *sul2* gene in the DNA from manure and manured soils [10,15] due to different conditions in various countries. The *sul3* gene was found at low frequencies in

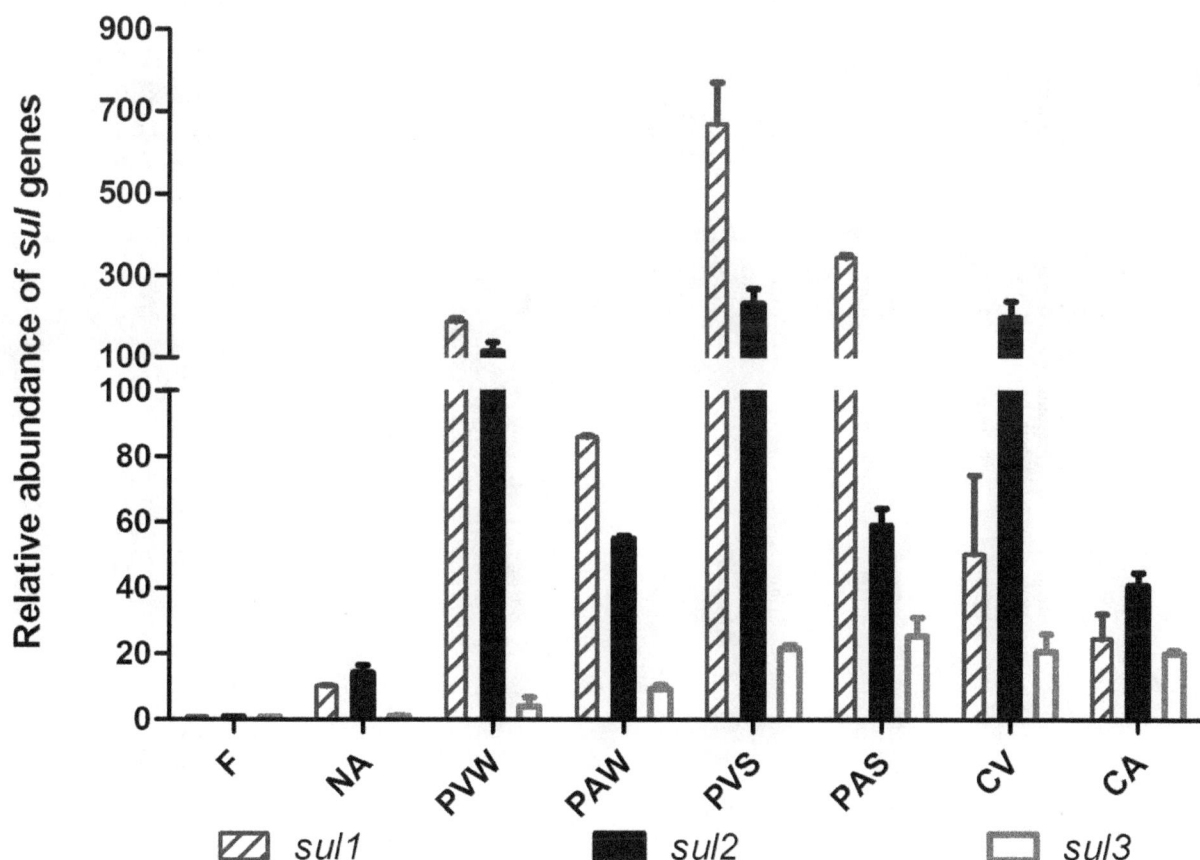

Figure 3. Relative quantity of sulfonamides resistant genes in soils with and without manure treatment.

Table 1. Distribution of *sul1*, *sul2* and *sul3* genes in genomic DNA and plasmid of SR isolates (in samples M, F, NA, CV and CA).

sul gene combination		M (n = 6[a]/6[b])		F (n = 1/0)		NA (n = 2/2)		CV (n = 20/20)		CA (n = 36/36)	
		NO. of isolates (%)		NO. of isolates (%)		NO. of isolates (%)		NO. of isolates (%)		NO. of isolates (%)	
		Genomic DNA	Plasmid DNA	Genomic DNA	Plasmid DNA	Genomic DNA	Plasmid DNA	Genomic DNA	Plasmid DNA	Genomic DNA	Plasmid DNA
Single genes	*sul1*	0 (0.0)	0 (0.0)	0 (0.0)	0 (0.0)	0 (0.0)	0 (0.0)	0 (0.0)	1 (5.0)	2 (5.6)	0 (0.0)
	sul2	0 (0.0)	0 (0.0)	1 (100.0)	0 (0.0)	2 (100.0)	0 (0.0)	2 (10.0)	3 (15.0)	3 (8.3)	2 (5.6)
	sul3	0 (0.0)	0 (0.0)	0 (0.0)	0 (0.0)	0 (0.0)	0 (0.0)	0 (0.0)	0 (0.0)	1 (2.8)	0 (0.0)
Two genes	*sul1+sul2*	0 (0.0)	0 (0.0)	0 (0.0)	0 (0.0)	0 (0.0)	0 (0.0)	4 (20.0)	7 (35.0)	10 (27.8)	2 (5.6)
	sul1+sul3	0 (0.0)	0 (0.0)	0 (0.0)	0 (0.0)	0 (0.0)	0 (0.0)	0 (0.0)	0 (0.0)	0 (0.0)	0 (0.0)
	sul2+sul3	0 (0.0)	0 (0.0)	0 (0.0)	0 (0.0)	0 (0.0)	1 (50.0)	0 (0.0)	0 (0.0)	1 (2.8)	9 (25.0)
Three genes	*sul1+sul2 +sul3*	6 (100.0)	6 (100.0)	0 (0.0)	0 (0.0)	0 (0.0)	1 (50.0)	12 (60.0)	9 (45.0)	19 (52.8)	23 (63.9)
None		0 (0.0)	0 (0.0)	0 (0.0)	0 (0.0)	0 (0.0)	0 (0.0)	0 (0.0)	0 (0.0)	0 (0.0)	0 (0.0)
Total	*sul1*	6 (100.0)	6 (100.0)	0 (0.0)	0 (0.0)	0 (0.0)	1 (50.0)	16 (80.00)	17 (85.0)	27 (75.0)	25 (69.4)
	sul2	6 (100.0)	6 (100.0)	1 (100.0)	0 (0.0)	2 (100)	2 (100.0)	19 (95.0)	19 (95.0)	29 (80.6)	36 (100.0)
	sul3	6 (100.0)	6 (100.0)	0 (0.0)	0 (0.0)	0 (0.0)	2 (100.0)	13 (65.0)	9 (45.0)	21 (58.3)	32 (88.9)
Total of SR isolate positive for *sul* genes		6 (100.0)	6 (100.0)	1 (100.0)	0 (0.0)	2 (100.0)	2 (100.0)	19 (95.0)	20 (100.0)	36 (100.0)	36 (100.0)

a = genomic DNA, b = plasmid.

Table 2. Distribution of *sul1*, *sul2* and *sul3* genes in genomic DNA and plasmid of SR isolates (in samples PVW, PAW, PVS and PAS).

sul gene combination		PVW (n = 65/47)		PAW (n = 57/43)		PVS (n = 25/22)		PAS (n = 25/22)	
		NO. of isolates (%)		NO. of isolates (%)		NO. of isolates (%)		NO. of isolates (%)	
		Genomic DNA	Plasmid DNA	Genomic DNA	Plasmid DNA	Genomic DNA	Plasmid DNA	Genomic DNA	Plasmid DNA
Single genes	*sul1*	0 (0.0)	23 (48.9)	0 (0.0)	19 (44.2)	15 (60.0)	0 (0.0)	24 (96.0)	0 (0.0)
	sul2	28 (43.1)	3 (6.4)	15 (26.3)	8 (18.6)	1 (4.0)	15 (68.2)	0 (0.0)	16 (72.7)
	sul3	0 (0.0)	1 (2.1)	0 (0.0)	2 (4.7)	0 (0.0)	0 (0.0)	0 (0.0)	0 (0.0)
Two genes	*sul1+sul2*	34 (52.3)	11 (23.4)	41 (71.9)	9 (20.9)	1 (4.0)	6 (27.3)	1 (4.0)	6 (27.3)
	sul1+sul3	0 (0.0)	2 (4.3)	0 (0.0)	0 (0.0)	0 (0.0)	0 (0.0)	0 (0.0)	0 (0.0)
	sul2+sul3	0 (0.0)	0 (0.0)	1 (1.8)	0 (0.0)	0 (0.0)	1 (4.5)	0 (0.0)	0 (0.0)
Three genes	*sul1+sul2+sul3*	1 (1.5)	0 (0.0)	0 (0.0)	0 (0.0)	0 (0.0)	0 (0.0)	0 (0.0)	0 (0.0)
None		2 (3.1)	7 (14.9)	0 (0.0)	5 (11.6)	8 (32.0)	0 (0.0)	0 (0.0)	0 (0.0)
Total	*sul1*	35 (53.8)	36 (76.6)	41 (71.9)	24 (55.8)	16 (64.0)	6 (27.3)	25 (100.0)	6 (27.3)
	sul2	63 (96.9)	14 (29.8)	57 (100.0)	17 (39.5)	2 (8.0)	21 (95.5)	1 (4.0)	22 (100.0)
	sul3	1 (1.5)	3 (6.4)	1 (1.8)	2 (4.7)	0 (0.0)	1 (4.5)	0 (0.0)	0 (0.0)
Total of SR isolate positive for *sul* genes		63 (96.9)	40 (85.1)	57 (100.0)	38 (88.4)	17 (68.0)	22 (100.0)	25 (100.0)	22 (100.0)

a = genomic DNA, b = plasmid.

our samples, whereas recently, Suzuki et al showed that sul3 was major sul in seawater [22]. Hoa et al. suggested that most of the *sul* genes are located on the chromosome [15]. However, there was no significant difference between the overall percentage of the isolates carrying the *sul* genes located on the genomic DNA and those on the plasmids in our study. It was interesting to note that the frequency order of the *sul1* and *sul2* genes from the isolates of the pig-manured soils for the genomic DNA was opposite that for the plasmids. In the isolates collected from the pig-manured soils in winter, *sul2* was the most prevalent gene located within the genomic DNA (96.9% and 100.0% in PVW and PAW, respectively) followed by *sul1* (53.8% and 71.9% in PVW and PAW, respectively); *sul1* was the most prevalent gene located on plasmids (76.6% and 55.8% in PVW and PAW, respectively) followed by *sul2* (29.8% and 39.5% in PVW and PAW, respectively). However, in the isolates collected from pig-manured soil in summer, the order of *sul1* (64.0% and 100.0% in PVW and PAW, respectively) > *sul2* (8.0% and 4.0% in PVW and PAW, respectively) in the genomic DNA and *sul2* (95.0% and 100.0% in PVW and PAW, respectively) > *sul1* (27.3% and 27.3% in PVW and PAW, respectively) in the plasmids was determined. We concluded that in most isolates, *sul1* and *sul2* were located in the different mobile elements and transferred at different rates.

Furthermore, the animal type was a significant factor influencing the expression frequency of *sul* genes, which showed that the frequency in the chicken-manured soil was higher than that in the pig-manured soil ($p < 0.05$), which was consistent with the data of the concentration of sulfonamides in the soil.

We also determined the co-presence of any two different *sul* genes on the chromosome and plasmids in a single isolate. The combination of *sul1* and *sul2* on the chromosome was the most frequent and was present in PVW, PAW, PVS, PAS, CV and CA (52.3%, 71.9%, 4.0%, 4.0%, 20.0% and 27.8%, respectively), and the *sul1*, *sul2* and *sul3* genes were highly co-present on the chromosomes of M, CV and CA (100%, 60.0% and 52.8%, respectively). The co-presence of *sul2* and *sul3* was only detected in two isolates from PAW and CA, respectively, and the co-existence of *sul1* and *sul3* was not detected in any SR isolates. The *sul1* and *sul2* genes were also frequently detected together in the plasmids (23.4%, 20.9%, 27.3%, 27.3%, 35.0% and 5.6% in PVW, PAW, PVS, PAS, CV and CA, respectively). In contrast, the co-presence of *sul2* and *sul3* was only detected in NA (50.0%), PVS (4.5%) and CA (25.0%), and the co-presence of *sul1* and *sul3* was not found in any plasmids. Furthermore, the three *sul* genes were co-present in the plasmids of M (100%), NA (50.0%), CV (45.0%), and CA (63.9%). We concluded that the combination of *sul1* and *sul2* was the most frequent and that the co-existence of *sul1* and *sul3* was not found in the genomic DNA or plasmids. Based on these results, the co-presence of the three *sul* genes was only in the isolates from manure and soil from chicken farms, suggesting that there was a positive correlation between the frequency of the co-presence of the three *sul* genes and the time and amount of repeated manure applications.

In summary, the *sul* genes, either individually or in combinations of two or three, were present in the SR isolates at high frequencies. Nearly all plasmids from the SR isolates contained the *sul* genes (with the exception of F). This observation suggests that the resistance that we observed in most cases was linked to plasmids or other mobile genetic elements, which theoretically have transfer potential. The SR isolates could possibly carry these *sul* genes through gene transfer under selection conditions, leading to an increase in antibiotic resistance among bacteria.

SR bacterial and *sul* genes

The distribution of *sul* genes in bacteria species is listed in Table 3. *Bacillus* was the most prevalent *sul*-positive genus in the soil samples of this study, carrying the *sul* genes in 43.88% of the total isolates; thus, this genus could be the main reservoir of the *sul* genes. This finding was not consistent with other studies that showed that *Acinetobacter* was the dominant genus in aquatic environments (wastewater and shrimp ponds of north Vietnam) and manured agricultural clay soils and slurry samples in the United Kingdom [15,16]. Except for different environments, what makes the difference of genus may be the different condition of culture, such as 28 or 30°C incubation in these two references, not 37°C. It was reported that *Bacillus* spp. have developed resistance to most antibiotic groups, but only a few species of *Bacillus* have been reported to be sensitive to sulfonamides [23]. *Pseudomonas* and *Shigella* were the second and third most prevalent, carrying the *sul* genes in 11.39% and 8.02% of all isolates, respectively. Ventilator-acquired pneumonia, respiratory tract infections in immunocompromised patients and chronic respiratory infections in cystic fibrosis patients were associated with the *Pseudomonas* species (especially *P. aeruginosa*) [24]. *Enterobacteriaceae* species including *Shigella*, *Klebsiella*, and *Escherichia* have represented some of the most dominant bacterial infections over the last 30 years [24]. In the Henan Province of China, 72.6% of infections were caused by *Shigella* strains in 2006 [25].

To the best of our knowledge, this report is the first on *sul* genes in *Chryseobacterium*, *Cupriavidus*, *Flavobacterium*, *Hydrogenophaga*, *Lysinibacillus*, *Massilia*, *Microbacterium*, *Microvirga*, *Pseudoxanthomonas*, *Rhizibium*, *Rhodococcus*, *Sphingopyxis*, *Staphylococcus*, *Streptococcus*, and *Streptomyces* from soils and the first that indicates the widespread presence of ARB in the arable soils of China. Previous studies demonstrated the co-presence of *sul1*, *sul2* and *sul3* in a single cell; this was detected in *Acinetobacter*, *Bacillus*, *Psychrobacter*, *Escherichia coli*, and *Salmonella* [15,16,26,27]. In our study, these three *sul* genes were simultaneously found in *Arthrobacter*, *Brevibacterium*, *Citrobacter*, *Cupriavidus*, *Flavobacterium*, *Lysinibacillus*, *Pseudomonas*, *Pseudoxanthomonas*, *Rhizibium*, *Sphingobacterium*, *Staphylococcus*, *Stenotrophomonas*, *Streptococcus*, and *Streptomyces*, with the exception of three genera (*Bacillus*, *Escherichia*, and *Shigella*). This result indicates that the three *sul* genes are common and widely distributed in ARB in soil. Additionally, the *sul3* gene was detected for the first time in *Achromobacter*, *Chryseobacterium*, *Citrobacter*, *Cupriavidus*, *Flavobacterium*, *Lysinibacillus*, *Pseudoxanthomonas*, *Rhizibium*, *Sphingobacterium*, *Staphylococcus*, *Streptococcus*, and *Streptomyces* from arable soils.

It was revealed that the manured soils could be a reservoir of sulfonamide ARBs and ARGs, according to the observation of high frequency of various combinations of the *sul* genes in bacteria of manured agricultural soils, which may bring potential hazards to human and ecosystem health. Therefore, the diversity of ARGs and ARB enriched at the farm level should be the focus of more attention.

Conclusion

A comprehensive study of sulfonamide ARB and ARGs in livestock and poultry farms in Jiangsu Province of China revealed that the fertilization with antibiotic-polluted manure had a significant influence on the selection of sulfonamide ARB and ARGs. The sample type, animal type and sampling time may affect the prevalence and distribution rule of SR genes. The results from the identification of the SR bacteria genus and the description of the genotypes in the genus revealed that resistant

Table 3. Summary of *sul* genotype of sul-positive bacterial species isolated.

Genus	No. of total sul-positive isolates (%)	Source of isolates	*sul* genotype	No. of sul-positive isolates
Achromobacter	4 (1.69)	NA, PVW, CV	sul2	2
			sul1 sul2	1
			sul2 sul3	1
Arthrobacter	3 (1.27)	CV, CA	sul1 sul2	1
			sul1 sul2 sul3	2
Bacillus	104 (43.88)	F, PVW, PAW, PVS, PAS, CV, CA	sul1	2
			sul2	23
			sul1 sul2	66
			sul2 sul3	1
			sul1 sul2 sul3	12
Brevibacterium	16 (6.75)	PVW, PAW, PVS, PAS	sul2	4
			sul1 sul2	11
			sul1 sul2 sul3	1
Chryseobacterium	2 (0.84)	PVW, PVS	sul1 sul2	2
Citrobacter	1 (0.42)	CA	sul1 sul2 sul3	1
Cupriavidus	3 (1.27)	CA	sul1 sul2 sul3	3
Escherichia	3 (1.27)	PVW, CA	sul1 sul2	1
			sul1 sul2 sul3	2
Flavobacterium	5 (2.11)	CV, CA	sul2	1
			sul1 sul2	1
			sul1 sul2 sul3	3
Hydrogenophaga	1 (0.42)	PVS	sul1 sul2	1
Klebsiella	2 (0.84)	PAS	sul1 sul2	2
Lysinibacillus	7 (2.95)	PVW, PAW, PAS	sul1	1
			sul1 sul2	4
			sul1 sul2 sul3	2
Massilia	1 (0.42)	PVW	sul2	1
Microbacterium	1 (0.42)	PAW	sul1 sul2	1
Microvirga	1 (0.42)	PAS	sul1 sul2	1
Pseudomonas	27 (11.39)	PVW, PAW, CV	sul2	1
			sul1 sul2	23
			sul1 sul2 sul3	3
Pseudoxanthomonas	7 (2.95)	PVW, PVS	sul2	2
			sul1 sul2	4
			sul1 sul2 sul3	1
Rhizobium	3 (1.27)	PVS, CV	sul1 sul2	2
			sul1 sul2 sul3	1
Rhodococcus	6 (2.53)	PVW, PAW, PVS, PAS	sul1 sul2	6
Shigella	19 (8.02)	CV, CA, M	sul1 sul2 sul3	19

Table 3. Cont.

Genus	No. of total sul-positive isolates (%)	Source of isolates	*sul* genotype	No. of sul-positive isolates
Sphingobacterium	4 (1.69)	VA	sul1 sul2	1
			sul1 sul2 sul3	3
Sphingopyxis	2 (0.84)	PVW, PAW	sul2	1
			sul1 sul2	1
Staphylococcus	3 (1.27)	PAW, CA	sul1 sul2 sul3	3
Stenotrophomonas	6 (2.53)	CV, CA, NA	sul2 sul3	1
			sul1 sul2 sul3	5
Streptococcus	1 (0.42)	CV	sul1 sul2 sul3	1
Streptomyces	5 (2.11)	PVW, PAW, CA	sul1 sul2	1
			sul1 sul2 sul3	4

opportunistic pathogens increased the risk of ARGs affecting public health. Overall, the high frequency of various combinations of the *sul* genes in manured agricultural soil samples of Southeastern China should be the focus of more attention, and the regulation and supervision of veterinary antibiotics are urgently needed in China.

Supporting Information

Figure S1 Prevalences of SR bacteria belonging to different genera identified in the studied soils.

File S1 Contains the following files: **Table S1**. Detailed information on sampling sites in present study. **Table S2**. Primers for PCR in this Study. **Table S3**. Primers for quantitative PCR in this Study.

Author Contributions

Conceived and designed the experiments: NW. Performed the experiments: XHY SJJ JZ. Analyzed the data: XHY NW. Contributed reagents/materials/analysis tools: NW BPY XHY. Contributed to the writing of the manuscript: NW XHY. Guided the experiment: BPY SXG.

References

1. Ghosh S, LaPara TM (2007) The effects of subtherapeutic antibiotic use in farm animals on the proliferation and persistence of antibiotic resistance among soil bacteria. ISME J 1: 191–203.
2. Popowska M, Rzeczycka M, Miernik A, Krawczyk-Balska A, Walsh F, et al. (2012) Influence of soil use on prevalence of tetracycline, streptomycin, and erythromycin resistance and associated resistance genes. Antimicrob Agents Chemother 56: 1434–1443.
3. Heuer H, Schmitt H, Smalla K (2011) Antibiotic resistance gene spread due to manure application on agricultural fields. Curr Opin Microbiol 14: 236–243.
4. Marshall BM, Levy SB (2011) Food animals and antimicrobials: impacts on human health. Clinical microbiology reviews 24: 718–733.
5. Hvistendahl M (2012) Public Health China Takes Aim at Rampant Antibiotic Resistance. Science 336: 795–795.
6. Kim K-R, Owens G, Kwon S-I, So K-H, Lee D-B, et al. (2011) Occurrence and environmental fate of veterinary antibiotics in the terrestrial environment. Water Air Soil Poll 214: 163–174.
7. Zhu YG, Johnson TA, Su JQ, Qiao M, Guo GX, et al. (2013) Diverse and abundant antibiotic resistance genes in Chinese swine farms. Proc Natl Acad Sci U S A 110: 3435–3440.
8. Ungemach F (1999) Figures on quantities of antibacterials used for different purposes in the EU countries and interpretation. Acta Vet Scand Suppl 93: 89–97; discussion 97–88, 111–117.
9. Kools SA, Moltmann JF, Knacker T (2008) Estimating the use of veterinary medicines in the European Union. Regul Toxicol Pharm 50: 59–65.
10. Heuer H, Smalla K (2007) Manure and sulfadiazine synergistically increased bacterial antibiotic resistance in soil over at least two months. Environ Microbiol 9: 657–666.
11. Skold O (2000) Sulfonamide resistance: mechanisms and trends. Drug Resist Updat 3: 155–160.
12. Perreten V, Boerlin P (2003) A new sulfonamide resistance gene (sul3) in Escherichia coli is widespread in the pig population of Switzerland. Antimicrob Agents Chemother 47: 1169–1172.
13. Yun MK, Wu Y, Li Z, Zhao Y, Waddell MB, et al. (2012) Catalysis and sulfa drug resistance in dihydropteroate synthase. Science 335: 1110–1114.
14. Heuer H, Szczepanowski R, Schneiker S, Pühler A, Top E, et al. (2004) The complete sequences of plasmids pB2 and pB3 provide evidence for a recent ancestor of the IncP-1β group without any accessory genes. Microbiology 150: 3591–3599.
15. Hoa PTP, Nonaka L, Hung Viet P, Suzuki S (2008) Detection of the sul1, sul2, and sul3 genes in sulfonamide-resistant bacteria from wastewater and shrimp ponds of north Vietnam. Sci Total Environ 405: 377–384.
16. Byrne-Bailey KG, Gaze WH, Kay P, Boxall AB, Hawkey PM, et al. (2009) Prevalence of sulfonamide resistance genes in bacterial isolates from manured agricultural soils and pig slurry in the United Kingdom. Antimicrob Agents Chemother 53: 696–702.
17. Sengeløv G, Agersø Y, Halling-Sørensen B, Baloda SB, Andersen JS, et al. (2003) Bacterial antibiotic resistance levels in Danish farmland as a result of treatment with pig manure slurry. Environ Int 28: 587–595.
18. Chun J, Lee JH, Jung Y, Kim M, Kim S, et al. (2007) EzTaxon: a web-based tool for the identification of prokaryotes based on 16S ribosomal RNA gene sequences. Int J Syst Evol Microbiol 57: 2259–2261.
19. Hoa PTP, Managaki S, Nakada N, Takada H, Anh DH, et al. (2010) Abundance of sulfonamide-resistant bacteria and their resistance genes in integrated aquaculture-agriculture ponds, North Vietnam. Interdisciplinary studies on environmental chemistry - biological responses to contaminants. Tokyo: TERRAPUB. 15–22.
20. Jechalke S, Kopmann C, Rosendahl I, Groeneweg J, Weichelt V, et al. (2013) Increased abundance and transferability of resistance genes after field application of manure from sulfadiazine-treated pigs. Appl Environ Microbiol 79: 1704–1711.
21. Hoa PTP, Managaki S, Nakada N, Takada H, Shimizu A, et al. (2011) Antibiotic contamination and occurrence of antibiotic-resistant bacteria in aquatic environments of northern Vietnam. Science of the Total Environment 409: 2894–2901.

22. Suzuki S, Ogo M, Miller TW, Shimizu A, Takada H, et al. (2013) Who possesses drug resistance genes in the aquatic environment sulfamethoxazole (SMX) resistance genes among the bacterial community in water environment of Metro-Manila, Philippines. Frontiers in Microbiology 4.

23. Valderas MW, Bourne PC, Barrow WW (2007) Genetic basis for sulfonamide resistance in Bacillus anthracis. Microb Drug Resist 13: 11–20.

24. Diene SM, Rolain J-M (2013) Investigation of antibiotic resistance in the genomic era of multidrug-resistant Gram-negative bacilli, especially Enterobacteriaceae, Pseudomonas and Acinetobacter. Expert Rev Anti-Infe 11: 277–296.

25. Xia S, Xu B, Huang L, Zhao JY, Ran L, et al. (2011) Prevalence and characterization of human Shigella infections in Henan Province, China, in 2006. J Clin Microbiol 49: 232–242.

26. Antunes P, Machado J, Sousa JC, Peixe L (2005) Dissemination of sulfonamide resistance genes (sul1, sul2, and sul3) in Portuguese Salmonella enterica strains and relation with integrons. Antimicrob Agents Chemother 49: 836–839.

27. Hammerum AM, Sandvang D, Andersen SR, Seyfarth AM, Porsbo LJ, et al. (2006) Detection of sul1, sul2 and sul3 in sulphonamide resistant Escherichia coli isolates obtained from healthy humans, pork and pigs in Denmark. Int J Food Microbiol 106: 235–237.

Selectively Adsorptive Extraction of Phenylarsonic Acids in Chicken Tissue by Carboxymethyl α-Cyclodextrin Immobilized Fe₃O₄ Magnetic Nanoparticles Followed Ultra Performance Liquid Chromatography Coupled Tandem Mass Spectrometry Detection

Jing Jia[2], Wei Zhang[1,3], Jing Wang[1,3], Peilong Wang[1,3]*, Ruohua Zhu[2]*

1 Key Laboratory of Agrifood Safety and Quality, Ministry of Agriculture, Beijing, P.R. China, **2** Department of Chemistry, Capital Normal University, Beijing, China, **3** Institute of Quality Standards and Testing Technology for Agriculture Products, China Agricultural Academy of Science, Beijing, P.R. China

Abstract

Carboxymethyl α-cyclodextrin immobilized Fe₃O₄ magnetic nanoparticles (CM-α-CD-Fe₃O₄) were synthesized for the selectively adsorptive extraction of five phenylarsonic acids including *p*-amino phenylarsonic acid, *p*-nitro phenylarsonic acid, *p*-hydroxy phenylarsonic acid, *p*-acylamino phenylarsonic acid and *p*-hydroxy-3-nitro phenylarsonic acid in chicken tissue. Using ultra performance liquid chromatography coupled with tandem mass spectrometry (UPLC-MS/MS), a highly sensitive analytical method was proposed for the determination of five phenylarsonic acids. It was shown that CM-α-CD-Fe₃O₄ could extract the five phenylarsonic acids in complex chicken tissue samples with high extraction efficiency. Under the optimal conditions, a high enrichment factor, ranging from 349 to 606 fold, was obtained. The limits of detection (LODs) (at a signal-to-noise ratio of 3) were in the range of 0.05–0.11 μg/kg for the five phenylarsonic acids. The proposed method was applied for the determination of five target phenylarsonic acids in chicken muscle and liver samples. Recoveries for the spiked samples with 0.2 μg/kg, 2.0 μg/kg and 20 μg/kg of each phenylarsonic acids were in the range of 77.2%–110.2%, with a relative standard deviation (RSD) of less than 12.5%.

Editor: Bing Xu, Brandeis University, United States of America

Funding: The authors would like to thank the NSFC (No. 31201832), Special Fund for Agro-scientific Research in the Public Interest (No. 201203094), and National International cooperation program (No. 2012DFA31140) for financially supporting this research. The funders had no role in study design, data collection and analysis, decision to publish, or preparation of the manuscript.

Competing Interests: The authors have declared that no competing interests exist.

* Email: wplcon99@163.com (PW); zhurh@mail.cnu.edu.cn (RZ)

Introduction

4-Hydroxy-3-nitrobenzenearsonic acid, also known as roxarsone (ROX), has been used since 1944 as a feed additive in the poultry industry to promote growth and to control coccidiosis, a parasitic disease that infects the intestinal tract of poultry [1]. Besides ROX, some other organic arsenic compounds, including p-amino phenylarsonic acid (p-APAA), p-nitro phenylarsonic acid (p-NPAA), p-hydroxy phenylarsonic acid (p-HPAA) and p-acylamino phenylarsonic acid (AAPAA) (**Fig. 1**) have been successively employed for the same purposes. Their slight structural difference, i. e. different substituent groups on the aromatic ring, results in different growth-promoting and disease-controlling effects [2]. Phenylarsonic acids have been approved as feed additives by many countries at levels of 25–50 mg/kg [3]. Recent studies showed that phenylarsonic acids in the environment might be converted into elemental arsenic and other inorganic arsenic compounds, which are known to be strongly carcinogenic [4]. Some countries in EU strictly control the use of

phenylarsonic acid additives, while in the U.S., Tyson Foods, the country's largest poultry producer, stopped the use of arsenic compounds in 2004. After the release of 2011 FDA report of elevated inorganic arsenic in the livers of chickens treated with ROX, Pfizer Animal Health, the US manufacturer of ROX, quickly suspended ROX sales [5]. In order to monitor the residues of phenylarsonic acid in animal products, it is of great significance to establish a convenient, sensitive and reliable method to analyze the organic arsenic in samples [6].

Several analytical methods for the determination of phenylarsonic acids in the environment have been reported, including liquid chromatography (LC) coupled to atomic absorption spectroscopy (AAS) [7], atomic emission spectroscopy (AES) [8] or atomic fluorescence spectrometry (AFS) [9] as well as gas chromatography-mass spectrometry (GC-MS) [10], capillary electrophoresis (CE) coupled to ultraviolet and visible light detector [11] or inductively coupled plasma-mass spectrometry (ICP-MS) [12]. LC has been demonstrated to be the most effective method in arsenic separation, and ICP-MS can provide low

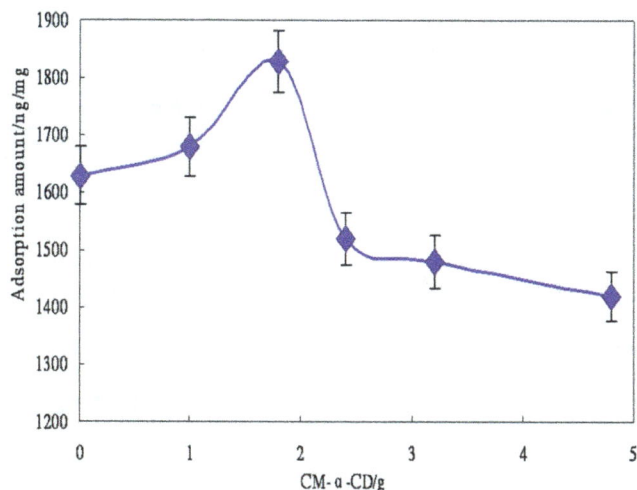

Figure 1. Effect of CM-α-CD amount used in modification on the adsorption efficiency of phenylarsonic acids (ROX). (n = 3, with RSD<6.5%).

detection limit. Therefore, LC-ICP-MS is one of the most powerful research means for analysis of organic arsenic in complex samples [2,13]. The LC-MS/MS is a powerful separation and detection platform in multi residues analysis. Pergantis et al [14] have developed a stable method for determination of 5 phenylarsonic acids including ROX, p-APAA, p-NPAA, AAPAA, p-HPAA and other organic arenics using LC-MS/MS in positive ionization mode. The LODs of the developed method achieved sub ng/g level, whereas the analytical time was too long. Furthermore, compared with other arsenic speciation methods, the LC-MS/MS could provide more structural information of the phenylarsonic acids.

After feeding, nearly all phenylarsonic acids are excreted unchanged to the environment through the disposal of poultry litter, so the residues in animal tissues are very low [15,16]. Some studies indicated that the approved conditions of use mandate a 5-day withdrawal period from the medicated feed before animals are slaughtered, and limits are in place for total residues of combined arsenic (As) in meat from ROX treated animals [0.5 mg kg^{-1} As in muscle tissue and eggs, and 2 mg kg^{-1} As in liver and kidney] [17]. To further lower the detection limits of phenylarsonic acids in complex biological samples, nano-materials have been used to selectively extract and concentrate arsenic compounds. The most commonly used nano-materials for the adsorption of arsenic compounds include goethite [18], titania [19], iron oxide [20,21], Fe_3O_4 nanoparticles [22,23] or modified Fe_3O_4 nanoparticles [24]. Among these nano-materials, Fe_3O_4 nanoparticles are well suited for arsenic analysis due to the following advantages. First, they can be easily isolated from solutions by applying an external magnetic field [25], which ensures simplified sample adsorption and elution processes. Second, Fe_3O_4 nanoparticles have been demonstrated to have higher affinity toward arsenic element than other nano-materials, resulting in higher arsenic extraction efficiency. Fe_3O_4 magnetic nanoparticles have been used for removing inorganic arsenic ions in water sample. For example, Fe_3O_4 nanoparticles dispersed in chelating resins or coated with adequate chelating agents have been used for the removal of a wide range of metal ions from wastewater, overall displaying higher adsorption capacity than traditional materials [26–29]. More recently, β-CD coated Fe_3O_4 nanoparticles have been

successfully applied for the removal of methylene blue and copper ions [30,31]. It is found that the cavity of cyclodextrin and its surface hydroxyl group can impart better binding capability and chemical stability to the magnetic particle [32].

In this work, a highly sensitive determination method was established to monitor five phenylarsonic acids in chicken tissues. In our first attempt, it was found that the adsorption of phenylarsonic acids by Fe_3O_4 magnetic nanoparticles was not as good as inorganic arsenic compounds. It was then decided to use CM-α-CD to couple on the surface of Fe_3O_4 providing CD cavities to fit the benzene rings in the structures of phenylarsonic acid compounds. In addition, hydrogen bonding and electrostatic interactions between hydroxyl/carboxyl groups of modified CM-α-CD and amino/nitro of phenyl arsenic acids were also increased. The adsorption properties of modified Fe_3O_4 magnetic nanoparticles to phenylarsonic acids were studied, and the interactions between nanoparticles and phenylarsonic acids were examined. The synthesized materials were successfully applied in the sample clean-up and pre-concentration of phenylarsonic acids in chicken muscle and liver samples, which were subsequently separated and detected by UPLC-MS/MS.

Materials and Methods

Apparatus

ICP-MS (Agilent 7500Ce, USA) was used to study the adsorption properties of synthesized material for phenylarsonic acids. The optimum operation parameters of ICP-MS were selected by tuning. The power was 1550W, the flow rates of cooling air, auxiliary air and carrier were 15.0 L/min, 1.0 L/min and 1.12 L/min, respectively. The sample rate of ascension by using peristaltic pump was set as 1.0 mL/min. The integration time for arsenic was 0.9 s/isotope. The operating parameters of UPLC-MS/MS (Waters Xevo TQ, USA) were as follows: capillary voltage = 2.8 kv, desolvation temperature = 450°C, desolvation gas flow rate = 600 L/Hr. The mobile phase was a mixture of acetonitrile (solvent A) and water containing 0.1% formic acid (solvent B) at a flow rate of 0.3 mL/min. All chromatographic separations were carried out in linear gradient mode as follows. In the first minute, solvent A was maintained at 98% (v/v). Solvent B quickly dropped to 30% from 1 to 3 min, followed by dramatic increase back to 98% from 3 to 5 min. MS parameters of UPLC-MS/MS are showed in **Table 1**.

The closed microwave digestion system (CEM MARS, American) was used to digest samples for the determination of the total arsenic in the crude samples. The homogenizer (IKA, Germany) was applied to sample pretreatment and rocking hammock bed was from Zhicheng, Shanghai, China.

Standard solutions and reagents

Five phenylarsonic acids (98%) were obtained from the Chinese Academy of Agricultural Sciences. The standard stock solutions were prepared by dissolving each arsenic species in pure water at an arsenic concentration equivalent to that of 1 mg/mL phenylarsonic acids and stored at 4°C in the dark.

Reagents for preparing magnetic nanoparticles: $FeCl_2 \cdot 4H_2O$ and $FeCl_3 \cdot 6H_2O$ (analytical reagent grade) were purchased from Tianjin Guangfu Fine Chemical Research Institute. Sodium chloroacetate (98%) was bought from Alfa Aesar. α-cyclodextrin (98%) was purchased from Beijing Dilang Biochemical Technology Co., Ltd., China. Other reagents including methanol, ethanol, acetone, toluene, formic acid and sodium hydroxide were of analytical reagent grade and all bought from Beijing Chemical

Table 1. Analytical parameters of MS/MS.

Compounds	Molecular weight	Precursor Ion (m/z)	Product Ion (m/z)	Cone voltage (V)	Collision voltage (V)
p-APAA	217	218	92	27	21
			109	27	16
p-HPAA	218	219	110	28	18
			201	28	13
p-NPAA	247	248	202	30	18
			230	30	14
AAPAA	260	261	244	29	14
ROX	263	264	246	33	14

Plant. The water used throughout the experiment was purified using a Milli-Q water purification system (Millipore, Germany).

Preparation of CM-α-CD stabilized magnetic nanoparticles

Synthesis of CM-α-CD. CM-α-CD was prepared according to the following procedures. α-CD (3.55 mmoL) and NaOH (90.2 mmoL) were first dissolved into 20 ml water. The solution was then heated at 90°C for 5 min, followed by the addition of 74.6 mmoL sodium chloroacetate. The solution was heated for 3 h at 90°C under stirring. Once cooled to room temperature, the pH of the solution was adjusted to 6–7 by hydrochloric acid. The nearly neutral solution obtained was then poured into about 500 mL methanol. CM-α-CD was precipitated out as white solids, which was filtered and washed with methanol for a few times and then dried under vacuum for 3 d at 50°C and 0.085 MPa. The melting point of the CM-α-CD product was about 245.5°C as determined by micro melting point apparatus.

Synthesis of Fe_3O_4 nanoparticles. Fe_3O_4 magnetic nanoparticles were prepared according to the conventional co-precipitation method [33]. A mixture of 2.0 g $FeCl_2 \cdot 4H_2O$ and 5.2 g $FeCl_3 \cdot 6H_2O$ was added into a 500 mL conical flask containing 200 ml 0.05 M HCl. After dissolution of the solids, 250 ml 0.75 M NaOH solution was poured into the flask under a blanket of N_2. The mixture was stirred for another 2 h at 80°C. The Fe_3O_4 nanoparticles were then obtained in the form of black precipitates, which were separated with a magnet and washed subsequently by water (3 times) and ethanol (twice). It should be noted that both the HCl and NaOH solutions were degassed by a sonicator for 20 min before use.

CM-α-CD modified Fe_3O_4 nanoparticles (CM-α-CD-Fe_3O_4). The Fe_3O_4 nanoparticles prepared in the previous step were added into 60 ml PBS buffer solution (pH = 6.6) containing 1.6 g of CM-α-CD. The suspension was sonicated for 3 min and then stirred for 3 h at 80°C. After cooling to room temperature, the nanoparticles were washed several times by PBS buffer solution to remove excess CM-α-CD. The CM-α-CD modified Fe_3O_4 nanoparticles were then dried at 80°C in a vacuum oven.

Adsorption procedure

Static adsorption. In a 10 mL centrifuge tube, 5 mg CM-α-CD-Fe_3O_4 nanoparticles were mixed with 8 mL standard solution of phenylarsenic acid with a given concentration. The centrifuge tube was placed on a rocking hammock bed at a rate of 270 rpm. After equilibrating for 30 min, the magnetic nanoparticles were separated from the solution with external magnetic field. The

nanoparticles were rinsed twice with ethanol and dried in N_2. To desorb target compounds, 1.0 mL pure water was added to the nanoparticles followed by equilibration for 10 min. The aqueous solution containing target compounds was filtered through a 0.22 μm Poly (ether sulfones) (PES) syringe filter and analyzed by ICP-MS. Standard phenylarsenic acid solutions of other concentrations were analyzed in the same way.

Dynamic adsorption. In a 10 mL centrifuge tube, 5 mg CM-α-CD-Fe_3O_4 nanoparticles were mixed with 8 mL standard solution of phenylarsenic acid with a given concentration. A number of centrifuge tubes were prepared in this manner for a given concentration of standard. The centrifuge tubes were placed on an orbital shaker at a rate of 270 r/min. At different time points, one tube was removed and the magnetic nanoparticles contained in the tube were separated from the solution with external magnetic field. The following washing, desorption and ICP-MS analysis procedures were the same as those in the static adsorption step. Standard phenylarsenic acid solutions of other concentrations were analyzed in the same way.

Sample analysis

Chicken tissues including meat and liver were bought from a supermarket in Beijing. Chicken meat and chicken liver samples were pulverized and freeze-dried for 24 h. The freeze-dried sample was homogenized by grinding and frozen until analysis. In a 50 mL centrifuge tube, 5.0 g freeze-dried chicken tissue sample and 10 mL ethanol were added. The extraction was repeated twice, each lasting 30 min. The combined extracts were equilibrated with 5 mg CM-α-CD-Fe_3O_4 nanoparticles for 10 min. Then the magnetic nanoparticles which adsorbed target analytes were separated under external magnetic field, the analytes adsorbed on the nanoparticles were then desorbed with 2 mL deionized water. The aqueous solution containing target compounds was filtered through a 0.22 μm PES syringe filter and analyzed by UPLC-MS-MS.

Determination of total arsenic

The total amount of arsenic in chicken tissue samples was determined by microwave digestion ICP-MS according to reference [34]. Each digestion can containing 0.5 g chicken tissues samples was added 5 mL 65% nitric acid. Stages digestion method by controlling temperature was used. The obtained digestion solution was diluted until the concentration of nitric acid fell below 5% and then subjected to ICP-MS analysis.

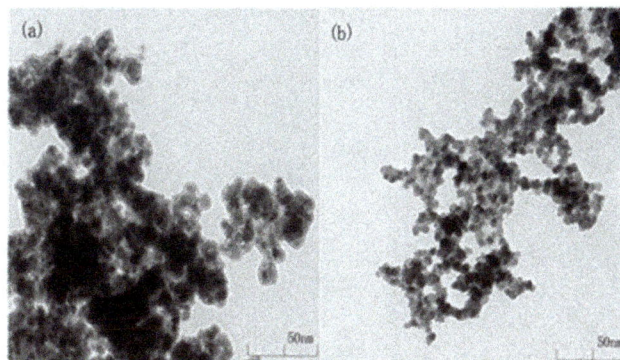

Figure 4. TEM images of magnetic Fe₃O₄ nanoparticles (a) and CM-α-CD-Fe₃O₄ nanoparticles (b).

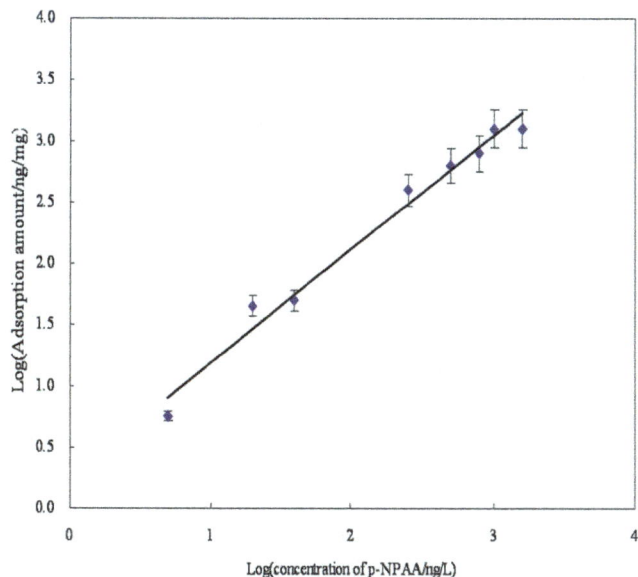

Figure 2. Recovery of different concentrations of phenylarsonic acids (ROX). (n = 3, with RSD<5.0%).

Results and Discussion

Optimization of adsorption efficiency

For optimal adsorption efficiency, the amounts of CM-α-CD for the modification of Fe₃O₄ nanoparticles were varied in six dosages of 0.8, 1.6, 2.4, 3.2, and 4.8 g to obtain 450 mL of Fe₃O₄ suspension as described in section 2.3.2. The adsorption was carried out by equilibrating 5 mg CM-α-CD-Fe₃O₄ in 8 mL ROX standard solution at a concentration of 1 μg/L. The adsorption efficiency was then determined by ICP-MS. As shown in **Fig. 1**, the highest adsorption efficiency corresponds to 1.6 g CM-α-CD. **Fig. 2** shows the adsorption behaviors of ROX at different concentrations by CM-α-CD-Fe₃O₄. The amount of

adsorbed ROX is linear over a large concentration range of ROX. To verify the important role of CM-α-CD in the adsorption of phenylarsonic acid, the adsorption efficiencies of modified and unmodified Fe₃O₄ nanoparticles for p-NPAA were compared. As shown in **Fig. 3**, modified material was obviously superior to unmodified material.

Characterization of magnetic CM-α-CD-Fe₃O₄

The transmission electron microscope (TEM), fourier transform infrared spectrometry (FTIR) and thermo gravimetric analyzer (TGA) were used to characterized the magnetic nanoparticles. The TEM images of Fe₃O₄ and CM-α-CD-Fe₃O₄ nanoparticles are shown in **Fig. 4 (a)** and **Fig. 4 (b)**, respectively. Unmodified Fe₃O₄ nanoparticles, approximately 10 nm in diameter, trend to aggregate because of size effect of Fe₃O₄ nanoparticles. On the other hand, CM-α-CD-Fe₃O₄ nanoparticles shown in **Fig. 4 (b)** are much better dispersed in an aqueous solution with diameters of about 5 nm, because the Fe₃O₄ nanoparticles were modified with CM-α-CD and the surface of Fe₃O₄ nanoparticles was protected by CM-α-CD. It can be observed in **Fig. 4 (b)** that the composite of modified nanoparticles was more compacted and displayed roughly spherical shapes.

The FTIR spectrums of Fe₃O₄, CM-α-CD and CM-α-CD-Fe₃O₄ nanoparticles respectively were scanned (See Fig. S1). In all three samples, a strong characteristic O-H absorption band at

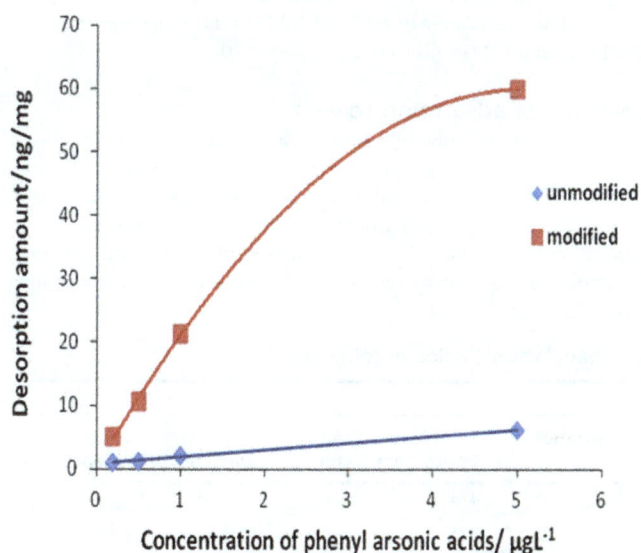

Fig. 3 Comparison of Fe3O4 nanoparticles (blue diamond) and CM-a-CD-Fe3O4 nanoparticles for adsorption of ROX (red square).

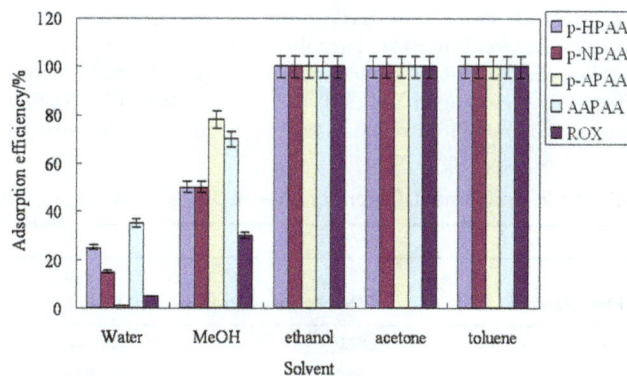

Figure 5. Effect of adsorption solvent on the adsorption efficiency of five phenylarsonic acids. The concentration of each phenylarsonic acids was 50 ng/mL. Adsorption time was 30 min. 5 mg adsorbent was used. (n = 3, with RSD<4.5%).

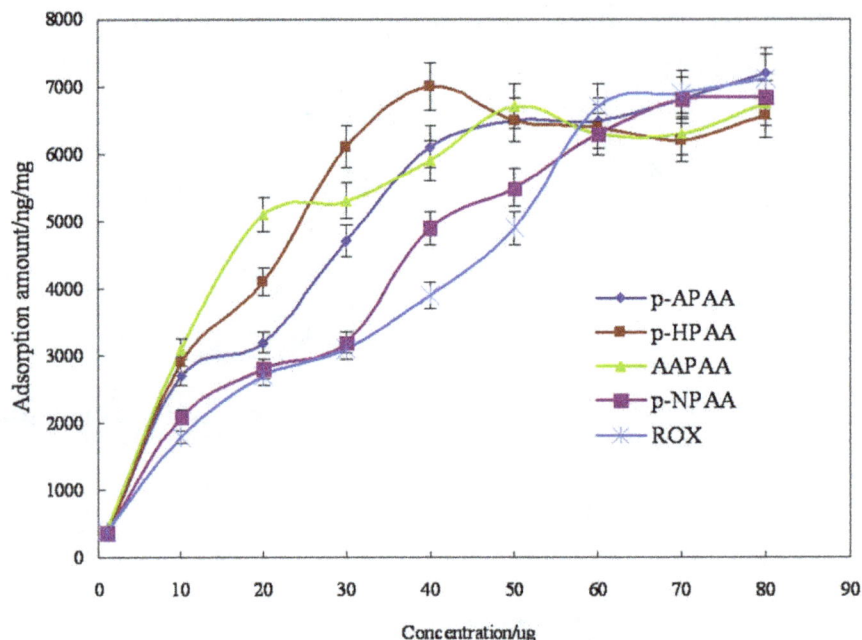

Figure 6. Adsorption saturation curve of CM-α-CD-Fe₃O₄ nanoparticles to five phenylarsonic acids. 5 mg adsorbent was used for each solution. Adsorption time was 30 min. Desorption was in water for 10 min. (n = 3, with RSD<5.7%).

around 3400 cm^{-1} is clearly visible. There is a strong Fe-O absorption peak at 580 cm^{-1} for both Fe_3O_4 and CM-α-CD-Fe_3O_4, suggesting the intactness of the Fe-O bond during the modification process. Compared with the innate Fe_3O_4 nanoparticles, new characteristic peaks at 1100 and 2900 cm^{-1} appeared in the spectrum of CM-α-CD- Fe_3O_4 nanoparticles, corresponding to C-O-C and C-H groups of CM-α-CD, which are also visible in the spectrum of pure CM-α-CD. This confirms the success of the modification of CM-α-CD to Fe_3O_4 nanoparticles.

The amount of CM-α-CD grafted on the surface of Fe_3O_4 and the number of CM-α-CD molecules immobilized on a single Fe_3O_4 nanoparticle can be estimated from the TGA results. It is known that CM-α-CD decomposes completely above 600°C. The TGA curves of Fe_3O_4 and CM- α-CD-Fe_3O_4 exhibit two steps of weight loss. In both cases, the first step can be attributed to the loss of residual water. The amounts of CM-α-CD coated on the surface of Fe_3O_4, calculated on the basis of the second-step loss, are 5.3% and 9.6% for Fe_3O_4 nanoparticles and CM-α-CD-Fe_3O_4 nanoparticles, respectively. The increase of the weight loss can be ascribed to CM-α-CD grafted on Fe_3O_4 nanoparticles (about 43 mg/g). According to Y. P. Wu [35], the number of CM-α-CD molecules immobilized on a single Fe_3O_4 can be calculated in equation (1):

$$N = \frac{4 \times (W_{CD} - W_{Fe_3O_4}) \times N_A \rho \pi R^3}{3 \times M_{CD} \times (1 - W_{CD}) \times 10^{21}} \qquad (1)$$

where N is the number of CM-α-CD molecules immobilized on each Fe_3O_4, R is the mean radius of CM-α-CD-Fe_3O_4 (5.0 nm based on the TEM results), ρ is the density of the nanoparticle (4.8 g/cm^3), N_A is Avogadro's number, W_{CD} and $W_{Fe_3O_4}$ are the weight losses of the CM-α-CD-Fe_3O_4 and Fe_3O_4 respectively, M_{CD} is the molar mass of CM-α-CD immobilized on CM-α-CD-Fe_3O_4. The calculated number of CM-α-CD molecules immobilized on each CM-α-CD-Fe_3O_4 is about 70.

Selection of adsorption solvent

The solvent for phenylarsonic acid solutions is an important factor affecting the adsorption efficiency. Five candidate solvents, i.e., water, methanol, ethanol, acetone and toluene, were compared (Fig. 5). The adsorption efficiency of all target arsenic species could reach 100% in ethanol, acetone and toluene. Ethanol was selected as the solvent in the following experiments

Table 2. Enrichment factors of CM-α-CD-Fe₃O₄ for 0.1 μg/L of each phenylarsonic acids in toluene solvent.

Phenylarsonic acids	Theory concentration (μg/L)	Experimental concentration (μg/L)	Enrichment factor	Enrichment Efficiency (%)
p-APAA	50.0	41.4	414	82.8
AAPAA	50.0	42.6	426	85.2
ROX	50.0	58.1	581	116.2
p-HPAA	50.0	60.6	606	121.2
p-NPAA	50.0	34.9	349	69.8

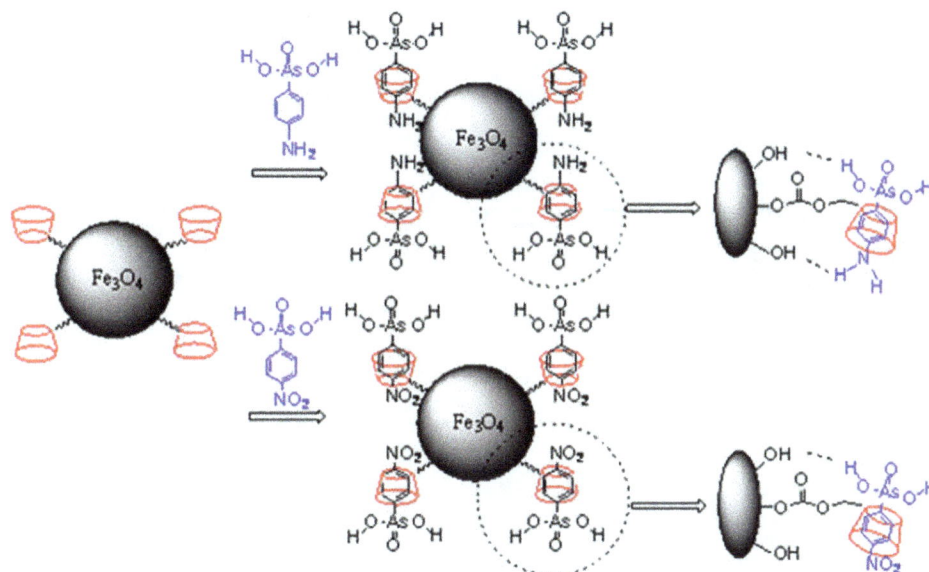

Figure 7. The likely interactions between CM-α-CD-Fe₃O₄ and the acids p-APAA and p-NPAA.

due to the relatively low toxicity. In water and methanol, the adsorption efficiencies of the five target compounds are very poor, which might be because of the water and methanol possess higher polarity than ethanol, acetone and toluene.

Static adsorption

Saturation of adsorption. The adsorption saturation curve is shown in **Fig. 6**. The point of saturation was reached at 40 µg for p-HPAA, p-APAA, and AAPAA, whereas it was 60 µg for p-NPAA and ROX, corresponding to about 0.2 µmol of each phenylarsonic acid.

The amount of CM-α-CD coated on the surface of Fe_3O_4 could be estimated from the saturation of adsorption. Since one phenylarsonic acid molecule was supposed to fit one CD cavity, there should be approximately equal amount of CD and phenylarsonic acid. Thus, the amount of CM-α-CD might be calculated through its molecular mass of 1029 according to equation (2). The estimated amount of grafted CM-α-CD was about 41.2 mg/g, which was similar to 43 mg/g calculated by TGA in 3.2.3.

$$m_{CM-a-CD} = \frac{n_{saturated} \times M_{CM-a-CD}}{m_{adsorbent}} \qquad (2)$$

Enrichment factors of low concentration. The lowest concentrations of phenylarsonic acids that could be absorbed by CM-α-CD-Fe₃O₄ as well as enrichment factors were investigated. Ultimately, the enrichment factors of p-APAA, AAPAA, ROX, p-HPAA and p-NPAA were 414, 426, 581, 606 and 349 (theoretical enrichment factor was 500) at the concentration of 0.1 µg/L. Enrichment efficiency was 69.8%-121.2% and experimental results are shown in **Table 2**.

Dynamic adsorption

Optimization of adsorption time and desorption time. The water was selected as desorption solvent. The adsorption efficiency was optimized by varying the adsorption time in the range of 1–40 min with all other parameters held constant. The adsorption efficiency increased with the adsorption time from 1 to 30 min followed by a plateau. The rate of adsorption was so high that 5 min was enough to adsorb the target compounds. Similarly, desorption efficiency was optimized by varying desorption time in the range of 1–10 min. The desorption efficiency increased with desorption time till 10 min, at which point it plateaued. Therefore, 10 min was selected as desorption time and desorption efficiency was achieved above 75%.

The tolerance of coexisting inorganic ions and organic analogues. By fixing the concentration of each phenylarsonic acids at 50 µg/L and 12 coexisting inorganic ions including Mn^{2+}, Cu^{2+}, K^+, Zn^{2+}, Ba^{2+}, Fe^{3+}, Mg^{2+}, Ca^{2+}, Pb^{2+}, Cr^{3+}, Cd^{2+}, and

Table 3. Concentration (µg/L) of coexisting analogues when tolerance factor less than 5%.

	phenol	benzoic acid	p-hydroxbenzoic acid	m-hydroxbenzoic acid
AAPAA	500	500	500	500
p-HPAA	500	250	250	250
p-NPAA	50	500	50	250
p-APAA	>500	500	500	250
ROX	250	500	500	250

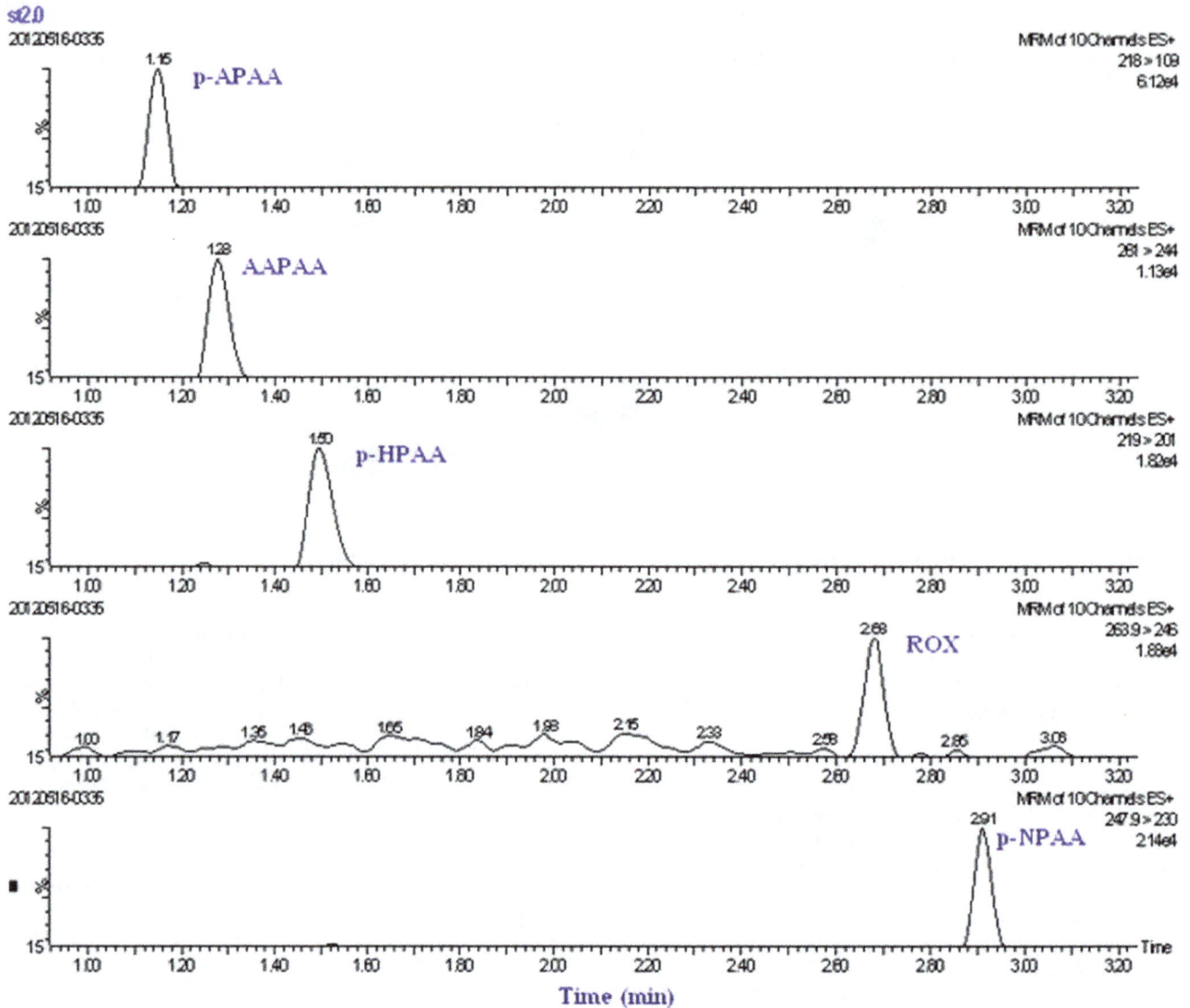

Figure 8. MRM chromatogram of five arsenical compounds standards at 2.0 μg/L.

Sn^{2+} at concentrations of 500, 5000, and 50000 μg/L for each inorganic ions respectively, interference of inorganic ions on the adsorption of phenylarsonic acids was studied. No significant interferences were observed under the optimum conditions described above in the presence of inorganic ions as high as 50000 μg/L. The influence of four organic analogues including phenol, benzoic acid, p-hydroxbenzoic and m-hydroxbenzoic acid at concentrations of 50, 250 and 500 μg/L for each organic

Table 4. Analytical performance data with UPLC-MS/MS.

Compounds	Linear range (μg/L)	Linear equation	R^2	LOD (μg/kg)
p-APAA	0.2–10.0	y = 7305.3x+2211.7	0.9952	0.05
AAPAA	0.2–10.0	y = 2287.2x+398.63	1.0000	0.05
p-HPAA	0.2–10.0	y = 5739.4x+935.79	0.9984	0.10
ROX	0.2–10.0	y = 3249.2x-144.89	1.0000	0.10
p-NPAA	0.2–10.0	y = 1712.3x-87.287	0.9988	0.11

x: the concentration of five phenylarsonic acids, y: peak area. LOD = 3 S/N of blank chicken tissues sample.

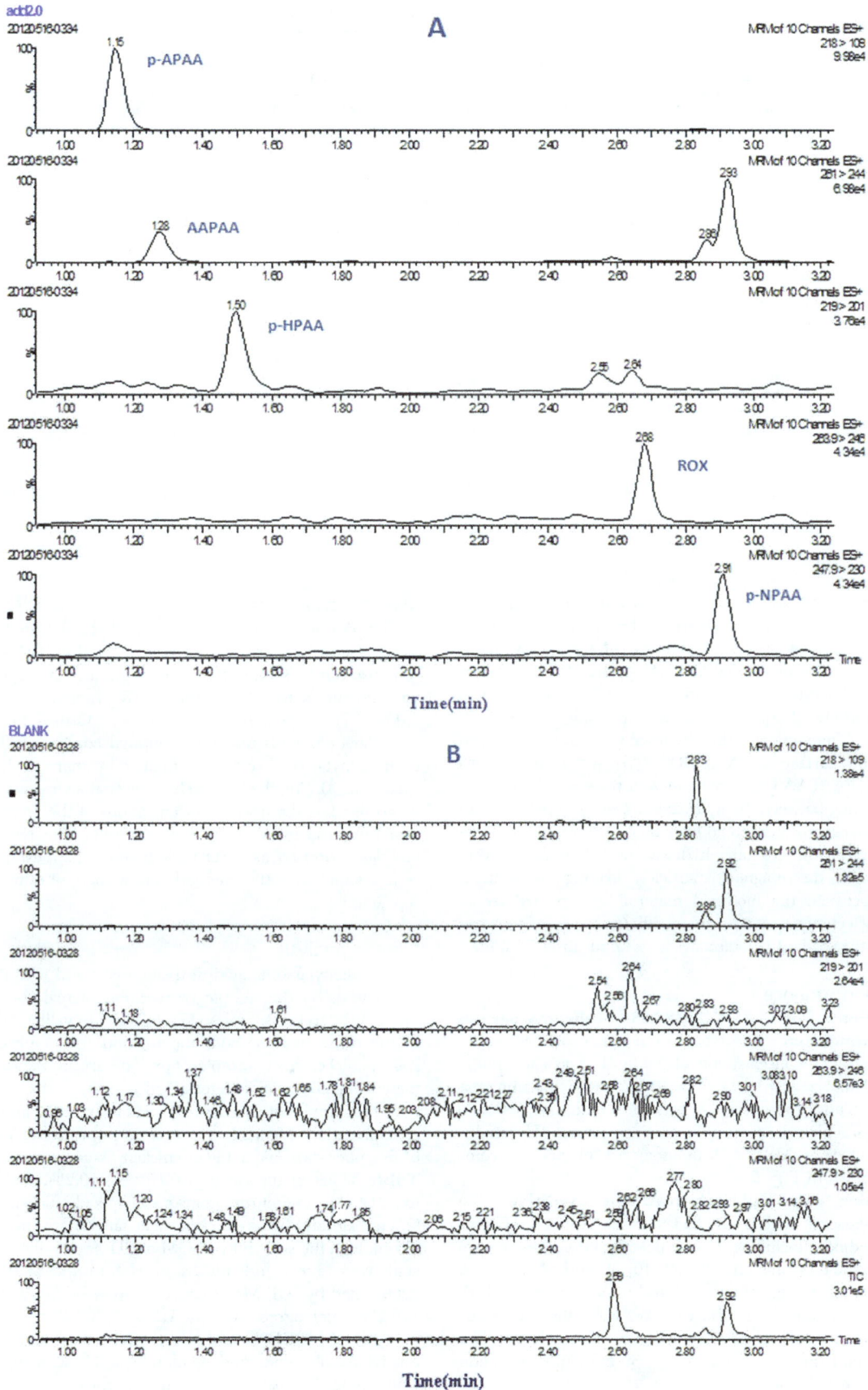

Figure 9. MRM chromatogram of spiked chicken meat (2.0 µg/kg) (A) and blank (B).

Table 5. Recoveries of spiked arsenics in chicken liver and muscle sample (n = 5).

Compounds	Chicken liver			Chicken meat		
	Added (μg/kg)	Recovery (%)	RSD (%)	Added (μg/kg)	Recovery (%)	RSD (%)
p-APAA	0.2	80.7	5.0	0.2	87.3	8.0
	2.0	89.5	6.7	2.0	86.1	7.6
	20	87.7	4.2	20	101.8	5.2
p-HPAA	0.2	94.3	4.9	0.2	77.2	9.1
	2.0	97.9	6.6	2.0	95.1	5.1
	20	93.7	12.5	20	79.9	5.2
p-NPAA	0.2	102.2	4.7	0.2	91.8	7.6
	2.0	110.2	5.4	2.0	98.8	5.7
	20	99.0	10.8	20	102.8	6.3
AAPAA	0.2	82.4	4.4	0.2	83.7	12.1
	2.0	82.6	5.4	2.0	82.6	8.1
	20	78.3	2.0	20	79.6	8.5
ROX	0.2	93.4	2.9	0.2	96.9	10.1
	2.0	104.2	7.3	2.0	104.2	6.4
	20	106.1	4.8	20	108.7	5.2

analogues respectively on the adsorption of target arsenic species was investigated. As shown in **Table 3**, the allowed concentrations of coexisting analogues were about 500 μg/L or 250 μg/L with a tolerance factor of less than 5%. However, p-NPAA was an exception. The tolerance concentration of p-hydroxybenzoic acid was only 50 μg/L. Possible explanation is shown in **Fig. 7**. Despite other interactions, the hydrogen bonding between hydroxyl on the surface of CM-α-CD-Fe$_3$O$_4$ and amino groups of p-APAA and AAPAA was very strong but there was hardly any hydrogen bonding between the hydroxyl and nitro of p-NPAA. In any case, phenylarsonic acid could form inclusion complex with α-CD and therefore the arsenate, hydroxyl or amino group were forced closer to the magnetic particles, leading to stronger interactions between the modified material and phenylarsonic acids. Both selectivity and adsorption efficiency were improved compared to that of Fe$_3$O$_4$ nanoparticles without modification.

Analytical performance

Optimization of UPLC-MS-MS chromatograph separation condition. A mixture of five arsenic compound standards, each at a concentration of 2.0 μg/L, was successfully separated and analyzed by UPLC-MS/MS in less than 3 min on a C18 column. Under the optimal separation conditions, baseline separation was achieved for every arsenic compound. The multi-reaction monitoring (MRM) chromatograms of arsenic compounds are shown in **Fig. 8**.

Optimization of sample pretreatment condition. To extract phenylarsonic acids from chicken tissue samples, different extraction methods (including ultrasonic extraction, microwave extraction), extraction time (10, 20, 30, 40, 50 and 60 min) and extraction solvents (methanol, ethanol and toluene) were studied. The optimal recovery was above 75% with the ultrasonic extraction method and toluene as the extraction solvent. The proper extraction time was found to be 30 min and extraction should be conducted twice.

Method evaluation

The developed method was validated by determining the linearity and LOD of arsenic species listed in **Table 4**. A linear response can be seen in the concentration range of 0.20–10 μg/L (enrichment factor 2.5), with the R^2 ranging from 0.9951 to 1.0000. The repeatability study was performed for each of the phenylarsonic acids under the optimal conditions. The LOD of each phenylarsonic acid was estimated by analyzing blank samples spiked at 0.2 μg/kg of each target analytes and they were determined as the lowest concentrations of the analyte for which signal-to-noise ratios were 3 respectively. The resultant repeatabilities expressed as RSD varied from 0.85% to 4.49%. These results show that the method has a high sensitivity and good repeatability.

Sample analysis

The phenylarsonic acid in tissue was stored with the prototype compound. So the sample preparation of real tissue samples is same as the spiked samples [17]. In order to validate the suitability of the developed method, the method was applied to analyze spiked chicken tissue samples. For comparison, the total arsenic in these samples was also determined.

In the samples of chicken meat and liver, 0.2, 2 and 20 μg/kg of each phenylarsonic acids were spiked, respectively. The recoveries of five phenylarsonic acids in chicken tissue samples, as shown in **Table 5**, fell in the ranges of 77.2%–110.2%, with a RSD less than 12.5%. The chromatograms of spiked chicken meat sample (2.0 μg/kg) and blank chicken meat sample are shown in **Fig. 9**. To confirm the selectivity of CM-α-CD-Fe$_3$O$_4$, five phenylarsonic acids were spiked individually in each sample at 0.2 μg/kg and determined by ICP-MS. Recoveries were in the range of 81.5%–119.2% which agreed with the UPLC-MS/MS results. This shows that the CM-α-CD-Fe$_3$O$_4$ nanoparticles have high selectivity for phenylarsonic acid and can be used for sample clean-up. Combining CM-α-CD-Fe$_3$O$_4$ nanoparticles clean-up procedure with UPLC-MS-MS technique would potentially result in practical application in the analysis of trace phenyl arsenic acids.

Conclusions

In this work, CM-α-CD-Fe$_3$O$_4$ nanoparticles were synthesized to selectively extract and enrich phenylarsonic acids. CM-α-CD-Fe$_3$O$_4$ nanoparticles exhibited excellent selectivity and adsorption efficiency for five phenylarsonic acids because of the size selectivity of α-CD and the affinity of Fe$_3$O$_4$ to arsenic. In the sub ppb level of phenylarsonic acids, the enrichment factor was higher than 400 and the extraction efficiency higher than 70%. Coupled with UPLC-MS/MS, a fast, selective and convenient analytical method for the determination of phenylarsonic acid was developed. Comparing with published method [14], the developed method showed satisfactory sensitivity due to the selectively adsorption of CM-α-CD-Fe$_3$O$_4$ nanoparticles for phenylarsonic acids in complex sample matrix.

Supporting Information

Figure S1 FTIR spectra of Fe$_3$O$_4$ magnetic nanoparticles (a), CM-α-CD (b) and CM-α- CD- Fe$_3$O$_4$ (c).

Author Contributions

Conceived and designed the experiments: PW RZ. Performed the experiments: JJ. Analyzed the data: JW PW. Contributed reagents/materials/analysis tools: WZ PW. Contributed to the writing of the manuscript: PW RZ.

References

1. Jaafar J (2001) Separation of phenylarsonic compounds by ion-pairing reversed phase-high performance liquid chromatography. Jurnal Teknologi C UTM 35: 71–83.

2. Wang PL, Zhao GL, Tian J, Su XO (2010) High-performance liquid chromatography-inductively coupled plasma mass spectrometry based method for the determination of organic arsenic feed additives and speciation of anionic arsenics in animal feed. J. Agric. Food Chem 58(9): 5263–70.

3. Chiou PW, Chen KL, Yu B (1997) Effects of roxarsone on performance, toxicity, tissue accumulation and residue of eggs and excreta in laying hens. J. Sci. Food Agric 74(2): 229–236.

4. Bartel M, Ebert F, Leffers L (2011) Toxicological characterization of the inorganic and organic arsenic metabolite thio-DMA(V) in cultured human lung cells. J. Toxicol 2011: 1–9.

5. FDA (2011) Study 275. 30 Arsenic speciation in broiler chickens. a) Summary Final Report. b) Amendments to final report. c) Analyst's report. d) Statistician's Report.

6. Thirunavukkarasu OS, Viraraghavan T, Subramanian KS, Tanjore S (2002) Organic arsenic removal from drinking water. Urban Water 4: 415–21.

7. Samanta G, Chowdhury TR, Mandal BK, Biswas BK, Chowdhury UK, et al. (1999) Flow Injection Hydride Generation Atomic Absorption Spectrometry for Determination of Arsenic in Water and Biological Samples from Arsenic-Affected Districts of West Bengal, India, and Bangladesh. Microchemical J 62(1): 174–91.

8. Ma MS, Le XC, Hu B, Jiang ZC (2000) High performance liquid chromatography with hydride generation atomic fluorescence detection for rapid urinary arsenic speciation. J. Ana. Sci 16: 89–96.

9. Liu JJ, Yu HX, Song HB, Qiu J, Sun FM, et al. (2008) Simultaneous determination of p-arsanilic acid and roxarsone in feed by liquid chromatography-hydride generation online coupled with atomic fluorescence spectrometry. J. Environ. Monit 10: 975–8.

10. Roerdink AR, Aldstadt JH (2004) Sensitive method for the determination of roxarsone using solid-phase microextraction with multi-detector gas chromatography. J. Chromatogr. A 1057 (1-2): 177–83.

11. Li P, Hu B (2011) Sensitive determination of phenylarsenic compounds based on a dual preconcentration method with capillary electrophoresis/UV detection. J. Chromatogr. A 1218 (29): 4779–87.

12. Rosal CG, Momplaisir GM, Heithmar EM (2005) Roxarsone and transformation products in chicken manure: determination by capillary electrophoresis-inductively coupled plasma-mass spectrometry. Electrophoresis 26(7-8) 1606–14.

13. Bednar AJ, Garbarino JR, Burkhardt MR, Ranville JF, Wildeman TR (2004) Field and laboratory arsenic speciation methods and their application to natural-water analysis. Water Res 38 (2): 355–64.

14. Pergantis SA, Winnik W, Betowski D (1997) Determination of ten organoarsenic compounds using microbore high-performance liquid chromatography coupled with electrospray mass spectrometry-mass spectrometry. J. Anal. At. Spectrom 12: 531–6

15. Chen SL, Yeh SJ, Yang MH, Lin TH (1995) Trace element concentration and arsenic speciation in the well water of a Taiwan area with endemic Blackfoot disease. Biol. Trace Elem. Res 48 (3): 263–74.

16. Kuehnelt D, Goessler W (2003) Organometallic compounds in the environment, ed. P. J. Craig, John Wiley & Sons Ltd., Chichester, pp. 223–75.

17. Conklin SD, Shockey N, Kubachka K, Howard KD, Carson MC (2012) Development of an ion chromatography-inductively coupled plasma-mass spectrometry method to determine inorganic arsenic in liver from chickens treated with roxarsone. J. Agric. Food Chem. 60(37): 9394–404.

18. Chen WR, Huang CH (2012) Surface adsorption of organoarsenic roxarsone and arsanilic acid on iron and aluminum oxides. J. Hazard. Mater 227–228: 378–85.

19. Mao XJ, Chen BB, Hu B (2011) Titania immobilized polypropylene hollow fiber as a disposable coating for stir bar sorptive extraction–high performance liquid chromatography–inductively coupled plasma mass spectrometry speciation of arsenic in chicken tissues. J.Chromatogr.A 1218 (1):1–9.

20. Mayo JT, Yavuz C, Yean S, Cong L, Shipley H, et al. (2007) The effect of nanocrystalline magnetite size on arsenic removal. Sci.Technol. Adv. Mater 8(1-2):71–5.

21. Akin I, Arslan G, Tor A, Ersoz M, Cengeloglu Y (2012) Arsenic(V) removal from underground water by magnetic nanoparticles synthesized from waste red mud. J. Hazard. Mater 235–236: 62–8.

22. Li RX, Liu SM (2011) Preparation and characterization of cross-linked β-cyclodextrin polymer/Fe$_3$O$_4$ composite nanoparticles with core-shell structures. Chin. Chem. Lett 22 (2): 217–20.

23. Kan XW, Zhao Q, Shao DL, Geng ZR, Wang ZL, et al. (2010) Preparation and Recognition Properties of Bovine Hemoglobin Magnetic Molecularly Imprinted Polymers. J. Phys. Chem. B 114 (11): 3999–4004.

24. Liu ZG, Zhang FS, Sasai R (2010) Arsenate removal from water using Fe$_3$O$_4$-loaded activated carbon prepared from waste biomass. Chem. Eng. J. 160 (1): 57–62.

25. Badruddoza AZM, Hidajat K, Uddin MS (2010) Synthesis and characterization of β-cyclodextrin-conjugated magnetic nanoparticles and their uses as solid-phase artificial chaperones in refolding of carbonic anhydrase bovine. J. Colloid. Interface Sci 346 (2): 337–46.

26. Zhang CC, Li X, Pang JX (2001) Synthesis and adsorption properties of magnetic resin microbeads with amine and mercaptan as chelating groups. J. Appl. Polym. Sci 82 (7) 1587–92.

27. Atia AA, Donia AM, El-Enein SA, Yousif AM (2007) Effect of Chain Length of Aliphatic Amines Immobilized on a Magnetic Glycidyl Methacrylate Resin towards the Uptake Behavior of Hg(II) from Aqueous Solutions. Sep. Sci. Technol. 42 (2): 403–20.

28. Atia AA, Donia AM, Yousif AM (2008) Removal of some hazardous heavy metals from aqueous solution using magnetic chelating resin with iminodiacetate functionality. Sep. Purif. Technol 61 (3): 348–57.

29. Mohan D, Pittman CU (2007) Arsenic removal from water/wastewater using adsorbents—A critical review. J. Hazard. Mater 142 (1-2): 1–53.

30. Badruddoza AZM, Hazel GSS, Hidajat K, Uddin MS (2010) Synthesis of carboxymethyl-β-cyclodextrin conjugated magnetic nano-adsorbent for removal of methylene blue. Colloids Surf. A 367 (1-3): 85–95.

31. Badruddoza AZM, Tay ASH, Tan PY, Hidajat K, Uddin MS (2011) Carboxymethyl-β-cyclodextrin conjugated magnetic nanoparticles as nano-adsorbents for removal of copper ions: Synthesis and adsorption studies. J. Hazard. Mater.185 (2-3): 1177–86.

32. Zhi J, Tian XL, Zhao W, Shen JB, Tong B, et al. (2008) Self-assembled film based on carboxymethyl-β-cyclodextrin and diazoresin and its binding properties for methylene blue. J. Colloid. Interface Sci. 319 (1): 270–6.

33. Wu W, He QG, Jiang CZ (2008) Magnetic Iron Oxide Nanoparticles: Synthesis and Surface Functionalization Strategies. Nanoscale Res. Lett 3(11): 397–415.

34. Krachler M, Radner H, Irgolic KJ (1996) Microwave digestion methods for the determination of trace elements in brain and liver samples by inductively coupled plasma mass spectrometry. Fresenius' Journal of Analytical Chemistry 355(2): 120–8.

35. Wu YP, Zuo F, Zheng ZH, Ding XB, Peng YX (2009) A Novel Approach to Molecular Recognition Surface of Magnetic Nanoparticles Based on Host–Guest Effect. Nanoscale Res. Lett. 4 (7): 738–47.

A Draft *De Novo* Genome Assembly for the Northern Bobwhite (*Colinus virginianus*) Reveals Evidence for a Rapid Decline in Effective Population Size Beginning in the Late Pleistocene

Yvette A. Halley[1], Scot E. Dowd[2], Jared E. Decker[3], Paul M. Seabury[4], Eric Bhattarai[1], Charles D. Johnson[5], Dale Rollins[6], Ian R. Tizard[1], Donald J. Brightsmith[1], Markus J. Peterson[7], Jeremy F. Taylor[3], Christopher M. Seabury[1]*

1 Department of Veterinary Pathobiology, College of Veterinary Medicine, Texas A&M University, College Station, Texas, United States of America, 2 Molecular Research LP, Shallowater, Texas, United States of America, 3 Division of Animal Sciences, University of Missouri, Columbia, Missouri, United States of America, 4 ElanTech Inc., Greenbelt, Maryland, United States of America, 5 Genomics and Bioinformatics Core, Texas A&M AgriLife Research, College Station, Texas, United States of America, 6 Rolling Plains Quail Research Ranch, Rotan, Texas, United States of America, 7 Department of Wildlife and Fisheries Sciences, Texas A&M University, College Station, Texas, United States of America

Abstract

Wild populations of northern bobwhites (*Colinus virginianus*; hereafter bobwhite) have declined across nearly all of their U.S. range, and despite their importance as an experimental wildlife model for ecotoxicology studies, no bobwhite draft genome assembly currently exists. Herein, we present a bobwhite draft *de novo* genome assembly with annotation, comparative analyses including genome-wide analyses of divergence with the chicken (*Gallus gallus*) and zebra finch (*Taeniopygia guttata*) genomes, and coalescent modeling to reconstruct the demographic history of the bobwhite for comparison to other birds currently in decline (i.e., scarlet macaw; *Ara macao*). More than 90% of the assembled bobwhite genome was captured within <40,000 final scaffolds (N50 = 45.4 Kb) despite evidence for approximately 3.22 heterozygous polymorphisms per Kb, and three annotation analyses produced evidence for >14,000 unique genes and proteins. Bobwhite analyses of divergence with the chicken and zebra finch genomes revealed many extremely conserved gene sequences, and evidence for lineage-specific divergence of noncoding regions. Coalescent models for reconstructing the demographic history of the bobwhite and the scarlet macaw provided evidence for population bottlenecks which were temporally coincident with human colonization of the New World, the late Pleistocene collapse of the megafauna, and the last glacial maximum. Demographic trends predicted for the bobwhite and the scarlet macaw also were concordant with how opposing natural selection strategies (i.e., skewness in the *r*-/*K*-selection continuum) would be expected to shape genome diversity and the effective population sizes in these species, which is directly relevant to future conservation efforts.

Editor: Axel Janke, BiK-F Biodiversity and Climate Research Center, Germany

Funding: The funders had no role in study design, data collection, data analysis, interpretation of the data or analyses, decision to publish, or drafting of the manuscript. This study was funded by private donations to CMS from Mr. Joe Crafton, members of Park Cities Quail, and the Rolling Plains Quail Research Ranch. DR directs the Rolling Plains Quail Research Ranch, which funded this study in part, but DR had no role in the primary analysis or interpretation of the data or analyses. DR provided reagents/materials/analysis tools and did make editorial comments and suggestions related to the final manuscript.

Competing Interests: The authors have the following competing interests to declare: SED and CDJ run sequencing service centers, SED is Owner of General Partner and CEO of Molecular Research LP, PMS is the brother of CMS and is also a collaborator and employee of ElanTech Inc. ElanTech Inc allows PMS to collaborate and participate in peer-reviewed publications. DR is now a retired Texas AgriLife Extension Wildlife Specialist who serves as the Director of the Rolling Plains Quail Research Ranch, which is a 501(c)(3) nonprofit organization.

* E-mail: cseabury@cvm.tamu.edu

Introduction

The northern bobwhite (*Colinus virginianus*; hereafter bobwhite) ranges throughout the United States (U.S.), Mexico and parts of the Caribbean, and is one of 32 species belonging to the family Odontophoridae (New World Quail) [1]. Within this family, the bobwhite is arguably the most diverse, with 22 named subspecies varying both in size (increasing from south to north) and morphology [1]. Specifically, the most overt morphological variation occurs on the head and underparts, which are marked by variable combinations of grey, brown, and white [1]. At

present, the bobwhite is one of the most broadly researched and intensively managed wildlife species in North America [2–4]. The suitability of the bobwhite as a model wildlife species for climate change, land use, toxicology, and conservation studies has also been well established [2–11].

Historically, the relative abundance of bobwhites across their native range has often been described as following a boom-bust pattern, with substantial variation in abundance among years [2,12–14]. Although broad scale declines in bobwhite abundance probably began somewhere between 1875 and 1905 [15–17], several better quantified studies of this long-term decline utilizing either breeding bird surveys or Christmas bird count data were reported beginning more than 20 years ago [5–6,18–21]. This range-wide decline in bobwhite abundance across most of the U.S. is still ongoing today [22–23]. The precise reasons for recent population declines in the U.S. appear to be a complex issue, and have been attributed to factors such as variation in annual rainfall [2,12–13], thermal tolerances of developing embryos within a period of global warming [24–25], shifts in land use and scale coupled with the decline of suitable habitat [2–3,14,20–21], red imported fire ants (Solenopsis invicta) [26–27], sensitivity to ecotoxins [28–29], and harvest intensity by humans [30–32], particularly during drought conditions [3,13]. Population declines have prompted intense recent efforts to translocate bobwhites to fragmented parts of their historic range where modern abundance is low. However, the results of these translocations have proven to be highly variable [33–35], with one such recent study demonstrating that bobwhites fail to thrive in historically suitable habitats that have since become fragmented [35]. Restocking via the release of pen-reared bobwhites has also been explored, with all such efforts achieving low survival rates [33,36–38], and those that do survive may potentially dilute local genetic adaptations via successful mating with remnant members of wild populations [38].

Historically, little genome-wide sequence and polymorphism data have been reported for many important wildlife species, thereby limiting the implementation of genomic approaches for addressing key biological questions in these species. However, the emergence of high-yielding, cost-effective next generation sequencing technologies in conjunction with enhanced bioinformatics tools have catalyzed a "genomics-era" for these species, with new avian genome sequence assemblies either recently reported or currently underway for the Puerto Rican parrot (Amazona vittata) [39], flycatchers (Ficedula spp) [40], budgerigar (Melopsittacus undulatus; http://aviangenomes.org/budgerigar-raw-reads/), saker and peregrine falcons (Falco peregrinus; Falco cherrug) [41], Darwin's finch (Geospiza fortis; http://gigadb.org/darwins-finch/), and the scarlet macaw (Ara macao) [42]. At present, the bobwhite is without an annotated draft genome assembly, thereby precluding genome-wide studies of extant wild bobwhite populations, and the utilization of this information to positively augment available management strategies. Likewise, utilization of the bobwhite as an experimental wildlife model cannot be fully enabled in the absence of modern genomic tools and resources.

Cytogenetic analyses have demonstrated that the bobwhite diploid chromosome number is 2n = 82, which includes 5 pairs of autosomal macrochromosomes and the sex chromosomes, 8 pairs of intermediately sized autosomes, and 27 pairs of autosomal microchromosomes [43–44]. Recent genomic efforts have focused on generating bobwhite cDNA sequences for the construction of a custom microarray (8,454 genes) to study the physiological effects of ecotoxicity [11], and for comparative studies with the annotated domestic chicken (Gallus gallus) genome [45]. However, no genome maps (i.e., linkage, radiation hybrid, BAC tiling paths) exist for the bobwhite. Consequently, we utilized >2.3 billion next generation sequence reads produced from paired-end (PE) and mate pair (MP) libraries to produce a draft de novo genome sequence assembly for a wild female bobwhite, and compared our assembly to other established and well-annotated avian reference genome assemblies [46–48]. We also used three in silico approaches to facilitate genome annotation, and assessed the genomic information content of the draft bobwhite assembly via comparative sequence alignment to the chicken (G. gallus 4.0) and zebra finch genomes (T. guttata 3.2.4) followed by a genome-wide analysis of divergence [42]. Finally, we inferred the population history of the bobwhite and compared it to the scarlet macaw using whole-genome

sequence data generated for both species. The results of this study facilitate genome-wide analyses for the bobwhite, and also enable modern genomics research in other evolutionarily related birds for which research funding is limited.

Results and Discussion

Genome Sequencing and de novo Assembly

Herein, we assembled a genome sequence for Pattie Marie, a wild, adult female bobwhite from Texas. All sequence data were generated with the Illumina HiSeq 2000 sequencing system (v2 Chemistry; Illumina Inc.; San Diego, CA). As previously described [42], we estimated the bobwhite nuclear genome size to be ≈1.19–1.20 Gigabase pairs (Gbp; See Methods). While this estimate does not fully account for the lack of completeness in all existing avian genome assemblies (i.e., collapsed repeats), it is useful for determining whether the majority of the bobwhite genome was captured by our de novo assembly. Collectively, more than 2.36 billion trimmed sequence reads derived from three libraries (see Methods) were used in the assembly process (Table 1), which yielded ≥142× theoretical genome coverage (1.19–1.20 Gbp) as input data, and ≥77× assembled coverage (Table 2). Summary and comparative data for major characteristics of the bobwhite draft de novo genome assembly are presented in Table 2, which also includes a comparison to the initial releases of two established and well annotated avian reference genomes from the order Galliformes [46–47].

To assess the consistency of our assembly and scaffolding procedures, and to facilitate fine-scale analyses of divergence as previously described, we produced a simple de novo (i.e. no scaffolding; hereafter NB1.0) and a scaffolded de novo assembly (hereafter NB1.1), with the scaffolding procedure using both PE and MP reads to close gaps and join contigs. The concordance between the two assemblies was profound, with >90% of the simple de novo contig sequences mapping onto the scaffolded assembly with zero alignment gaps (Table 2, Table S1). Our first generation scaffolded assembly contained 1.172 Gbp (including N's representing gaps; 1.047 Gbp of unambiguous sequence) distributed across 220,307 scaffolds, with a N50 contig size of 45.4 Kbp (Table 2). Moreover, >90% of the assembled genome was captured within <40,000 scaffolds (Fig. 1). Importantly, these results meet or exceed similar quality benchmarks and summary statistics initially described for several other avian genome assemblies (i.e., Puerto Rican parrot, scarlet macaw, chicken, turkey) [39,42,46–47], but do not exceed summary statistics (i.e., scaffold N50, etc) for some recent assemblies (i.e., Flycatcher, Peregrine and Saker Falcons) that utilize either ultra-large insert mate pair libraries and/or available maps for enhanced scaffolding [40–41].

Comparative Genome Alignment, Predicted Repeat Content, and Genome-Wide Variant Detection

Both bobwhite genome sequence assemblies (NB1.0; NB1.1) were aligned to the available chicken (G. gallus 4.0) and zebra finch (T. guttata 3.2.4) reference genomes via blastn (Tables S2 and S3), which allowed for orientation of most de novo contigs to their orthologous genomic positions, additional quality control investigations regarding our scaffolding procedure (Table S1), and a genome-wide analysis of divergence with quality control analyses as previously described [42]. Examination of the NB1.0 blastn alignments (E-value and bitscore top hits) across all chicken nuclear chromosomes revealed very stable levels of nucleotide divergence (overall percent identity, Median = 83.20%, Mean = 82.94%), with alignments to GGA24 and GGA16

Table 1. Summary of Illumina sequence data used for *de novo* assembly of the bobwhite genome.

Data Source	Total Reads[a]	Library Type	Insert Size PD Dist. (bp)[b]	Average Read Length (bp)[c]
Illumina HiSeq	1,575,625,135	Small Insert Paired End	230–475[c]	84
Illumina HiSeq	510,031,444	Mate Pair (Small)	2100–3100[c]	49
Illumina HiSeq	276,134,302	Mate Pair (Medium)	4600–6000[c]	50

[a]Total usable reads after quality and adapter trimming (n = 2,361,790,881).
[b]Insert size and corresponding range of paired distances for each Illumina sequencing library.
[c]Averages for quality and adapter trimmed reads, rounded to the nearest bp.

producing the highest (Median = 85.08%, Mean = 85.05%) and lowest (Median = 76.88%, Mean = 75.48%) percent identities, respectively (Table S2). Evaluation of the NB1.0 blastn alignments (E-value and bitscore top hits) across all zebra finch nuclear chromosomes also revealed stable but greater overall levels of nucleotide divergence (overall percent identity, Median = 77.30%, Mean = 79.04%), with alignments to TGU-LGE22 as well as TGU28 producing the highest (Median ≥ 81.62%, Mean ≥ 81.76%), and TGU16 the lowest (Median = 74.48%, Mean = 75.41%) percent identities, respectively (Table S2). Similar trends in nucleotide divergence were also observed for the NB1.1 blastn alignments to the chicken and zebra finch nuclear chromosomes (Table S3), with greater nucleotide divergence from the zebra finch genome being compatible with larger estimated divergence times (100–106 MYA), as compared to the chicken (56–62 MYA; http://www.timetree.org/) [49–50].

The minimum estimated repetitive DNA content (excluding N's) for the scaffolded bobwhite genome was approximately 8.08%, as predicted by RepeatMasker (RM; Table 3; Table S4). This estimate was greater than those reported for the Puerto Rican parrot, saker and peregrine falcon, scarlet macaw, turkey, and zebra finch genomes using RM [39,41–42,47–48], but less than that reported for the chicken genome [46]. However, read-based scaffolding involving the insertion of "N's" into gaps is known to result in the underestimation of genome-wide repetitive content [42]. Nevertheless, a common feature of the bobwhite, scarlet macaw, chicken, turkey, and zebra finch genomes is the high proportion of LINE-CR1 interspersed repeats [42,46–48] that are conserved across these divergent avian lineages. In fact, the majority of the predicted repeat content in the bobwhite genome consisted of interspersed repeats, of which most belong to four groups of transposable elements including SINEs, L2/CR1/Rex non-LTR retrotransposons, retroviral LTR retrotransposons, and at least three DNA transposons (hobo Activator, Tc1-IS630-Pogo, PiggyBac). Similar to the chicken, the bobwhite genome was predicted to contain about one third as many retrovirus-derived LTR elements as the zebra finch [48], but more SINEs than the chicken [46,48]. To further evaluate the repetitive content within the bobwhite genome, we utilized PHOBOS (v3.3.12) [51] to predict and characterize genome-wide tandem repeats (microsatellite loci) for the purpose of identifying loci that could be utilized for population genetic studies. Collectively, we identified 3,584,054 tandem repeats (Table S5) consisting of 2 to 10 bp sequence motifs that were repeated at least twice, which is greater than 50% more tandem repeats than was recently predicted for the scarlet macaw [42]. Bobwhite tandem repeats were characterized as follows: 644,064 di-, 997,112 tri-, 577,913 tetra-, 518,315 penta-, 552,957 hexa-, 143,590 hepta-, 93,583 octa-, 35,260 nona-, and 21,260 decanucleotide microsatellites (Table S5). Importantly, microsatellite genotyping as a means to assess parentage, gene flow, population structure, and covey composition within and between bobwhite populations has historically been limited to very few genetic markers [38,52–53], and therefore, the

Table 2. Summary data for the bobwhite *de novo* genome assembly with comparison to the initial turkey and chicken genome assemblies.

Genome Characteristics	Simple *de novo* Bobwhite 1.0[a]	Scaffolded Bobwhite 1.1[b]	Turkey 2.01	Chicken 1.0
Total Contig Length[c]	1.042 Gbp	1.047 Gbp	0.931 Gbp	1.047 Gbp
Total Contigs >1 Kb	198,672	65,833	128,271	98,612
N50 Contig Size	6,260 bp	45,400 bp	12,594 bp	36,000 bp
Largest Contig	163,812 bp	600,691 bp	90,000 bp	442,000 bp
Total Contigs	374,224	220,307	152,641	NA[d]
Contig Coverage	≥100×[e]	≥77×[f]	17×	7×
Cost (M = million)	<$0.020M[g]	<$0.020M[g]	<$0.250M	>$10M

[a]No scaffolding procedure implemented (NB1.0).
[b]Scaffolding based on paired reads (NB1.1); no genome maps or BACs were available.
[c]Excluding gaps; scaffolded assembly with gaps (i.e., N's) = 1.172 Gbp.
[d]Not provided; see [46].
[e]Median and average coverage, excluding contigs with coverage >300× (n = 4,293).
[f]Median and average coverage, excluding scaffolds with coverage >300× (n = 3,717).
[g]The one-time cost of sequencing also reflects all library costs.

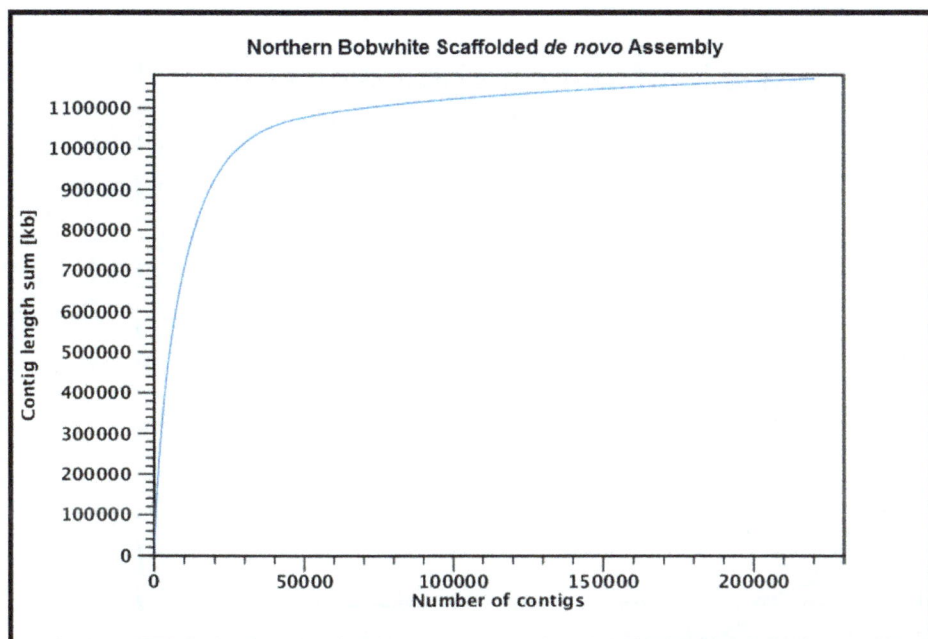

Figure 1. Relationship Between Total Contig Length (Kbp) and Total Contig Number for the Scaffolded Bobwhite (Colinus virginianus) Genome (NB1.1). The y-axis represents total contig length, expressed in kilobase pairs (Kbp), and the x-axis represents the total number of scaffolds. The bobwhite genome was estimated to be 1.19–1.20 Gbp. For NB1.1 (1.172 Gbp), >90% of the assembled genome was captured within <40,000 scaffolds.

resources described herein will directly enable genome-wide population genetic studies for the bobwhite.

To provide the first characterization of genome-wide sequence variation for a wild bobwhite, we investigated the frequency and distribution of putative single nucleotide polymorphisms (SNPs) and small insertion-deletion mutations resulting from biparental inheritance of alternative alleles (heterozygosity) within the repeat-masked scaffolded *de novo* assembly (NB1.1). Collectively, 3,503,457 SNPs and 268,981 small indels (Coverage $\geq 10\times$ and $\leq 572\times$) were predicted (Fig. 2), which corresponds to an average genome-wide density (i.e., intra-individual variation) of approxi-

mately 3.22 heterozygous polymorphisms per Kbp for the autosomes. Considering only high quality putative SNPs, the bobwhite heterozygous SNP rate was approximately 2.99 SNPs per Kbp. This estimate is four times greater than that reported for the peregrine falcon, more than three times greater than for the scarlet macaw and saker falcon, approximately twice that of the zebra finch and turkey, and is second only to the chicken and the flycatcher, which are most similar to the bobwhite in terms of putative heterozygous SNPs per Kbp [40–42,47–48,54]. Despite evidence for recent population declines across the majority of the bobwhite's historic U.S. range [5–6,20–23], our wild Texas bobwhite possesses extraordinary levels of genome-wide variation as compared to most other avian species for which draft *de novo* genome assemblies are currently available.

Bobwhite Population History as Inferred From Whole-Genome Sequence Data

Using high-quality autosomal SNP density data, we implemented a pairwise sequentially Markovian coalescent (PSMC) model [55] to reconstruct the demographic history of our wild bobwhite (Pattie Marie), and for comparison, we also produced a PSMC analysis for a wild female scarlet macaw (Neblina; Fig. 3) [42]. For both species, we inferred their demographic history using the per-site pairwise sequence divergence to represent time, and the scaled mutation rate to represent population size [55]. Importantly, many biological characteristics associated with the bobwhite are largely typical of an *r*-selected avian species, whereas the scarlet macaw clearly exhibits characteristics of *K*-selection [56–59]. However, despite the fundamental biological differences in how these two avian species achieve reproductive success within their respective habitats, both species experienced pronounced bottlenecks which were predicted to begin approximately 20–58 thousand years ago (kya), with the range in timing of this interval being a product of modeling a range of underlying mutation rates (Fig. 3; See

Table 3. Major classes of repetitive content predicted by RepeatMasker within the bobwhite NB1.1 scaffolded *de novo* assembly.

Repeat Type	Total	Total bp (% of Genome)[a]
Predicted	Elements[a]	
SINEs	4,425	545,252 (0.047%)
LINEs (L2/CR1/Rex)	172,398	44,762,255 (3.818%)
LTR Retroviral	31,766	8,987,247 (0.767%)
DNA Transposons	22,793	6,863,495 (0.585%)
Unclassified Interspersed Repeats	2,096	337,844 (0.0288%)
Small RNA	757	70,666 (0.006%)
Satellites	3,624	580,253 (0.050%)
Low Complexity & Simple Repeats	403,599	32,608,785 (2.781%)
Totals	**641,458**	**94,755,797 (8.08%)**

[a]Scaffolded *de novo* assembly NB1.1 (1.17 Gb including gaps with N's).

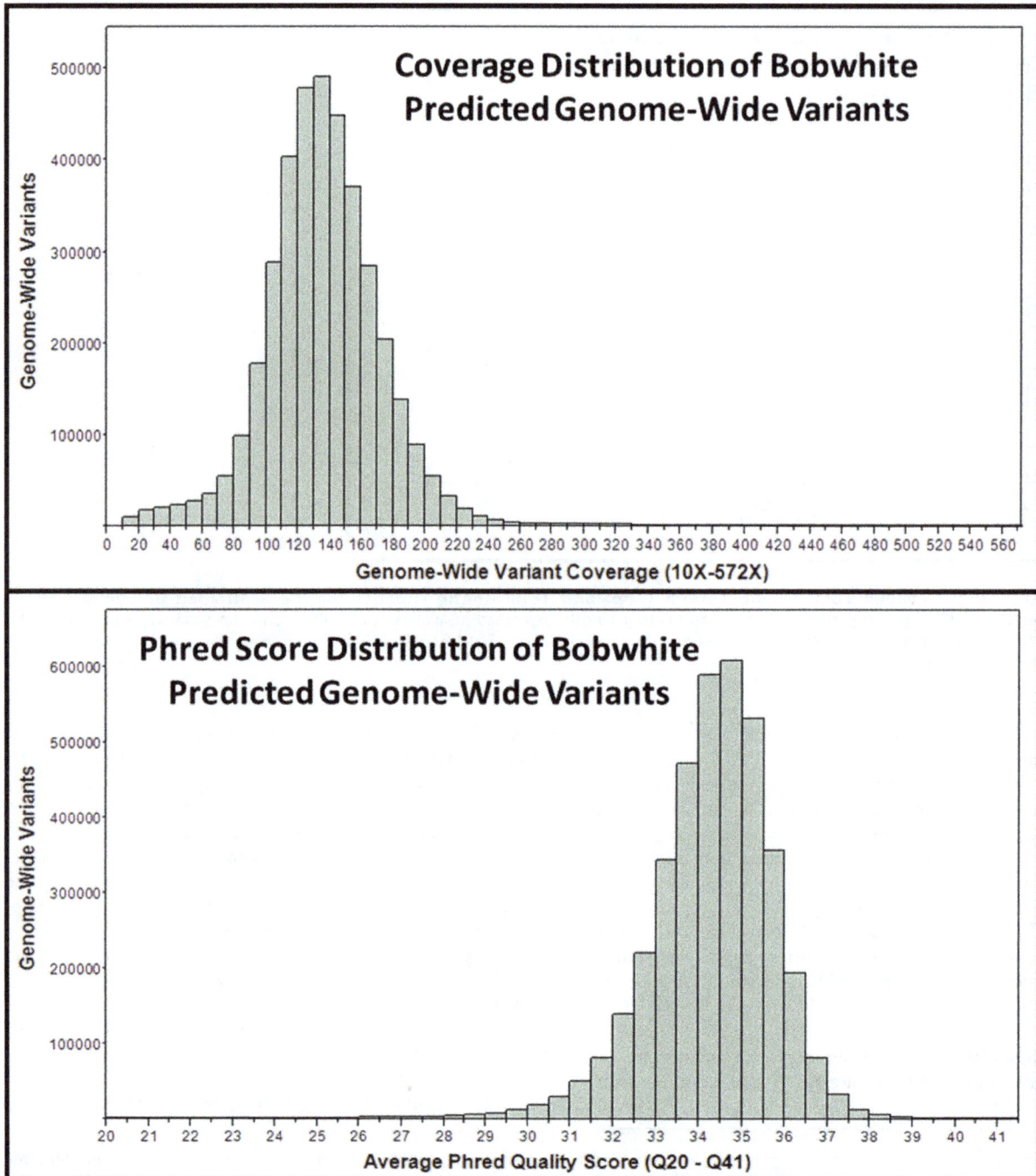

Figure 2. Autosomal Coverage and Quality Score Distributions for Variants Predicted in the Scaffolded Bobwhite (*Colinus virginianus*) Genome (NB1.1). Total genome-wide variants predicted within NB1.1 appears on the y-axis, with coverage and quality scores presented on the x-axis, respectively. Total variants include putative single nucleotide polymorphisms and small insertion deletion mutations (≤5 bp) that were predicted within the repeat masked NB1.1 assembly.

Methods). The temporal synchronicity of these bottlenecks for the bobwhite and the scarlet macaw became more coincident as the assumed mutation rate approached the human mutation rate (PSMC default $\mu = 2.5 \times 10^{-8}$). Beginning approximately 20 kya, the bobwhite (generation time = 1.22 yrs; Fig. 3) and the scarlet macaw (generation time = 12.7 yrs; Fig. 3; See Methods) demonstrate synchronous declines in their estimated effective population sizes (N_e), with this trend persisting up until about 9–10 kya, which

is coincident with the timing of modern human colonization of the New World (15,500–40,000 years ago) [60–63], the collapse of the megafauna [64–66], and the last glacial maximum (LGM) [67–68]. The geographic expansion of modern man has previously been proposed (i.e., subsistence hunting; overkill) as one highly efficient mechanism for the late Pleistocene collapse of the megafauna in the Americas, and to a lesser degree, in Eurasia [64,66]. Both the bobwhite and the scarlet macaw were hunted by indigenous peoples of the Americas [1,69–71]. However, the peregrine falcon also experienced a bottleneck at about the same time as the bobwhite and the scarlet macaw, possibly due to climate-driven habitat diminution [41], which may also explain some or even most aspect(s) of the predicted declines that we detected. Moreover, the peregrine falcon previously used for PSMC modeling was not sampled from the New World [41], which further confirms the possibility for the LGM [67–68] being explanatory for temporally relevant global declines of many animal populations, with recent evidence of swine population declines (i.e., European and Asian wild boar; *Sus scrofa*) [72] during the same time intervals as the bobwhite and scarlet macaw declines (Fig. 3).

Relevant to modern conservation biology and conservation genetics, it is clear that the estimated N_e of the bobwhite remained large even after a historic bottleneck (i.e., up to about 9–10 kya), with a historic peak N_e which was more than 6.6 times larger than the scarlet macaw (Fig. 3). This result was relatively unsurprising given the high autosomal SNP rate predicted for the bobwhite in this study (2.99 SNP per Kbp). When avian mutation rates (i.e., bobwhite, scarlet macaw) were modeled according to the human

mutation rate (PSMC default $\mu = 2.5 \times 10^{-8}$), as was also assumed for the wild boar [72], peak N_e for the bobwhite was estimated at approximately 95,000 about 20 kya, with a subsequent decline to approximately 72,000 by 9–10 kya (Fig. 3). The most recent bobwhite peak which arises near 10^{-4} on the "Time" x-axis (scaled in units of $2\mu T$) appears to be an artifact due to PSMC being unable to model a continued decline in N_e until the present, with a similar statistical signature and corresponding overestimation of N_e detected prior to a population decrease that was predicted in the Denisovan genome analysis [73]. Estimates of modern N_e in the bobwhite will require multiple sequenced individuals [74] to adequately estimate the severity of the predicted decline. Relevant to modern bobwhite declines observed across the majority of their U.S. range [5–6,20–23], our demographic analysis indicates that the *r*-selection strategy employed by the bobwhite can be very effective with respect to rapid increases in N_e (i.e., see the increase at 4×10^{-3} $2\mu T$ in Fig. 3). Therefore, it is apparent that these recent bobwhite declines may potentially be reversed at least to some degree (i.e., boom-bust pattern) in regions with suitable habitats, ample annual rainfall, and low harvest intensity. In striking contrast to the

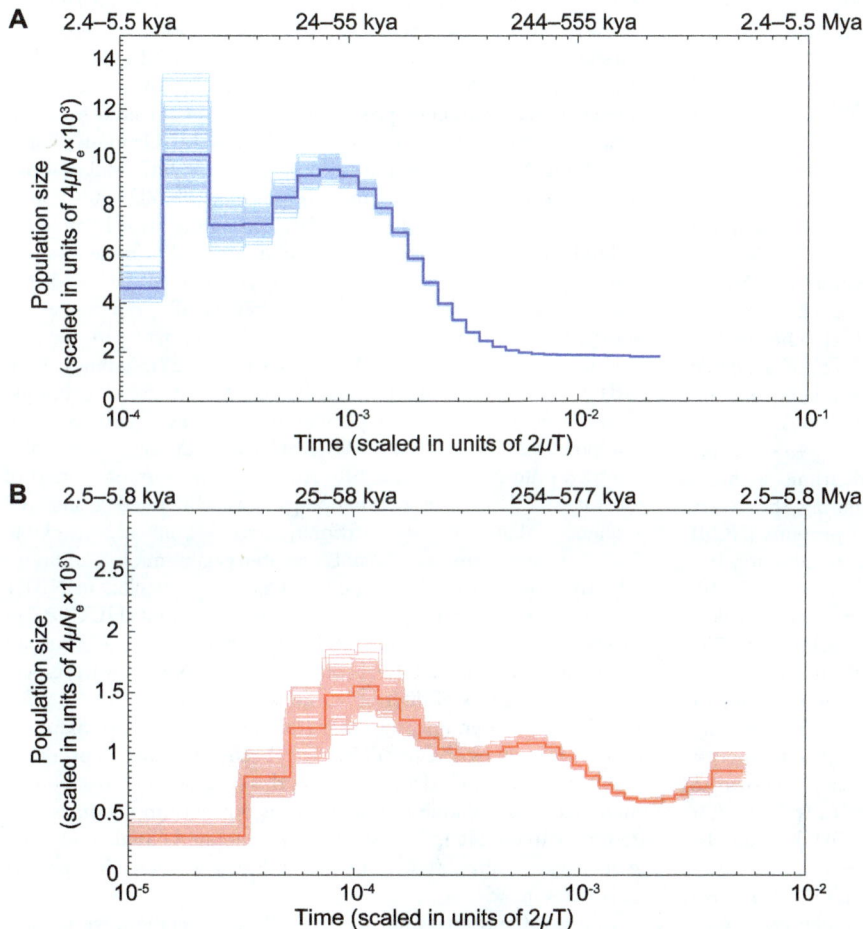

Figure 3. Comparative Demographic History Analysis and PSMC Effective Population Size Estimates for Bobwhite (*Colinus virginianus*) (A) and Scarlet Macaw (*Ara macao*) (B). Estimates of effective population size are presented on the y-axis as the scaled mutation rate. The bottom x-axis represents per-site pairwise sequence divergence and the top x-axis represents years before present, both on a log scale. Generation intervals of 1.22 years for the bobwhite (*Colinus virginianus*) and 12.7 years for the scarlet macaw (*Ara macao*) were used (See Methods). In the absence of known per-generation *de novo* mutation rates for the bobwhite and the scarlet macaw, we used the two human mutation rates (μ) of 1.1×10^{-8} and 2.5×10^{-8} per generation [124,125] (see Methods). Darker lines represent the population size inference, and lighter, thinner lines represent 100 bootstraps to quantify uncertainty of the inference.

bobwhite, peak N_e for the scarlet macaw (assuming $\mu = 2.5 \times 10^{-8}$) was never as large, and was estimated at approximately 15,500 about 25 kya, with a subsequent collapse to approximately 3,000 by 2.5 kya (Fig. 3); despite the fact that Neblina is from Brazil (i.e., wild caught) and was part of the population found in the Amazon Basin and adjacent lowlands, with an estimated population habitat range that exceeds 5 million km^2. Our analysis of these data strongly underscores the importance of conservation biology and conservation genetics in the scarlet macaw and other related pscittacines that rely heavily on K-selection [56–58]. Notably, the disparities in peak N_e as well as the more recent estimates (10 kya) for the bobwhite and the scarlet macaw are likely to reflect long-term, opposing differences in the r-/K- selection continuum [56–58], and suggest that species which rely heavily on facets of K-selection for success, like the scarlet macaw, could be at higher risk of experiencing more rapid and dramatic declines in N_e that are likely to prolong recovery. In fact, even under the perception of relatively ideal biological conditions in the field, N_e for large K-selected avian species like the scarlet macaw may be much lower than presumed based on the amount of available habitat, and the estimated total population size. Our findings highlight the need to conserve large populations of scarlet macaws and similar species in order to maintain genomic diversity and corresponding N_e to avoid unmasking deleterious alleles by way of increasing homozygosity, as observed for the highly endangered Spix's Macaws [75–76]. However, caution is necessary when interpreting the results of PSMC, as population size reductions and population fragmentation may not always be easily differentiated [55].

Annotation of the Bobwhite Genome

Three *in silico* methods were used to annotate the scaffolded bobwhite genome (NB1.1). Initially, we used GlimmerHMM [77–78] to comparatively predict putative exons within the NB1.1 assembly, with algorithm training conducted using all annotated chicken genes (*G gallus* 4.0) as recently described [42]. The chicken was chosen for training based on the superior level of available annotation and the lowest estimated time since divergence (56–62 MYA), as compared to the zebra finch (100–106 MYA) and the turkey (56–62 MYA; http://www.timetree.org/) [49–50]. All GlimmerHMM predicted exons were filtered using a high-throughput distributed BLAST engine implementing the blastx algorithm in conjunction with all available bird proteins (NCBI non-redundant avian protein sequences), and the E-value top hits to known avian proteins were retained and summarized [42,79]. Collectively, this comparative *in silico* approach produced statistical evidence for 37,851 annotation models, of which 15,759 represented unique genes and corresponding proteins (Table S6). Similar to the first-generation comparative annotation reported for the scarlet macaw, the number of unique annotation models that are reported here were based on blastx assignments to unique protein hit definitions (i.e. unique accessions), which is known to underestimate the total unique annotation models produced (for review see [42]). As one example, within the NB1.1 assembly, 3,532 genome-wide annotation models were predicted for eight unique protein accessions representing non-LTR retrovirus reverse transcriptases and/or reverse transcriptase-like genes (i.e., *pol*-like ORFs; RT-like RNA-dependent DNA-polymerases) which have also been predicted in large copy numbers in the chicken nuclear genome (Table S6; GenBank Accessions AAA49022.1, AAA49023.1, AAA49024.1, AAA49025.1, AAA49026.1, AAA49027.1, AAA49028.1; AAA58720.1). Moreover, the prediction of multi-copy genes within all avian genomes routinely utilizes naming schemes which include "like" or "similar to" a specific GenBank accession [42]. Our initial comparative annotation

procedure culminated with a blastx hit definition representing the highest scoring avian protein curated by NCBI. Therefore, some loci predicted to encode very similar putative proteins, including multi-copy loci such as those representing gene family members, may be assigned to the same specific protein accession(s) by the blastx algorithm. As occurred for the scarlet macaw genome [42], the absence of bobwhite genome maps and cDNA sequences to guide our initial annotation process also precluded the generation of complete *in silico* models for most bobwhite nuclear genes. Nevertheless, this procedure was successful at identifying bobwhite scaffolds predicted to contain genes encoding moderate to large proteins, which also included some multi-exonic genes distributed across large physical distances (i.e., *TLR2, TNRC18, NBEA*, respectively; Table S6). Investigation of the blastn comparative alignment data for NB1.1 (Table S3) revealed that all or most of the scaffolds predicted to possess exons encoding these genes (*TLR2, TNRC18, NBEA*) aligned to their orthologous genomic locations in the chicken (*G. gallus* 4.0) and zebra finch (*T. guttata* 3.2.4) genomes. Overall, the results of our comparative annotation for the bobwhite using GlimmerHMM and blastx were similar to those reported for the scarlet macaw [42], but with more annotation models predicted by way of higher genome coverage, and substantially less time since divergence from the chicken.

In a second approach to NB1.1 annotation, we used the Ensembl Galgal4.71 (*G. gallus*) cDNA refseqs (n = 16,396) and *ab initio* (GENSCAN) sequences (n = 40,571) in an iterative, sequence-based alignment process specifically engineered for transcript mapping and discovery (see Methods; CLC Genomics Large Gap Read Mapper Algorithm, [42]). Of the 56,967 total putative transcripts utilized in this analysis pipeline, 39,603 (70%) were successfully mapped onto the NB1.1 assembly, which included redundant annotation models. Approximately 59% of the mapped transcripts contained gaps which corresponded to predicted intron-exon boundaries and/or species-specific differences in transcript composition (i.e. regions with no match to NB1.1). Specifically, 12,290 Galgal4.71 cDNA refseq mappings onto NB1.1 were produced, with 10,959 of these possessing unique Ensembl gene names and protein descriptions (Table S7). An additional 27,309 *ab initio* (GENSCAN) transcripts were also mapped onto NB1.1 (Table S8). An exhaustive summarization of all Galgal4.71 transcript mappings was generated using the sequence alignment map format, and is publicly available (http://vetmed.tamu.edu/faculty/cseabury/genomics). Additionally, the positions of all mapped Galgal4.71 transcripts in NB1.1 and the corresponding gene descriptions (Ensembl, HUGO) are provided in Table S7. Our analysis of these data, including an examination of the scaffolded contig positions (NB1.1) with respect to annotated genes of interest within the chicken genome (*G. gallus* 4.0; Table S7), demonstrates that comparative transcript mapping onto the genomes of more distantly related avian species produces viable annotation models. However, this result and corresponding inference is not unique to our study, as other avian genomes (i.e., zebra finch) are often at least partially annotated based on chicken sequences (http://www.ncbi.nlm.nih.gov/genome/367?project _id = 32405).

In a third and final approach to NB1.1 annotation, we utilized the few, low-coverage cDNA sequences that were previously produced for the bobwhite to generate species-specific annotation models. Specifically, we obtained and trimmed 478,142 bobwhite cDNA sequences previously utilized in the construction of a custom bobwhite cDNA microarray [11] (SRA: SRR036708), and subsequently used the quality and adaptor trimmed reads (n = 325,569; average length = 232 bp) for a strict *de novo* assembly of putative bobwhite transcripts (See Methods). Altogether, 21,367 *de novo* contigs were generated, and of these, 21,011 (98%) were produced from two or more overlapping reads, with most of these contigs (n = 18,135; 85%) possessing ≤5× average coverage. Using the same iterative, sequence alignment process (CLC Genomics Large Gap Read Mapper) described for the Galgal4.71 comparative annotation, we successfully mapped 98% of the assembled bobwhite transcripts (n = 21,002) onto NB1.1. Approximately 31% of the mapped transcripts produced gapped alignments that were considered putative intron-exon boundaries. All *de novo* contigs representing bobwhite transcripts were characterized using a high-throughput distributed BLAST engine implementing blastx in conjunction with all available bird proteins (NCBI non-redundant avian protein sequences), and the top ranked hits (i.e., E-value, bitscore) to known avian proteins were retained and summarized [79]. Altogether, 8,708 *de novo* contigs (i.e. bobwhite putative transcripts) produced statistical evidence for assignment to at least one known or predicted avian protein (Table S9). Further evaluation of the top hits also revealed some evidence for redundancy across the blastx protein assignments (i.e. same protein; similar alignment length, E-value, and bitscore for two or more avian species). An exhaustive summary of all bobwhite transcript mappings to NB1.1 was also generated using the sequence alignment map format, and is available online (http://vetmed.tamu.edu/faculty/cseabury/genomics). Likewise, the positions of all bobwhite transcripts in NB1.1 are provided in Table S10.

A comparison of all three annotation methods revealed evidence for both novel and redundant annotation models. For example, 8,463 assembled (*de novo*) bobwhite transcripts could be mapped directly onto the Ensembl Galgal4.71 transcripts by sequence similarity and alignment, and of these, 5,537 were redundant with 3,728 unique annotations produced by mapping the Ensembl Galgal4.71 transcripts directly onto NB1.1. Importantly, the overall utility and impact of the previously generated bobwhite cDNA sequences [11] could not be fully realized in the absence of a draft *de novo* genome assembly. Similar to the scarlet macaw genome project [42], both of our bobwhite assemblies (NB1.0, NB1.1) were successful at reconstructing a complete mitochondrial genome at an average coverage of 159×, which resulted in the annotation of 13 mitochondrial protein coding genes (*ND1, ND2, COX1, COX2, ATP8, ATP6, COX3, ND3, ND4L, ND4, ND5, ND6, CYTB*), two ribosomal RNA genes (*12S, 16S*), 21 tRNA genes, and a predicted D-loop (Table S6). Despite the effectiveness of our mitochondrial and nuclear gene predictions, it should also be noted that even three annotation approaches applied to NB1.1 were not sufficient to exhaustively predict every expected bobwhite nuclear gene. For example, studies of the avian major histocompatibility complex (MHC) have established expectations for gene content among several different bird species, with our approaches providing evidence for many (i.e., *HLA-A, TAP1, TAP2, C4, HLA-DMA, HLA-B2, TRIM7, TRIM27, TRIM39, GNB2L1, CSNK2B, BRD2, FLOT1, CIITA, TNXB, CLEC2D*) but not all previously described avian MHC genes (Table S6) [46–48,80–84]. While the limitations of our three annotation methods were not surprising, the results were sufficient to facilitate informed genome-wide analyses for the bobwhite. Moreover, even well-established avian genomes, such as the chicken and zebra finch genomes, have yet to be exhaustively annotated. Nevertheless, the results of our annotation analyses provide a foundation for implementing interdisciplinary research initiatives ranging from ecotoxicology to molecular ecology and population genomics in the bobwhite.

Whole-Genome Analysis of Divergence and Development of Candidate Genes

One of the most interesting scientific questions to be directed toward the interpretation of new genome sequences is: "What makes each species unique?". We used the percentile and composite variable approach as well as the validation and quality control procedures previously described [42] to identify *de novo* contigs (NB1.0) displaying evidence of extreme nucleotide conservation and divergence (i.e. outliers) relative to the chicken (*G. gallus* 4.0) and zebra finch (*T. guttata* 3.2.4) genomes (Fig. 4; See Methods). The *de novo* contigs (NB1.0) are useful for this purpose because they provide a shotgun-like fragmentation of the bobwhite genome that is nearly devoid of N's (i.e. intra-contig gaps), which facilitates fine-scale comparative nucleotide alignments that often span large portions, the majority, or even the entire length of the contig sequences. A genome-wide nucleotide sequence comparison of the bobwhite and chicken genomes revealed outlier contigs harboring coding and noncoding loci that were characterized either on the basis of known function and/or the results of human genome wide-association studies (GWAS) (Fig. 4; Table 4; Table S11). Two general trait classes (cardiovascular, pulmonary) were routinely associated with loci predicted within or immediately flanking the aligned positions of bobwhite contigs (NB1.0) classified as outliers for extreme conservation with the chicken genome (Table 4; Table S11). This result is compatible with the supposition that loci modulating cardiovascular and pulmonary traits are often highly conserved across divergent avian lineages [42]. One plausible explanation for this is that birds are unique within the superclass Tetrapoda because they are biologically equipped for both bipedalism and powered flight [85], which may place larger and different demands on the cardiovascular and pulmonary systems than for organisms where mobility is limited to a single terrestrial method (i.e., bipedalism, quadrupedalism). In addition to cardiovascular and pulmonary traits, one bobwhite outlier contig (NB1.0) for extreme conservation with the chicken genome also included a gene (*LDB2*) that is known to be strongly associated with body weight and average daily gain in juvenile chickens [86]. This result is compatible with the fact that both the chicken and bobwhite are gallinaceous birds which produce precocial young, and therefore, are likely to share some genetic mechanisms governing early onset juvenile growth and development. Examination of all bobwhite contigs (NB1.0) classified as outliers for divergence with the chicken revealed relatively few predicted genes, with sequences of unknown orthology and noncoding regions being the most common results observed (Table 4; Table S11). This is concordant with the hypothesis that noncoding regions of the genome (i.e., promoters, noncoding DNA possessing functional regulatory elements including repeats) are likely to underlie differences in species-specific genome regulation and traits [87–90]. Some of the most interesting bobwhite contigs (NB1.0) displaying evidence for extreme divergence were predicted to contain putative introns for *CSMD2* as well as *TNIK*, and to flank *LPHN3* (intergenic region; Table 4; Table S11). These three genes have all been associated with human brain-related traits including heritable differences in brain structure (*CSMD2*, voxel measures) [91], measures of activation within the dorsolateral prefrontal cortex (*TNIK*) [92] and working memory in schizophrenia patients receiving the drug Quetiapine [93]. Our whole genome-wide analysis of divergence between the bobwhite and the chicken provides further evidence that noncoding regions of the genome are likely to play a tangible role in the developmental manifestation of species-specific traits [87–90], including both neurocognition and behavior [91–93].

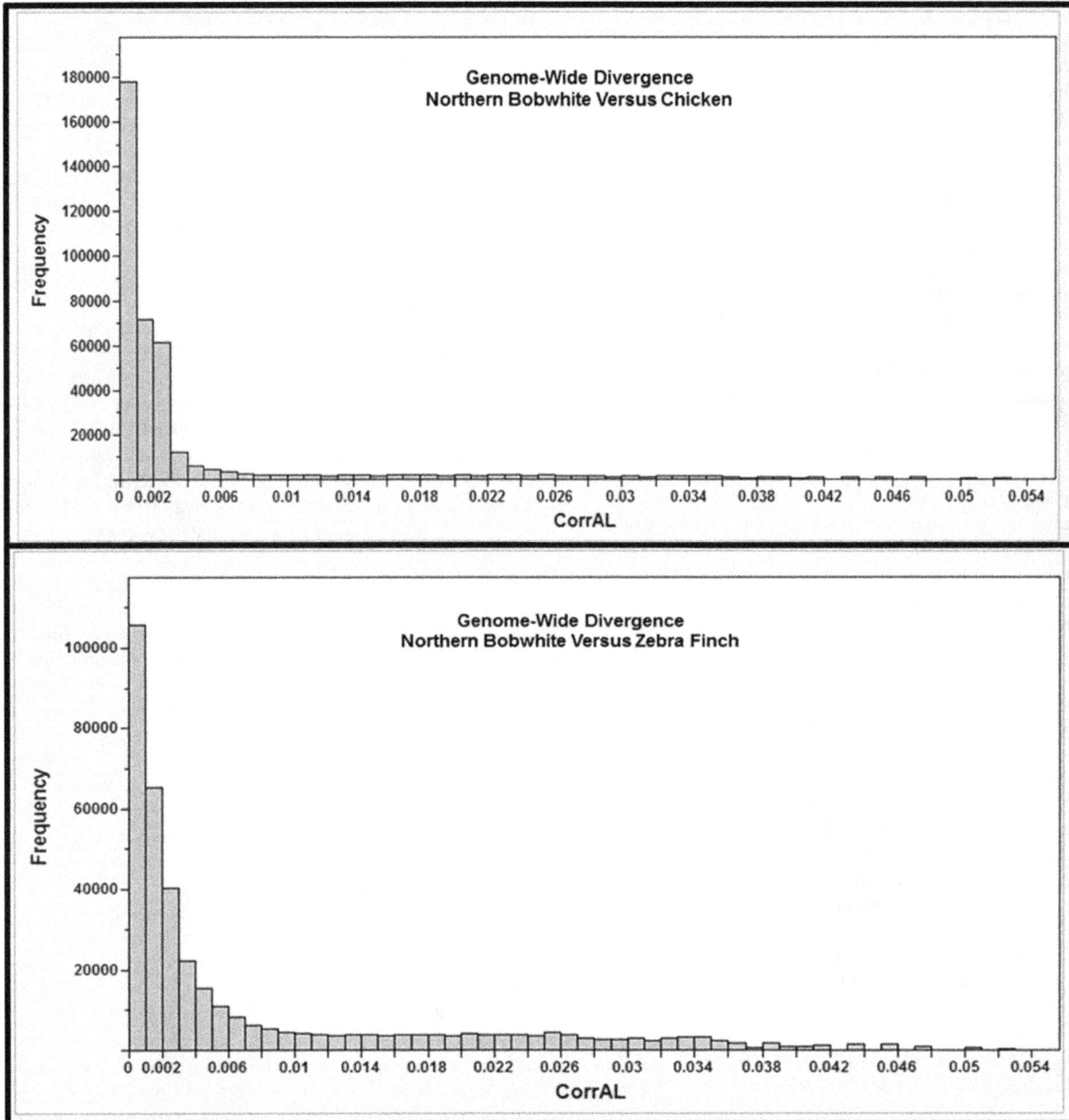

Figure 4. Whole Genome Analysis of Divergence. (Top) Genome-wide nucleotide-based divergence (CorrectedForAL) between the bobwhite (*Colinus virginianus*; NB1.0; simple *de novo* assembly) and the chicken genome (*Gallus gallus* 4.0). (Bottom) Genome-wide nucleotide-based divergence (CorrectedForAL) between the bobwhite (*Colinus virginianus*; NB1.0; simple *de novo* assembly) and the zebra finch genomes (*Taeniopygia guttata* 1.1, 3.2.4). Each histogram represents the full distribution of the composite variable defined as: $CorrectedForAL = \dfrac{(\frac{PercentID}{100})}{AlignmentLength}$ [42]. The left edges of the distributions represent extreme conservation, whereas the right edges indicate extreme putative divergence. The observed ranges of the composite variable were 2.19545E-05 – 0.052631579 (chicken), and 4.28493E-05 – 0.052631579 (zebra finch). Distributional outliers were predicted using a percentile-based approach (99.98th and 0.02th) to construct interval bounds capturing >99% of the total data points in each distribution.

Comparison of the bobwhite (NB1.0) and zebra finch genomes (*T. guttata* 3.2.4) also revealed evidence for extreme nucleotide conservation and divergence (Fig. 4; Table 5; Table S11). In comparison to the zebra finch genome, two general trait classes (osteogenic, cardiovascular) were routinely associated with loci predicted within or immediately flanking the aligned positions of bobwhite contigs (NB1.0) classified as outliers for extreme

conservation (Table 5; Table S11). Within these contigs, the presence of orthologous gene sequences previously associated with human cardiovascular traits (or their proximal noncoding flanking regions) was relatively unsurprising, as this result also occurred during our analysis of divergence with the chicken genome (Table 4; Table 5; Table S11), and in a previous study of the scarlet macaw genome [42]. Therefore, it is apparent that some

Table 4. Biologically relevant bobwhite NB1.0 simple *de novo* outliers from a genome-wide analysis of divergence with the chicken genome (*G. gallus* 4.0).

Predicted Outlier Contig Genes[a,b,c]	Known Function or GWAS Trait Classification	References
BCL11B[a]	Aortic Stiffness	[126]
ALPK3[a]	Cardiac Heath and Development	[127]
SETBP1[a], *FAF1*[a]	Heart Ventricular Conduction	[128]
MEF2A[a], *LPL*[a],	Cardiomyopathy	[129–130]
KCNJ2[a]	Heart Q-wave T-wave Interval Length	[131]
LDB2[a], *PTPRF*[a], *ATP10B*[a]	Coronary Artery Disease	[132–134]
ZNF652[a], *FIGN*[a], *CHIC2*[a]	Blood Pressure	[135–136]
CFDP1[a], *KCNJ2*[a]	Pulmonary Function and Health	[137,138]
GRM3[a], *RELN*[a], *RORB*[a],	Cognitive Abilities	[139–141]
CSMD2[b]	Brain Structure	[91]
TNIK[b]	Brain Imaging	[92]
LPHN3[b]	Working Memory	[93]

[a]Outlier for extreme nucleotide-based conservation.
[b]Outlier for extreme nucleotide-based divergence.
[c]See Table S11 for an exhaustive list of outlier contigs with annotation.

loci associated with cardiovascular and pulmonary traits in humans appear to be extremely conserved across multiple avian species, including some of the same loci identified by similar analyses involving the scarlet macaw, chicken, and zebra finch genomes (Table S11) [42]. Among the bobwhite contigs classified as outliers for extreme conservation with the zebra finch, we also observed orthologous gene sequences (or their proximal noncoding flaking regions) which were previously associated with human bone density, strength, regeneration, and spinal development as well as human height and waist circumference (Table 5; Table S11). Interestingly, the overall size and stature of the bobwhite (i.e. height or length, wingspan) is actually more similar to the zebra finch than to the chicken [94–96], which is compatible with these results. Additionally, while the temporal order of ossification for avian skeletal elements is known to be conserved across divergent bird species (i.e., duck, quail, zebra finch) [97], some aspects of wild bobwhite medullary bone formation (i.e., annual frequency of occurrence) are arguably far more similar to the zebra finch than to domesticated chickens, which have been bred and utilized for continuous egg production [98–100]. Therefore, some similarities in the underlying biology of these two bird species were reconciled with the genomic information content found within several bobwhite outlier contigs displaying evidence for extreme conservation with the zebra finch genome. At the opposite end of the distribution (Fig. 4), and across all diverged outliers with respect to the zebra finch genome, one of the most intriguing results was a bobwhite contig predicted to contain an *LDB2* intron (Table 5; Table S11). Notably, *LDB2* was implicated as an outlier for extreme conservation with the chicken genome (Table 4; Table S11), and is known to be strongly associated with body weight and average daily gain in precocial juvenile chickens [86]. The observation of this same putative gene (a different NB1.0 contig) with respect to extreme divergence with the zebra finch genome (Table 5; Table S11) may potentially reflect the different developmental strategies associated with the bobwhite and the zebra finch (i.e., precocial versus altricial) [101–103]. Two additional contigs classified as outliers for divergence were also predicted to be proximal to genes implicated by human GWAS studies for age at menarche (*NR4A2*) [104] and reasoning in

schizophrenia patients receiving the drug Quetiapine (*ZNF706*; Table 5; Table S11) [93]. Interestingly, both wild and domesticated zebra finches reach sexual maturity earlier than do bobwhites, with hypersexuality in the zebra finch considered to be an adaptation to arid environments [105–106]. However, any potential relationships between *ZNF706* and specific underlying

Table 5. Biologically relevant bobwhite NB1.0 simple *de novo* outliers from a genome-wide analysis of divergence with the zebra finch genome (*T. guttata* 3.2.4).

Predicted Outlier Contig Genes[a,b,c]	Known Function or GWAS Trait Classification	References
CDH13[a], *CXADR*[a],	Blood Pressure	[142–143]
VTI1A[a], *KLF12*[a]	Heart Ventricular Conduction	[128]
BCL11B[a]	Aortic Stiffness	[126]
GJA1[a]	Resting Heart Rate	[144]
JAG1[a]	Bone Density	[145]
VPS13B[a]	Bone Strength	[146]
SALL1[a]	Bone Mineral Density	[147]
STAU2[a]	Spinal Development	[148]
SATB2[a]	Osteogenic Differentiation And Regeneration	[149]
ZFHX4[a], *BNC2*[a],	Height	[150–151]
STX16[a], *APCDD1L*[a]	Waist Circumference	[152]
GRIA1[a]	Anthropometric Traits	[153]
LDB2[b]	Body Weight	[86]
LDB2[b]	Average Daily Gain	[86]
NR4A2[b]	Age of onset of Menarche	[104]
ZNF706[b]	Reasoning	[93]

[a]Outlier for extreme nucleotide-based conservation.
[b]Outlier for extreme nucleotide-based divergence.
[c]See Table S11 for an exhaustive list of outlier contigs with annotation.

biological differences between the bobwhite and zebra finch were not apparent, especially since no studies have comparatively evaluated a battery of cognitive traits in these two species using standardized methods.

Quality Control Investigation for Analyses of Divergence

All NB1.0 contigs classified as putative outliers for divergence (Fig. 4; right tail) shared one unifying feature: A 19–20 bp alignment with 100% identity to a reference genome (i.e., chicken or zebra finch) regardless of contig size (Range = 300 bp to 1,471 bp; Median = 385 bp; Mean = 438 bp). These short alignments had variable sequences, with the common feature being the short length (19–20 bp), and produced values for the composite variable ($CorrectedForAL = \frac{(\frac{PercentID}{100})}{AlignmentLength}$) that ranged from 0.050 to 0.053 (i.e., $\frac{(\frac{100}{100})}{20bp} or \frac{(\frac{100}{100})}{19bp}$). This was expected based upon previous observations [42], and at least three plausible explanations for this result include: 1) The orthologous sequences are simply missing from the chicken and/or zebra finch genome assemblies; 2) The NB1.0 contigs are misassembled; or 3) The NB1.0 contigs represent true outliers for nucleotide divergence and include species-specific insertion-deletion mutations. Some sequences are invariably missing from every draft genome assembly (i.e., unassembled). Therefore, we searched five databases curated by NCBI (i.e., refseq_genomic, refseq_rna, nr/nt, traces-WGS, traces-other DNA) for nucleotide alignments that would facilitate NB1.0 contig characterization and/or help refute the diverged outlier status of these contigs, and in all cases found little or no evidence for a conclusively better blastn alignment to the chicken or zebra finch genomes (See Methods). However, some of these contigs actually produce better blastn alignments (i.e., E-value, bitscore) to other vertebrate species, including other avian species, which is not compatible with outlier status (diverged) resulting solely from contig misassembly (Table S2; Table S11).

Regarding our whole-genome analyses of divergence, all NB1.0 contigs classified as outliers for extreme conservation (Fig. 4; extreme left edge) were moderately large (Range = 9,647 bp to 89,591 bp; Median = 22,792 bp; Mean = 25,196 bp) in comparison to outliers for divergence (Range = 300 bp to 1471 bp; Median = 385 bp; Mean = 438 bp). Again, this trend was expected and has been previously described [42]. Therefore, we conducted several quality control (QC) analyses that were designed to assess whether factors other than nucleotide sequence divergence were responsible for our results. First, we used summary data from the two comparative genome alignments performed using blastn to estimate pairwise correlations among the following: NB1.0 contig size (bp), contig percent GC, contig percent identity, and contig alignment length (bp). Moderate correlations between NB1.0 contig alignment length and contig size were observed with respect to the chicken (r = 0.649, Nonparametric τ = 0.656) and zebra finch genome alignments (r = 0.490, Nonparametric τ = 0.492), whereas weak correlations were observed between percent identity and alignment length (chicken: r = 0.127, Nonparametric τ = 0.071; zebra finch: r = −0.371, Nonparametric τ = −0.469). Weak correlations were also observed for all other investigated parameters. This result is important because the two parameters that drive our analysis of divergence are the percent identity and the alignment length, which were jointly used to construct a composite variable (CorrectedForAL) representing percent identity normalized for alignment length across all NB1.0 contigs which produced blastn alignments to the chicken and zebra finch genomes. In a second QC analysis, we applied the same percentile based approach (Percentiles = 99.98[th] and 0.02[th]) used in our whole-genome analyses of divergence to examine the full, ordered distribution of NB1.0 contig sizes, and determined that only 2 contigs (chicken analysis; contigs 4309, 7216) were in common with the 244 implicated as outliers for conservation or divergence (Table S11). This result argues against contig size being deterministic for outlier status. Finally, for larger contigs, such as those classified as outliers for conservation, the blastn procedure often produces multiple meaningful alignments, which are appended below the most "significant" hit (i.e., E-value and bitscore top ranked hit). These appended alignments include both noncontiguous (i.e., gaps due to insertion-deletion mutations) and less "significant" comparative alignments (i.e., increasing nucleotide sequence divergence). To assess the reliability of utilizing only the top ranked hit (i.e., E-value and bitscore) as a proxy for larger contigs which may produce multiple, syntenic, noncontiguous hits spanning either the majority or even the entire contig length, we used the additional (i.e., appended) non-overlapping alignment data (percent identity, alignment length) for the conserved outlier contigs to recalculate our composite variable (Table S12). Across all 145 unique contigs categorized as conserved outliers, the new (recalculated) composite variable only further confirmed the original outlier status (i.e., extreme conservation), which is in agreement with the results of a similar study involving the scarlet macaw genome (Table S12) [42]. Moreover, the NB1.0 contigs classified as outliers for extreme conservation are actually highly conserved genomic regions for which extended nucleotide conservation persists for the two compared species, which cannot occur in the presence of species-specific genomic rearrangements, copy number variants whereby one or more amplification-deletion boundaries are traversed, or in the presence of frequent and complex repetitive elements. Nevertheless, only NB1.0 contigs which produced blastn results (>99%) could be included in our analyses of divergence and quality control analyses, as they provided the data required to construct the composite variable. All NB1.0 contigs for which no alignments were achieved with respect to the chicken or zebra finch genomes are provided in Table S2.

Conclusions

The ability to rapidly generate low-cost, high quality avian draft *de novo* genome assemblies in conjunction with coalescent models to reconstruct the demographic histories of species which are currently in decline provides a foundation for understanding and monitoring both historic and recent population trends. Although the bobwhite has clearly declined across much of its native range [5–6,20–23], our estimates of N_e up until about 9–10 kya demonstrate that genomic diversity has remained quite high despite a substantial, historic bottleneck (Fig. 3). The same cannot be said for the scarlet macaw (Fig. 3), with our analyses indicating that N_e for the scarlet macaw was never as large as the bobwhite (Fig. 3), and with the large disparity in effective population sizes between these two highly divergent species most likely a product of their opposing natural selection strategies (i. e., r- versus K-selection). Short generation times and large clutches in the bobwhite provide more opportunities for the creation of genomic diversity via meiotic recombination and new mutation than do the long generation times, small clutches, and very small broods for the scarlet macaw [56–59,107–108]. Therefore, our observations are concordant with genomic signatures of selection created by how opposing selection strategies (i.e., skewness in the r- versus K-selection continuum) would be expected to shape genomic diversity and the corresponding effective population sizes in these species [56,58]. Considering the findings of human GWAS studies (i.e., genes, noncoding regions), the results of our whole-genome analyses of divergence were often consistent with several fundamental biological differences noted between three divergent avian species, with independent replication of some outlier loci and trait classes that were previously suggested to be important among avian species [42]. We also identified several potential candidate genes and noncoding regions which coincide with human GWAS studies for biological traits that appear disparate among the three investigated bird species, but also found previously reported evidence for purifying selection operating on some of the same genes we identified within our conserved outlier contigs (Table S11). As described for a recent analysis of the scarlet macaw genome, the overwhelming majority of the bobwhite contigs (NB1.0) classified as outliers for divergence with the chicken and zebra finch were determined to contain noncoding sequences, which is consistent with the hypothesis that noncoding regions of the genome are likely to underlie differences in species-specific genome regulation and traits [42,87–90].

Methods

Source of Bobwhite (Colinus virginianus) Genomic DNA

We utilized skeletal muscle derived from the legs of a wild, female bobwhite ("Pattie Marie") from Fisher county Texas to isolate high molecular weight genomic DNA using the MasterPure DNA Purification Kit (Epicentre Biotechnologies, Inc., Madison, WI). Ethical clearance is not applicable to samples obtained from lawfully harvested wild bobwhites. The protocol for isolating genomic DNA followed the manufacturer's recommendations, and we confirmed the presence of high molecular weight genomic DNA by agarose gel electrophoresis, with subsequent initial quantification of multiple individual isolates performed using a Nano Drop 1000 (Thermo Fisher Scientific, Wilmington, DE).

Genome Sequencing Strategy

Prior to library construction, bobwhite genomic DNA was quantitated using the Qubit DNA HS assay and Qubit 2.0 flourometer (Life Technologies Inc., Carlsbad, CA), with further evaluation by agarose gel electrophoresis. All samples contained high molecular weight DNA >15 kb, with little or no degradation, thereby making them suitable for PE and MP library preparation. For creation of a small insert PE library, approximately 1.0 μg of DNA was normalized to 40 μl and fragmented to approximately 300 bp using the QSonica plate sonication system (Qsonica Inc., Newton CT). The fragmented DNA was blunt-end repaired, 3' adenylated and ligated with multiplex compatible adapters using the NEXTflex DNA Sequencing Kit for Illumina (Bioo Scientific cat # 514104) prior to size selection (200–400 bp fragments) using SPRI beads (Agencourt Inc, Brea CA). PCR enrichment was performed to selectively amplify bobwhite DNA fragments with adapters on both ends as follows: 98°C for 30 sec, 10cycles [98°C for 10 sec, 65°C for 30 sec, 72°C for 60 sec], 72°C for 5 minutes, 10°C hold. Bobwhite PE library validation was performed using the Bioanalyzer 2100 High Sensitivity DNA assay (Agilent Inc., Santa Clara, CA), with quantitation performed using the Qubit HS DNA assay. Thereafter, two MP sequencing libraries (Table 1) were created by following the Illumina Mate Pair v2 Library Preparation procedure for 2–5 Kbp fragments (Part #15008135 Rev A; Illumina Inc., San Diego, CA) as recently described [42]. The final PE and MP libraries were diluted to 10 nM in preparation for sequencing on a HiSeq 2000 genetic analysis system (Illumina Inc., San Diego, CA). The bobwhite PE library was processed using PE-100 cycle runs (2×100 bp), and the MP libraries were processed using MP-50 cycle runs (2×50 bp), with

data generation (i.e., image processing and base calling) occurring in real time on the instrument. All clustering and base-calling was performed as recommended by the manufacturer. A summary of Illumina reads for all libraries is provided in Table 1. Prior to assembly, we used knowledge of avian genome size (nuclear DNA content, C-value) [154] in conjunction with physical knowledge of modern avian genome assemblies (bp) to estimate the size of the bobwhite nuclear genome [42].

Genome Assembly

Prior to assembly, all Illumina sequence reads were first trimmed for quality and adapter sequences using the CLC Genomics Workbench. Briefly, Phred quality base scores (Q) were converted into error probabilities, read-based running sums for quality were calculated, and reads were trimmed as recently described [42]. Following initial quality trimming, a second algorithm was used to trim ambiguous nucleotides (N) from the ends of every sequence read by referring to a user-specified maximum number of ambiguous nucleotides allowed (n = 2) at each end of the sequence, with subsequent removal of all other ambiguous bases. Finally, we also used the Workbench (i.e. Smith-Waterman algorithm) to specify, identify, and remove all sequencing adapters that could potentially be present in our sequence reads.

For the simple de novo (NB1.0) and the scaffolded assemblies (NB1.1) we used the CLC de novo assembler (v4.9), which has also been utilized for the generation of the scarlet macaw and Norway spruce genome assemblies [42,109]. Briefly, the CLC assembler implements the following general procedures: 1) Creation of a table of "words" observed in the sequence data, with retention and utilization of "word" frequency data; 2) Creation of a de Bruijn graph from the "word" table; 3) Utilization of the sequence reads to resolve paths through bubbles caused by SNPs, read errors, and small repeats; 4) Utilization of paired read information (i.e., paired distances and orientation of reads) to resolve more complex bubbles (i.e., larger repeats and/or structural variation); 5) Output of final simple de novo contigs (NB1.0) derived from a preponderance of evidence supporting discrete "word" paths, and also supported by the mapping-back process. For the scaffolded de novo assembly (NB1.0), the CLC assembler implemented one additional step in which paired reads spanning two contigs were used to estimate the distance between them, determine their relative orientation, and join them where appropriate using "N's"; the number of which reflect the estimated intercontig distance. Notably, not all de novo contigs can be joined to another by read-based scaffolding (i.e., in the absence of map data), and therefore, we use the term scaffolds to collectively refer to the final set of contigs for which read-based scaffolding was attempted. For both assemblies we utilized the same strict assembly parameters in conjunction with all trimmed, unmasked sequence reads (Table 1) as previously described [42], but with the following exceptions: minimum contig length = 300 bp; minimum read length fraction = 0.95; minimum fraction of nucleotide identity (similarity) = 0.95. Paired distances within the Workbench are user-specified, with incorrect specification (i.e., range too narrow or too wide) negatively impacting de novo genome assembly. Therefore, using knowledge from library construction and characterization (i.e., agarose gel electrophoresis; Agilent Bioanalyzer) as a guide, we initially assembled the sequence reads multiple times (iteratively), each with incremental increases in the specified paired distances, until the observed paired distances for each library resembled a bell shaped curve centered about a mean that was compatible with library construction and assessment data. For both bobwhite genome assemblies (NB1.0, NB1.1), the user-specified paired distances for all libraries are presented in Table 1. To further suppress genome misassembly, the CLC assembler (i.e., NB1.0, NB1.1) was instructed to break paired reads exhibiting the wrong distance or orientation(s), and only utilize those reads as single reads within the assembly process. This approach is conservative and favors the creation of more contigs with smaller N50 over the creation of larger and fewer contigs that are likely to contain more assembly errors. Assembly statistics for NB1.0 and NB1.1 are provided in Tables S13 and S14.

Estimating Concordance Between Genome Assemblies

Treating all NB1.0 contig sequences as individual sequence reads, we used the CLC Large Gap Read Mapper algorithm to iteratively search the scaffolded genome assembly (NB1.1) for the best matches (v2.0 beta 10) as previously described [42]. A single, initial round of iterative searching resulted in 91% of the NB1.0 contigs mapping onto the NB1.1 assembly, with 99% of these mappings containing no gaps. Thereafter, a SAM output was created, which was then used to parse out the coordinates of all mapped NB1.0 contigs for the purpose of creating a reference table summarizing the concordance between the two assemblies (Table S1). All parsing and joining was performed using Microsoft SQL Server 2008 R2.

Comparative Genome Alignment, Characterization of Repeat Content, and Variant Prediction

The NB1.0 and NB1.1 genome assemblies were aligned to the chicken (*G. gallus* 4.0) and zebra finch (*T. guttata* 1.1, 3.2.4) reference genome assemblies (including ChrUN, unplaced) using the blastn algorithm (version 2.2.26+). To minimize disk space and enable continuous data processing we used an E-value step-down procedure as recently described [42]. After each step, we exported the results and parsed out the top hit (E-value, bitscore) for each bobwhite contig (NB1.0, NB1.1). E-value ties were broken by bitscore. All parsing was performed using Microsoft SQL Server 2008 R2.

To estimate the minimum repetitive content within the bobwhite genome (NB1.1), we processed all of the scaffolds with RepeatMasker (http://www.repeatmasker.org/; RepBase16.0.1). As described for the scarlet macaw genome [42], we conducted a two-stage, composite analysis which consisted of masking the NB1.1 contigs with both the chicken and zebra finch repeat libraries to cumulatively estimate the detectable repetitive content. Additionally, we used PHOBOS (v3.3.12) [51] to detect and characterize genome-wide microsatellite loci with the following settings: Extend exact search; Repeat unit size range from 2 to 10; Maximum successive N's allowed in a repeat = 2; Recursion depth = 5; Minimum and maximum percent perfection = 80% and 100%, respectively [42]. Finally, the average coverage and total number of comparative blastn hits for each *de novo* contig (NB1.0, NB1.1) also provided insight regarding unmasked repeats when cross referenced with the results of RepeatMasker (Tables S4, S13, S14).

Following a two-stage RepeatMasker analysis (chicken+zebra finch repeat libraries), the masked NB1.1 scaffolds became the reference sequences used for SNP and indel prediction as previously described [42,110–111]. After reference mapping all the trimmed sequence reads onto the double-masked NB1.1 assembly using the same assembly parameters described above, we used the CLC probabilistic variant detection algorithm (v6.0.4) to predict and estimate genome-wide variation (i.e., SNPs, indels) with the following settings: ignore nonspecific matches = yes; ignore broken read pairs = no; minimum coverage = 10; variant probability ≥0.95; require variant in both forward and reverse

reads = yes; maximum expected variants = 2; ignore quality scores = no. Histograms representing the NB1.1 coverage distribution of predicted genome-wide variants and their corresponding phred score distribution were produced using JMP Pro 10.0.1 (SAS Institute Inc., Cary, NC).

"In silico" Annotation of the Bobwhite Genome

Initially, we used GlimmerHMM [47,77–78] to predict exons and putative gene models within NB1.1. GlimmerHMM was trained using all annotated chicken genes (*G gallus* 4.0) as recently described [42], which is similar to an approach used for annotation of the turkey genome [47]. Thereafter, we characterized, assessed support, and filtered GlimmerHMM predictions via blastx [79] in conjunction with all available bird proteins (NCBI non-redundant avian protein sequences), with the top hits (E-value, bitscore; minimum E-value = 1E-04) to known avian proteins retained and summarized as previously described [42].

In a second approach to annotation, we used the Ensembl Galgal4.71 (*G. gallus*) cDNA refseqs (n = 16,396) and *ab initio* (GENSCAN) sequences (n = 40,571) in an iterative, sequence-based alignment process for comparative transcript mapping and discovery. Galgal4.71 transcript length ranged from 108 bp to 93,941 bp. Briefly, we used the CLC large gap read mapper (v2.0 beta 10) to iteratively search the NB1.1 assembly for the best Galgal4.71 nucleotide matches. The CLC large gap read mapper was utilized as previously described [42], but with the following exceptions: maximum distance from seed = 100,000; minimum fraction of identity (similarity) = 0.80; minimum read length fraction = 0.001. Our settings for minimum read length fraction were necessary to facilitate mapping for large Galgal4.71 transcripts. However, this setting did not impede or nullify the stringency of mapping smaller transcripts, as the best matches (i.e. longest length fraction and highest similarity) were sought and reported. A SAM file representing all Galgal4.71 mappings was created using the CLC Genomics Workbench. Gene names (HUGO), descriptions, and protein information for the Ensembl Galgal4.71 cDNA refseqs were obtained from BioMart-Ensembl (http://useast.ensembl.org/biomart/martview/) and NCBI (http://www.ncbi.nlm.nih.gov/sites/batchentrez).

In a third approach to annotation, we obtained 478,142 bobwhite cDNA sequences (Roche 454) previously used to construct a microarray [11] (SRA: SRR036708) and trimmed them for quality and adaptors. Thereafter, the remaining sequences (n = 325,569; average length = 232 bp) were assembled using the CLC *de novo* assembler (v6.0.4) and the same strict assembly parameters utilized for NB1.0 and NB1.1. *De novo* contigs (50 bp to 6466 bp) generated from bobwhite cDNA sequences were mapped onto NB1.1 using the CLC large gap read mapper as described above for the Galgal4.71 transcripts, but with the following modifications: minimum fraction of identity (similarity) = 0.90; minimum read length fraction = 0.01. All *de novo* contigs generated from bobwhite cDNA sequences were characterized using blastx [79] in conjunction with all available bird proteins (NCBI non-redundant avian protein sequences) as previously described [42]. A SAM file representing all bobwhite cDNA *de novo* contig mappings was created using the CLC Genomics Workbench.

The bobwhite contig containing the mitochondrial genome (NB1.0, NB1.1) was manually annotated using the chicken as a guide (GenBank Accession HQ857212), and several available BLAST tools (blastn, bl2seq, blastp; http://blast.ncbi.nlm.nih.gov/). Thereafter, we used tRNAscan-SE (http://lowelab.ucsc.edu/tRNAscan-SE/) to predict tRNA genes, with one tRNA manually predicted by comparative sequence analysis.

Whole-Genome Analyses of Divergence and Development of Candidate Genes

For all NB1.0 contigs that produced blastn hits to the chicken (*G. gallus* 4.0) or zebra finch genomes (*T. guttata* 3.2.4), we normalized the observed percent identity for differences in alignment length across both comparative genome alignments using the following formula:

$$CorrectedForAL = \frac{(\frac{PercentID}{100})}{AlignmentLength}$$ [42]. This method is

mathematically similar and related to the p-distance [112], and allows for genome-wide nucleotide by nucleotide comparison of both coding and noncoding DNA, with a previous investigation supporting the use of alignment based sequence comparison and distance estimation for conserved genomes [113]. Thereafter, we visualized the full distribution of this composite variable by producing histograms within JMP Pro 10.0.1 (SAS Institute Inc., Cary, NC). The full distribution of observed "CorrectedForAL values" produced from each comparative genome alignment is highly skewed and resistant to standard transformation methods [42]. Therefore, we used a percentile approach to identify outlier contigs based on establishing interval bounds within the ordered distributions (at the 99.98th and 0.02th percentiles). All analytical procedures including outlier definition, detection by percentile-cutoff locations, and quality control analyses followed methods previously described [42]. All NB1.0 contigs implicated as outliers for divergence were scrutinized by searching five databases curated by NCBI (i.e., refseq_genomic, refseq_rna, nr/nt, traces-WGS, traces-other DNA) for blastn alignments that would further confirm or refute their outlier status. Trace alignments (i.e., WGS; other) with bitscores ≥15% larger than the original bitscore were considered false positives for extreme divergence, and were removed from the final list of putative outliers. NB1.0 contigs classified as outliers for extreme conservation were annotated based on the individual reference genome from which they were identified (i.e., *G. gallus* 4.0; *T. guttata* 3.2.4; See Table S11). Established knowledge of gene function (i.e., among outliers) in combination with the human GWAS literature were used to identify potential candidate genes for biological traits among the avian species compared.

Effective population size estimation

The bobwhite and scarlet macaw were chosen for comparison using PSMC [55] because they occupy opposing positions on the *r-/K*-selection continuum [56–58], with bobwhites being largely typical of an *r*-selected avian species, and the scarlet macaw clearly exhibiting characteristics of *K*-selection [56–59]. This allowed us to test the hypothesis that historic effective population size estimates for an *r*-selected avian species should theoretically exceed that of a *K*-selected avian species, and to compare the magnitude by which they differed. The input file for PSMC [55] was prepared according to the PSMC author's recommendations. For the bobwhite, variants with less than 46× coverage or more than 280× coverage were filtered from the diploid consensus. For the scarlet macaw, variants with less than 4× coverage or more than 26× coverage were filtered from the diploid consensus. Only NB1.1 and scarlet macaw (SMAC 1.1) [42] scaffolds aligning to autosomes were used. The maximum $2N_0$ coalescent time (parameter –t) was varied until at least 10 recombinations per atomic interval were observed. PSMC was run for 25 iterations, with –t10 –r5 –p "4+25*2+4+6" options used for the bobwhite and –t6 –r5 –p "4+25*2+4+6" used for the scarlet macaw. One hundred bootstraps were used to calculate confidence intervals. We used the per-site pairwise sequence divergence to represent time and the scaled mutation rate to represent population size [55]. To estimate generation time for the bobwhite, we evaluated long-term survivorship studies from across their U.S. range that did not rely on radio telemetry [114–119]. Radio telemetry studies often greatly underestimate survivorship, so generation time based on such studies would also be underestimated [120]. Bobwhite generation time (*g*) was estimated as: $g = a+[s/(1-s)]$ [41,121], where *a* = age of sexual maturity (~1 yr), and *s* = adult survival rate, as reported across the survivorship studies evaluated. We used the median generation time (1.22 yrs; range = 1.17–1.39 yrs) estimated across all studies for the bobwhite. At present, little is known about generation times in the scarlet macaw, with one source proposing a generation time of 12.7 years (http://www.birdlife.org/datazone/speciesfactsheet.php?id = 1551&m = 1). By considering an expected (*s*) of at least 90% across the scarlet macaw's range (i.e., in protected and unprotected regions), and (*a*) equivalent to 4 yrs, we estimated generation time for the scarlet macaw as approximately 13 yrs. Therefore, we used *g* = 12.7 in our PSMC analysis. Notably, our assumptions regarding *s* = 0.90 and *a* = 4.0 were both biologically feasible and reasonable, as evidenced by previous studies [107,122–123]. Similar to recent PSMC analyses for the pig (*Sus scrofa*) genome [72], there are also no convincing data available regarding a different mutation rate in our birds (i.e., bobwhite, scarlet macaw) as compared to humans ($1.1-2.5 \times 10^{-8}$ mutations per generation) [124–125]. In fact, we initially estimated the substitution rate for the bobwhite and the scarlet macaw using autosomal genome alignment data and estimated divergence times as previously described [41], but found that these estimates produced unreasonable PSMC results due to underestimation of the per-generation *de novo* mutation rate, as has been predicted by using the substitution rate [155]. The most likely reasons for this are the relatively large estimated divergence times between the bobwhite and scarlet macaw as compared to other available, well annotated bird genomes (i.e., chicken, zebra finch, turkey), a very short generation interval for the bobwhite, a potential bias that is introduced by estimating the mutation rate via whole genome alignment (i.e., conserved regions align more stringently and more frequently), and the fact that the substitution rate only accounts for those mutations in lineages that persist in the face of drift and selection, which is not the same as the per-generation mutation rate observed from parent genome to offspring [155]. For these reasons, we used two reasonable estimates for the mutation rate (i.e., 1.1×10^{-8} and the PSMC default value of 2.5×10^{-8} mutations per generation) to calibrate sequence divergence to years [55].

Data Access

This Whole Genome Shotgun project has been deposited at DDBJ/EMBL/GenBank under the accessions AWGT00000000 and AWGU00000000. The versions described in this paper are the first versions: AWGT01000000; AWGU01000000. data and other project materials are also available at the bobwhite genome project website: http://vetmed.tamu.edu/faculty/cseabury/genomics.

Supporting Information

Table S1 NB1.0 Contig Map Positions in NB1.1.

Table S2 NB1.0 Comparative Genome Alignment to Chicken (S2a) and Zebra Finch (S2b).

Table S3 NB1.1 Comparative Genome Aligment to Chicken (S3a) and Zebra Finch (S3b).

Table S4 Summary of all Repeat Masker Analyses.

Table S5 Summary of All PHOBOS Repeat Analyses.

Table S6 Summary of Putative Nuclear Annotation Models via GlimmerHMM and Blastx with Manual Annotation of the Mitochondria and a Synopsis of MHC Annotations.

Table S7 Galgal4.71 cDNA Refseq Mappings onto NB1.1.

Table S8 Galgal 4.71 *ab initio* (GENSCAN) Transcript Mappings to NB1.1.

Table S9 Bobwhite *de novo* cDNA Contigs-Blastx to all Avian Proteins.

Table S10 Bobwhite cDNA Contig Map Positions in NB1.1.

Table S11 Bobwhite *de novo* Outlier Contigs (NB1.0) from Genome-Wide Analyses of Divergence with the Chicken and Zebra Finch Genomes.

Table S12 NB1.0 QC Analysis on Conserved Outliers Using Additional (Appended) Non-overlapping Blastn data (Chicken S12a, Zebra Finch S12b) To Recalculate the Composite Variable.

Table S13 NB1.0 Simple *de novo* Assembly Stats.

Table S14 NB1.1 Scaffolded *de novo* Assembly Stats.

Acknowledgments

We thank the Texas AgriLife Genomics and Bioinformatics Core, Texas A&M University, and the Missouri Sequencing Core (Nathan Bivens; Sean Blake) at the University of Missouri for high quality sequencing services. CMS is thankful to Joe Crafton for his enthusiasm and commitment to bobwhite management and restoration. CMS also thanks Jeff Skelton, Rick Young, Enrique Terrazas, Stuart Slattery, and Nathan Brown of the TAMU CVM computer support group for I.T. support, maintenance, and the freedom to explore innovative computing solutions to data processing and storage.

Author Contributions

Conceived and designed the experiments: CMS. Performed the experiments: YH CMS. Analyzed the data: CMS YH SED JED MJP. Contributed reagents/materials/analysis tools: DR IRT DJB CDJ. Wrote the paper: YH CMS. Assembled the Bobwhite Genome (Iteratively): YH CMS. Performed Comparative Genome Alignments: YH. Predicted Repeat Content: YH CMS. Performed Genome Wide Variant Detection: CMS. Estimated Bobwhite and Scarlet Macaw Generation Times: MJP CMS. Performed Coalescent Modeling: JED. Interpreted Coalescent Modeling: CMS JED JFT. Performed Bobwhite Genome Annotation: YH CMS SED. Compiled, Parsed, and Scripted Annotation Tables and Genomic Data: CMS PMS SED. Performed Whole Genome Analyses of Divergence: YH CMS. Performed Quality Control Analyses: YH CMS EB PMS. Managed Data: YH CMS PMS. Provided important comments and suggestions for the manuscript: SED JED PMS EB CDJ DR IRT DJB MJP JFT.

References

1. Del Hoyo J, Elliot A, Sargatal J (1997) Handbook of the Birds of the World (Vol. 2): New World Vultures to Guineafowl. Barcelona: Lynx Edicións. 413–425 pp.
2. Lusk J, Guthery FS, George RR, Peterson MJ, DeMason SJ (2002) Relative abundance of bobwhites in relation to weather and land use. J Wildlife Manage 66: 1040–1051.
3. Williams CK, Guthery FS, Applegate RD, Peterson MJ (2004) The northern bobwhite decline: scaling our management for the twenty-first century. Wildlife Soc Bull 32: 861–869.
4. Quinn MJ, Hanna TL, Shiflett AA, McFarland CA, Cook ME, et al. (2012) Interspecific effects of 4A-DNT (4-amino-2,6-dinitrotoluene) and RDX (1,3,5-trinitro-1,3,5-triazine) in Japanese quail, Northern bobwhite, and Zebra finch. Ecotoxicology 22: 231–239.
5. Brennan LA (1991) How can we reverse the Northern Bobwhite population decline? Wildlife Soc Bull 19: 544–555.
6. Sauer JR, Link WA, Nichols JD, Royle JA (2004) Using the North American Breeding Bird Survey as a tool for conservation: a critique of Bart et al. (2004). J Wildlife Manage 69: 1321–1326.
7. Johnson MS, Michie W, Bazar MA, Gogal RM (2005) Influence of oral 2, 4-dinitrotoluene exposure to the Northern Bobwhite (*Colinus virginianus*). Int J Toxicol 24: 265–274.
8. Quinn MJ Jr, Bazar MA, McFarland CA, Perkins EJ, Gust KA, et al. (2007) Effects of subchronic exposure to 2, 6-dinitrotoluene in the northern bobwhite (*Colinus Virginianus*). Environ Toxicol Chem 26: 2202–2207.
9. Quinn MJ Jr, McFarland CA, LaFiandra EM, Johnson MS (2009) A preliminary assessment of relative sensitivities to foreign red blood cell challenges in the northern bobwhite for potential evaluation of immunotoxicity. J Immunotoxicol 6: 171–173.
10. Brausch JM, Blackwell BR, Beall BN, Caudillo C, Kolli V, et al. (2010) Effects of polycyclic aromatic hydrocarbons in northern bobwhite quail (*Colinus virginianus*). J Toxicol Env Health 73: 540–551.
11. Rawat A, Gust KA, Deng Y, Garcia-Reyero N, Quinn MJ, et al. (2010) From raw materials to validated system: The construction of a genomic library and microarray to interpret systemic perturbations in Northern bobwhite. Physiol Genomics 42: 219–235.
12. Bridges AS, Peterson MJ, Silvy NJ, Smeins FE, Wu XB (2001) Differential influence of weather on regional quail abundance in Texas. J Wildl Manage 65: 10–18.
13. Hernández F, Hernández F, Arredondo JA, Bryant FC, Brennan LA, et al. (2005) Influence of precipitation on demographics of northern bobwhites in southern Texas. Wildlife Soc Bull 33: 1071–1079.
14. Hernández F, Peterson MJ (2007) Northern bobwhite ecology and life history. In: Brennan LA editor. Texas quails: Ecology and management. College Station: Texas A&M University Press. pp. 40–64.
15. Leopold A (1931) Report on a game survey of the north central states. Madison: Democrat Printing Company.
16. Errington PL, Hamerstrom FN Jr (1936) The northern bob-white's winter territory. Iowa State College of Agriculture and Mechanical Arts Research Bulletin 201: 305–443.
17. Lehmann VW (1937) Increase quail by improving their habitat. Austin: Texas Game, Fish and Oyster Commission. 44 p.
18. Droege S, Sauer JR (1990) Northern bobwhite, gray partridge, and ring-necked pheasant population trends (1966–1988) from the North American Breeding Bird Survey. In Church KE, Warner RE, Brady SJ, editors. Perdix V: gray partridge and ring-necked pheasant workshop. Emporia: Kansas Department of Wildlife and Parks. pp. 2–20.
19. Church KE, Sauer JR, Droege S (1993) Population trends of quails in North America. Proceedings of the National Quail Symposium 3: 44–54.
20. Brady SJ, Flather CH, Church KE (1998) Range-wide declines of northern bobwhite (Colinus virginianus): land use patterns and population trends. Gibier Faune Sauvage; Game and Wildlife 15: 413–431.
21. Peterson MJ, Wu XB, Rho P (2002) Rangewide trends in landuse and northern bobwhite abundance: An exploratory analysis. Proc Nat Quail Sym 5: 35–44.
22. Sauer JR, Hines JE, Fallon JE, Pardieck KL, Ziolkowski DJ Jr, et al. (2012) The North American Breeding Bird Survey, results and analysis 1966–2011. Version 07.03.2013. USGS Patuxent Wildlife Research Center, Laurel, Maryland, Available from http://www.mbr-pwrc.usgs.gov/bbs/bbs.html. Accessed 23 October 2013.
23. Hernández F, Brennan LA, DeMaso SJ, Sands JP, Wester DB (2013) On reversing the northern bobwhite population decline: 20 years later. Wildl Soc Bull 37: 177–188.

24. Guthery FS, Forrester ND, Nolte KR, Cohen WE, Kuvlesky WP Jr (2000) Potential effects of global warming on quail populations. In Brennan LA, Palmer WE, Burger LW Jr., Pruden TL, editors. Quail IV: Proceedings of the Fourth National Quail Symposium. Tallahassee: Tall Timbers Research Station. pp. 198–204.

25. Reyna KS, Burggren WW (2012) Upper lethal temperatures of Northern Bobwhite embryos and the thermal properties of their eggs. Poultry Sci 91: 41–46.

26. Mueller JM, Dabbert CB, Demarais S, Forbes AR (1999) Northern bobwhite chick mortality caused by red imported fire ants. J Wildlife Manage 63: 1291–1298.

27. Allen CR, Willey RD, Myers PE, Horton PM, Buffa J (2000) Impact of red imported fire ant infestation on northern bobwhite quail abundance trends in southeastern United States. J Agric Urban Entomo 17: 43–51.

28. Ottinger MA, Quinn MJ Jr, Lavoie E, Abdelnabi MA, Thompson N, et al. (2005) Consequences of endocrine disrupting chemicals on reproductive endocrine function in birds: establishing reliable end points of exposure. Domest Anim Endocrin 29: 411–419.

29. Kitulagodage M, Isanhart J, Buttemer WA, Hooper MJ, Astheimer LB (2011) Fipronil toxicity in northern bobwhite quail Colinus virginianus: reduced feeding behavior and sulfone formation. Chemosphere 83: 524–530.

30. Peterson MJ, Perez RM (2000) Is quail hunting self regulatory?: Northern bobwhite and scaled quail abundance and quail hunting in Texas. Proc Nat Quail Sym: 85–91.

31. Peterson MJ (2001) Northern bobwhite and scaled quail abundance and hunting regulation: A Texas example. J Wildl Manage 65: 828–837.

32. Williams CK, Lutz SR, Applegate RD (2004) Winter survival and additive harvest in northern bobwhite coveys in Kansas. J Wildlife Manage 68: 94–100.

33. DeVos T Jr, Speake DW (1995) Effects of releasing pen-raised northern bobwhites on survival rates of wild populations of northern bobwhites. Wildlife Soc Bull 23: 267–273.

34. Terhune TM, Sisson DC, Palmer WE, Faircloth BC, Stribling HL, et al. (2010) Translocation to a fragmented landscape: survival, movement, and site fidelity of northern bobwhites. Ecol Appl 20: 1040–1052.

35. Scott JL, Hernández F, Brennan LA, Ballard BM, Janis M, et al. (2012) Population demographics of translocated northern bobwhites on fragmented habitat. Wildlife Soc Bull: 10.1002/wsb.239.

36. Baumgartner FM (1944) Dispersal and survival of game farm bobwhite quail in north central Oklahoma. J Wildlife Manage 8: 112–118.

37. Buechner HK (1950) An evaluation of restocking with pen-reared bobwhite. J Wildlife Manage 14: 363–377.

38. Evans KO, Smith MD, Burger LW Jr, Chambers RJ, Houston AE, et al. (2006) Release of pen-reared bobwhites: Potential consequences to the genetic integrity of resident wild populations. In: Cederbaum SB, Faircloth BC, Terhune TM, Thompson JJ, Carroll JP, editors. Gamebird. Georgia: University of Georgia. pp. 121–133.

39. Oleksyk TK, Pombert JF, Siu D, Mazo-Vargas A, Ramos B, et al. (2012) A locally funded Puerto Rican parrot (Amazona vittata) genome sequencing project increases avian data and advances young researcher education. GigaScience 1: 14.

40. Ellegren H, Smeds L, Burri R, Olason PI, Backström N, et al. (2012) The genomic landscape of species divergence in Ficedula flycatchers. Nature 491: 756–760.

41. Zhan X, Pan S, Wang J, Dixon A, He J, et al. (2013) Peregrine and saker falcon genome sequences provide insights into evolution of a predatory lifestyle. Nat Genet 45: 563–566.

42. Seabury CM, Dowd SE, Seabury PM, Raudsepp TR, Brightsmith DJ, et al. (2013) A Multi-Platform Draft de novo Genome Assembly and Comparative Analysis for the Scarlet Macaw (Ara macao). PLoS ONE 8: e15811. doi: 10.1371/journal.pbio.1000475.

43. Beçak ML, Beçak W, Roberts FL, Shoffner RN, Volpe EP (1971) Aves. In: Benirschke K, Hsu TC, editors. Chromosome atlas: fish, amphibians, reptiles, and birds (Vol.1). New York: Springer-Verlag. p. AV-3.

44. Hale DW, Ryder EJ, Sudman PD, Greenbaum IF (1988) Application of synaptonemal complex techniques for determination of diploid number and chromosomal morphology of birds. The Auk 105: 776–779.

45. Rawat A, Gust KA, Elasri MO, Perkins EJ (2010) Quail Genomics: a knowledgebase for Northern Bobwhite. BMC Bioinformatics 11: S313.

46. Hillier LW, Miller W, Birney E, Warren W, Hardison RC, et al. (2004) Sequence and comparative analysis of the chicken genome provide unique perspectives on vertebrate evolution. Nature 432: 695–716.

47. Dalloul RA, Long JA, Zimin AV, Aslam L, Beal K, et al. (2010) Multi-platform next-generation sequencing of the domestic turkey (Meleagris gallopavo): genome assembly and analysis. PLoS Biol 9. doi: 10.1371/journal.pbio.1000475.

48. Warren WC, Clayton DF, Ellegren H, Arnold AP, Hillier LW, et al. (2010) The genome of a songbird. Nature 464: 757–762.

49. Hedges SB, Dudley J, Kumar S (2006) TimeTree: a public knowledge-base of divergence times among organisms. Bioinformatics 22: 2971–2972.

50. Kumar S, Hedges SB (2011) TimeTree2: species divergence times on the iPhone. Bioinformatics 27: 2023–2024.

51. Mayer C, Leese F, Tollrian R (2010) Genome-wide analysis of tandem repeats in Daphnia pulex-a comparative approach. BMC Genomics 11: 277.

52. Schable NA, Faircloth BC, Palmer WE, Carroll JP, Burger LW, et al. (2004) Tetranucleotide and dinucleotide microsatellite loci from the northern bobwhite (Colinus virginianus). Molec Ecol Notes 4: 415–419.

53. Faircloth BC, Terhune TM, Schable NA, Glenn TC, Palmer WE, et al. (2009) Ten microsatellite loci from Northern Bobwhite (Colinus virginianus). Conserv Genet 10: 535–538.

54. Wong GKS, Liu B, Wang J, Zhang Y, Yang X, et al. (2004) A genetic variation map for chicken with 2.8 million single-nucleotide polymorphisms. Nature 432: 717–722.

55. Li H, Durbin R (2011) Inference of human population history from individual whole-genome sequences. Nature 475: 493–496.

56. Dobzhansky T (1950) Evolution in the tropics. Amer Sci 38: 209–221.

57. MacArthur RH, Wilson EO (1967) The theory of island biogeography. Princeton: Princeton University Press.

58. Pianka ER (1970) On r- and K-Selection. Am Nat 104: 592–597.

59. Brennan LA (2007) Texas Quails: Ecology and Management. College Station: Texas A&M University Press.

60. Eshleman JA, Malhi RS, Smith DG (2003) Mitochondrial DNA studies of Native Americans: conceptions and misconceptions of the population prehistory of the Americas. Evol Anthropol 12: 7–18.

61. Gilbert MTP, Jenkins DL, Götherstrom A, Naveran N, Sanchez JJ, et al. (2008) DNA from pre-clovis human coprolites in Oregon, North America. Science 320: 786–789.

62. Waters MR, Forman SL, Jennings TA, Nordt LC, Driese SG, et al. (2011) The buttermilk creek complex and origins of Clovis at the Debra L. Friedkin site, Texas. Science 331: 1599–1603

63. Waters MR, Stafford TW Jr, McDonald HG, Gustafson C, Rasmussen M, et al. (2011) Pre-Clovis mastodon hunting 13,800 years ago at the Manis site, Washington. Science 334: 351–353.

64. Alroy J (2001) A multispecies overkill simulation of the end-Pleistocene megafaunal mass extinction. Science 292: 1893–1896.

65. Firestone RB, West A, Kennett JP, Becker L, Bunch TE, et al. (2007) Evidence for an extraterrestrial impact 12,900 years ago that contributed to the megafaunal extinctions and the Younger Dryas cooling. Proc Natl Acad Sci USA 104: 16016–16021.

66. Pushkina D, Raia P (2008) Human influence on distribution and extinctions of the late Pleistocene Eurasian megafauna. J Hum Evol 54: 769–782.

67. Yokoyama Y, Lambeck K, De Deckker P, Johnston P, Fifield LK (2000) Timing of the last glacial maximum from observed sea-level minima. Nature 406: 713–716.

68. Clark PU, Dyke AS, Shakun JD, Carlson AE, Clark J, et al. (2009) The last glacial maximum. Science 325: 710–714.

69. Redford KH, Robinson JG (1987) The game of choice: Patterns of Indian and Colonist Hunting in the Neotropics. Am Anthropol 89: 650–667.

70. Jackson HE, Scott SL (1995) The faunal record of the southeastern elite: The implications of economy, social relations, and ideology. Southeastern Archaeology 14: 103–119.

71. Kricher JC (1999) A neotropical companion: an introduction to the animals, plants, and ecosystems of the New World tropics (2nd Edition). Princeton: Princeton Univ Press.

72. Groenen MAM, Archibald AL, Uenishi H, Tuggle CK, Takeuchi Y, et al. (2012) Analyses of pig genomes provide insight into porcine demography and evolution. Nature 491: 393–398.

73. Meyer M, Kircher M, Gansauge MT, Li H, Racimo F, et al. (2012) A high-coverage genome sequence from an archaic Denisovan individual. Science 338: 222–226.

74. Sheehan S, Harris K, Song YS (2013) Estimating variable effective population sizes from multiple genomes: a sequentially markov conditional sampling distribution approach. Genetics 194: 647–62.

75. Caparroz R, Miyaki CY, Bampi MI, Wajntal A (2001) Analysis of the genetic variability in a sample of the remaining group of Spix's Macaw (Cyanopsitta spixii, Psittafores: Aves) by DNA fingerprinting. Biol Conserv 99: 307–311.

76. Hemmings N, West M, Birkhead TR (2012) Causes of hatching failure in endangered birds. Biol Lett doi:10.1098/rsbl.2012.0655.

77. Delcher AL, Harmon D, Kasif S, White O, Salzberg SL (1999) Improved microbial gene identification with GLIMMER. Nucleic Acids Res 27.23: 4636–4641.

78. Majoros WH, Pertea M, Salzberg SL (2004) TigrScan and GlimmerHMM: two open-source ab initio eukaryotic gene-finders. Bioinformatics 20: 2878–2879.

79. Dowd SE, Zaragoza J, Rodriguez JR, Oliver MJ, Payton PR (2005) Windows.NET network distributed basic local alignment search toolkit (W.ND-BLAST). BMC Bioinformatics 6: 93.

80. Kaufman J, Milne S, Göbel TW, Walker BA, Jacob JP, et al. (1999) The chicken B locus is a minimal essential major histocompatibility complex. Nature 401: 923–925.

81. Hughes CR, Miles S, Walbroehl JM (2008) Support for the minimal essential MHC hypothesis: a parrot with a single, highly polymorphic MHC class II B gene. Immunogenetics 60: 219–231.

82. Balakrishnan CN, Ekblom R, Völker M, Westerdahl H, Godinez R, et al. (2010) Gene duplication and fragmentation in the zebra finch major histocompatibility complex. BMC Biol 8: 29.

83. Ekblom R, Stapley J, Ball AD, Birkhead T, Burke T, et al. (2011) Genetic mapping of the major histocompatibility complex in the zebra finch (*Taeniopygia guttata*). Immunogenetics 63: 523–530.

84. Monson MS, Mendoza KM, Velleman SG, Strasburg GM, Reed KM (2013) Expression profiles for genes in the turkey major histocompatibility complex B-locus. Poultry Sci 92: 1523–1534.

85. Casinos A, Cubo J (2001) Avian long bones, flight and bipedalism. Comp Biochem Phys A 131: 159–167.

86. Gu X, Feng C, Ma L, Song C, Wang Y, et al. (2011) Genome-wide association study of body weight in chicken F2 resource population. PLoS ONE 6: e21872. doi: 10.1371/journal.pone.0021872.

87. Meisler MH (2001) Evolutionarily conserved noncoding DNA in the human genome: How much and what for? Genome Res 11: 1617–1618.

88. Prabhakar S, Noonan JP, Pääbo S, Rubin EM (2006) Accelerated evolution of conserved noncoding sequences in humans. Science 314: 786.

89. Pheasant M, Mattick JS (2007) Raising the estimate of functional human sequences. Genome Res 17: 1245–1253.

90. Johnson R, Samuel J, Keow C, Leng N, Jauch R, et al. (2009) Evolution of the vertebrate gene regulatory network controlled by the transcriptional repressor REST. Mol Biol Evol 26: 1491–1507.

91. Stein L, Hua X, Lee S, Ho AJ, Leow D, et al. (2010) Voxelwise genome-wide association study (vGWAS). Neuroimage 53: 1160–1174.

92. Potkin SG, Guffanti G, Lakatos A, Turner JA, Kruggel F, et al. (2009) Hippocampal atrophy as a quantitative trait in a genome-wide association study identifying novel susceptibility genes for Alzheimer's disease. PloS ONE 4: e6501. doi: 10.1371/journal.pone.0006501.

93. McClay JL, Adkins DE, Åberg K, Bukszár J, Khachane AN, et al. (2011) Genome-wide pharmacogenomic study of neurocognition as an indicator of antipsychotic treatment response in schizophrenia. Neuropsychopharmacol 36: 616–626.

94. Madge S, McGowan PJ, Kirwan GM (2002) Plate 59: Bobwhite. In Pheasants, partridges and grouse: a guide to the pheasants, partridges, quails, grouse, guineafowl, buttonquails and sandgrouse of the world. Princeton: Princeton University Press.

95. Higgins PJ, Peter JM, Cowling SJ (2006) Handbook of Australian, New Zealand and Antartic Birds, Boatbills to Starlings (Vol. 6). Melbourn; Oxford University Press. 1132 p.

96. Del Hoyo J, Elliott A, Christie DA (2010) Handbook of the Birds of the World (Vol. 15): Weavers to New World Warblers. Barcelona: Lynx Edicións. 357 p.

97. Mitgutsch C, Wimmer C, Sánchez-Villagra MR, Hahnloser R, Schneider RA (2011) Timing of ossification in duck, quail, and zebra finch: intraspecific variation, heterochronies, and life history evolution. Zool Sci 28: 491.

98. Ringoen AR (1945) Deposition of medullary bone in the female English sparrow, *Passer domesticus* (*Linnaeus*), and the Bobwhite quail, *Colinus virginianus*. J Morphol 77: 265–283.

99. Dacke CG, Arkle S, Cook DJ, Wormstone IM, Jones S, et al. (1993) Medullary bone and avian calcium regulation. J Exp Biol 184: 63–88.

100. Reynolds SJ (1997) Uptake of ingested calcium during egg production in the zebra finch (*Taeniopygia guttata*). The Auk: 562–569.

101. Starck JM, Ricklefs RE (1998) Patterns of development: the altricial-precocial Spectrum. In: Starck JM, Ricklefs RE, editors. Avian growth and development: evolution within the altricial-precocial spectrum. New York: Oxford university press. pp. 3–30.

102. Blom J, Lilja C (2005) A comparative study of embryonic development of some bird species with different patterns of postnatal growth. Zoology 108: 81–95.

103. Murray JR, Varian-Ramos CW, Welch ZS, Saha MS (2013) Embryological staging of the zebra finch, *Taeniopygia guttata*. J Morphol 274: 1090–1110.

104. Elks CE, Perry JR, Sulem P, Chasman DI, Franceschini N, et al. (2010) Thirty new loci for age at menarche identified by a meta-analysis of genome-wide association studies. Nat Genet 42: 1077–1085.

105. Guthery FS (2006) On Bobwhites (Issue 27, W. L. Moody Jr. Natural History Series). College Station: Texas A&M University Press. 124 p.

106. Nager RG, Law G (2010) The Zebra Finch. In: Hubrecht R, Kirkwood J, editors. The UFAW handbook on the care and management of laboratory and other research animals. Ames: Wiley-Blackwell. pp. 674–685.

107. Brightsmith DJ, Hilburn J, del Campo A, Boyd J, Frisius M, et al. (2005) The use of hand-raised psittacines for reintroduction: a case study of scarlet macaws (*Ara macao*) in Peru and Costa Rica. Biol Conserv 121: 465–472.

108. Vigo G, Williams M, Brightsmith DJ (2011) Growth of Scarlet Macaw (*Ara macao*) chicks in southeastern Peru. Neotropical Ornithology 22: 143–153.

109. Nystedt B, Street NR, Wetterbom A, Zuccolo A, Lin YC, et al. (2013) The Norway spruce genome sequence and conifer genome evolution. Nature 497: 579–584.

110. Sanchez CC, Smith TPL, Wiedman RT, Vallejo RL, Salem M, et al. (2009) Single nucleotide polymorphism discovery in rainbow trout by deep sequencing of a reduced representation library. BMC Genomics 10: 559.

111. Seabury CM, Bhattarai EK, Taylor JF, Viswanathan GG, Cooper SM, et al. (2011) Genome-wide polymorphism and comparative analyses in the white-tailed deer (*Odocoileus virginianus*): a model for conservation genomics. PLoS ONE 6: e15811. doi: 10.1371/journal.pone.0015811.

112. Nei M, Kumar S (2000) Molecular evolution and phylogenetics: Evolutionary Change of DNA Sequences. New York: Oxford University Press. 33 p.

113. Rosenberg MS (2005) Evolutionary distance estimation and fidelity of pair wise sequence alignment. BMC Bioinformatics 6: 102.

114. Marsden HM, Baskett TS (1958) Annual mortality in a banded bobwhite population. J Wildlife Manage 22: 414–419.

115. Kabat C, Thompson DR (1963) Wisconsin quail, 1834–1962: Population dynamics and habitat management. Tech. Bull. Wis. Conserv. Dep. No.30.

116. Speake DW (1967) Ecology and management studies of the bobwhite quail in the Alabama Piedmont. Ph.D. Dissertation, Auburn University, Alabama.

117. Roseberry JL, Klimstra WD (1984) Population ecology of the bobwhite. Carbondale: Southern Illinois University Press.

118. Pollock KH, Moore CT, Davidson WR, Kellogg FE, Doster GL (1989) Survival rates of bobwhite quail based on band recovery analyses. J Wildlife Manage 53: 1–6.

119. Folk TH, Holmes RR, Grand JB (2007) Variation in northern bobwhite demography along two temporal scales. Popul Ecol 49: 211–219.

120. Guthery FS, Lusk JJ (2004) Radiotelemetry studies: Are we radio-handicapping northern bobwhites? Wildl Soc Bull 32: 194–201.

121. Lande R, Engen S. Sæther BE (2003) Stochastic Population Dynamics in Ecology and Conservation. New York: Oxford Univ. Press.

122. Vaughan C, Nemeth NM, Cary J, Temple S (2005) Response of a Scarlet Macaw (*Ara macao*) population to conservation practices in Costa Rica. Bird Conserv Int 15: 119–30.

123. Strem RI, Bouzat JL (2012) Population viability analysis of the blue-throated macaw (*Ara glaucogularis*) using individual-based and cohort-based PVA programs. The Open Conservation Biology Journal 6: 12–24.

124. Nachman MW, Crowell SL (2000) Estimate of the mutation rate per nucleotide in humans. Genetics 156: 297–304.

125. Roach JC, Glusman G, Smit AF, Huff CD, Hubley R, et al. (2010) Analysis of genetic inheritance in a family quartet by whole-genome sequencing. Science 328: 636–9.

126. Mitchell GF, Verwoert GC, Tarasov KV, Isaacs A, Smith AV, et al. (2012) Common Genetic Variation in the 3'-BCL11B Gene Desert Is Associated With Carotid-Femoral Pulse Wave Velocity and Excess Cardiovascular Disease Risk Clinical Perspective. Circulation: Cardiovascular Genetics 5: 81–90.

127. Van Sligtenhorst I, Ding ZM, Shi ZZ, Read RW, Hansen G, et al (2012) Cardiomyopathy in α-Kinase 3 (ALPK3)–Deficient Mice. Vet Patho Online 49: 131–141.

128. Sotoodehnia N, Isaacs A, de Bakker PI, Dörr M, Newton-Cheh C, et al. (2010) Common variants in 22 loci are associated with QRS duration and cardiac ventricular conduction. Nat Genet 42: 1068–1076.

129. Companioni O, Rodríguez Esparragón F, Medina Fernández-Aceituno A, Rodríguez Pérez JC (2011) Genetic variants, cardiovascular risk and genome-wide association studies. Rev Esp Cardiol 64: 509–514.

130. Middelberg R, Ferreira M, Henders A, Heath A, Madden P, et al. (2011) Genetic variants in LPL, OASL and TOMM40/APOE-C1-C2-C4 genes are associated with multiple cardiovascular-related traits. BMC Med Genet 12: 123.

131. Pfeufer A, Sanna S, Arking DE, Müller M, Gateva V, et al. (2009) Common variants at ten loci modulate the QT interval duration in the QTSCD Study. Nat Genet 41: 407–414.

132. Hägg S, Skogsberg J, Lundström J, Noori P, Nilsson R, et al. (2009) Multiorgan expression profiling uncovers a gene module in coronary artery disease involving transendothelial migration of leukocytes and LIM domain binding 2: The Stockholm atherosclerosis gene expression (STAGE) study. PloS Genet 5: e1000754. doi: 10.1371/journal.pgen.1000754.

133. Menzaghi C, Paroni G, De Bonis C, Coco A, Vigna C, et al. (2008) The protein tyrosine phosphatase receptor type f (PTPRF) locus is associated with coronary artery disease in type 2 diabetes. J Intern Med 263: 653–654.

134. Nolan DK, Sutton B, Haynes C, Johnson J, Sebek J, et al. (2012) Fine mapping of a linkage peak with integration of lipid traits identifies novel coronary artery disease genes on chromosome 5. BMC Genet 13: 12.

135. Newton-Cheh C, Eijgelsheim M, Rice KM, de Bakker PI, Yin X, et al. (2009) Common variants at ten loci influence QT interval duration in the QTGEN Study. Nat Genet 41: 399–406.

136. Wain LV, Verwoert GC, O'Reilly PF, Shi G, Johnson T, et al. (2011) Genome-wide association study identifies six new loci influencing pulse pressure and mean arterial pressure. Nat Genet 43: 1005–1011.

137. Artigas MS, Loth DW, Wain LV, Gharib SA, Obeidat ME, et al. (2011) Genome-wide association and large-scale follow up identifies 16 new loci influencing lung function. Nat Genet 43: 1082–1090.

138. Hancock DB, Artigas MS, Gharib SA, Henry A, Manichaikul A, et al. (2012) Genome-wide joint meta-analysis of SNP and SNP-by-smoking interaction identifies novel loci for pulmonary function. PLoS Genet 8: e1003098. doi: 10.1371/journal.pgen.1003098.

139. Egan MF, Straub RE, Goldberg TE, Yakub I, Callicott JH, et al. (2004) Variation in GRM3 affects cognition, prefrontal glutamate, and risk for schizophrenia. Proc Natl Acad Sci USA 101: 12604–12609.

140. Kramer PL, Xu H, Woltjer RL, Westaway SK, Clark D, et al. (2011) Alzheimer disease pathology in cognitively healthy elderly: A genome-wide study. Neurobiol Aging 32: 2113–2122.

141. Ersland KM, Christoforou A, Stansberg C, Espeseth T, Mattheisen M, et al. (2012) Gene-based analysis of regionally enriched cortical genes in GWAS data sets of cognitive traits and psychiatric disorders. PLOS ONE, 7: e31687. doi: 10.1371/journal.pone.0031687.

142. Levy D, Larson MG, Benjamin EJ, Newton-Cheh C, Wang TJ, et al. (2007) Framingham Heart Study 100K Project: genome-wide associations for blood pressure and arterial stiffness. BMC Med Genet 8: S3.

143. Shetty PB, Tang H, Tayo BO, Morrison AC, Hanis CL, et al. (2012) Variants in CXADR and F2RL1 are associated with blood pressure and obesity in African-Americans in regions identified through admixture mapping. J hypertens 10: 1970–1976.

144. Eijgelsheim M, Newton-Cheh C, Sotoodehnia N, de Bakker PI, Müller M, et al. (2010) Genome-wide association analysis identifies multiple loci related to resting heart rate. Hum Mol Genet 19: 3885–3894.

145. Kung AW, Xiao SM, Cherny S, Li GH, Gao Y, et al. (2010) Association of Stochastic Population Dynamics in Ecology and JAG1 with bone mineral density and osteoporotic fractures: a genome-wide association study and follow-up replication studies. Am J Hum Genet 86: 229.

146. Deng FY, Zhao LJ, Pei YF, Sha BY, Liu XG, et al. (2010) Genome-wide copy number variation association study suggested VPS13B gene for osteoporosis in Caucasians. Osteoporosis Int 21: 579–587.

147. Estrada K, Styrkarsdottir U, Evangelou E, Hsu YH, Duncan EL, et al. (2012) Genome-wide meta-analysis identifies 56 bone mineral density loci and reveals 14 loci associated with risk of fracture. Nat Genet 44: 491–501.

148. Lebeau G, Miller LC, Tartas M, McAdam R, Laplante I, et al. (2011) Staufen 2 regulates mGluR long-term depression and Map1b mRNA distribution in hippocampal neurons. Learn Memory 18: 314–326.

149. Zhang J, Tu Q, Grosschedl R, Kim MS, Griffin T, et al. (2011) Roles of SATB2 in osteogenic differentiation and bone regeneration. Tissue Eng 17: 1767–1776.

150. Gudbjartsson DF, Walters GB, Thorleifsson G, Stefansson H, Halldorsson BV, et al. (2008) Many sequence variants affecting diversity of adult human height. Nat Genet 40: 609–615.

151. Allen HL, Estrada K, Lettre G, Berndt SI, Weedon MN, et al. (2010) Hundreds of variants clustered in genomic loci and biological pathways affect human height. Nature 467: 832–838.

152. Smith EN, Chen W, Kähönen M, Kettunen J, Lehtimäki T, et al. (2010) Longitudinal genome-wide association of cardiovascular disease risk factors in the Bogalusa heart study. PLoS Genet 6: e1001094. doi: 10.1371/journal.pgen.1001094.

153. Polašek O, Marušić A, Rotim K, Hayward C, Vitart V, et al. (2010) Genome-wide association study of anthropometric traits in Korčula Island, Croatia. Croat Med J 50: 7–16.

154. Tiersch TR, Wachtel SS (1991) On the Evolution of Genome Size in Birds. J Hered 82: 363–368. doi: 10.1590/s1415-47571998000200006.

155. Barrick JE, Lenski RE (2013) Genome dynamics during experimental evolution. Nat Rev Genet 14(12):827–39. doi: 10.1038/nrg3564.

Parents and Early Life Environment Affect Behavioral Development of Laying Hen Chickens

Elske N. de Haas[1]*, **J. Elizabeth Bolhuis**[1], **Bas Kemp**[1], **Ton G. G. Groothuis**[2], **T. Bas Rodenburg**[3]

1 Adaptation Physiology Group, Department of Animal Science, Wageningen University and Research, Wageningen, The Netherlands, **2** Behavioural Biology, Centre for Behaviour and Neuroscience, University of Groningen, Groningen, The Netherlands, **3** Behavioural Ecology Group, Department of Animal Sciences, Wageningen University and Research, Wageningen, The Netherlands

Abstract

Severe feather pecking (SFP) in commercial laying hens is a maladaptive behavior which is associated with anxiety traits. Many experimental studies have shown that stress in the parents can affect anxiety in the offspring, but until now these effects have been neglected in addressing the problem of SFP in commercially kept laying hens. We therefore studied whether parental stock (PS) affected the development of SFP and anxiety in their offspring. We used flocks from a brown and white genetic hybrid because genetic background can affect SFP and anxiety. As SFP can also be influenced by housing conditions on the rearing farm, we included effects of housing system and litter availability in the analysis. Forty-seven rearing flocks, originating from ten PS flocks were followed. Behavioral and physiological parameters related to anxiety and SFP were studied in the PS at 40 weeks of age and in the rearing flocks at one, five, ten and fifteen weeks of age. We found that PS had an effect on SFP at one week of age and on anxiety at one and five weeks of age. In the white hybrid, but not in the brown hybrid, high levels of maternal corticosterone, maternal feather damage and maternal whole-blood serotonin levels showed positive relations with offsprings' SFP at one week and offsprings' anxiety at one and five weeks of age. Disruption and limitation of litter supply at an early age on the rearing farms increased SFP, feather damage and fearfulness. These effects were most prominent in the brown hybrid. It appeared that hens from a brown hybrid are more affected by environmental conditions, while hens from a white hybrid were more strongly affected by parental effects. These results are important for designing measures to prevent the development of SFP, which may require a different approach in brown and white flocks.

Editor: William Barendse, CSIRO, Australia

Funding: This work was supported by the Division of Earth and Life Sciences with financial aid from the Netherlands Organization for Scientific Research and the Ministry of Economic Affairs within the program "The Value of Animal Welfare" for the specific project named "Preventing feather pecking in laying hens: from principle to practice" (http://www.nwo.nl/en/research-and-results/research-projects/46/2300154446.html). The funders had no role in study design, data collection and analysis, decision to publish, or preparation of the manuscript.

Competing Interests: The authors have declared that no competing interests exist.

* E-mail: elske.dehaas@wur.nl

Introduction

In mammals, but also in avian and fish species, mothers can affect the behavioral development of their offspring both before and after birth or hatch (e.g. humans [1,2], rodents [3,4], fish [5], wild birds [6] and domesticated birds [7]; for reviews see: [8–10], farm animals [11], birds [12,13]). Mechanisms by which birds may pass information to their offspring are through hormone transfer to the egg [12,14] and/or via epigenetic pathways [15–17]. By these mechanisms the developing embryo may be better prepared for its future environment; this is also referred to as a "predictive adaptive response" [18,19]. In poultry, yolk-hormone levels can vary according to stressful environmental conditions [20]. Exposure to repeated, unpredictable events (Japanese quail [21], domestic chicken [16]) and daily exposure to humans (Japanese quail [22]) can alter egg-hormone levels. Stress experienced by the hen can also reduce her own body weight [7] and egg weight [23,24], and in this way influence offspring development too. Such maternal effects may underlie the repeated finding that offspring of stressed birds have higher anxiety levels compared with offspring from non-stressed birds [7,21,25–27].

These maternal effects may have important implications for the poultry industry, but have so far been overlooked. In commercial laying hens, feather pecking (FP), the plucking of- and pecking at feathers of conspecifics [28], is a maladaptive behavior. The severe form of FP (severe feather pecking: SFP) has serious consequences for animal welfare as it causes pain and stress in the recipient and can lead to mortality due to cannibalism. Counter measures against FP, such as beak trimming, adjustments of light intensity or supply of foraging materials [29], are only partially successful and we studied the possibility that maternal effects play a role. The tendency to develop SFP seems to be related to anxiety-related behavioral and physiological traits [30–33]. For example, chicks which show high anxiety in an Open Field test (social isolation in a novel environment) have stronger tendencies to perform SFP [30,31,33,34]. Also, birds with high anxiety levels show high post-stress plasma corticosterone levels whilst having low whole-blood serotonin levels, which were linked to feather pecking tendencies [32,33]. The predisposition to be more anxious and develop FP has a genetic component, as birds of a white ancestor origin are generally more anxious than birds of a brown origin [24,34–38]. The predisposition for anxiety can be affected by level of stress of the parents [7,39]. Therefore, it is important to assess this relationship under commercial conditions where it can affect millions of laying hens. In the poultry industry, parental flocks (parent stock: PS) are flocks which contain thousands of breeder hens and roosters housed together. They produce a multitude of offspring flocks (rearing flocks) which themselves contain thousands per flock. Additionally, the housing conditions during the offspring's early life can affect development of behavior [40,41] including FP [42,43]. Factors such as a large group size [44,45], a high stocking density [46,47] and a lack of litter or unsuitable litter [48–50] have been shown to increase the development of FP.

In this study, we examined in two crosses of laying hens (Dekalb White: DW and ISA Borwn: ISA) whether parent stock had an effect on the development of FP and anxiety in their offspring. To understand the relation between parents and offspring, we studied which behavioral and physiological parameters (feather damage, plasma corticosterone levels and serotonin levels) of the parent stock coincided with high levels of SFP and anxiety in their offspring. In addition, we studied how litter supply and housing conditions during rearing affected the development of FP. Commercial PS flocks had an impact on the development of anxiety and SFP in their offspring, especially for the DW hybrid. Litter conditions and housing system also showed to have a substantial effect on SFP and anxiety, especially for the ISA hybrid.

Materials and Methods

As one-on-one relations between parents and offspring cannot be determined under commercial conditions - due to the impossibility of individual recognition within large flocks of birds - data were assessed on flock level for both PS and rearing flocks.

Ethics Statement

This study comprises an on-farm longitudinal follow-up study on commercial laying hens, conducted between August 2010 and March 2012, which was approved by the Institutional Animal Care and Use Committee of Wageningen University, The Netherlands (permit number for parental flocks: DEC 2010042, permit number for rearing flocks: DEC 2010083).

Parent stock

Experimental animals and housing. Ten commercial flocks of parent stock (PS) of the rearing company Ter Heerdt BV, Babberich, The Netherlands were studied. Five of these were ISA Brown (ISA) parent stock (white hens, brown roosters) and five were Dekalb White (DW) parent stock (white hens and roosters). ISA Brown PS chickens originate from a Rhode Island Red and a Rhode Island White founder line. Dekalb White chickens originate from two White Leghorn founder lines. The ten PS flocks were situated at 7 different breeding farms, meaning that 3 farms had both hybrids while the remaining had either DW or ISA only. Flocks of different hybrids from the same breeding farm were taken as separate flocks. Rooster/hen ratio was approximate 1:10 for all flocks. Flocks were kept on commercial propagator farms with floor housing, partly slatted floors, and litter. For details on housing see [24].

Measurements. At 40 weeks of age, levels of feather damage, basal plasma-corticosterone and whole-blood serotonin levels of parental hens were assessed. For a detailed description of the measurements, see [24]. For 20 hens per flock, blood samples were drawn from the wing vein within two min after capturing the hen. Blood samples were analyzed for plasma-corticosterone (CORT) and whole-blood serotonin (5-HT) levels (for details, see [24]).

Each hen was individually taken from a random location in the chicken house (left or right; front or middle; floor or slats or nest boxes) to an adjacent room. After blood sampling, feather damage on neck, back and belly was assessed, and scored on a 3-point scale: no damage (a), slight damage (b), severe damage (c). Scores per area were summed to give a total body score [51] between 0 (no damage) and 2 (most severe damage). Fertilized eggs were collected daily and were incubated in a commercial incubator of the hatchery of Ter Heerdt BV, Zevenaar, The Netherlands. Fertilized eggs were collected per farm and hybrid. The pooled data per farm and hybrid are referred to as parent stock (PS).

Rearing flocks

Experimental animals and housing. Per PS flock (n = 10) between three to seven rearing flocks were studied, of which 23 were DW and 24 were ISA (n = 47 rearing flocks in total). The 47 rearing flocks were situated at 25 different rearing farms. Age of the parents at time of incubation varied from 30 to 60 weeks of age, with a majority around 40 weeks. The rearing flocks contained only hen-chicks. At one day after hatch chicks arrived at the rearing farm on which they stayed until approximate 17 weeks of age. All rearing flocks were housed in a tier-system of which 39 flocks were housed in an aviary system and 8 flocks in a floor system to which levels were gradually added (level system). All systems provided tiers, a litter area, slatted area, perches, multiple nipple drinkers and feeding troughs at different levels but no nest boxes or outdoor area. During the first five weeks of life, in the aviary system adjacent cages were either closed, restricting the number of chicks within the same enclosure (between 30–60), or partly-open (between 30–100). Chicks in the level system were placed in one large flock which varied between 10.000 and 30.000 chicks. Upon arrival, chicks were housed under temperatures ranging between 30 and 33°C with humidity levels between 50 and 65%. Temperature was gradually decreased to approximately 19°C at 10 weeks of age, which was maintained from 10 to 17 weeks of age. Chicks were kept under artificial light either with or without additional LED light with intensities ranging from 1 – 25 LUX measured with a Voltcraft MS-1300 light meter (Conrad Electric Benelux, Oldenzaal, The Netherlands) on bird level. Light regime was a 4-h light/2-h dark cycle for the first seven days of life. After seven days, light regime was adjusted to a 16-h light/8-h dark cycle and light was subsequently decreased gradually from 16 to 9 consecutive hours per day. Each week, one hour of light was removed from the schedule, until 9 hours per day was reached (at 10 weeks of age). Chicks received a commercial diet: mashed starter 1 from one until four weeks of age; semi mashed starter 2 from four until ten weeks of age; and crumbled pre-lay diet from 10 until 17 weeks of age. Chicks were placed within the aviary system on cardboard paper (also called chick paper: [48]) varying from 50 to 90 grams per square meter. This cardboard paper prevented the chicks getting stuck or falling through the mesh wire of the system due to their small body size. It also enabled the accumulation of spilled food, excretions and/or litter and thus provided a foraging substrate. Around five weeks of age, exposure to the litter area within the system was enabled for all flocks. In the aviary system, all walls of the cage tiers were opened and the corridor between tiers became litter area. In the open level system the side walls of the system were opened, and the outside corridor became litter area. Litter supply could, however, be disrupted from seven to 10 days prior to opening the system by the removal of cardboard paper without additional litter being supplied (hereafter named litter disruption). Farmers use this approach to accustom chicks to their new flooring condition (i.e. wire or plastic surface without cardboard paper). Also, litter supply could be limited by supplying the cardboard paper remnants without additional flooring substrate such as wood-shavings or alfa-alfa (hereafter named litter limitation). The code of practice of maximum stocking densities was applied, enabling sufficient space per bird in the chicken house. Birds were vaccinated according to the standard vaccination protocol used by the rearing company. Extra specific vaccinations could be requested by the laying hen farm for which the birds were reared.

Measurements. At four age points during the rearing period behavioral observations were conducted: week one, five, ten and fifteen weeks of age (see Figure 1).

Anxiety related tests. Tests related to fear and anxiety were conducted at one, five and 10 weeks of age. Fear of humans was assessed by exposure to either a human arm in their home cage (at one and five weeks of age) or a human standing in the litter area (at 10 weeks of age). In the level system, fear of humans was assessed only by a human standing in the litter area at all ages. Fear of novelty was assessed by exposure to a novel wooden box (5*5*2 cm) with colored tape (red, yellow, white and green) at one and five weeks of age, and a novel stick with colored tape (a 50 cm PVC tube with colored tape) at 10 weeks of age [51]. In both tests, birds were exposed for two min to the human observer and the novel object separately. Every ten seconds, we counted the number of birds within close proximity (i.e. 25 cm). For the novel object test, we calculated at which time point at least three birds approached. As birds often did not approach within 25 cm during the human observer test, we estimated the minimal distance in cm of hens that approached over the total test duration. For each flock, tests were repeated four times at different locations in the chicken house (front, middle-front, middle-back, back) always under a light source to limit lack of visibility. A preliminary analysis was performed to assess the effect of location and as location did not affect the latency to approach the novel object or the minimal distance to the human observer, we averaged all values over our four tests. Separation anxiety was measured by a social isolation/novel environment test. Individual chicks, selected from random locations in the chicken house (n = 20 in week one, n = 15 in week five), were tested. Chicks were positioned inside a round orange bucket (30 cm Ø, with 30 cm height) at one week of age and a round white bucket (40 cm Ø, with 50 cm height) at five weeks of age for a duration of one min. At five weeks of age a larger bucket was needed to prevent chicks from jumping out the smaller bucket. The observer was out of sight of the chick while testing, but was able to record high pitched vocalizations; i.e. latency to vocalize and number of vocalizations. High pitched

vocalizations are referred to as alarm or distress calls [52,53]. They are interpreted as an attempt to reinstate contact with conspecifics and as indicating separation anxiety [54].

Feather pecking and feather damage. At one, five and 10 weeks of age feather pecking (FP) behavior was recorded during two 20-min observations in each flock. For each observation, FP was recorded by means of behavior sampling at a predetermined location of approximately 1 m^2 within the chicken house, covering all resources (feeding through, drinking nipples, litter area, tiers and perches). FP was recorded as the frequency of pecks/20 min observation time. Gentle FP (GFP) was recorded as nibbling and gentle feather pecks without a reaction in the receiver, while severe FP (SFP) was recorded as forceful pecks with attempts to pull feathers out to back of the recipient body generally leading to a withdrawal response of the receiver [28,42]. Aggressive pecks to neck and head, were also recorded but due to limited observation numbers, these data were not further analyzed. Prior to observations, the observer waited until birds were habituated to her presence by the criterion that 80% of chicks present were not directing their attention to the observer. The number of chicks within the observation area could vary between 15 and 50 chicks due to unrestricted physical boundaries. Feather damage was assessed at five, 10 and 15 weeks of age. At each age point, 20 chicks per flock, chosen selectively from random locations within the chicken house, were assessed for feather damage to the neck, back and belly region, similar to feather damage scoring in PS hens [51]. However, the wing and tail area were included as extra areas of measurement using a 0/1 scale, as slight damage to the tips of the feathers in these regions early in life possibly indicates the presence of SFP before severe damage is perceived. Total body score (FS) was the sum of values for all body regions, similar to the scoring system for PS hens, but damage to the tips of wings was added to the total body score as a value of 0.5.

Blood parameters. At 15 weeks of age, prior to assessment of feather damage, 20 hens per flock were blood sampled. Samples were always collected around 11–12 a.m. before feeding. An identical procedure was applied for blood sampling and analysis as with the PS hens (for details, see [24]). In short, individual hens were chosen selectively from random locations (floor, tier, perch, front and middle) in the chicken house and sampled within two minutes after capture. Blood (2.5-mL) was stored in 4-mL EDTA tubes and immediately put on ice. For whole-blood serotonin (5-HT) analysis, 1.1 mL of blood was pipetted out of the total amount and stored at −80°C. 1 mL of blood was used for analysis (see [32] for detailed description). 5-HT concentrations (nmol/mL)

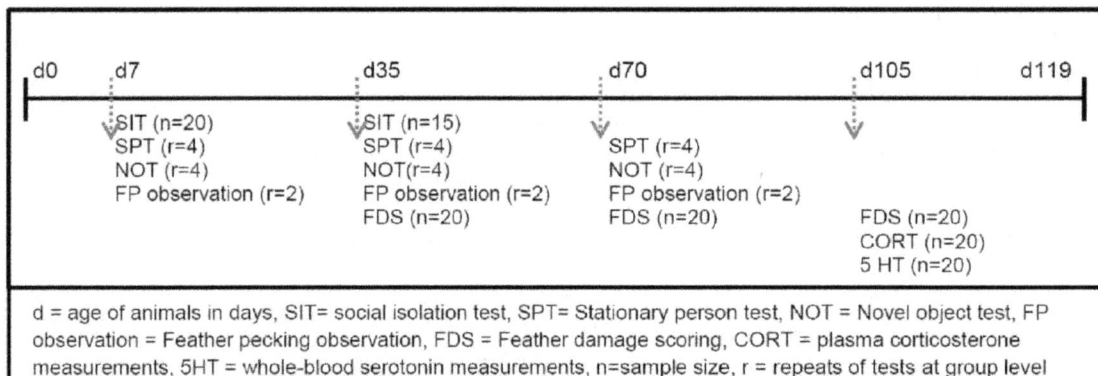

d0	d7		d35		d70		d105	d119

SIT (n=20)   SIT (n=15)
SPT (r=4)   SPT (r=4)   SPT (r=4)
NOT (r=4)   NOT (r=4)   NOT (r=4)
FP observation (r=2) FP observation (r=2) FP observation (r=2)
                    FDS (n=20)   FDS (n=20)   FDS (n=20)
                                                            CORT (n=20)
                                                            5 HT (n=20)

d = age of animals in days, SIT= social isolation test, SPT= Stationary person test, NOT = Novel object test, FP observation = Feather pecking observation, FDS = Feather damage scoring, CORT = plasma corticosterone measurements, 5HT = whole-blood serotonin measurements, n=sample size, r = repeats of tests at group level

Figure 1. Time line of age of birds in days (d) with tests executed at specific ages.

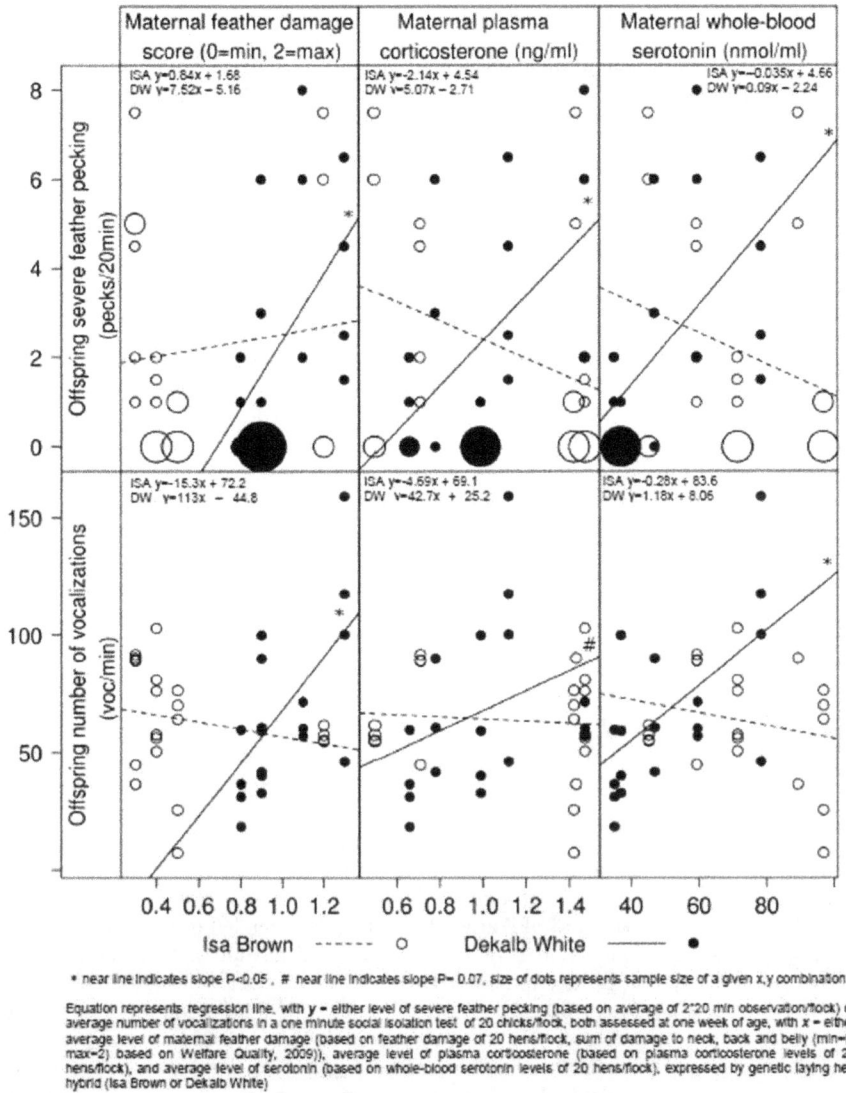

Figure 2. Average level of maternal feather damage [left panel], average level of maternal plasma-corticosterone [middle panel] and average level of whole-blood serotonin levels [right panel] with their offsprings' average level of severe feather pecking at week one of age [upper panels] and the number of vocalizations in a social isolation at one week of age [lower panels].

were assessed by fluorescence assay and compared with a standard curve of 5-HT stock of increasing dilutions. A Perkin-Elmer 2000 Fluorescence spectrophotometer was used to determine fluorescence at 283 and 540 nm. For basal plasma corticosterone (CORT) analysis, 1.4 mL of blood was centrifuged at 2,095 × g at 21°C for 6 min to obtain plasma. Plasma was stored at −20°C before CORT was analyzed at the Faculty of Bio Engineer Science, University of Leuven (Belgium). For the determination of corticosterone concentrations, a competitive radio-immunoassay was performed with the ImmuChem Double Antibody Corticosterone 125I RIA Kit for Rats and Mice of MP Biomedicals LLC (Bio-Connect Diagnostics BV, The Netherlands) with appropriately diluted plasma specimens (for details see [33]).

Statistical analysis. Data were analyzed with SAS 9.2. For each flock, flock averages were calculated. A general linear model (GLM) included the fixed effects of PS, hybrid (DW vs. ISA) and housing system (open, partly open, closed). For the variables which

showed an effect of PS, an additional analysis was conducted to investigate the underlying factors. The average level of CORT, 5-HT and feather damage of the PS hens and age of the PS were added separately as a covariate in the model, which substituted the factor PS, and were tested with its interaction with factor hybrid. For the variables measured from five weeks of age onwards, the effects of limitation of litter (yes/no), disruption of litter supply (yes/no) and the interaction between limitation and disruption of litter supply, and their single interaction with hybrid were added to the model. Post-hoc least square means were used to assess pairwise differences. Correlations between the residuals of the variables (based on a GLM with PS) were assessed, by hybrid, to determine relations between variables related to anxiety and FP. Plots were examined for outliers to confirm the calculated R-values. The normality of the distribution of the residuals was checked, and no transformations were needed. All data is expressed as means ± SEM.

Results

Parental effects

SFP at one week of age was affected by parent stock (PS) ($F_{8,39} = 4.09$, $P = 0.002$). Additional analysis revealed that for the DW hybrid, but not for the ISA hybrid, offspring' SFP at one week of age was related to high maternal plasma-CORT (CORT*hybrid: $F_{1,39} = 6.25$, $P = 0.02$), high maternal whole-blood 5-HT (5-HT*hybrid: $F_{1,39} = 7.72$, $P = 0.01$) and high maternal feather damage score (FS*hybrid: $F_{1,39} = 5.02$, $P = 0.03$, see Figure 2 [top panel]. For the ISA hybrid, no effects of maternal CORT, 5-HT or feather damage was found on offsprings' SFP at one week of age. PS affected the number of vocalizations in the social isolation test at five weeks of age ($F_{8,43} = 2.56$, $P = 0.03$) and tended to affect the number of vocalizations at one week of age ($F_{8,39} = 2.21$, $P = 0.06$). PS did not affect the latency to vocalize at one week ($F_{8,39} = 0.22$, $P = 0.98$) or five weeks of age ($F_{8,43} = 1.48$, $P = 0.20$). Additional analysis revealed that for the DW hybrid but not for the ISA hybrid, a high level of vocalizations in the social isolation test at one week of age were related to high levels of maternal whole-blood 5-HT (5-HT*hybrid: $F_{1,39} = 9.18$, $P = 0.005$) and high maternal feather damage (FS*hybrid: $F_{1,39} = 9.16$, $P = 0.005$) and tended to relate to high levels of maternal plasma-CORT (CORT*hybrid: $F_{1,39} = 3.48$, $P = 0.07$) see Figure 2 [bottom panel]. High number of vocalizations at five weeks of age were related to high maternal feather damage in the DW hybrid (FS*hybrid: $F_{1,43} = 5.98$, $P = 0.02$: DW y = 38.4x − 26.98). For the ISA hybrid, no effects of maternal CORT, 5-HT or feather damage was found on number of vocalizations of the offspring at one or five week of age. Neither PS age nor its interaction with hybrid affected SFP, or vocalizations in the social isolation test at one week of age (SFP$_{week1}$: PS age: $F_{1,39} = 0.75$, $P = 0.39$, PS age * hybrid $F_{1,39} = 2.19$, $P = 0.15$; vocalizations$_{week1}$: PS age: $F_{1,39} = 0.26$, $P = 0.61$; PS age * hybrid $F_{1,39} = 0.09$, $P = 0.76$). PS did not affect SFP and GFP at five or ten weeks of age, feather damage, fearfulness at any other age.

Housing effects

See table 1 for differences and pair-wise comparisons of housing system for FP, fear and feather damage. SFP at ten weeks and GFP at one and ten weeks was highest, and SFP at five weeks tended to be highest, in the open level system compared to the closed and partly-open aviary system (housing-system effect: SFP$_{week1}$: $F_{2,39} = 1.93$, $P = 0.16$, SFP$_{week5}$: $F_{2,43} = 2.62$, $P = 0.10$; SFP$_{week10}$: $F_{2,45} = 11.55$, $P = 0.002$; GFP$_{week1}$: $F_{2,38} = 4.09$, $P = 0.03$, GFP$_{week5}$: $F_{2,44} = 0.38$, $P = 0.69$, GFP$_{week10}$: $F_{2,45} = 4.48$, $P = 0.02$, see Table 1). Feather damage score at ten weeks, but not at five or fifteen weeks, was highest for flocks that were housed in an open level system compared to an aviary system (housing-system effect: FS$_{week5}$: $F_{2,45} = 1.81$, $P = 0.18$, FS$_{week10}$: $F_{2,45} = 3.14$, $P = 0.05$, FS$_{week15}$: $F_{2,42} = 1.26$, $P = 0.30$). At one and five weeks of age, the latency of at least three birds to approach a novel object (NOT) was shortest in the open level system compared to the open and partly-open aviary system (housing-system effect: NOT$_{week1}$: $F_{2,39} = 17.02$, $P < 0.0001$, NOT$_{week5}$ $F_{2,45} = 4.81$, $P = 0.01$, NOT$_{week10}$: $F_{2,43} = 0.65$, $P = 0.53$). In the fear for humans test at one week of age, the effect of housing-system was significant ($F_{2,39} = 16.7$, $P < 0.0001$: open: 96.1 ± 26 cm, closed: 29 ± 3.2 cm, partly-open: 23.6 ± 1.8 cm). This effect is, however, an artifact caused by the different spatial dimensions of the systems on the test variable (minimal distance, i.e. the minimal distance can be larger in an open system vs. the other systems purely due to the systems' spatial dimension) and the setting of the test (i.e. in the aviary systems response to a human arm, while in the level system

response to a standing person is measured). Therefore, this results is not reported in Table 1. Housing system did not affect minimal distance to the human observer at five or ten weeks of age (housing-system effect: SPT$_{week5}$: $F_{2,44} = 0.13$, $P = 0.87$; SPT$_{week10}$: $F_{2,44} = 0.51$, $P = 0.60$).

Genetic effects

GFP tended to be higher for DW than for ISA birds at one week of age (GFP$_{week1}$: $F_{1,38} = 3.69$, $P = 0.06$: DW: 16.8 ± 3.2 pecks/20 min vs. ISA: 11.4 ± 1.7 pecks/20 min). At five and ten weeks of age GFP did not differ between hybrids (GFP$_{week5}$: $F_{1,44} = 0.11$, $P = 0.73$, GFP$_{week10}$: $F_{1,45} = 0.01$, $P = 0.94$). SFP was not affected by hybrid at one or ten weeks of age (SFP$_{week1}$: $F_{1,39} = 0.00$, $P = 0.97$: SFP$_{week10}$: $F_{1,45} = 1.16$, $P = 0.29$). SFP at week 5 of age was affected by the interaction of hybrid with litter limitation, which will be explained further-on under litter effects. At ten weeks of age, but not at one or five weeks of age, DW birds kept a greater distance to the human observer than ISA birds (SPT$_{week1}$: $F_{1,28} = 0.77$, $P = 0.39$; SPT$_{week5}$: $F_{1,28} = 0.09$, $P = 0.76$; SPT$_{week10}$ $F_{1,28} = 12.15$, $P = 0.002$: DW: 152.9 ± 17.8 cm vs. ISA: 57.9 ± 11.0 cm). Whole-blood serotonin (5-HT) was higher for ISA birds than for DW birds ($F_{1,44} = 64.03$, $P < 0.001$: DW: 60.8 ± 1.26 nmol/ml vs. ISA: 88.6 ± 2.54 nmol/ml). Plasma CORT was not affected by hybrid ($F_{1,44} = 0.00$, $P = 0.96$: DW: 1.85 ± 0.06 ng/ml vs. ISA: 2.05 ± 0.15 ng/ml).

Litter effects

The combination of both litter disruption and litter limitation resulted in the highest levels of SFP at five weeks of age (litter disruption * litter limitation: $F_{1,43} = 4.12$, $P = 0.05$, Figure 3a) and a similar but non-significant trend for GFP at five weeks (litter disruption * litter limitation: $F_{1,44} = 1.13$, $P = 0.30$, Figure 3b). GFP and SFP at week 10 of age also tended to be affected by the interaction between limitation and disruption (litter limitation * litter disruption: GFP$_{week10}$: $F_{1,45} = 3.12$, $P = 0.08$; SFP$_{week10}$: $F_{1,45} = 3.32$, $P = 0.08$, Figure 3a,b). Limitation of litter alone increased SFP at five weeks in the ISA hybrid but not in the DW hybrid (hybrid * limitation: $F_{1,43} = 7.36$, $P = 0.01$, see Figure 4a) while GFP did not differ between hybrids (hybrid * limitation: $F_{1,44} = 0.04$, $P = 0.84$, Figure 4b). Disruption of litter alone increased feather damage score at week 5 and 10 but not at 15 weeks of age (disruption: FS$_{week5}$: $F_{1,45} = 18.55$, $P = 0.002$, FS$_{week10}$: $F_{1,45} = 6.55$, $P = 0.02$, FS$_{week15}$: $F_{1,45} = 0.48$, $P = 0.51$, Table 2). These effects were most strong for the DW hybrid at five weeks of age (hybrid * disruption: FS$_{week5}$: $F_{1,45} = 4.21$, $P = 0.05$, FS$_{week10}$: $F_{1,45} = 0.34$, $P = 0.56$, FS$_{week15}$: $F_{1,45} = 0.79$, $P = 0.35$, Figure 5). Independent of hybrid, in flocks which experienced a litter disruption, birds tended to keep a greater distance to the human observer (litter disruption: $F_{1,44} = 3.00$, $P = 0.09$: disruption: 126.7 ± 15.7 cm vs. no disruption: 63.9 ± 17.0 cm) and tended to approach a novel object later (litter disruption: $F_{1,43} = 3.78$, $P = 0.06$, disruption: 31.1 ± 5.0 s. vs. no disruption: 17.0 ± 2.4 s.) in comparison to flocks that did not experience litter disruption. Whole-blood 5-HT was higher when litter was disrupted then when litter was not disrupted (litter disruption: $F_{1,44} = 4.24$, $P = 0.05$; disruption: 64.2 ± 3.6 nmol/ml vs. no disruption: 57.0 ± 3.5 nmol/ml). Plasma-corticosterone was not affected by litter supply (litter disruption: $F_{1,44} = 0.49$, $P = 0.48$, litter limitation: $F_{1,44} = 0.18$, $P = 0.67$). Disruption in access to litter affected the response to social isolation at five weeks differently between the hybrids; ISA birds that had a disruption in litter supply vocalized less than ISA birds that did not have a disruption in litter supply (hybrid * disruption: $F_{1,43} = 4.08$, $P = 0.05$) and had a longer

Table 1. Means ± SEM of response variables of the behavioral tests, feather pecking observations and feather damage scoring of rearing flocks housed in an open, closed or party-open system.

Variables			System		
Tests	Age	Response variables	Open (n = 8)	Closed (n = 25)	Partly open (n = 14)
Stationary person test					
	Week 1	Minimal distance (cm)	-	-	-
	Week 5	Minimal distance (cm)	71.7±19.2	78.2±13.7	74.7±22.9
	Week 10	Minimal distance (cm)	45.9±17.4	113.7±17.8	117.5±22.6
Novel object test					
	Week 1	Latency of 3 birds to approach (s)	**33.2±6.5ᵃ**	**87.6±6.0ᵇ**	**94.2±8.0ᵇ**
	Week 5	Latency of 3 birds to approach (s)	**17.2±2.6ᵃ**	**69.5±8.5ᵇ**	**68.1±11.5ᵇ**
	Week 10	Latency of 3 birds to approach (s)	14.3±1.7	30.0±5.6	24.8±4.6
Social isolation test					
	Week 1	Number of vocalizations/min	55.3±7.0	70.6±6.8	61.2±8.8
	Week 1	Latency to vocalise (s)	9.8±1.5	11.6±1.4	10.4±2.0
	Week 5	Number of vocalizations/min	24.5±8.6	21.3±2.7	15.1±1.5
	Week 5	Latency to vocalise (s)	23.6±5.1	24.0±2.8	27.8±4.0
Feather pecking behaviour (pecks/20 min)					
	Week 1	Gentle feather pecking	**24.4±1.9ᵃ**	**9.7±1.7ᵇ**	**16.3±4.1ᶜ**
	Week 5	Gentle feather pecking	70.6±27.4	74.8±17.1	41.9±9.7
	Week 10	Gentle feather pecking	**71.1±14.6ᵃ**	**23.8±5.2ᵇ**	**42.9±11.8ᶜ**
	Week 1	Severe feather pecking	4.0±1.0	1.7±0.5	2.1±0.9
	Week 5	Severe feather pecking	**15.4±5.8ˣ**	**9.6±1.6ˣ**	**4.0±1.6ʸ**
	Week 10	Severe feather pecking	**7.3±1.9ᵃ**	**1.6±0.5ᵇ**	**2.2±0.8ᶜ**
Feather damage scoring (min = 0, max = 2)					
	Week 5	Average feather score	0.24±0.07	0.28±0.05	0.31±0.05
	Week 10	Average feather score	**0.29±0.08ᵃ**	**0.23±0.03ᵇ**	**0.22±0.05ᵇ**
	Week 15	Average feather score	0.23±0.04	0.14±0.02	0.13±0.02

Bold number with superscripts a,b,c indicate P-value of <0.05; bold numbers with superscripts x,y,z indicate P-value <0.1>0.05 (different superscript letters indicate pair-wise differences), " –" indicate non determined effect due to effects of an artifact of the system on the response variable.

latency to vocalize (hybrid * disruption: $F_{1,43} = 3.63$, P = 0.04) while the opposite was the case for the DW birds (Figure 6).

Relations between anxiety and feather pecking

For both hybrids, average feather damage score at five weeks was higher when the latency to vocalize in the social isolation test at one week of age was higher ($r = 0.46$, P < 0.003, Figure 7). In the ISA birds, whole-blood serotonin levels were higher if the latency to vocalize in the social isolation test at one week was higher ($r_{ISA} = 0.67$, P < 0.001, Figure 7), but this was not significant in the DW birds ($r_{DW} = 0.22$, P = 0.37). As 5-HT was higher for birds which experienced a litter disruption, we assessed the correlation within litter disruption groups within the ISA hybrid. For litter disruption the correlation between 5-HT and vocalizations at one week was positive ($r_{limitation} = 0.72$, P = 0.02), while without litter disruption the correlation was not significant ($r_{no\ limitation} = 0.17$, P = 0.70).

Discussion

This is the first on-farm study in which maternal effects on the behavioral development of offspring are described for laying hens. We explored and examined which maternal and environmental effects act on the development of feather pecking (FP) from one

until fifteen weeks of age in two hybrids: Dekalb White (DW) and ISA Brown (ISA). As FP is related to anxiety [30,31], we also assessed this relationship under commercial conditions.

Maternal effects

In the DW hybrid, high maternal plasma-corticosterone (CORT), whole-blood serotonin (5-HT) and feather damage were positively related to offsprings' severe FP (SFP) at one week of age and offsprings' vocalizations upon social isolation at one and five weeks of age. The latter are indicative of fearfulness and anxiety [55–57]. These results suggest that within the DW hybrid, maternal state can affect behavioural development of the offspring and thereby cause high fearfulness [55,56] and SFP. These maternal effects may derive from high levels of stress (affecting CORT) and feather pecking in the maternal birds (affecting 5-HT and feather damage, for details see [24]). Offspring of mothers with high CORT have repeatedly shown to have high levels of fearfulness (hens [7], quail [22,58,59]) and emotional reactivity (hens [15,16], quail [21,26,60]). Altered deposition of nutrients and hormones in the egg may underlie the maternal effects we found (for review see [12]). High CORT of the mother, due to living in a stressful environment, can affect yolk-hormones such as testosterone [7,21,26,60,61], progesterone [22,60] and oestrogens [15,20] which can influence offspring behavior [12,14]. Addition-

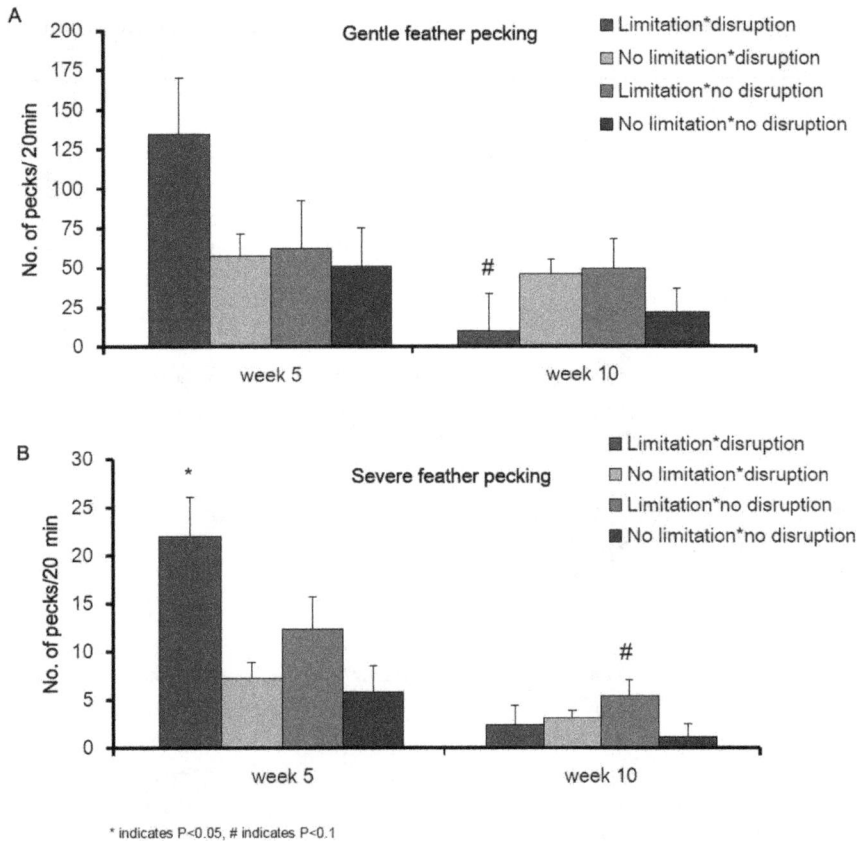

Figure 3. A. Gentle feather pecking at 5 and 10 weeks of age in relation to litter disruption and litter limitation B. Severe feather pecking at 5 and 10 weeks of age in relation to litter disruption and litter limitation.

ally, high maternal CORT has been related to low egg weight [23,24] and chick weight post hatch [7,16,22]. It is known that ISA and DW birds with high maternal CORT induced by CORT implants differ in yolk-steroid levels and yolk-mass [23]. Offsprings' fearfulness and SFP in our study may thus be influenced by egg hormone and nutrient content as affected by maternal physiology. In the present study these maternal effects depended on genotype. Breed-dependent differences in epigenetic programming (similar gene-expression patterns over generations and other non-genetic inheritable traits, see reviews on epigenetic studies in mice and chickens [62–65]) have also been identified as a putative mechanism of maternal effects [15–17]. These epigenetic changes may even be induced by altered egg-hormone content [15,60,66]. Differences in genetic and epigenetic inheritance between laying hen lines may be the reason why we only recorded maternal effects in the DW hybrid and not in the ISA hybrid.

Environmental effects

Housing effects. In the open level system, chicks had a shorter latency to approach a novel object, but also had the highest gentle FP (GFP) at one and ten weeks of age, highest SFP at five (a tendency $P < 0.1$) and ten weeks of age, and highest feather damage at ten weeks of age compared to chicks housed in a closed or partly-open aviary system. Although GFP and SFP originate from different behavioral needs [28,42] and involvement of different genes [67] and gene-expression patterns [68], one does not necessarily lead to the other [31,42,69], but the co-existence of

both may result in feather damage. In the open level system chicks are placed together with thousands of other individuals inside a large area from day one. In both aviary systems group size is substantially smaller than in the level system as the (partially) closed walls of the aviary system limit the space nor group size to extent to over hundreds. Effects of system are therefore likely to partly be group size related. Social transmission of behavior [70,71], such as the approach of novel objects and FP, may have occurred more readily in a large group [72] as there are more birds from which to copy and synchronize behavior. Previous studies suggest that FP is socially transmitted within a group (SFP [34,73,74], GFP [74–76]). In the closed aviary system we recorded a peak in GFP at five weeks of age. GFP seems to stem from social exploration [28,76] and presumably underlies this result. Birds are mixed at around four to five weeks, and this may elicit social exploration, which presumably would have already occurred in the other systems. These results indicate that housing system (possibly related to differences in group size which affect social exploration and social transmission) influences the development of FP and feather damage on-farm.

Litter effects

Litter disruption (taking away foraging substrate for a period of 7–10 days) and litter limitation (limited supplementation in the form of remnant of chicken paper) had a substantial effect on FP and fear responses (Table 3). Especially at five weeks of age, disruption of litter led to high SFP, GFP and feather damage. During litter disruption, three factors are at play: 1) disturbances

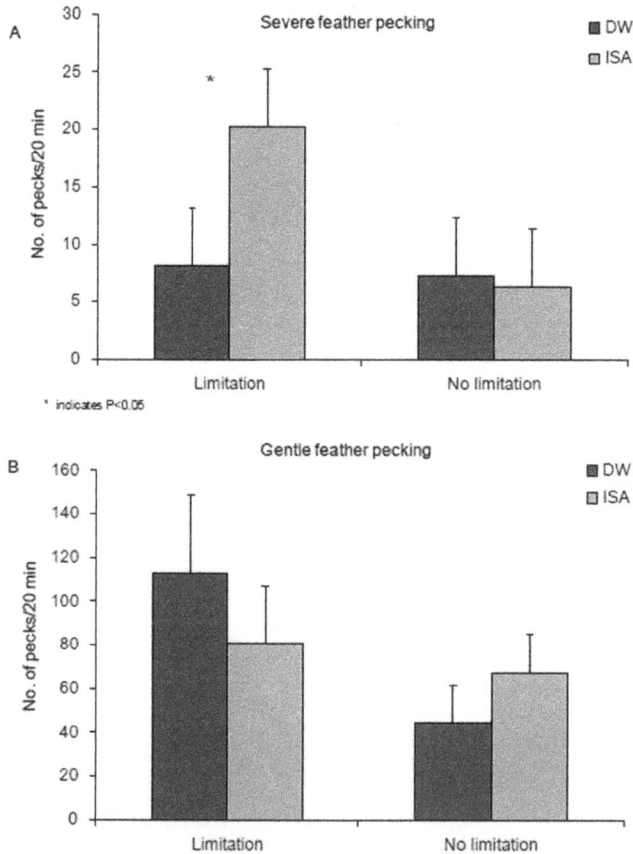

Figure 4. A. Severe feather pecking at 5 weeks of age in Dekalb White (DW) and ISA brown (ISA) birds in relation to litter limitation B. Gentle feather pecking at 5 weeks of age in Dekalb White (DW) and ISA brown (ISA) birds in relation to litter limitation.

by the farmer, who takes out cardboard paper, handles and mixes birds, 2) removal of cardboard paper and thus removal of foraging material, and 3) disrupted uptake of fibers or excretions from cardboard paper.

The first factor, disturbance by the farmer, may elicit stress related to fear of humans, as indicated by the greater distance to the human observer in flocks in which access to litter was disrupted. Additionally, absence of litter may induce frustration which can result in SFP [77]. The act of SFP itself (pecking and pulling feathers) causes distress in the victims i.e. withdrawal, escape attempts, vocalizations [42] and can lead to disturbances in

Table 2. Feather damage score at week five, ten and fifteen of age in relation to litter disruption.

Feather damage score	Disruption	
	Yes	No
week 5	0.45(0.06)[a]	0.18(0.05)[b]
week 10	0.29(0.06)[a]	0.15(0.06)[b]
week 15	0.16(0.03)	0.15(0.03)

Numbers with superscripts a,b indicate P-value of <0.05.

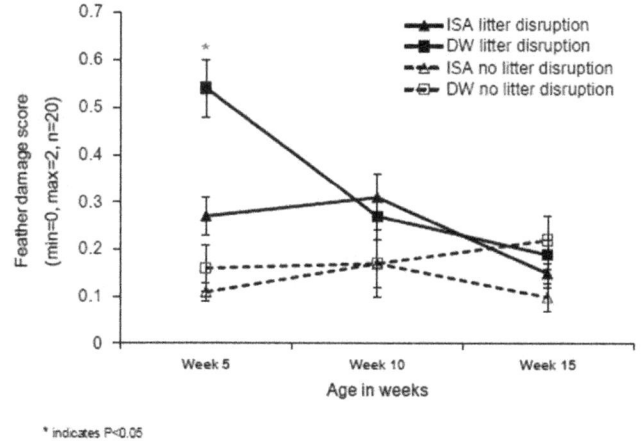

Figure 5. Feather damage score of Dekalb White (DW) and ISA brown (ISA) at 5, 10 and 15 weeks of age in relation to litter disruption.

the flock [78,79]. Taken together, litter disruption can, either directly or indirectly via SFP, increase a flock's fear level.

The second factor, removal of foraging substrate, has most probably the largest influence on the occurrence of FP. FP is considered redirected foraging pecking [80], and increases when

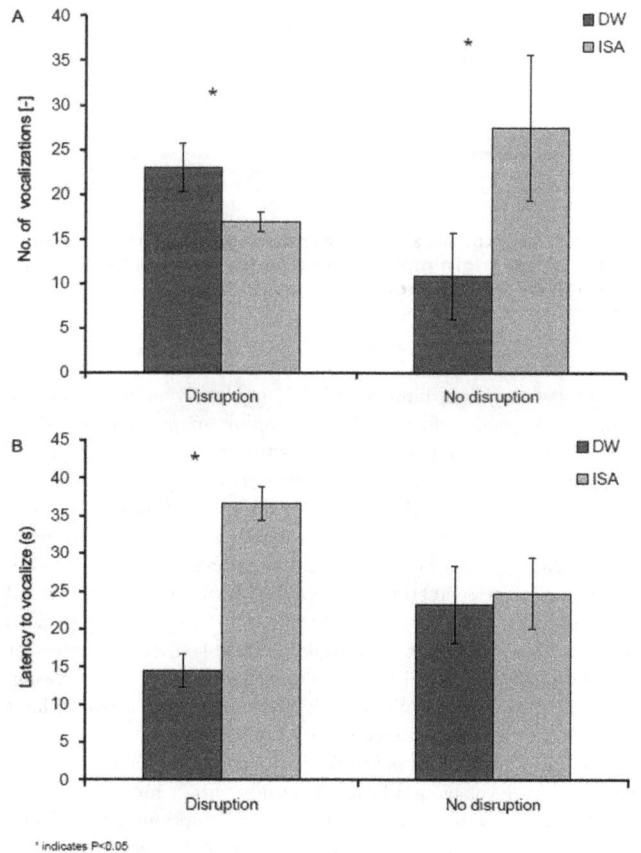

Figure 6. A. Number of vocalizations upon social isolation at 5 weeks of age in Dekalb White (DW) and ISA Brown (ISA) chicks in relation to litter disruption B. Latency to vocalize upon social isolation at 5 weeks of age in Dekalb White (DW) and ISA Brown (ISA) chicks in relation to litter disruption.

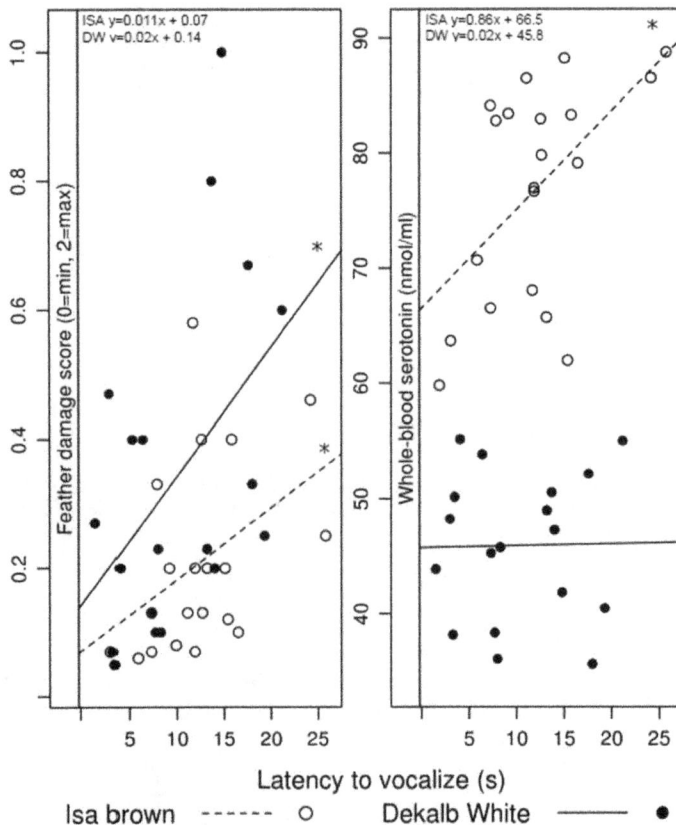

ISA y=0.011x + 0.07
DW y=0.02x + 0.14

ISA y=0.86x + 66.5
DW y=0.02x + 45.8

Feather damage score (0=min, 2=max)

Whole-blood serotonin (nmol/ml)

Latency to vocalize (s)

Isa brown   ------   O   Dekalb White   ————   ●

* near line indicates slope P<0.05

Equation represents regression line, with y = average latency per flock to vocalize in social isolation test at one week of age based on 20 chicks/flock at one week of age, with x = either average level of feather damage per flock at five weeks of age based on 20 hens/flock, sum of damage to neck, back and belly (min=0, max=2) based on Welfare Quality, 2009)) and average level of whole-blood serotonin (based on 20 hens/flock), expressed by genetic laying hen hybrid (Isa Brown or Dekalb White)

Figure 7. Average level of feather damage score at five weeks of age [left panel] and average level of whole-blood serotonin at 15 weeks of age [right panel] related to the average latency to vocalize in a social isolation at one week of age in flocks of Dekalb White (DW) and ISA Brown (ISA) laying hens.

foraging material is limited [43,81], especially at an early age [48,49]. As said before, lack of foraging material can induce frustration when the need to forage is thwarted [77] and results in SFP which in turn can lead to feather damage, as shown in this study. On top of litter disruption, a subsequent limitation of litter brought an additive effect in the development of SFP. At any given time, foraging material is important in prevention of SFP [81]. Feather damage seemed to reduce when birds age, irrespective of litter supply. This may be influenced by the molting periods around 10 weeks of age [82], making loose feathers available for ingestion from the floor [83]. Feather pecking during early rearing, as affected by litter supply, may however still yield a risk of later outbreaks of feather damage during lay [42,84].

The third factor, lack of uptake of fibers or excretions, probably affected the level of 5-HT, as our study shows increased whole blood 5-HT levels in flocks with litter disruption. Litter (often wood shavings, alfalfa or remaining cardboard paper) contains fibers, excretions and feather particles. Uptake of these large particles can stimulate gut motility [85], alter gut micro biota [86,87] and activate immunity in various ways [88]. Particularly feather eating, which is linked with FP [83,90,91] has been associated with increased gut motility [92]. The enterochromaffin cells in the gut contain 5-HT which are released upon stimulation

of the intestinal tract [89]. As a result of a temporary lack of litter birds may have a strong need to forage, possible enhancing feather and litter uptake afterwards as over-compensation [83] which altogether affects 5-HT release. Our study shows that whole-blood 5-HT can be influenced by litter disruption.

In the ISA hybrid, especially under disruption of litter, a positive correlation between fear-response at one week of age and 5-HT at fifteen weeks of age was detected. In a previous study, in Rhode Island Red birds (RIR), one of the founder lines of ISA, the correlation between fear responses and brain 5-HT was also dependent on the environment. RIR birds mixed with birds of another line showed a negative correlation between fear and 5-HT while RIR birds which we kept in non-mixed groups showed a positive correlation [38]. The positive correlation between fear-responses and 5-HT under litter disruption in our study could be influenced by effects of mixing and substrate intake but probably also by the high levels of SFP occurring under litter disruption. 5-HT activity has been suggested to relate to the development of SFP [42] (brain 5-HT young [90] and adult birds [91], peripheral 5-HT [32,33] and both brain and peripheral 5-HT [38]). Both brain and peripheral 5-HT have also been associated with fearfulness [32,33,92], and, in our study, peripheral 5-HT was influenced by litter disruption.

Table 3. Effects of litter supply (litter disruption and litter limitation) on feather pecking, fearfulness and whole-blood serotonin in laying hens.

Effects of litter supply	Severe feather pecking	Gentle feather pecking	Feather damage test	Stationary person test	Novel object test test	Social isolation test	Whole-blood 5-HT	Correlations
Disruption	↑ week 5 (ISA)*		↑ week 5** ↑ week 10#	↑ week 10#	↑ week 10#	↑ week 5 (ISA)*	↑ week 15*	5-HT at 15 weeks & social isolation test at 5 weeks $r=0.67$**
Limitation	↑ week 5 (ISA)*	↑ week 5*						
Disruption* limitation	↑ week 5* ~ week 10#	↑ week 5 ~ week 5NS						
Effects displayed in:	Figure 3	Figure 3	Figure 5	text	text	Figure 6	text	Figure 7

↑ increase due to described effect, ~ effects dependent on interaction, NS: not significant, ** indicates effects $P<0.01$, * indicates effects $P<0.05$, # indicates effects $P<0.1$, (ISA) effects only valid for ISA hybrid, r = correlation coefficient, week = age of hens in weeks

Within the ISA hybrid, disruption of litter also caused higher anxiety in the social isolation test at five weeks of age. ISA birds appear to be more strongly affected by their (social) environment than DW birds [38]. In comparison to other hybrids, birds from a brown origin (in the PS [24] and founder lines [36]) repeatedly show higher fear in response to social isolation [93–95] and novel items in their home environment. ISA birds are also more affected by social factors such as group size [24] and mixing [34,37] than DW birds. Taken together with other studies, it appears that ISA birds are more strongly affected by their (social) environment in comparison to DW birds who are more sensitive to maternal effects.

Fear and feather pecking

For both hybrids we found that latency to vocalize during social isolation at one week of age was related to feather damage at five weeks of age, which complements the relationship between anxiety traits in social isolation tests and FP [30,31]. This may also explain why we still see FP under optimal conditions with regard to litter. In DW birds, fear of humans was higher than in ISA birds, which was similar to the study of the PS [24]. DW birds are more easily frightened by exposure to humans [34] as indicated by higher fear-responses and plasma-CORT after human handling [35,36]. DW birds also have relatively low levels of whole-blood 5-HT compared to ISA birds (shown in this study, in the PS [24] and founder lines [38]), which may represent a risk in the development of FP [42]. In addition, the maternal effects on fearfulness in early life may predispose DW birds to develop SFP. The predisposition to develop FP may thus stem from different origins depending on genotype.

Conclusion

This study shows for the first time that maternal effects in commercial laying hens play an important role in early life behavioral development of their offspring. Our study indicates that maternal stress can create a risk for the development of anxiety and maladaptive behavior such as feather pecking (FP) in laying hens. These maternal effects depend on genotype, with birds from a White Leghorn origin being sensitive. Litter availability is of utmost importance for laying hens, and reduced the risk of FP, especially for birds from a Rhode Island Red origin who also become more anxious and fearful as a result of disruption in litter supply. These results provide new knowledge that is important for preventing the development of anxiety and FP in laying hens.

Acknowledgments

This research would have been impossible if it was not for the cooperation of Ter Heerdt BV hatchery (Babberich, The Netherlands) and it's farmers for which we are very grateful. We thank Linda Persoon and Lourdes Icalla, MSc students of the Wageningen University for their help in the behavioral observations. Gratefully acknowledged are the lab-technicians of the Adaptation Physiology Group (Wageningen UR, The Netherlands): Ger de Vries-Reilingh also on-farm and Rudie Koopmanschap for the 5-HT assay. Daniel Vermeulen from the Faculty of Bio engineer science, University of Leuven (Leuven, Belgium) for the extensive plasma CORT analysis. Gus Rose of Animal Breeding and Genomics Centre (Wageningen UR, The Netherlands) is thanked for its creativity in designing the graphs.

Author Contributions

Conceived and designed the experiments: EDH JEB TBR TGG BK. Performed the experiments: EDH TBR. Analyzed the data: EDH TBR JEB BK. Contributed reagents/materials/analysis tools: JEB BK. Wrote the paper: EDH JEB BK TGG TBR. Grant application: TBR.

References

1. Lumey LH, Stein AD, Kahn HS, van der Pal-de Bruin KM, Blauw GJ, et al. (2007) Cohort Profile: The Dutch Hunger Winter Families study. International Journal of Epidemiology 36: 1196–1204.

2. Viltart O, Vanbesien-Mailliot CCA (2007) Impact of prenatal stress on neuroendocrine programming. The Scientific World Journal 7: 1493–1537.

3. Champagne FA, Rissman EF (2011) Behavioral epigenetics: A new frontier in the study of hormones and behavior. Hormones and Behavior 59: 277–278.

4. Weaver ICG, Cervoni N, Champagne FA, D'Alessio AC, Sharma S, et al. (2004) Epigenetic programming by maternal behavior. Nature Neuroscience 7: 847–854.

5. Eriksen MS, Faerevik G, Kittilsen S, McCormick MI, Damsgard B, et al. (2011) Stressed mothers - troubled offspring: a study of behavioural maternal effects in farmed Salmo salar. Journal of Fish Biology 79: 575–586.

6. Groothuis TGG, Muller W, Von Engelhardt N, Carere C, Eising C (2005) Maternal hormones as a tool to adjust offspring phenotype in avian species. Neuroscience and Biobehavioral Reviews 29: 329–352.

7. Janczak AM, Torjesen P, Palme R, Bakken M (2007) Effects of stress in hens on the behaviour of their offspring. Applied Animal Behaviour Science 107: 66–77.

8. Gudsnuk KMA, Champagne FA (2011) Epigenetic Effects of Early Developmental Experiences. Clinics in Perinatology 38: 703–+.

9. Charil A, Laplante DP, Vaillancourt C, King S (2010) Prenatal stress and brain development. Brain Research Reviews 65: 56–79.

10. Brunton PJ, Russell JA (2010) Prenatal Social Stress in the Rat Programmes Neuroendocrine and Behavioural Responses to Stress in the Adult Offspring: Sex-Specific Effects. Journal of Neuroendocrinology 22: 258–271.

11. Rutherford KMD, Donald RD, Arnott G, Rooke JA, Dixon L, et al. (2012) Farm animal welfare: assessing risks attributable to the prenatal environment. Animal Welfare 21: 419–429.

12. Henriksen R, Rettenbacher S, Groothuis TGG (2011) Prenatal stress in birds: Pathways, effects, function and perspectives. Neuroscience and Biobehavioral Reviews 35: 1484–1501.

13. Richard-Yris MA, Michel N, Bertin A (2005) Nongenomic inheritance of emotional reactivity in Japanese quail. Developmental Psychobiology 46: 1–12.

14. Groothuis TGG, Schwabl H (2008) Hormone-mediated maternal effects in birds: mechanisms matter but what do we know of them? Philosophical Transactions of the Royal Society B-Biological Sciences 363: 1647–1661.

15. Natt D, Lindqvist N, Stranneheim H, Lundeberg J, Torjesen PA, et al. (2009) Inheritance of Acquired Behaviour Adaptations and Brain Gene Expression in Chickens. PLoS One 4: e6405.

16. Goerlich VC, Nätt D, Elfwing M, Macdonald B, Jensen P (2012) Transgenerational effects of early experience on behavioral, hormonal and gene expression responses to acute stress in the precocial chicken. Hormones and Behavior 61: 711–718.

17. Lindqvist C, Janczak AM, Natt D, Baranowska I, Lindqvist N, et al. (2007) Transmission of stress-induced learning impairment and associated brain gene expression from parents to offspring in chickens. PLoS One 2: e364.

18. Bateson P (2007) Developmental plasticity and evolutionary biology. Journal of Nutrition 137: 1060–1062.

19. Gluckman PD, Hanson MA, Spencer HG (2005) Predictive adaptive responses and human evolution. Trends in Ecology & Evolution 20: 527–533.

20. Janczak AM, Torjesen P, Rettenbacher S (2009) Environmental effects on steroid hormone concentrations in laying hens' eggs. Acta Agriculturae Scandinavica Section a-Animal Science 59: 80–84.

21. Guibert F, Richard-Yris M-A, Lumineau S, Kotrschal K, Bertin A, et al. (2011) Unpredictable mild stressors on laying females influence the composition of Japanese quail eggs and offspring's phenotype. Applied Animal Behaviour Science 132: 51–60.

22. Bertin A, Richard-Yris M-A, Houdelier C, Lumineau S, Moestl E, et al. (2008) Habituation to humans affects yolk steroid levels and offspring phenotype in quail. Hormones and Behavior 54: 396–402.

23. Henriksen R, Groothuis TG, Rettenbacher S (2011) Elevated Plasma Corticosterone Decreases Yolk Testosterone and Progesterone in Chickens: Linking Maternal Stress and Hormone-Mediated Maternal Effects. Plos One 6.

24. de Haas EN, Kemp B, Bolhuis JE, Groothuis T, Rodenburg TB (2013) Fear, stress, and feather pecking in commercial white and brown laying hen parent-stock flocks and their relationships with production parameters. Poultry Science 92: 2259–2269.

25. Janczak AM, Heikkila M, Valros A, Torjesen P, Andersen IL, et al. (2007) Effects of embryonic corticosterone exposure and post-hatch handling on tonic immobility and willingness to compete in chicks. Applied Animal Behaviour Science 107: 275–286.

26. Guibert F, Richard-Yris M-A, Lumineau S, Kotrschal K, Guemene D, et al. (2010) Social Instability in Laying Quail: Consequences on Yolk Steroids and Offspring's Phenotype. Plos One 5.

27. Davis KA, Schmidt JB, Doescher RM, Satterlee DG (2008) Fear responses of offspring from divergent quail stress response line hens treated with corticosterone during egg formation. Poultry Science 87: 1303–1313.

28. Savory CJ (1995) Feather pecking and cannibalism. World's Poultry Science Journal 51: 215–219.

29. Nicol CJ, Bestman M, Gilani A-M, De Haas EN, De Jong IC, et al. (2013) The prevention and control of feather pecking: application to commercial systems. World's Poultry Science Journal 69: 775–788.

30. Jones RB, Blokhuis HJ, Beuving G (1995) Open-field and tonic immobility responses in domestic chicks of two genetic lines differing in their propensity to feather peck. British Poultry Science 36: 525–530.

31. Rodenburg TB, Buitenhuis AJ, Ask B, Uitdehaag KA, Koene P, et al. (2004) Genetic and phenotypic correlations between feather pecking and open-field response in laying hens at two different ages. Behavior Genetics 34: 407–415.

32. Bolhuis JE, Ellen ED, Van Reenen CG, De Groot J, Ten Napel J, et al. (2009) Effects of genetic group selection against mortality on behaviour and peripheral serotonin in domestic laying hens with trimmed and intact beaks. Physiology & Behavior 97: 470–475.

33. Rodenburg TB, Bolhuis JE, Koopmanschap RE, Ellen ED, Decuypere E (2009) Maternal care and selection for low mortality affect post-stress corticosterone and peripheral serotonin in laying hens. Physiology & Behavior 98: 519–523.

34. Uitdehaag KA, Rodenburg TB, Bolhuis JE, Decuypere E, Komen H (2009) Mixed housing of different genetic lines of laying hens negatively affects feather pecking and fear related behaviour. Applied Animal Behaviour Science 116: 58–66.

35. Fraisse F, Cockrem JF (2006) Corticosterone and fear behaviour in white and brown caged laying hens. British Poultry Science 47: 110–119.

36. Uitdehaag K, Kornen H, Rodenburg TB, Kemp B, van Arendonk J (2008) The novel object test as predictor of feather damage in cage-housed Rhode Island Red and White Leghorn laying hens. Applied Animal Behaviour Science 109: 292–305.

37. Uitdehaag KA, Rodenburg TB, van Hierden YM, Bolhuis JE, Toscano MJ, et al. (2008) Effects of mixed housing of birds from two genetic lines of laying hens on open field and manual restraint responses. Behavioural Processes 79: 13–18.

38. Uitdehaag KA, Rodenburg TB, Van Reenen CG, Koopmanschap RE, Reilingh GD, et al. (2011) Effects of genetic origin and social environment on behavioral response to manual restraint and monoamine functioning in laying hens. Poultry Science 90: 1629–1636.

39. Janczak AM, Braastad BO, Bakken M (2006) Behavioural effects of embryonic exposure to corticosterone in chickens. Applied Animal Behaviour Science 96: 69–82.

40. Rodenburg TB, Komen H, Ellen ED, Uitdehaag KA, van Arendonk JAM (2008) Selection method and early-life history affect behavioural development, feather pecking and cannibalism in laying hens: A review. Applied Animal Behaviour Science 110: 217–228.

41. Rogers LJ (1995) Behavioural transitions in early posthatching life. The development of brain and behaviour in the chicken. Wallingford, United Kingdom: CAB International. pp. 157–183.

42. Rodenburg TB, van Krimpen MM, de Jong IC, de Haas EN, Kops MS, et al. (2013) The prevention and control of feather pecking in laying hens: identifying the underlying principles. Worlds Poultry Science Journal 69: 361–373.

43. Gilani AM, Knowles TG, Nicol CJ (2012) The effect of dark brooders on feather pecking on commercial farms. Applied Animal Behaviour Science 142: 42–50.

44. Bilcík B, Keeling LJ (2000) Relationship between feather pecking and ground pecking in laying hens and the effect of group size. Applied Animal Behaviour Science 68: 55–66.

45. Kjaer JB (2004) Effects of stocking density and group size on the condition of the skin and feathers of pheasant chicks. The Veterinary Record 154: 556–558.

46. Rodenburg TB, Koene P (2007) The impact of group size on damaging behaviours, aggression, fear and stress in farm animals. Applied Animal Behaviour Science 103: 205–214.

47. Zimmerman PH, Lindberg AC, Pope SJ, Glen E, Bolhuis JE, et al. (2006) The effect of stocking density, flock size and modified management on laying hen behaviour and welfare in a non-cage system. Applied Animal Behaviour Science 101: 111–124.

48. de Jong IC, Reuvekamp BFJ, Gunnink H (2013) Can substrate in early rearing prevent feather pecking in adult laying hens? Animal Welfare 22: 305–314.

49. Huber-Eicher B, Wechsler B (1998) The effect of quality and availability of foraging materials on feather pecking in laying hen chicks. Animal Behaviour 55: 861–873.

50. HuberEicher B, Wechsler B (1997) Feather pecking in domestic chicks: its relation to dustbathing and foraging. Animal Behaviour 54: 757–768.

51. Welfare Quality (2009) Welfare Quality® assessment protocol for poultry (broilers, laying hens). Welfare Quality® Consortium, Lelystad, the Netherlands.

52. Collias NE (1987) The vocal repertoire of the Red Junglefowl: a spectrographic classification and the code of communication. The Condor 89: 510–524.

53. Sufka KJ, Hughes RA (1991) Differential effects of handling on isolation-induced vocalizations, hypoalgesia and hyperthermia in domestic fowl. Physiology & Behavior 50: 129–133.

54. Warnick JE, Huang CJ, Acevedo EO, Sufka KJ (2009) Modelling the anxiety-depression continuum in chicks. Journal of Psychopharmacology 23: 143–156.

55. Suarez SD, Gallup GG (1983) Social reinstatement and Open-Field testing in chickens. Animal Learning & Behavior 11: 119–126.

56. Gallup GG, Suarez SD (1980) An ethological analysis of open-field behaviour in chickens. Animal Behaviour 28: 368–378.

57. Faure JM, Jones RB, Bessei W (1983) Fear and social motivation as factors in open-field behaviour of the domestic chick. Biology of Behaviour 8: 103–116.

58. Houdelier C, Lumineau S, Bertin A, Guibert F, De Margerie E, et al. (2011) Development of Fearfulness in Birds: Genetic Factors Modulate Non-Genetic Maternal Influences. Plos One 6.

59. Bertin A, Richard-Yris MA, Houdelier C, Richard S, Lumineau S, et al. (2009) Divergent selection for inherent fearfulness leads to divergent yolk steroid levels in quail. Behaviour 146: 757–770.

60. Guibert F, Lumineau S, Kotrschal K, Mostl E, Richard-Yris MA, et al. (2013) Trans-generational effects of prenatal stress in quail. Proceedings of the Royal Society B-Biological Sciences 280.

61. Schwabl H (1997) The contents of maternal testosterone in house sparrow Passer domesticus eggs vary with breeding conditions. Naturwissenschaften 84: 406–408.

62. Champagne FA (2010) Early Adversity and Developmental Outcomes: Interaction Between Genetics, Epigenetics, and Social Experiences Across the Life Span. Perspectives on Psychological Science 5: 564–574.

63. Champagne FA (2012) Interplay Between Social Experiences and the Genome: Epigenetic Consequences for Behavior. In: Sokolowski MB, Goodwin SF, editors. Gene-Environment Interplay. pp. 33–57.

64. Curley JP, Mashoodh R, Champagne FA (2011) Epigenetics and the origins of paternal effects. Hormones and Behavior 59: 306–314.

65. Berghof T, Parmentier H, Lammers A (2013) Transgenerational epigenetic effects on innate immunity in broilers: An underestimated field to be explored? Poultry Science 92: 2904–2913.

66. Ho DH, Burggren WW (2010) Epigenetics and transgenerational transfer: a physiological perspective. Journal of Experimental Biology 213: 3–16.

67. Buitenhuis AJ, Rodenburg TB, Wissink PH, Visscher J, Koene P, et al. (2004) Genetic and phenotypic correlations between feather pecking behavior, stress response, immune response, and egg quality traits in laying hens. Poultry Science 83: 1077–1082.

68. Hughes AL, Buitenhuis AJ (2010) Reduced variance of gene expression at numerous loci in a population of chickens selected for high feather pecking. Poultry Science 89: 1858–1869.

69. Newberry RC, Keeling LJ, Estevez I, Bilcik B (2007) Behaviour when young as a predictor of severe feather pecking in adult laying hens: The redirected foraging hypothesis revisited. Applied Animal Behaviour Science 107: 262–274.

70. Nicol CJ (1995) The social transmission of information and behavior. Applied Animal Behaviour Science 44: 79–98.

71. Tolman CW (1964) Social Facilitation Of Feeding Behaviour In Domestic Chick. Animal Behaviour 12: 245–251.

72. Croney CC, Newberry RC (2007) Group size and cognitive processes. Applied Animal Behaviour Science 103: 215–228.

73. Zeltner E, Klein T, Huber-Eicher B (2000) Is there social transmission of feather pecking in groups of laying hen chicks? Animal Behaviour 60: 211–216.

74. McAdie TM, Keeling LJ (2002) The social transmission of feather pecking in laying hens: effects of environment and age. Applied Animal Behaviour Science 75: 147–159.

75. McAdie TM, Keeling LJ (2000) Effect of manipulating feathers of laying hens on the incidence of feather pecking and cannibalism. Applied Animal Behaviour Science 68: 215–229.

76. Riedstra B, Groothuis TGG (2002) Early feather pecking as a form of social exploration: the effect of group stability on feather pecking and tonic immobility in domestic chicks. Applied Animal Behaviour Science 77: 127–138.

77. Rodenburg TB, Koene P, Spruijt BM (2004) Reaction to frustration in high and low feather pecking lines of laying hens from commercial or semi-natural rearing conditions. Behavioural Processes 65: 179–188.

78. Bright A (2008) Vocalisations and acoustic parameters of flock noise from feather pecking and non-feather pecking laying flocks. British Poultry Science 49: 241–249.

79. Koene P, Zimmerman P, Bokkers E, Rodenburg B (2001) Vocalisation due to frustration in layer and broiler chickens. In: Oester H, Wyss C, editors. pp. 95–100.

80. Blokhuis HJ (1986) Feather-pecking in poultry: its relation with ground-pecking. Applied Animal Behaviour Science 16: 63–67.

81. Nicol CJ, Lindberg AC, Phillips AJ, Pope SJ, Wilkins LJ, et al. (2001) Influence of prior exposure to wood shavings on feather pecking, dustbathing and foraging in adult laying hens. Applied Animal Behaviour Science 73: 141–155.

82. Savory CJ, Mann JS (1997) Behavioural development in groups of pen-housed pullets in relation to genetic strain, age and food form. British Poultry Science 38: 38–47.

83. Harlander-Matauschek A, Benda I, Lavetti C, Djukic M, Bessei W (2007) The relative preferences for wood shavings or feathers in high and low feather pecking birds. Applied Animal Behaviour Science 107: 78–87.

84. Bright A (2009) Time course of plumage damage in commercial layers. Veterinary Record 164: 334–335.

85. Amerah AM, Ravindran V, Lentle RG, Thomas DG (2007) Feed particle size: Implications on the digestion and performance of poultry. Worlds Poultry Science Journal 63: 439–455.

86. Meyer B, Bessei AW, Vahjen W, Zentek J, Harlander-Matauschek A (2012) Dietary inclusion of feathers affects intestinal microbiota and microbial metabolites in growing Leghorn-type chickens. Poultry Science 91: 1506–1513.

87. Meyer B, Zentek J, Harlander-Matauschek A (2013) Differences in intestinal microbial metabolites in laying hens with high and low levels of repetitive feather-pecking behavior. Physiology & Behavior 110: 96–101.

88. Mossner R, Lesch KP (1998) Role of serotonin in the immune system and in neuroimmune interactions. Brain Behavior and Immunity 12: 249–271.

89. Ebert-Zavos E, Horvat-Gordon M, Taylor A, Bartell PA (2013) Biological Clocks in the Duodenum and the Diurnal Regulation of Duodenal and Plasma Serotonin. Plos One 8.

90. van Hierden YM, de Boer SF, Koolhaas JM, Korte SM (2004) The Control of Feather Pecking by Serotonin. Behavioral Neuroscience 118: 575–583.

91. Kops MS, de Haas EN, Rodenburg TB, Ellen ED, Korte-Bouws GAH, et al. (2013) Effects of feather pecking phenotype (severe feather peckers, victims and non-peckers) on serotonergic and dopaminergic activity in four brain areas of laying hens (Gallus gallus domesticus). Physiology and Behavior 120: 77–82.

92. Kops MS, de Haas EN, Rodenburg TB, Ellen ED, Korte-Bouws GAH, et al. (2013) Selection for low mortality in laying hens affects catecholamine levels in the arcopallium, a brain area involved in fear and motor regulation. Behavioural Brain Research 257: 54–61.

93. Ghareeb K, Niebuhr K, Awad WA, Waiblinger S, Troxler J (2008) Stability of fear and sociality in two strains of laying hens. British Poultry Science 49: 502–508.

94. Hocking PM, Channing CE, Waddington D, Jones RB (2001) Age-related changes in fear, sociality and pecking behaviours in two strains of laying hen. British Poultry Science 42: 414–423.

95. Ghareeb K, Awad WA, Niebuhr K, Böhm J, Troxler J (2008) Individual differences in fear and social reinstatement behaviours in laying hens. International Journal of Poultry Science 7: 843–851.

6

Activation of Duck RIG-I by TRIM25 Is Independent of Anchored Ubiquitin

Domingo Miranzo-Navarro, Katharine E. Magor*

Department of Biological Sciences and the Li Ka Shing Institute of Virology, University of Alberta, Edmonton, Alberta, Canada

Abstract

Retinoic acid inducible gene I (RIG-I) is a viral RNA sensor crucial in defense against several viruses including measles, influenza A and hepatitis C. RIG-I activates type-I interferon signalling through the adaptor for mitochondrial antiviral signaling (MAVS). The E3 ubiquitin ligase, tripartite motif containing protein 25 (TRIM25), activates human RIG-I through generation of anchored K63-linked polyubiquitin chains attached to lysine 172, or alternatively, through the generation of unanchored K63-linked polyubiquitin chains that interact non-covalently with RIG-I CARD domains. Previously, we identified RIG-I of ducks, of interest because ducks are the host and natural reservoir of influenza viruses, and showed it initiates innate immune signaling leading to production of interferon-beta (IFN-β). We noted that K172 is not conserved in RIG-I of ducks and other avian species, or mouse. Because K172 is important for both mechanisms of activation of human RIG-I, we investigated whether duck RIG-I was activated by TRIM25, and if other residues were the sites for attachment of ubiquitin. Here we show duck RIG-I CARD domains are ubiquitinated for activation, and ubiquitination depends on interaction with TRIM25, as a splice variant that cannot interact with TRIM25 is not ubiquitinated, and cannot be activated. We expressed GST-fusion proteins of duck CARD domains and characterized TRIM25 modifications of CARD domains by mass spectrometry. We identified two sites that are ubiquitinated in duck CARD domains, K167 and K193, and detected K63 linked polyubiquitin chains. Site directed mutagenesis of each site alone, does not alter the ubiquitination profile of the duck CARD domains. However, mutation of both sites resulted in loss of all attached ubiquitin and polyubiquitin chains. Remarkably, the double mutant duck RIG-I CARD still interacts with TRIM25, and can still be activated. Our results demonstrate that anchored ubiquitin chains are not necessary for TRIM25 activation of duck RIG-I.

Editor: Pierre Boudinot, INRA, France

Funding: This study was supported by Canadian Institutes of Health Research grant CIHR MOP 125865 to KEM, and Natural Sciences and Engineering Research Council Discovery Grant RGPIN 228035 to KEM. The funders had no role in study design, data collection and analysis, decision to publish, or preparation of the manuscript.

Competing Interests: The authors have read the journal's policy and have the following conflict. The authors have issued a patent on the duck RIG-I gene used in the studies: 2009071 "Transgenic Chickens Expressing RIG-I and Viral Resistance", United States Provisional Patent.

* E-mail: kmagor@ualberta.ca

Introduction

RIG-I is an intracellular detector of 5′ triphosphate RNA that activates a signaling pathway leading to the production of type I interferon and initiation of the antiviral state [1]. The three dimensional structures of RIG-I from several species [2,3,4,5] provide a molecular model for RIG-I activation [6]. The pathway starts upon sensing of viral RNA by the RIG-I helicase and regulatory domains, which undergo a conformational change, acting as a molecular camshaft that uses energy from ATP hydrolysis to expose the two caspase activator recruitment domains (CARDs) to the cytoplasm [7]. Activated CARD domains of RIG-I interact with CARD domains of MAVS (VISA, CARDIF, IPS-I) [8,9,10], which aggregate in a prion-like structure [11]. This conformational change allows MAVS to serve as a platform to recruit the other components of the pathway in a multiprotein signaling complex [12]. Downstream, activation of IRF3 and NF-κβ transcription factors induce type I interferon and production of proinflamatory cytokines [13].

RIG-I is normally found in an auto-repressed state in a closed conformation [14] through constitutive phosphorylation by PKC-α and PKC-β kinases [15]. Upon binding of RNA and subsequent conformational change, RIG-I is dephosphorylated by PP1 phosphatase [16], triggering the activation process. TRIM25, an E3 ubiquitin ligase is critically important in RIG-I activation [17]. The carboxy terminal SPRY domain of TRIM25 interacts with the first CARD of RIG-I involving T55, and attaches K63-linked ubiquitin chains to K172 within the second RIG-I CARD domain [17]. More recently, the need for attached ubiquitin has been disputed, as human CARD domains can be activated *in vitro* through interaction with unanchored K63-linked polyubiquitin chains produced by TRIM25, a process that also requires K172 [18,19]. RIPLET, another E3 ubiquitin ligase, is also involved in RIG-I activation through ubiquitination of RIG-I CARD domains [20,21]. The importance of RIG-I ubiquitination is underscored by the fact that influenza nonstructural protein 1 (NS1) blocks TRIM25 and RIPLET in a species-specific manner [22,23].

We are investigating RIG-I function and regulation in ducks, the natural host of influenza viruses, with the aim to examine influenza interference in this pathway. Waterfowl are the natural reservoir of most influenza A strains [24,25], and have a central role in the ecology and evolution of influenza A viruses. Some highly pathogenic avian influenza strains can replicate in ducks not showing disease signs, making waterfowl the "Trojan horses" of influenza infection [26]. Highly pathogenic avian influenza strains can be lethal in chickens, and are of concern because they occasionally infect humans. In the last decade H5N1 strains have infected 641 individuals with a mortality rate close to 60% [27]. We discovered a critical difference in pathogen detection between ducks and chickens, giving a plausible explanation for their difference in susceptibility. RIG-I of ducks is functional and highly expressed upon influenza infection, while chickens apparently lack RIG-I [28]. RIG-I initiates a robust and immediate induction of IFN stimulated genes in ducks within the first 24 hours after infection [29], contributing to duck survival.

Phosphorylation–dephosphorylation [16,30] and ubiquitination [17] are key processes in regulation of human RIG-I. The three-dimensional structure has been determined for duck RIG-I [31] facilitating structure-function comparison with human RIG-I, but the modifications required for regulation have not been examined. From an alignment of duck RIG-I with other sequences, the phosphorylation sites for inactivation are conserved (S8 and S168, corresponding to human S8 and T170, respectively) and thus we anticipate regulation by phosphorylation is conserved. However, we noted that K172, the site of activating ubiquitination, was not conserved in RIG-I of ducks, zebrafinch and other species, including mouse. Thus, we wondered whether duck RIG-I was modified by TRIM25 ubiquitination, and whether ubiquitin was attached at residues other than K172. Chickens have TRIM25, but RIPLET appears to be missing from the genome [23,32]. Here we demonstrate TRIM25 interaction is essential for activation of duck RIG-I expressed in chicken cells. However, while TRIM25 mediated ubiquitination occurs at alternate residues, the activation of RIG-I is independent of anchored ubiquitination, suggesting duck TRIM25 is involved in production of unanchored K63-linked ubiquitin chains.

Results

Duck RIG-I(N) Induces Interferon-beta in Chicken DF1 Cells

Chickens lack the intracellular RNA detector RIG-I, which is present in ducks [28], but have MDA5 [33], which converges on the same signalling pathway downstream of MAVS. We previously reported the reconstitution of the pathway, and induction of an innate immune response in chicken DF1 cells transfected with duck RIG-I and RIG-I ligand [28,34]. The N-terminal part of human RIG-I, RIG-I(N), is constitutively active and induces an innate immune response in the absence of the helicase and regulatory domains of RIG-I [17]. Exploring the possibility of similar behavior of duck RIG-I(N), we generated a GST-d2CARD construct and transfected it into the chicken embryonic fibroblast cell line, DF1. Using the dual luciferase assay with a chicken IFN-β promoter in a luciferase reporter plasmid (pGL3-chIFN-β) [28] we detected increased chicken IFN-β promoter activity in the cells transfected with the GST-d2CARD compared to the GST plasmid (Fig. 1A). We compared the expression of three interferon-stimulated-genes (ISG) in cells expressing GST-d2CARD to those with GST alone, and detected vast upregulation of *MX1*, *IFIT5* and *OASL* (Fig. 1B), suggesting that duck CARD domains constitutively activate the RIG-I pathway.

Duck RIG-I is Ubiquitinated in Chicken DF1 Cells

Activation of human RIG-I involves polyubiquitin chains attached at lysine 172 of the second CARD domain [17], but this residue is absent in duck RIG-I. To determine whether duck RIG-I is modified by attached ubiquitin, we examined ubiquitina-

tion of full-length duck Flag-RIG-I using HA-tagged ubiquitin (HA-Ub) and immunoprecipitation with anti-Flag antibodies. We co-transfected DF1 cells with pcDNA3.1 Hygro+ Flag-RIG-I and pcDNA HA-Ub [35] and used cell extracts for immunoprecipitation and western blotting. We detected ubiquitination as a smear of larger forms of RIG-I in the anti-HA immunoblot (Fig. 1C). To determine whether ubiquitination within the CARD domains was involved in regulation, we performed GST-pulldown from extracts of cells transfected with GST-d2CARD and GST control plasmids. We detected ubiquitinated bands in the anti-GST and anti-HA western blots (Fig. 1D). These results demonstrate that duck RIG-I CARD domains lacking the Lys 172 residue are nonetheless ubiquitinated in chicken cells under conditions where the RIG-I pathway is active.

Human TRIM25 Increases Ubiquitination of Duck CARD Domains

TRIM25, an E3 ubiquitin ligase, is involved in activation of human RIG-I through attachment of polyubiquitin chains [17]. To determine whether duck RIG-I is modified by human TRIM25 in chicken DF-1 cells, we used a GST-pulldown assay. We observed larger bands corresponding to ubiquitinated duck CARD domains on the immunoblot, when DF1 cells were co-transfected with human TRIM25-V5 and GST-d2CARD plasmids (Fig. 1E). To investigate whether duck RIG-I and human TRIM25 interact, we immunoblotted for the V5 epitope on TRIM25, and the pulldown of human TRIM25-V5 with duck CARD domains indicated a strong interaction between these proteins. Surprisingly, we detected no larger ubiquitinated forms, nor interaction between human CARD domains and human TRIM25 in extracts of chicken DF1 cells transfected with GST-h2CARD, suggesting that an additional factor needed for human TRIM25 and h2CARD interaction is not conserved in chicken cells (Fig. 1E). We confirmed that h2CARD was ubiquitinated and interacts with human TRIM25 in HEK293 cells demonstrating these constructs were functional (data not shown). We observed increased activation of the chicken IFN-β promoter upon co-transfection of human TRIM25-V5 and GST-d2CARD, and no chicken IFN-β promoter induction with h2CARD and hTRIM25-V5 (Fig. 1F). These results together suggest that duck CARD domains are activated by TRIM25 activity through modification involving ubiquitin chains.

Duck TRIM25 Interacts with and Increases Activation of Duck CARD Domains

Our results above suggest that chicken and human TRIM25 can function in the activation of duck RIG-I, however, the function of duck TRIM25 is unknown. We cloned duck TRIM25 into an expression vector with a C-terminal V5 tag to investigate its interaction with duck RIG-I. GST-pulldown clearly showed that duck TRIM25 interacts with duck RIG-I CARD domains but not with human CARD domains (Fig. 1G). We observed increased RIG-I activity in a dose dependent manner, with transfection of increasing amounts, up to 150 ng of dTRIM25 plasmid, indicated by induction of the chicken IFN-β promoter (Fig. 1H). Higher amounts of duck TRIM25 (500 ng) abrogated the induction of the RIG-I pathway (data not shown), as previously described for human TRIM25 [36]. These results demonstrate that duck TRIM25 has a role in the activation of the RIG-I pathway.

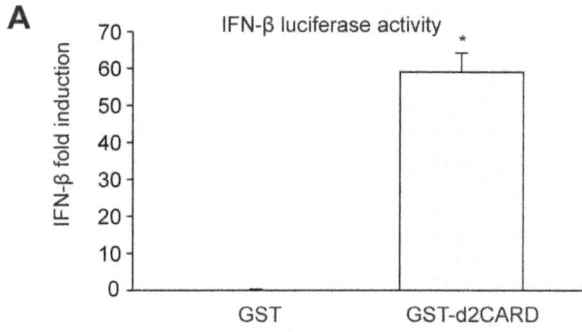

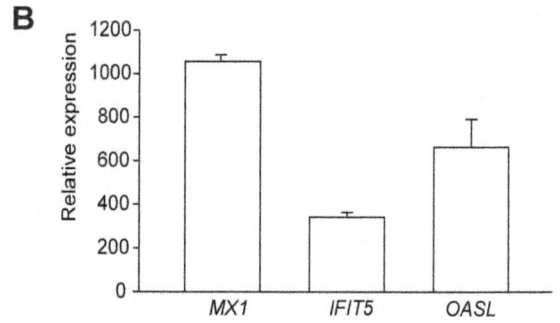

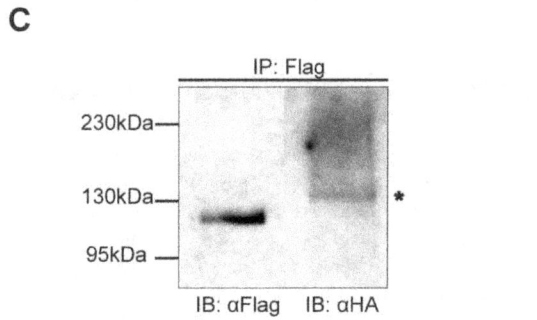

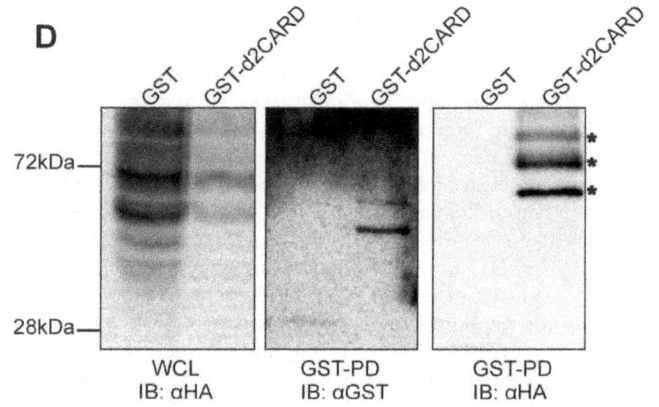

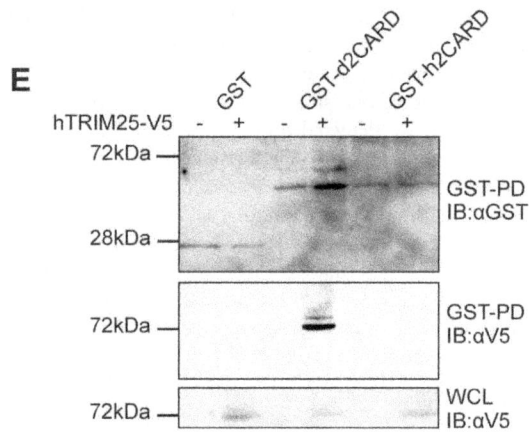

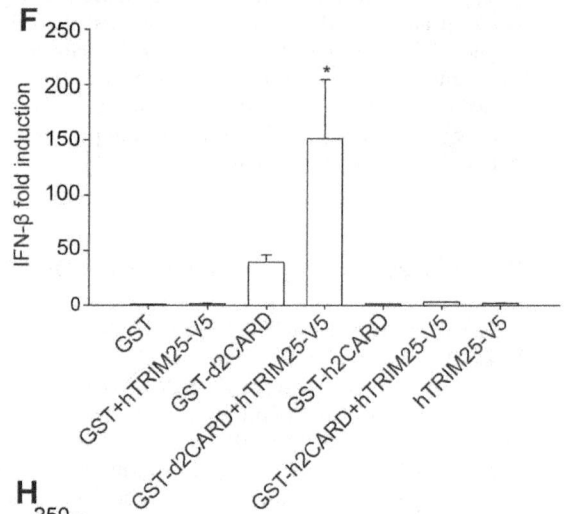

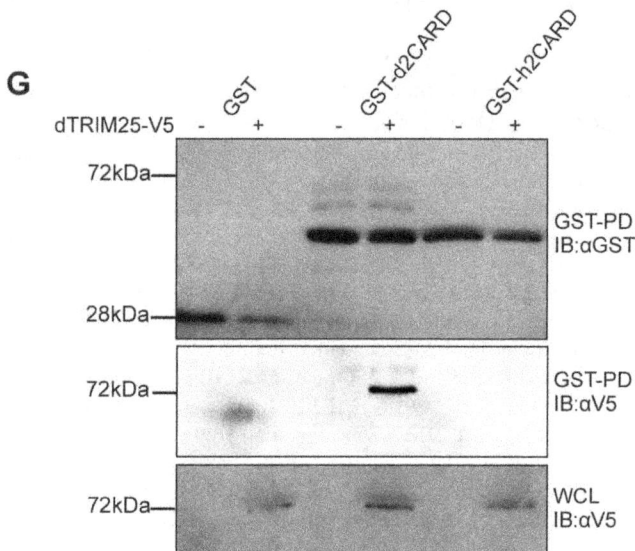

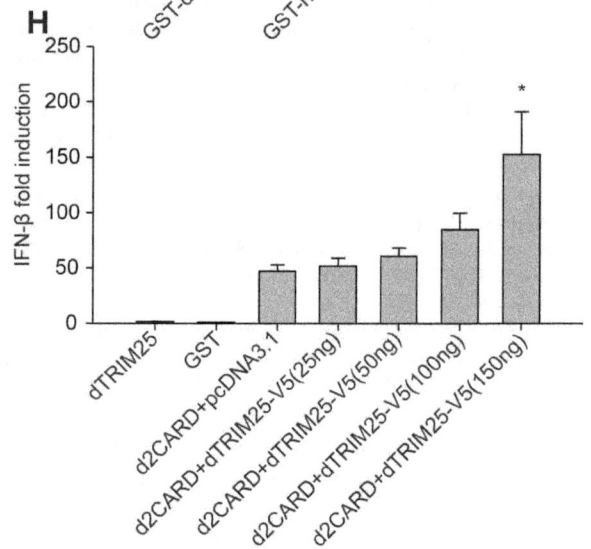

Figure 1. RIG-I 2CARD induces innate immune genes and ubiquitination by TRIM25 increases activation. A. Chicken IFN-β promoter activity in DF-1 cells transfected with GST or GST-d2CARD, shown as mean fold induction (±SD) from three independent experiments (n = 3) (* indicates P<0.001). B. Expression of innate immune genes (*MX1*, *IFIT5*, and *OASL*) upon signaling by duck 2CARD is shown relative to cells transfected with GST alone. Results are representative of three independent experiments and error bars show RQ$_{min/max}$ at a 95% confidence level. C. Extracts of DF1 chicken cells transfected with duck FLAG-RIG-I together with HA-ubiquitin were used for immunoprecipitation of RIG-I with anti-FLAG and immunoblotting with anti-FLAG and anti-HA. Asterisk indicates ubiquitinated RIG-I. D. GST or GST-d2CARD and HA-Ubiquitin were transfected into DF-1 cells and used for GST pulldown and immunoblotting. GST-d2CARD is ubiquitinated as indicated by presence of larger bands in anti-GST or anti-HA blots (indicated with *). E. Chicken DF1 cells transfected with GST or GST 2CARD fusion constructs from duck (d2CARD) or human (h2CARD) and V5-epitope tagged human TRIM25 were used for GST pulldown and immunoblotting with anti-GST and anti-V5 antibodies. Duck 2CARD is ubiquitinated and associates with human TRIM25. F. Chicken IFN-β promoter activity in DF-1 cells transfected with GST-d2CARD or GST-h2CARD and human TRIM25-V5, show GST-d2CARD is further activated by the presence of hTRIM25-V5 (* indicates P<0.001). Human 2CARD is less active and not ubiquitinated in chicken cells. Data are the mean ± SD (n = 3). G. GST pulldown and immunoblot demonstrating interaction of duck TRIM25-V5 with duck CARD domains but not with human CARD domains. H. Duck TRIM25 significantly activates d2CARD compared to d2CARD alone (* indicates P<0.001). Luciferase assay was performed using the chIFN-β promoter luciferase reporter and increasing amounts of duck TRIM25-V5 plasmid (between 25 ng and 150 ng) with a fixed amount of GST-d2CARD plasmid (5 ng). All d2CARD transfections show a statistically significant activation of the chIFN-β promoter compared to the GST control (P<0.005). Data are the mean ± SD (n = 3).

A Splice Variant of Duck RIG-I Cannot Interact with TRIM25

A splice variant deleting exon 2, the region necessary for interaction with TRIM25 including the residue T55 required for interaction and ubiquitination of TRIM25, was identified in human RIG-I [37]. The production of the splice variant, which is a dominant inhibitor of the RIG-I mediated antiviral IFN response, was proposed to be a mechanism of down regulation of RIG-I activity [37]. We previously noted that ducks have alanine instead of threonine at position 55 [28], which functions suboptimally in the human RIG-I T55A mutant [37]. To determine whether a similar splice variant exists in ducks and whether it forms a non-functional RIG-I, we examined transcripts present in tissues from influenza-infected ducks from a previous experiment [28]. RNA from lung tissues taken at 3 dpi was obtained from three groups of 3 ducks -mock infected with PBS, infected with low pathogenic avian influenza A/British Columbia 500/2005 (H5N2) (BC500), or highly pathogenic avian influenza A/Vietnam 1203/2004 (H5N1) (VN1203), and reverse-transcription PCR was used to amplify across RIG-I. We observed accumulation of the splice variant in tissues from ducks infected with both viruses (Fig. 2A). The duck splice variant was confirmed by sequencing to be equivalent to its human counterpart, missing a portion of the first CARD domain (exon 2) that is essential for TRIM25 interaction [37](Fig. 2B).

Duck RIG-I Splice Variant Cannot Activate Innate Immune Signaling

The human RIG-I splice variant does not interact with TRIM25 and is not ubiquitinated; therefore we examined the duck splice variant by GST-pulldown and immunoblotting. We observed a decrease in the interaction between the duck splice variant CARD domains (SVCARD) and human TRIM25 (Fig. 2C) or duck TRIM25 (Fig. 2D), compared with the intact GST-d2CARD fusion protein. We examined the attachment of HA-tagged ubiquitin to the SVCARD, in transfection followed by anti-HA immunoblotting. The ubiquitination of the splice variant was severely decreased in transfected DF1 cells (Fig. 2E). The activation of chIFN-β promoter for the intact CARD domains and the splice variant, relative to GST alone were compared. We observed total abrogation chIFN-β activation for the splice variant, even with addition of human TRIM25 (Fig. 2F). These results demonstrate that ducks have a splice variant of RIG-I expressed after infection. The abrogation of the innate immune response by the splice variant supports the idea that duck TRIM25 and RIG-I interact similarly to those of humans, and suggests the splice variant accumulates to down regulate the innate immune

response. We explored this possibility by cotransfecting GST-d2CARD with increasing amounts of SVCARD The SVCARD does not bind to TRIM25, thus does not compete with the interferon response produced by GST-d2CARD (Fig. S1A.). We expected that the full-length SVRIG-I would bind ligand, but being unable to signal, would dampen the interferon response, as the human SVRIG-I does in competition with native RIG-I in HEK293T cells infected with Sendai virus [37]. Because duck cell lines are not available, we transfected Flag-RIG-I and increasing amounts of the splice variant RIG-I in chicken DF-1 cells, and stimulated with RIG-I ligand or control RNA. We were unable to demonstrate significant interference in RIG-I signaling by the RIG-I splice variant (Fig. S1B).

Duck RIG-I is Ubiquitinated at Lysine 167 and Lysine 193

To determine the location of attached ubiquitin in duck CARD domains, the ubiquitinated forms of GST-d2CARD modified by human TRIM25-V5 were purified and analyzed by mass spectrometry. Three bands, larger than the GST-d2CARD protein, were detected in a coomassie stained polyacrylamide gel (Fig. 3A). These bands were equivalent in size to the ubiquitinated bands detected by western blot. We detected ubiquitin and GST-d2CARD fusion protein in all three bands, confirming ubiquitination of the duck CARD domains. Within the human CARD domains six lysines can be ubiquitinated (K99, K169, K172, K181, K190, and K193) in cells with active TRIM25 [37]. In duck CARD domains, only three lysines are conserved, K98, K167 and K193, equivalent to human K99, K169 and K193, respectively (Fig. 2B). In all three bands, we recovered Lys 193 with a diglycine modification tag indicating anchored ubiquitination (Fig. 3B). Lys 167 is the Lys nearest to the missing Lys 172, and its localization within the three-dimensional structure of duck RIG-I CARD domains [31] is close to Lys 172 (Fig. 3C). Lys 167 was ubiquitinated in band 1 (Fig. 3D). K98 was not ubiquitinated in any band. Also no ubiquitinated forms of K154 and K164, important for RIG-I activation by RIPLET [38], were present in any of the analyzed samples. These results demonstrate that duck CARD domains are ubiquitinated at Lys 167 and Lys 193 under activation conditions, and these Lys appear to be the only anchored ubiquitination sites within the CARD domains with a potential role in duck RIG-I activation.

K63-linked Polyubiquitin Chains were Isolated with Duck RIG-I CARD

Human RIG-I is activated through the attachment of anchored K63-linked polyubiquitin chains to the second CARD domain at K172 [17]. To examine the nature of the ubiquitin bound to RIG-

A

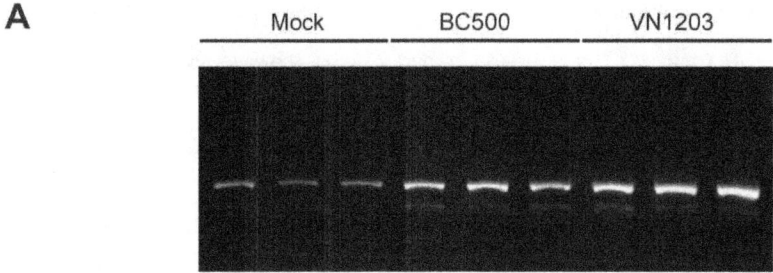

B

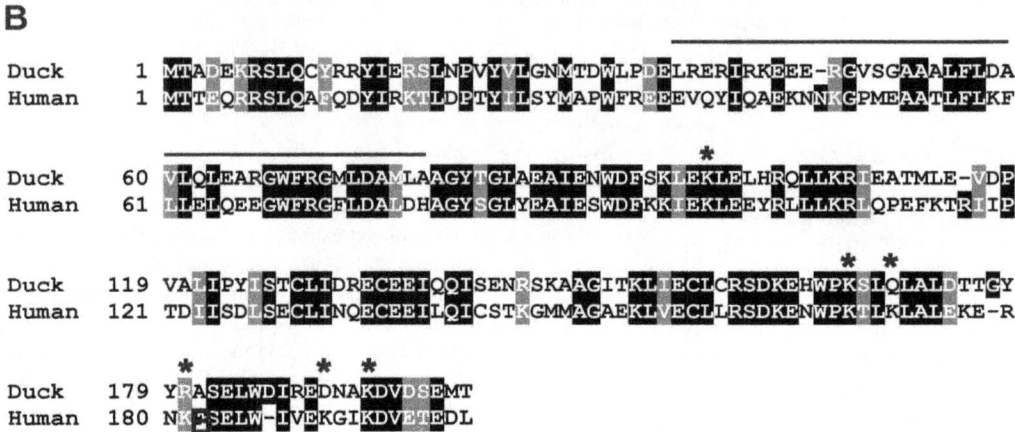

C 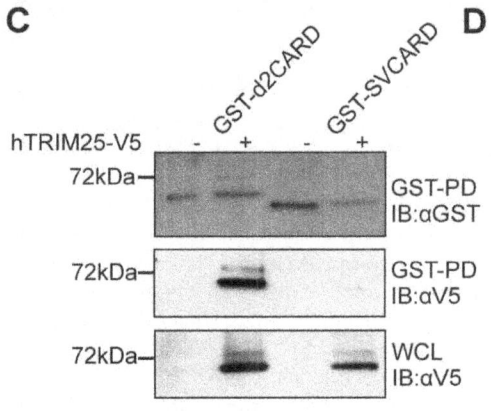 **D**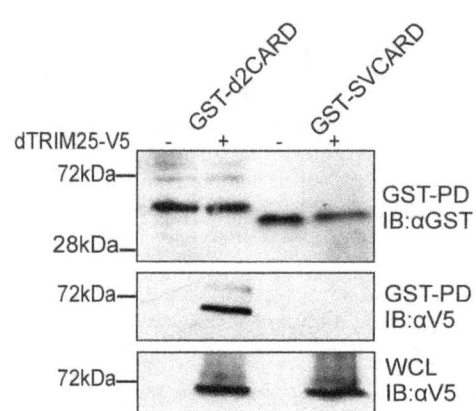

E 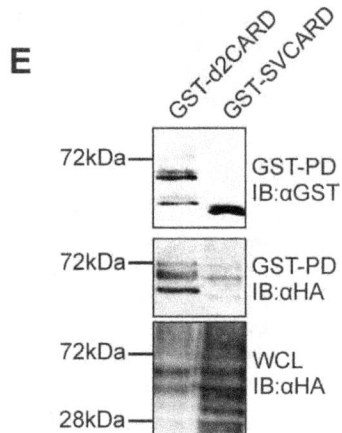 **F**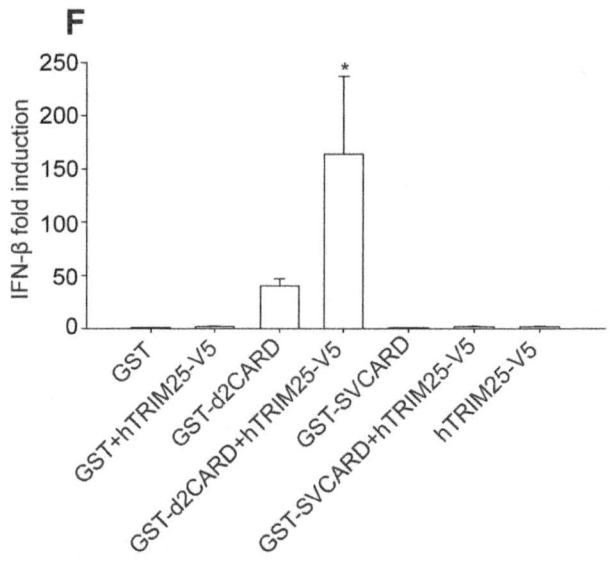

Figure 2. RIG-I splicing variant is not ubiquitinated and cannot activate innate immune signaling. A. Reverse transcription PCR showing amplification across exon 2 of duck RIG-I in lung tissue of ducks that were mock challenged, or infected with influenza A virus strains, BC500 or VN1203. B. Alignment of sequences of duck and human RIG-I CARD domains. Exon 2 sequence missing in splicing variant is overlined and Lys residues ubiquitinated in human RIG-I are marked with an asterisk. C. GST pulldown followed by immunoblotting indicates that GST-d2CARD interacts with human TRIM25-V5, while this interaction is greatly reduced for the splicing variant SVCARD. D. GST pulldown and immunoblotting showing no interaction between duck RIG-I CARD splice variant (SVCARD) and duck TRIM25-V5. E. GST-d2CARD is ubiquitinated while splicing variant is not. F. GST-2CARD activates the chIFN-β promoter in DF-1 cells, while splicing variant does not. Human TRIM25 significantly increased the chIFN-β promoter activity of d2CARD, but not SVCARD, compared to d2CARD alone (* indicates P<0.001). Data are the mean ± SD (n = 3).

I CARD domains, we co-transfected the GST-d2CARD and HA-ubiquitin wildtype and mutants [39]. The GST pulldown using HA-ubiquitin-WT shows the same ubiquitinated bands detected in the coomasie stained gel above. The HA-ubiquitin-K0 mutant, with all lysines mutated, had 2 bands in the HA blot of the pulldown that likely correspond to monoubiquitin attached at K167, at K193 or at both lysines (larger band) (Fig. 3E). The HA-ubiquitin-K63 mutant, with only the K63 lysine intact allowing formation of K63-linked polyubiquitin, has two additional bands, the lower part of the doublet in band 2 and band 3. Analyzing the MS/MS results we detect a strong signal for K63-linked polyubiquitin chains in band 2 (Fig. 3F). The size of the lower part of band 2 is consistent with the linkage of two ubiquitin molecules in a polyubiquitin chain.

Duck RIG-I CARD K167R/K193R Double Mutant is not Ubiquitinated

To ascertain the importance of sites for attachment of anchored polyubiquitin chains in the activation of duck CARD domains in DF1 chicken cells, we created a set of mutants and examined their ubiquitination. We first created a mutant, Q170K, to restore the equivalent of the human RIG-I lysine 172. This mutant did not show an increased ability to activate the RIG-I pathway and ubiquitinated bands look identical to those of the wild type construct. We then mutated each of the lysines, K167, and K193 independently, and together in a double mutant. Both of the single mutants (K167R, K193R) showed identical patterns of bands demonstrating ubiquitination, but the K167R/K193R double mutant lost all larger bands indicating it was not ubiquitinated, nor were K63 polyubiquitin chains attached (Fig. 4A). These results demonstrate that both K167 and K193 are attachment sites for bound ubiquitin and polyubiquitin chains in duck RIG-I CARD domains.

Duck RIG-I CARD K167R/K193R Mutant can be Activated by TRIM25

To determine whether the duck CARD domains with the mutated lysines could interact with TRIM25, we cotransfected the mutant with duck TRIM25 in DF-1 cells. We confirmed the interaction of duck CARD domains bearing K167 and K193 mutations with duck TRIM25 using GST-pulldown (Fig. 4B), suggesting these lysines are not essential for duck TRIM25 interaction. No larger bands were detected in the GST-pulldown samples in the presence of duck TRIM25 indicating that duck TRIM25 does not attach ubiquitin or polyubiquitin chains at any lysine other than K167 and K193 (Fig. 4B). Exploring the importance of K167 and K193 in the activation of RIG-I, we used the dual luciferase assay and examined chIFN-β promoter induction in cells transfected with the different mutants. None of the mutants lost its ability to activate the RIG-I pathway, as indicated by 400-fold to 800-fold induction of the chicken IFN-β promoter (Fig. 4C). These results suggest that ubiquitination of K167 and K193 is not essential for activation of RIG-I CARD domains. We show the ability of duck TRIM25 to increase the

activity of the double mutant K167R/K193R (Fig. 4D). Taken together these results demonstrate that duck TRIM25 is activating duck RIG-I through a mechanism that does not require anchored K63-linked polyubiquitin chains.

Discussion

Here we examine regulation of RIG-I in the natural host of avian influenza. Human RIG-I is activated by K63-linked polyubiquitin attached at lysine 172 in the second CARD domain [17], or interaction with unanchored K63-linked polyubiquitin chains, also involving lysine 172 [19], a residue that is missing in ducks and other birds. We demonstrate that chicken and duck TRIM25 are functional, and duck CARD domains are ubiquitinated. The duck TRIM25-RIG-I interaction plays a role in the activation of IFN-β production as demonstrated by the nonfunctional splice variant of RIG-I, which is unable to interact with TRIM25, and does not induce IFN-β. Using mass spectrometry to analyze ubiquitinated CARD domains, we determined that K167 and K193 were the ubiquitination sites of duck CARD domains. We showed that mutation of each site independently did not disrupt ubiquitination, however a double mutant with both K167R and K193R, was unable to undergo ubiquitination. Surprisingly, this double mutant retained activity and the ability to be activated by duck TRIM25. Thus, duck TRIM25 can activate RIG-I through a process that does not depend on attachment of ubiquitin chains to the CARD domains of RIG-I.

Duck CARD domains undergo robust ubiquitination and activate IFN-β in DF1 chicken cells indicating that chicken cells contain all the proteins required for the activation and downstream signaling of RIG-I. Previously, Rajsbaum et al. [23] showed that knockdown of chicken TRIM25 decreased IFN production in chicken cells. Since chickens lack RIG-I, this unexpected observation highlights that MDA5 and RIG-I share the same activation pathway, and likely TRIM25 is also involved in activation of MDA5. However, Gack et al., [17] showed MDA5 CARD domains are not modified by bound ubiquitin. Recent *in vitro* studies have shown that MDA5 and RIG-I interact with unanchored K63-linked polyubiquitins chains generated by TRIM25 [19] and are activated by the same dephosphatase [16]. The demonstration that RIG-I CARD domains are ubiquitinated in chicken cells suggests that chicken TRIM25 is activating duck RIG-I in the transfected DF-1 cells, and potentially also activates chicken MDA5.

Human TRIM25 ubiquitinated duck CARD domains and augmented their activation in chicken DF-1 cells. Surprisingly, human TRIM25 does not interact, ubiquitinate, or activate human CARD domains co-transfected in chicken DF-1 cells. This implies that some factor needed for this interaction is missing or diverged in chicken cells. Potentially, a factor needed in the ubiquitination steps is not conserved. The E2 ligases Ubc13-Uev1a and UbcH5a are needed for the activation of the RIG-I signaling pathway reconstituted *in vitro* [18]. Both E2 ligases interact with TRIM25 [17,40]. These two E2 ligases are also partners of the E3 ligase CHIP, and in this context Ubc13-Uev1a produces K63-

A

GST-d2CARD
+hTRIM25-V5

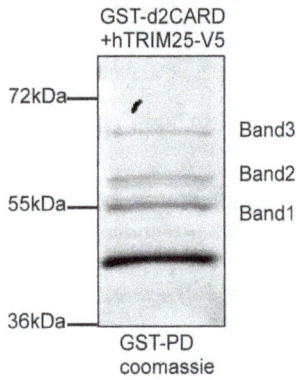

72kDa —

Band3
Band2
55kDa — Band1

36kDa —

GST-PD
coomassie

B   K 193 ubiquitination

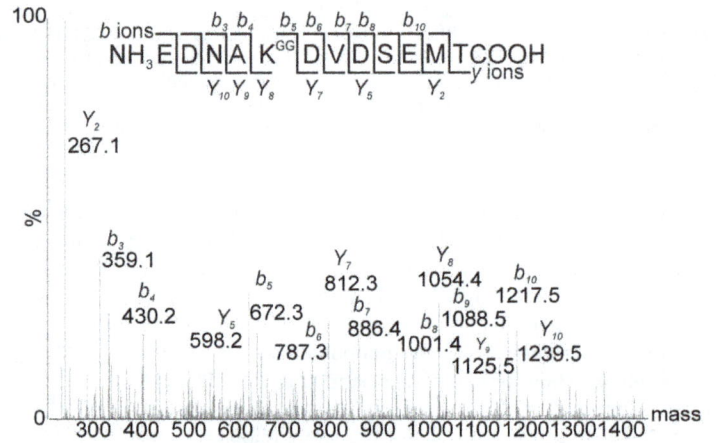

b ions
NH₃─E─D─N─A─K^GG─D─V─D─S─E─M─T─COOH

C

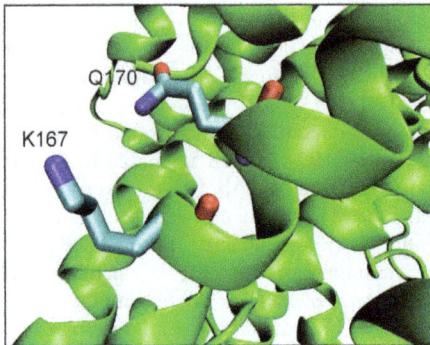

Q170

K167

D   K 167 ubiquitination

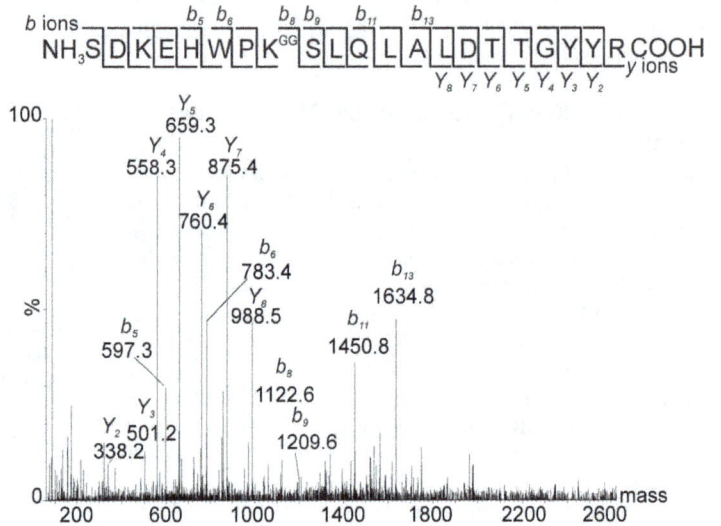

b ions
NH₃─S─D─K─E─H─W─P─K^GG─S─L─Q─L─A─L─D─T─T─G─Y─Y─R─COOH

E

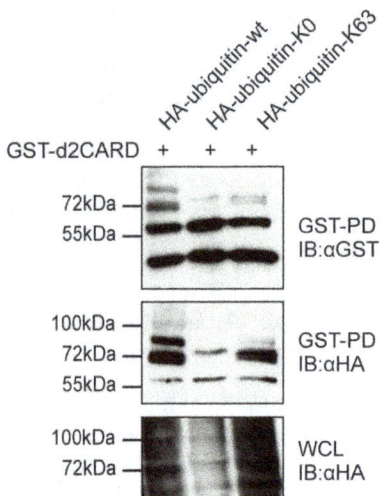

GST-d2CARD   +   +   +

72kDa —
55kDa —
GST-PD
IB:αGST

100kDa —
72kDa —
55kDa —
GST-PD
IB:αHA

100kDa —
72kDa —
WCL
IB:αHA

F   K 63 ubiquitin chains

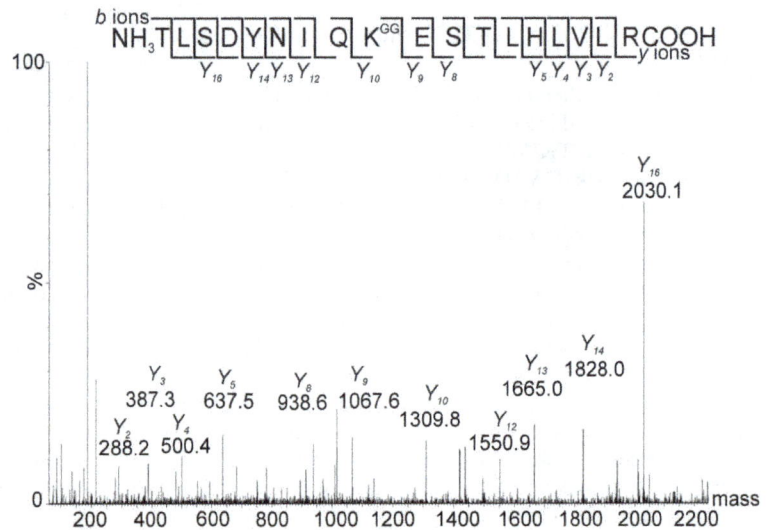

b ions
NH₃─T─L─S─D─Y─N─I─Q─K^GG─E─S─T─L─H─L─V─L─R─COOH

Figure 3. Duck RIG-I CARD domains are ubiquitinated at K167 and K193. A. Coomassie stained gel showing the different ubiquitinated forms of the duck RIG-I CARD domains. B. MaxEnt3 deconvoluted MS/MS spectra of a peptide bearing the typical diglycine signal for ubiquitination at Lys 193, showing the singly-charged forms of the B and Y ions detected. C. Three-dimensional detail of duck RIG-I structure in the region of Q170 (red), corresponding to human K172, and the closest lysine, K167 (blue) (PDB code:4a2w) [31]. Figure was created using Protean 3D from DNA Lasergene 9. D. MaxEnt3 deconvoluted MS/MS spectra of a peptide showing the ubiquitination signal for K167. E. GST pulldown of duck CARD domains from cells co-transfected with HA-ubiquitin, or HA-ubiquitin-K0 (all lysines mutated) or HA-ubiquitin K63 (only lysine 63 intact). Band 2 in the GST pulldown is a doublet and the lower band is missing in the absence of K63-linked ubiquitin chains. F. MaxEnt3 deconvoluted MS/MS spectra of a peptide of ubiquitin recovered from band 2 showing the diglycine signal at Lys 63, indicating the presence of Lys 63-linked polyubiquitin.

linked unanchored polyubiquitin chains, while UbcH5a catalyzes the formation of covalently attached polyubiquitin chains [41]. Both E2 ligases are highly conserved in chickens, and the human and duck TRIM25 are interchangeable for activation of duck RIG-I. An alternative hypothesis is that human CARD domains cannot interact with chicken MAVS, and this interaction is needed prior to attached monoubiquitination. While this idea is contrary to the existing models for RIG-I activation, the divergence of

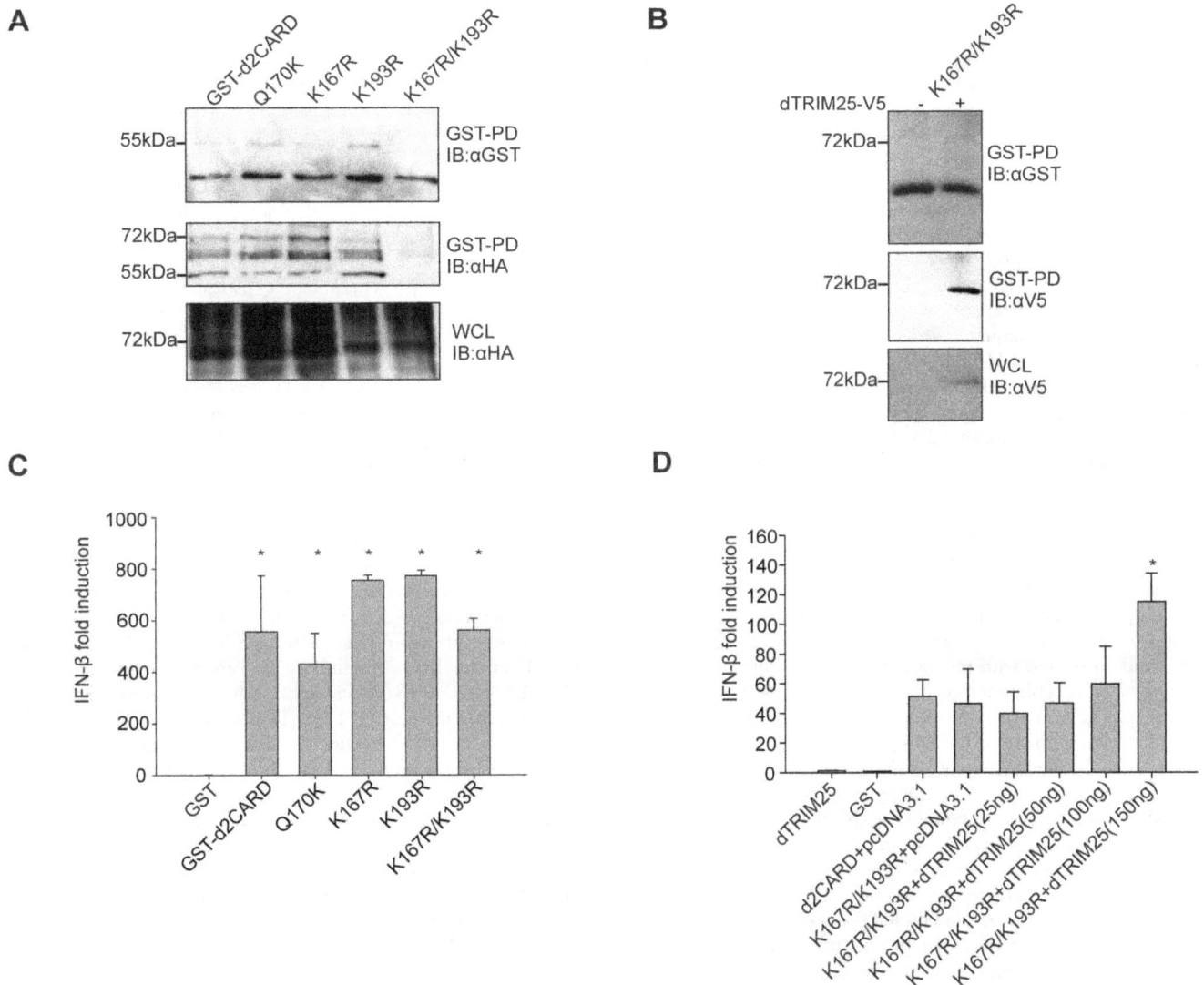

Figure 4. Ubiquitination but not activity of duck 2CARD is lost in double K-R mutant. A. GST pulldown and immunoblotting with anti-GST and anti-HA of samples from cells transfected with different GST-d2CARD mutants (Q170K, K167R, K193R, K167R/K193R) and HA-Ub plasmids showing a loss of ubiquitination only in the double mutant. B. GST pulldown and immunoblotting of K167/K193R mutant showing interaction with duck TRIM25-V5. C. GST-d2CARD and mutants Q170K, K167R, K193R, K167/193R activate the chIFN-β promoter when co-transfected into chicken DF-1 cells. Data represent the mean ± SD (n = 3). All mutants tested significantly activate the chIFN-β promoter compared with the GST control (P<0.05). None of the CARD mutants are statistically different compared to d2CARD. D. Luciferase assay was performed using the chIFN-β promoter and increasing amounts of duck TRIM25-V5 plasmid (between 25 ng and 150 ng) with a fixed amount of K167R/K193R mutant CARD domain plasmid (5 ng). Results are the mean±SD (n = 3) and dTRIM25 significantly increased activation of chIFN-β compared to d2CARD (* indicates P<0.005). All d2CARD samples show a statistically significant activation of the chIFN-β promoter compared with the GST control (P<0.05).

chicken and human MAVS also explains the previously observed inability of chicken MDA5 to function in human cells (VERO cells) or human MDA5 to function in chicken DF-1 cells [33]. It is worth noting that ducks, and other birds, have an alanine residue at D122, and the D122A mutant of human RIG-I(N) does not activate the IFN pathway even though it interacts with polyubiquitin chains [18], presumably because it cannot engage MAVS.

The conservation of a RIG-I splice variant present in tissue samples of influenza-infected ducks provides indirect evidence of the importance of the TRIM25 RIG-I physical interaction, and indicates the same region of duck RIG-I is involved. The RIG-I SV, lacking exon 2 cannot interact with duck TRIM25. This region of human RIG-I includes the residue T55, which is critically involved in the interaction, as T55I or T55E mutants are inactive [37]. Duck RIG-I does not have the residue T55, but has the conservative substitution alanine. The human RIG-I mutant T55A still interacts with human TRIM25 [37], but is suboptimal compared to wild type. Finally, the presence of this splice variant in tissues at 3 dpi, when the response to infection is winding down, is consistent with the hypothesis that the splice variant functions as a dominant inhibitor of full length RIG-I to prevent MAVS activation, as demonstrated for the human RIG-I splice variant [37]. However, we are unable to demonstrate inhibition of duck RIG-I signaling by the mutant SVRIG-I lacking exon 2, when both are equally expressed in chicken DF-1 cells, at amounts above the natural abundance of the transcript.

Using mass spectrometry we demonstrated that TRIM25 attaches anchored ubiquitin to the duck RIG-I CARD domains at residues K167 and K193. In band 1, we saw evidence of K167 and K193 ubiquitination, and also evidence for unubiquitinated K167, consistent with monoubiquitin attached to either of these residues, as the size of this band would predict. In band 2, we saw ubiquitination at K193 only, and a strong signal for K63-linked polyubiquitin chains. Band 2 was always seen as a doublet, and the lower band disappeared when the ubiquitin lacking all lysine residues was used, precluding production of polyubiquitin chains. The size of the upper band 2 is consistent with ubiquitin attached at both K167 and K193, while the lower band is a K63-linked polyubiquitin chain attached at one residue. Finally, in the third band we detected only ubiquitin attached at K193, and the size of this band correlated with the size of a chain of three ubiquitins. The pattern of ubiquitinated bands appeared identical to the bands arising from ubiquitination of human RIG-I, despite the ubiquitination of different residues.

We created a double mutant K167R/K193R form of the RIG-I card domains that could not be ubiquitinated by human or duck TRIM25. Independent mutation of each site (K167R or K193R) did not disrupt the attachment of ubiquitin, suggesting that attachment can occur at either of the identified sites. K167 lies close to the predicted site of phosphorylation at S168 in duck (corresponds to T170 in human RIG-I). Dephosphorylation of S168 may be a necessary step for the ubiquitination at either K167 or K193, and ubiquitination may prevent re-phosphorylation. No other lysines were found modified, and it appears that no other sites can be modified, since mutation of both lysines abrogates attachment of any ubiquitin. We also saw no attachment of ubiquitin at sites typically modified by RIPLET, as expected given the lack of RIPLET in the chicken genome [23,32].

The double mutant form of RIG-I, that cannot be ubiquitinated, indirectly demonstrates that activation of the duck CARD domains by duck or human TRIM25 does not require any anchored ubiquitin chains. TRIM25 produces unanchored ubiquitin chains that activate RIG-I *in vitro* [18], but the importance of this has not been demonstrated *in vivo*. The double mutant of duck RIG-I CARD domains, still interacts with human or duck TRIM25 and can be activated. The most likely explanation is that duck RIG-I CARD domains are associating with unanchored polyubiquitin produced by TRIM25. This is in direct contrast to the K172R mutation of human RIG-I CARD domains that abrogates ubiquitination and activation [17], including activation by interaction with unanchored polyubiquitin [18,19,42].

The role of anchored ubiquitin on K172 in the activation of human RIG-I has been questioned recently [18,19] and it was suggested that only unanchored K63-linked polyubiquitin chains are important for RIG-I activation. We demonstrated that the attachment of ubiquitin at duck RIG-I K167 and K193 is not necessary for activation of the CARD domains, but we suggest that duck RIG-I is unlikely to have ubiquitination sites without function in the cell. Indeed, the function of attached ubiquitin at K172 (human), or K167 and K193 (ducks) remains elusive, but perhaps this function is required for conformational changes leading to the activation of the intact RIG-I. The ability to generate anchored ubiquitin could be evidence of the ancestral activity of TRIM25, while activation by unanchored polyubiquitin chains was later derived. In addition, TRIM25 can undergo autoubiquitinylation with polyubiquitin chains [43]. Recent enzymatic studies suggest that TRIM25 generates the ubiquitin chains anchored to the E2 ligase and then they are transferred to the substrate [44], generating a plausible scenario for the coexistence of the two types of ubiquitin chains: some of the transferred chains will be attached to RIG-I and some will remain unanchored. Another possibility is that TRIM25 produces K63-linked polyubiquitin chains attached to RIG-I and an unknown deubiquitinating enzyme produces the unanchored chains [45]. Our data, with activation of duck RIG-I CARD domains in the absence of ubiquitin attachment sites makes the latter model unlikely.

In an alignment of RIG-I CARD domain amino acid sequences (Fig. S2), it is noteworthy that the lysines equivalent to human K169 and K193 are highly conserved across species, while the lysine equivalent to human K172 is not present in birds or rodents. RIG-I from avian species, including ducks, geese and zebrafinch, all have the equivalent lysine K169 and K193, and have the motif KSLQ, in the RIG-I regulatory site equivalent to the human $KT^{170}LK^{172}$. Turtle (*Chelonia mydas*) has the conserved K169 and K193, and has the motif KTFH. Teleost species, such as zebrafish (*Danio rerio*) have the sequence KVLK, and the K193 is not conserved. Mouse and rat have K169 and the motif KVLQ. Among RIG-I sequences for all species to date, K169 is conserved in all, and K193 is missing only in wild boar and rat. Serine or threonine at the site of phosphorylation of human RIG-I (T170) is well conserved among most species, but is missing in rodents and bats. Clearly, the RIG-I regulatory motif has changed over evolutionary time, possibly due to selection pressure from viruses intent on disrupting this regulation.

To suggest a model for duck RIG-I activation consistent with our observations, dephosphorylation at serine residues S8 and S168 allows RIG-I to interact with MAVS, where it undergoes monoubiquitination of lysine residues K167 or K193 in the RIG-I CARD domain by TRIM25, and attachment or association with polyubiquitin chains. Alternatively, additional ubiquitin moieties are attached to the monoubiquitin bound at either site. The activation of duck RIG-I in the absence of bound ubiquitin, which is in contrast to human RIG-I lacking K172, may reflect different specificity for duck RIG-I in binding unanchored polyubiquitin chains for activation. Indeed, the activation of duck RIG-I in the absence of attached polyubiquitin chains may be a mechanism to evade viral subversion of ubiquitin pathways [23], such as NS1 from H5N1 avian influenza, which preferentially interacts with TRIM25 in chicken cells, and decreases IFN production [23], even in the absence of RIG-I [28].

Materials and Methods

Ducks and Avian Influenza Virus Infections

cDNA samples prepared from lung tissue taken from ducks infected with avian influenza A/Vietnam 1203/04 (H5N1) (VN1203) or A/British Columbia 500/05 (H5N2) (BC500) were obtained from a previous study done in collaboration with Dr. Robert G. Webster at St. Jude Children's Research Hospital, Memphis, TN [28]. All animal experiments in that study were approved by the Animal Care and Use Committee of St. Jude Children's Research Hospital and performed in compliance with relevant institutional policies, National Institutes of Health regulations, and the Animal Welfare Act.

Cell Culture and Transfection

DF-1, a chicken embryonic fibroblast cell line derived from East Lansing strain eggs [46], was maintained in DMEM plus 10% FBS. Cells were seeded overnight in 24-well (2×10^5) or 6-well (8×10^5) plates and 24 h later cells were transfected with 1 µg/well (24-well plates) or 1.5 µg/well (6-well plates) of each of the indicated DNA constructs using Lipofectamine 2000™ reagent (Invitrogen) (ratio 1:2.5).

Plasmids

pcDNA3.1 (Hygro+) (Invitrogen) was the backbone plasmid used in all the constructs presented in this work and Phusion® High-Fidelity PCR Master Mix (NEB) was used in all PCR and site-directed mutagenesis reactions. To express the GST protein (pEBG plasmid) in DF1 chicken cells, a 716 bp PCR fragment containing the ORF coding for GST and the multicloning site from pEBG were amplified and cloned into pcDNA3.1 (Hygro+) using NheI-XhoI restriction sites encoded in the primers. The pcDNA3.1 (Hygro+)-GST construct was used for cloning the CARD domains (600 bp) of human RIG-I [17] and duck RIG-I [28] using BglII-ClaI restriction sites encoded in the primers to produce GST-CARD fusion constructs, GST-hCARD- and GST-d2CARD in pcDNA3.1 (Hygro+)-GST using BamHI-ClaI sites.

Duck TRIM25 was amplified and cloned into pCR 2.1-TOPO (Invitrogen) using cDNA from lung samples collected at 1 day post infection with H5N1 virus A/Vietnam/1203/04 [28]. The duck TRIM25 was cloned into pcDNA3.1 (Hygro+) using NheI-NotI containing primers with a V5 epitope coding sequence in the reverse primer.

The duck RIG-I splice variant was detected in cDNA prepared from a lung sample collected at 3 dpi with A/Vietnam/1203/04 (H5N1) [34]. PCR products across the intron were fully sequenced to confirm the same deletion of residues corresponding to exon 2 seen in human RIG-I [37]. For construction of the GST-splice variant CARD domain construct, the duck CARD domains cloned in pCR 2.1-TOPO with added sites BglII-ClaI (TOPO-d2CARD) was used as template for a PCR with primers flanking the second exon of RIG-I and facing outwards. The PCR product was treated with T4 polynucleotide kinase (NEB) to generate PCR fragments susceptible to ligation. The TOPO-SVCARD was digested with BglII-ClaI and the fragment containing SVCARD domains was cloned into pcDNA3.1 (Hygro+)-GST using BamHI-ClaI. The full length Flag-RIG-I was cloned into pcDNA3.1 (Hygro+) using NheI-NotI containing primers with a Flag epitope

coding sequence in the forward primer using pcDNA-RIG-I [28] as template. For construction of the splice variant version of the full-length Flag-RIG-I, the duck full length Flag-RIG-I cloned in pCR 2.1-TOPO with added sites NheI-NotI (TOPO-Flag-RIG-I) was used as template for a PCR with primers flanking the second exon of RIG-I facing outwards. The PCR product was treated with T4 polynucleotide kinase (NEB) to generate PCR fragments susceptible to ligation. The TOPO-Flag-RIG-I was digested with NheI-NotI and the fragment containing full length Flag-SVRIG-I was cloned into pcDNA3.1 using NheI-NotI. For induction of full length Flag-RIG-I and Flag-SVRIG-I constructs, 5'ppp-dsRNA and dephosphorylated dsRNA control (InvivoGen) were used.

All CARD mutants used in this work (Q170K, K167R, K193R, K167R/K193R) were made using an adapted protocol for site-directed mutagenesis [47] using TOPO-dCARD-GST as template. Mutations were confirmed by sequencing and the mutant CARD domains were cloned into the pcDNA3.1 (Hygro+)-GST construct using ΔBamHI/BglII-ClaI sites. Constructs were confirmed by complete sequencing. Human GST-CARD and human TRIM25-V5 were kindly provided by Dr. Michaela U. Gack. pcDNA3.1-HAUB (Addgene plasmid 18712) [35] was used to detect ubiquitinated forms of RIG-I. pRK5-HA-Ubiquitin-WT (Addgene plasmid 17608), pRK5-HA-Ubiquitin-K0 (Addgene plasmid 17603), pRK5-HA-Ubiquitin-K63 (Addgene plasmid 17606) [39] were used to study the nature of the ubiquitinated forms of RIG-I.

Immunoprecipitation, GST Pulldown, and Immunoblotting

Flag® Immunoprecipitation kit (Sigma-Aldrich) was used for immunoprecipitation experiments. GST pulldown was performed as previously described [17]. Briefly, DF1 cells transfected with DNA constructs (one confluent 6-well plate per sample) were lysed in 1200 µl of lysis buffer (50 mM TRIS pH 7.2, 150 mM NaCl, 1% [vol/vol] Triton X-100, protease inhibitor cocktail [Roche]), followed by centrifugation at 13,000 rpm for 5 min. For GST-pulldown assays, supernatants were mixed with 50 µl of glutathione Sepharose 4B resin (GE Healthcare) equilibrated with lysis buffer, and the binding reaction mix was incubated for 3 to 4 h at 4°C. The GST-pulldown was washed three times with ice-cold lysis buffer and eluted with 25 µl 4X Laemmli buffer, followed by boiling for 10 min. For immunoblotting, proteins were separated by SDS–polyacrylamide gel electrophoresis (SDS–PAGE) and transferred to a nitrocellulose membrane. Immunodetection was achieved with anti-V5 (1:5,000) (Invitrogen), anti-Flag (1:2,000) (Sigma), anti-HA (1:1000) (Sigma-Aldrich) or anti-GST (1:1000) (Dana-Farber Monoclonal Core Facility (DG-122) antibodies and proteins were visualized by chemiluminescence using the ECL kit (GE- Healthcare).

Q PCR and Luciferase Assay

Cells (2×10^5) were seeded overnight in 24-well plates. To quantify gene expression from transfected chicken DF1 cells, primers and probes specific for *Mx1*, *IFIT5* and *OASL* were used for Q-PCR as previously described [34]. Reverse transcription PCR for detection of RIG-I splice variant was carried out on cDNA samples previously prepared [28] from lung tissues of ducks infected with A/British Columbia 500/2005 (H5N2) and A/Vietnam 1203/2004 (H5N1), using forward (SVRIG-I.fw 5'-GAG CCT CAA CCC GGT CTA C-3') and reverse primer (SVRIG-I.rev 5'-GGT AGC TCC GAG CCT TCT TT-3') flanking duck RIG-I exon 2. Luciferase activity was measured using the Dual-Luciferase Reporter Assay System (Promega) 24 h after transfection with the chicken IFN-β promoter luciferase reporter plasmid (pGL3-chIFNβ) derived from the chicken *IFN2* gene as previously described [28,33,48]. Briefly, DF1 cells were transfected with fixed amounts of pGL3-chIFNβ (150 ng), the synthetic Renilla luciferase reporter construct phRTK as internal control (10 ng) and GST-d2CARD or K167R/K193R mutant (5 ng). Cells were also transfected with increasing amounts of duck TRIM25-V5 construct (between 25 ng and 150 ng) and variable amounts of pcDNA3.1 (Hygro+) (Invitrogen) to normalize the amount of transfected DNA.

Mass Spectrometry

The GST resin purified protein was separated by SDS-PAGE, and the bands corresponding to the ubiquitinated forms of duck RIG-I GST-CARD domains fusion protein were in-gel digested with trypsin (Promega) using the adapted protocol of Shevchenko et al. [49,50]. Samples were analyzed using a hybrid quadrupole orthogonal acceleration time-of-flight mass spectrometer (Waters, UK) equipped with a nanoACQUITY Ultra Performance liquid chromatography system (Waters, Milford, MA) as previously described [50]. Briefly, 2 µl of the peptide solution was injected into a VanGuard micro precolumn C18 cartridge connected to a 75 µm i.d.×150 µm Atlantis dC18 column (Waters, Milford, MA). Solvent A was 0.1% formic acid in water, and solvent B was 0.1% formic acid in acetonitrile. After 1 min trap wash in the precolumn with solvent A at flow rate of 10 µl/min, peptides were separated using solvent gradient and electrosprayed to the mass spectrometer at a flow rate of 350 nl/min. The collision energy used to perform MS/MS depended on the mass and charge state of the eluting peptides. The instrument was calibrated every 1 min with GFP and LecErK using the LockSpray. MassLynx (Waters MassLynx V4.1) was used for the data acquisition and analysis.

Supporting Information

Figure S1   The RIG-I splice variant is not a dominant inhibitor of RIG-I. A. The SVCARD is not active, and is not acting as a dominant inhibitor of RIG-I. The dual luciferase assay was performed using the chIFN-β promoter and increasing amounts of SVCARD plasmid (25 ng to 150 ng) with a fixed amount of GST-d2CARD plasmid (150 ng). Data are mean ± SD (n = 5). GSTdCARD activates the chIFN-β promoter compared

with the GST control (P<0.001). B. SVRIG-I does not act as a dominant inhibitor of duck RIG-I. Luciferase assay was performed using the chIFN-β promoter and increasing amounts of Flag-SVRIG-I plasmid (25 ng to 150 ng) with fixed amount Flag-SVRIG-I plasmid (150 ng). No significant decrease of the activation of the chIFN-β promoter was observed when increasing amounts of Flag-SVRIG-I plasmid were added. Data are mean ± SD (n = 3). All samples show activation of RIG-I by ligand compared to the pcDNA3.1+ control RNA sample (P<0.05).

Figure S2   Alignment of RIG-I CARD domains from vertebrate species. Amino acid alignment of selected sequences available in Genbank, including human (AAI07732), squirrel monkey (XP_003939778), horse (XP_001497895), wild boar (NP_998969), cat (XP_003995589), ferret (XP_004765417), Brandt's bat (EPQ03535), black flying fox (AEW46678), rabbit (XP_002708086), mouse (BAC37205), rat (XP_216380, duck (ACA61272), goose (AEG75816), zebrafinch (XP_002194560), green turtle (EMP30788), zebrafish (XP_002666571). Asterisks indicate ubiquitinated residues of human RIG-I CARD domains in the presence of active TRIM25. The plus symbol indicates the D122 residue and the circumflex indicates the regulatory phosphorylation sites.

Acknowledgments

We would like to acknowledge the technical assistance of Ximena Fleming-Canepa. For mass spectrometry help we thank Jing Zheng, Fatima Garcia-Quintanilla, Matias Musumeci and Jeremy A. Iwashkiw. We also thank Gustavo Fuertes-Vives for help with the 3D visualization of duck RIG-I. We thank Jerry Aldridge and Patrick Seiler for help with animal work in BL3 in the laboratory of Dr. Robert Webster as part of a previous study. We also thank two anonymous reviewers for excellent experimental suggestions.

Author Contributions

Conceived and designed the experiments: DMN KEM. Performed the experiments: DMN KEM. Analyzed the data: DMN KEM. Wrote the paper: DMN KEM.

References

1. Schlee M, Hartmann E, Coch C, Wimmenauer V, Janke M, et al. (2009) Approaching the RNA ligand for RIG-I? Immunol Rev 227: 66–74.
2. Civril F, Bennett M, Moldt M, Deimling T, Witte G, et al. (2011) The RIG-I ATPase domain structure reveals insights into ATP-dependent antiviral signalling. EMBO Rep 12: 1127–1134.
3. Jiang FG, Ramanathan A, Miller MT, Tang GQ, Gale M, et al. (2011) Structural basis of RNA recognition and activation by innate immune receptor RIG-I. Nature 479: 423–U184.
4. Kowalinski E, Lunardi T, McCarthy AA, Louber J, Brunel J, et al. (2011) Structural Basis for the Activation of Innate Immune Pattern-Recognition Receptor RIG-I by Viral RNA. Cell 147: 423–435.
5. Luo D, Ding SC, Vela A, Kohlway A, Lindenbach BD, et al. (2011) Structural insights into RNA recognition by RIG-I. Cell 147: 409–422.
6. Kolakofsky D, Kowalinski E, Cusack S (2012) A structure-based model of RIG-I activation. RNA 18: 2118–2127.
7. O'Neill LA, Bowie AG (2011) The powerstroke and camshaft of the RIG-I antiviral RNA detection machine. Cell 147: 259–261.
8. Kawai T, Takahashi K, Sato S, Coban C, Kumar H, et al. (2005) IPS-1, an adaptor triggering RIG-I- and Mda5-mediated type I interferon induction. Nat Immunol 6: 981–988.
9. Meylan E, Curran J, Hofmann K, Moradpour D, Binder M, et al. (2005) Cardif is an adaptor protein in the RIG-I antiviral pathway and is targeted by hepatitis C virus. Nature 437: 1167–1172.
10. Xu LG, Wang YY, Han KJ, Li LY, Zhai Z, et al. (2005) VISA is an adapter protein required for virus-triggered IFN-beta signaling. Mol Cell 19: 727–740.
11. Hou F, Sun L, Zheng H, Skaug B, Jiang QX, et al. (2011) MAVS forms functional prion-like aggregates to activate and propagate antiviral innate immune response. Cell 146: 448–461.
12. Takamatsu S, Onoguchi K, Onomoto K, Narita R, Takahasi K, et al. (2013) Functional characterization of domains of IPS-1 using an inducible oligomerization system. PLoS One 8: e53578.
13. Wang C, Liu X, Wei B (2011) Mitochondrion: an emerging platform critical for host antiviral signaling. Expert Opin Ther Targets 15: 647–665.
14. Saito T, Hirai R, Loo YM, Owen D, Johnson CL, et al. (2007) Regulation of innate antiviral defenses through a shared repressor domain in RIG-I and LGP2. Proc Natl Acad Sci U S A 104: 582–587.
15. Maharaj NP, Wies E, Stoll A, Gack MU (2012) Conventional protein kinase C-alpha (PKC-alpha) and PKC-beta negatively regulate RIG-I antiviral signal transduction. J Virol 86: 1358–1371.
16. Wies E, Wang MK, Maharaj NP, Chen K, Zhou S, et al. (2013) Dephosphorylation of the RNA Sensors RIG-I and MDA5 by the Phosphatase PP1 Is Essential for Innate Immune Signaling. Immunity 38: 437–449.
17. Gack MU, Shin YC, Joo CH, Urano T, Liang C, et al. (2007) TRIM25 RING-finger E3 ubiquitin ligase is essential for RIG-I-mediated antiviral activity. Nature 446: 916–920.
18. Zeng W, Sun L, Jiang X, Chen X, Hou F, et al. (2010) Reconstitution of the RIG-I pathway reveals a signaling role of unanchored polyubiquitin chains in innate immunity. Cell 141: 315–330.
19. Jiang X, Kinch LN, Brautigam CA, Chen X, Du F, et al. (2012) Ubiquitin-induced oligomerization of the RNA sensors RIG-I and MDA5 activates antiviral innate immune response. Immunity 36: 959–973.

20. Oshiumi H, Matsumoto M, Hatakeyama S, Seya T (2009) Riplet/RNF135, a RING finger protein, ubiquitinates RIG-I to promote interferon-beta induction during the early phase of viral infection. J Biol Chem 284: 807–817.

21. Oshiumi H, Miyashita M, Inoue N, Okabe M, Matsumoto M, et al. (2010) The ubiquitin ligase Riplet is essential for RIG-I-dependent innate immune responses to RNA virus infection. Cell Host Microbe 8: 496–509.

22. Gack MU, Albrecht RA, Urano T, Inn KS, Huang IC, et al. (2009) Influenza A virus NS1 targets the ubiquitin ligase TRIM25 to evade recognition by the host viral RNA sensor RIG-I. Cell Host Microbe 5: 439–449.

23. Rajsbaum R, Albrecht RA, Wang MK, Maharaj NP, Versteeg GA, et al. (2012) Species-specific inhibition of RIG-I ubiquitination and IFN induction by the influenza A virus NS1 protein. PLoS Pathog 8: e1003059.

24. Webster RG, Bean WJ, Gorman OT, Chambers TM, Kawaoka Y (1992) Evolution and ecology of influenza A viruses. Microbiol Rev 56: 152–179.

25. Spackman E (2008) A brief introduction to the avian influenza virus. Methods Mol Biol 436: 1–6.

26. Kim JK, Negovetich NJ, Forrest HL, Webster RG (2009) Ducks: the "Trojan horses" of H5N1 influenza. Influenza Other Respi Viruses 3: 121–128.

27. World Health Organization (2013) Cumulative number of confirmed human cases of avian influenza A/(H5N1) reported to the World Health Organization. Available: http://www.who.int/influenza/human_animal_interface/EN_GIP_20131008CumulativeNumberH5N1cases.pdf. Accessed 2013 Oct 8.

28. Barber MR, Aldridge JR Jr, Webster RG, Magor KE (2010) Association of RIG-I with innate immunity of ducks to influenza. Proc Natl Acad Sci U S A 107: 5913–5918.

29. Vanderven HA, Petkau K, Ryan-Jean KE, Aldridge JR Jr, Webster RG, et al. (2012) Avian influenza rapidly induces antiviral genes in duck lung and intestine. Mol Immunol 51: 316–324.

30. Nistal-Villan E, Gack MU, Martinez-Delgado G, Maharaj NP, Inn KS, et al. (2010) Negative role of RIG-I serine 8 phosphorylation in the regulation of interferon-beta production. J Biol Chem 285: 20252–20261.

31. Kowalinski E, Lunardi T, McCarthy AA, Louber J, Brunel J, et al. (2011) Structural basis for the activation of innate immune pattern-recognition receptor RIG-I by viral RNA. Cell 147: 423–435.

32. Magor KE, Miranzo Navarro D, Barber MR, Petkau K, Fleming-Canepa X, et al. (2013) Defense genes missing from the flight division. Dev Comp Immunol. 41: 377–88.

33. Childs K, Stock N, Ross C, Andrejeva J, Hilton L, et al. (2007) mda-5, but not RIG-I, is a common target for paramyxovirus V proteins. Virology 359: 190–200.

34. Barber MR, Aldridge JR Jr, Fleming-Canepa X, Wang YD, Webster RG, et al. (2013) Identification of avian RIG-I responsive genes during influenza infection. Mol Immunol 54: 89–97.

35. Kamitani T, Kito K, Nguyen HP, Yeh ET (1997) Characterization of NEDD8, a developmentally down-regulated ubiquitin-like protein. J Biol Chem 272: 28557–28562.

36. Versteeg GA, Rajsbaum R, Sanchez-Aparicio MT, Maestre AM, Valdiviezo J, et al. (2013) The E3-ligase TRIM family of proteins regulates signaling pathways triggered by innate immune pattern-recognition receptors. Immunity 38: 384–398.

37. Gack MU, Kirchhofer A, Shin YC, Inn KS, Liang C, et al. (2008) Roles of RIG-I N-terminal tandem CARD and splice variant in TRIM25-mediated antiviral signal transduction. Proc Natl Acad Sci U S A 105: 16743–16748.

38. Gao D, Yang YK, Wang RP, Zhou X, Diao FC, et al. (2009) REUL is a novel E3 ubiquitin ligase and stimulator of retinoic-acid-inducible gene-I. PLoS One 4: e5760.

39. Lim KL, Chew KC, Tan JM, Wang C, Chung KK, et al. (2005) Parkin mediates nonclassical, proteasomal-independent ubiquitination of synphilin-1: implications for Lewy body formation. J Neurosci 25: 2002–2009.

40. Markson G, Kiel C, Hyde R, Brown S, Charalabous P, et al. (2009) Analysis of the human E2 ubiquitin conjugating enzyme protein interaction network. Genome Res 19: 1905–1911.

41. Windheim M, Peggie M, Cohen P (2008) Two different classes of E2 ubiquitin-conjugating enzymes are required for the mono-ubiquitination of proteins and elongation by polyubiquitin chains with a specific topology. Biochem J 409: 723–729.

42. Feng M, Ding Z, Xu L, Kong L, Wang W, et al. (2013) Structural and biochemical studies of RIG-I antiviral signaling. Protein Cell 4: 142–154.

43. Marblestone JG, Larocque JP, Mattern MR, Leach CA (2012) Analysis of ubiquitin E3 ligase activity using selective polyubiquitin binding proteins. Biochim Biophys Acta 1823: 2094–2097.

44. Streich FC Jr, Ronchi VP, Connick JP, Haas AL (2013) Tripartite Motif Ligases Catalyze Polyubiquitin Chain Formation through a Cooperative Allosteric Mechanism. J Biol Chem 288: 8209–8221.

45. Maelfait J, Beyaert R (2012) Emerging role of ubiquitination in antiviral RIG-I signaling. Microbiol Mol Biol Rev 76: 33–45.

46. Schaefer-Klein J, Givol I, Barsov EV, Whitcomb JM, VanBrocklin M, et al. (1998) The EV-O-derived cell line DF-1 supports the efficient replication of avian leukosis-sarcoma viruses and vectors. Virology 248: 305–311.

47. Zheng L, Baumann U, Reymond JL (2004) An efficient one-step site-directed and site-saturation mutagenesis protocol. Nucleic Acids Res 32: e115.

48. Sick C, Schultz U, Munster U, Meier J, Kaspers B, et al. (1998) Promoter structures and differential responses to viral and nonviral inducers of chicken type I interferon genes. J Biol Chem 273: 9749–9754.

49. Shevchenko A, Jensen ON, Podtelejnikov AV, Sagliocco F, Wilm M, et al. (1996) Linking genome and proteome by mass spectrometry: large-scale identification of yeast proteins from two dimensional gels. Proc Natl Acad Sci U S A 93: 14440–14445.

50. Iwashkiw JA, Fentabil MA, Faridmoayer A, Mills DC, Peppler M, et al. (2012) Exploiting the Campylobacter jejuni protein glycosylation system for glycoengineering vaccines and diagnostic tools directed against brucellosis. Microb Cell Fact 11: 13.

7

A Mixture of Cod and Scallop Protein Reduces Adiposity and Improves Glucose Tolerance in High-Fat Fed Male C57BL/6J Mice

Hanne Sørup Tastesen[1,2], Alexander Krokedal Rønnevik[1,2], Kamil Borkowski[1], Lise Madsen[1,2], Karsten Kristiansen[1]*, Bjørn Liaset[2]*

1 Department of Biology, University of Copenhagen, Copenhagen, Denmark, 2 National Institute of Nutrition and Seafood Research, Bergen, Norway

Abstract

Low-protein and high-protein diets regulate energy metabolism in animals and humans. To evaluate whether different dietary protein sources modulate energy balance when ingested at average levels obesity-prone male C57BL/6J mice were pair-fed high-fat diets (67 energy percent fat, 18 energy percent sucrose and 15 energy percent protein) with either casein, chicken filet or a mixture of cod and scallop (1:1 on amino acid content) as protein sources. At equal energy intake, casein and cod/scallop fed mice had lower feed efficiency than chicken fed mice, which translated into reduced adipose tissue masses after seven weeks of feeding. Chicken fed mice had elevated hepatic triglyceride relative to casein and cod/scallop fed mice and elevated 4 h fasted plasma cholesterol concentrations compared to low-fat and casein fed mice. In casein fed mice the reduced adiposity was likely related to the observed three percent lower apparent fat digestibility compared to low-fat, chicken and cod/scallop fed mice. After six weeks of feeding an oral glucose tolerance test revealed that despite their lean phenotype, casein fed mice had reduced glucose tolerance compared to low-fat, chicken and cod/scallop fed mice. In a separate set of mice, effects on metabolism were evaluated by indirect calorimetry before onset of diet-induced obesity. Spontaneous locomotor activity decreased in casein and chicken fed mice when shifting from low-fat to high-fat diets, but cod/scallop feeding tended ($P=0.06$) to attenuate this decrease. Moreover, at this shift, energy expenditure decreased in all groups, but was decreased to a greater extent in casein fed than in cod/scallop fed mice, indicating that protein sources regulated energy expenditure differently. In conclusion, protein from different sources modulates energy balance in C57BL/6J mice when given at normal levels. Ingestion of a cod/scallop-mixture prevented diet-induced obesity compared to intake of chicken filet and preserved glucose tolerance compared to casein intake.

Editor: Zane Andrews, Monash University, Australia

Funding: This study was part of the Lean Seafood in the Prevention of the Metabolic Syndrome project which is financially supported by the Norwegian Research Council (Grant no. 200515/I30) and the National Institute of Nutrition and Seafood Research. Part of this work was also financially supported by the Danish Council for Strategic Research (Grant no. 2101-08-0053), the Danish Dairy Research Foundation, the Danish Natural Science Research Council, the Novo Nordisk Foundation and the Carlsberg Foundation. The funders had no role in study design, data collection and analysis, decision to publish, or preparation of the manuscript.

Competing Interests: The authors have declared that no competing interests exist.

* Email: kk@bio.ku.dk (KK); Bjorn.Liaset@nifes.no (BL)

Introduction

Identifying nutritional strategies to alleviate the obesity pandemic are of great interest. Diet-induced thermogenesis, i.e. the regulated liberation of energy in the form of heat [1], could lower food efficiency, and thereby diminish obesity development. Already in 1939, induction of adaptive thermogenesis by feeding rats very low (4–8 weight percent) or very high (54 weight percent) protein diets was described [2]. Later, the increment in thermogenesis by low-protein (LP) diets was verified in rats [3,4], in baby pigs [5], and similar effects were observed in young human subjects [6]. Thus, intake of LP diets induces thermogenesis, but instead of resulting in decreased body mass, the reduced food efficiency is compensated for by a higher food intake [7].

Whereas LP diets may increase energy intake, high-protein (HP) diets are more satiating than an isoenergetic amount of

carbohydrates or fat [8,9]. Moreover, HP intake induces higher postprandial thermogenesis than high-carbohydrate ingestion does [10,11]. It is likely that both reduced energy intake and elevated thermogenesis might be underlying mechanisms explaining, at least in part, the reduction in body mass observed in mice [12–15] and humans [9,16,17] by replacing carbohydrates with protein.

Despite the known effects of LP and HP diets on thermogenesis, limited information exists on whether varying protein sources affect body mass and composition differently [18]. From studies in rodents, we know that consumption of hydrolyzed rather than intact proteins reduces body mass gain, adipose tissue mass and hepatic and plasma lipid concentrations [19–21]. Moreover, whey ingestion decreased fat mass relative to casein intake in mice [22–24], and an intervention study with free-living overweight and obese subjects indicated that intake of whey protein, but not soy protein (both approximately 56 g protein/day for 23 weeks)

Table 1. Composition of the experimental diets.

	LF[a]	Casein[b]	Chicken[c]	Cod/scallop[d]
Composition (g/kg)				
Casein	190	215	–	–
Chicken	–	–	240	–
Cod	–	–	–	114
Scallop	–	–	–	133
L-Cystine	2.8	3	3	3
KCl	–	10.2	5.4	–
Corn starch	299	–	–	–
Maltodextrin 10	33	–	–	–
Sucrose	332	242	223	221
Lard	19	198	198	198
Vegetable oil[e]	23.7	198	198	198
Cellulose	47.4	50	50	50
AIN-76 mineral mix	9.5[f]	67	67	67
AIN-76 vitamin mix	9.5[g]	14	14	14
Choline bitartrate	1.9	2	2	2
Butylated hydroxytoluene	–	0.4	0.4	0.4
DiCalcium Phosphate	12.3	–	–	–
Calcium Carbonate	5.2	–	–	–
Potassium Citrate, 1 H$_2$O	15.6	–	–	–
Analyzed (g/kg)				
Crude protein[h]	170	190	190	200
Ash	31	48	60	75
Fat	44	390	400	390
Gross energy kJ/g	17.4	26.2	26.0	25.5

[a] OpenSource diet no. D12450B (Research Diets, Inc. NJ, USA).
[b] Casein (cat. no. C8654, lot BCBC 3986, Sigma-Aldrich, MO, USA).
[c] Chicken breast fillets (Kyllingfilet naturell, Ytterøykylling AS, Norway).
[d] Cod fillets (Wildcaught in the Northeastern Atlantic) and Canadian scallops, (Wild North Atlantic scallops, 20–30 ct, Placopecten magellanicus, Clearwater Seafoods Limited, NS, Canada).
[e] LF: soybean oil. Casein, chicken and cod/scallop: corn oil.
[f] Mineral Mix S10026.
[g] Vitamin Mix V100001.
[h] Crude protein, N $^\times$ 6.15 for casein; N $^\times$ 5.6 for chicken filet and cod/scallop.

resulted in a significant reduction in body mass, fat mass and waist circumference, relative to the carbohydrate (maltodextrin) control treatment [25]. In a randomized, double-blinded intervention study with cross-over design, ingestion of a liquid test meal consisting of 50% whey protein, 40% carbohydrate and 10% fat, induced a higher postprandial thermic effect than equal amounts of casein and soy protein [26]. Thus, studies in both rodents and humans indicate that different protein sources might affect body weight gain and adiposity differently.

The average (hereafter referred to as normal) protein intake in humans has been estimated to be 15–16 energy percent (E%) both in the US [27] and in the UK [28]. We have recently shown that obesity-prone C57BL/6J mice exhibited distinct metabolic responses to intake of various dietary protein sources, given as 15 E% protein in a high-fat (HF) background-diet [29]. Mice fed scallop muscle as the sole protein source were protected against diet-induced obesity, enlarged liver mass and hyperlipidemia as compared to mice fed chicken or cod filets. However, the scallop fed mice also had lower *ad libitum* feed intake, suggesting different satiating effects of the protein sources [29]. Therefore, the present study was undertaken in order to elucidate whether the protein sources casein (a commonly used standard reference protein), chicken breast filet or a mixture of cod filet and scallop muscle, would affect diet-induced obesity during equal energy intake (pair-feeding) in HF diets fed to male C57BL/6J mice for seven weeks. Furthermore, to evaluate instant differences in metabolism independent of the development of obesity indirect calorimetry was performed during the first 72 h of feeding on the HF diets containing protein from different sources. Protein from different sources at normal level was found to modulate energy balance in C57BL/6J mice, and consumption of a cod/scallop-mixture prevented HF diet-induced development of obesity compared to chicken and preserved glucose tolerance compared to casein.

Materials and Methods

Ethics Statement

The animal experiments were approved by the Norwegian National Animal Health Authorities (Experiment 1 (Expt. 1) performed in Norway, permit number 3421) and the Danish

National Animal Experiments Inspectorate (Experiment 2 (Expt. 2) performed in Denmark, permit number 2012-15-2934), and care and handling were carried out in strict compliance with the ethical standards of the 1964 Helsinki Declaration, as revised in 1983. No adverse events were observed. To ameliorate suffering the mice were anaesthetized by inhalation of isoflurane (4%, Isoba Vet) before being euthanized by exsanguination by cardiac puncture.

Experimental diets

Low-fat (LF) diet (10 E% fat, 70 E% carbohydrate and 20 E% protein, OpenSource Diet no. D12450B, Research Diets, NJ, USA) was used to feed mice during acclimatization periods and as a reference diet. Three isoenergetic experimental HF diets (67 E% fat, 18 E% carbohydrate (sucrose) and 15 E% protein) were made with protein from different sources; casein sodium salt from bovine milk (casein, Sigma C-8456) chicken breast filets (chicken) and a mixture of wild caught cod filets and Canadian scallop muscles (cod/scallop, 1:1 on amino acid content) as previously described [29,30] with the modification that three g cystine/kg diet were added to all diets in the present study. Raw chicken breast filets, raw cod filets and raw scallop muscle were minced, freeze dried and powdered before being added into the diets and apart from the supplemented cystine, the protein sources casein, chicken filet or cod filet and scallop represented the sole amino acid source in the respective diets. The final compositions of the diets are shown in Table 1 and Table 2.

Animal studies

Male C57BL/6JBomTac mice (Taconic, Ejby, Denmark) 8 to 9 weeks old and weighing approximately 25 g at arrival were housed individually at thermoneutrality ($28 \pm 1°$) under a 12 h light-dark cycle. The mice were fed LF diet during acclimatization to the animal facility. Two experiments were carried out as follows; Expt. 1 encompassed 32 mice ($n = 8$/group) which were assigned into four experimental groups by bodyweight after five days acclimatization and fed either LF diet or HF diets (Table 1 and 2) for seven weeks. At week six the mice were subjected to an oral glucose tolerance test (O-GTT). After seven weeks the mice were terminated and the following tissues were dissected out, weighed and frozen at $-80°C$; liver, kidneys, heart, skeletal muscle soleus, epididymal white adipose tissue (eWAT), perirenal/retroperitoneal white adipose tissue (p/rWAT), inguinal white adipose tissue (iWAT), and interscapular brown adipose tissue (iBAT). Feed efficiency was calculated as body mass gain per energy intake (g BM/MJoule), based on the data from week six, i.e., prior to the O-GTT. Expt. 2 encompassed 30 mice ($n = 10$/group). After seven days acclimatization on LF diet the mice were placed in indirect calorimetry cages for 72 hours of baseline indirect calorimetry and activity measurements while still on LF diet. Based on body weight and baseline measurements of total activity and RER in light and dark phases the mice were divided into three groups and fed the experimental HF diets for another 72 hours of measurements and subsequently terminated. The mice were in both experiments anaesthetized by inhalation of isoflurane (4%, Isoba Vet) and euthanized by exsanguination by cardiac puncture after a 4 h fast. The blood was heparinized (20.2 units sodium heparin/mL blood), centrifuged ($4°C$, 2500 g, 5 min) and plasma fractions were stored at $-80°C$ until analysis. To ensure comparability between Expt. 1 and Expt. 2. handling and care of the mice were standardized between the two experiments and mice from the same supplier and feed from the same batch were used for both experiments.

Pair-feeding

LF fed mice had free access to feed, whereas HF fed mice were pair-fed for the duration of both Expt. 1 and Expt. 2 to obtain equal energy intake. In Expt. 1 leftover diet was collected from every cage three times per week and intake calculated accordingly. On a group basis casein fed mice had slightly lower feed intake than chicken and cod/scallop fed mice and as a consequence hereof the mice in the chicken and cod/scallop fed groups were mildly feed restricted compared to the casein fed mice to obtain similar energy intakes between groups. The average daily food intake in grams for Expt. 1 was 3.01 g/d in LF fed mice, 2.06 g/d in casein fed mice, 2.20 g/d in chicken fed mice and 2.31 g/d in cod/scallop fed mice. The daily energy intake was 52.4 kJ/d in LF fed mice, 53.9 kJ/d in casein fed mice, 57.2 kJ/d (+6.0% vs. casein) in chicken fed mice and 58.8 kJ/d (+9.1% vs. casein) in cod/scallop fed mice. The pair-feeding during Expt. 2 was essentially done in the same way, however leftovers were collected and the mice fed every day to keep the energy intake similar between groups during this shorter study.

Diet composition analyses

Diets were analyzed as previously described [29]. In short; energy contents were determined by bomb calorimetry (Parr Instruments, Moline, IL, USA). For total amino acid analysis norvaline was added to samples as internal standard, samples were hydrolyzed (6 M HCl, $110 \pm 2°C$, 22 h) and derivatized (AccQ-Tag Ultra Derivatization Kit, Waters, MA, USA). Amino acids were separated and detected on the ACQUITY UPLC System (Waters, MA, USA), identified using Pierce Amino Acid Standard H (Thermo Fisher Scientific Inc., IL, USA) to which norvaline, taurine and hydroxy-proline were added and finally quantified by internal and external standard regression. For tryptophan analysis the samples were hydrolyzed ($Ba(OH)_2$, $110 \pm 2°C$, 20 h), pH adjusted to 6.2, separated by HPLC (Shimadzu 6A/6B) equipped with a SUPELCOSILTM LC-18 HPLC-column, detected in UV-spectrophotometer (Shimadzu SPD 6A) at 280 nm and quantified using a standard curve of L-Tryptophan (T-0254, Sigma-Aldrich). Total cysteine was determined, at the Norwegian Institute of Food, Fishery and Aquaculture, after oxidation of cysteine/cystine (9:1) performic acid (88%): H_2O_2 (30%) (v/v) to yield cysteic acid.

Feces collection

After six weeks of feeding (Expt. 1) the mice were placed in cages with the standard wood chip layer replaced by paper lining for the purpose of collecting feces for one week. Feces left behind in cages were collected, weighted and frozen at $-80°C$ until analyses for nitrogen and total fat content. Based on feces measurements and diet-intake data apparent digestibility of fat and nitrogen was calculated as follows: $100 \times$ (intake (mg) - fecal output (mg))/(intake (mg)).

Nitrogen and fat content in diets and feces

Nitrogen content was determined by the Dumas method using Leco FP-528 nitrogen analyzer (Leco Corp, MI, USA). The crude protein content in the diets was calculated as nitrogen content multiplied by 6.15 for casein and 5.6 for chicken filet and cod/scallop [31]. Total fat content was determined gravimetrically after extraction with organic solvents before and after acidic hydrolysis as described previously [29].

Plasma measurements

MaxMat PL II analyzer (MAXMAT S.A., Montpellier, France) and conventional kits were used to measure 4 h fasted plasma

Table 2. Amino acid composition of the experimental diets.

mmol/kg	LF	Casein	Chicken	Cod/scallop
Ala	60	68	138	119
Arg	31	36	66	73
Asx[a]	100	109	154	149
Cys	28	36	48	42
Glx[b]	274	307	212	191
Gly	40	47	109	192
His*	29	34	43	21
Ile*	66	77	75	61
Leu*	120	142	126	107
Lys*	99	112	136	116
Met*	29	34	36	34
Phe*	51	59	48	41
Pro	153	185	61	47
Ser	94	107	78	73
Thr*	63	70	78	64
Trp	9	11	11	8
Tyr	36	44	28	24
Val*	94	111	88	69
Hyp	<0.1	<0.1	3	3
Tau	<0.1	<0.1	1	61
EAA[c]	551	640	630	512
BCAA[d]	280	330	290	237
Total AA[e]	1376	1590	1540	1494

* essential amino acids.
[a] Asx: sum of Asp + Asn.
[b] Glx: sum of Glu + Gln.
[c] EAA: sum of essential amino acids.
[d] BCAA: sum of branched-chain amino acids.
[e] Total AA: total sum of amino acids.

lactate (Sentinel Diagnostics, Italy), triglyceride (TG), total cholesterol, LDL cholesterol, glucose and alanine aminotransferase (MaxMat, France), hydroxyl-butyrate (OH-butyrate) and glycerol (Randox, UK), free fatty acids (FFA), HDL cholesterol and total bile acids (Dialab, Austria) concentrations. 4 h fasted plasma insulin concentrations were analyzed using DRG mouse insulin ELISA kit (DRG Diagnostics, Germany).

Liver lipid analysis

Total liver lipids were extracted with chloroform:methanol (2:1, v:v). Lipid classes were analyzed via automated Camaq HPTLC system and separated on HPTLC silica gel 60 F plates as previously described [32].

qRT-PCR

Gene expression analysis was performed as described previously [21]. In short, total RNA was isolated from tissue samples with TRIzol Reagent (Invitrogen, Thermo Fisher Scientific, Carlsbad, CA, USA). Qualities and concentrations of the purified RNA were assessed using NanoDrop ND-1000 UV-Vis spectrophotometer (NanoDrop Technologies, Wilmington, DE, USA). Using GeneAmp PCR 9700 (Applied Biosystems, Thermo Fisher Scientific, Carlsbad, CA, USA), TaqMan RT buffer, dNTP, oligo(dT)primers, RNase inhibitor, Multiscribe Reverse Transcriptase (N808-0234, Applied Biosystems) and RNase-free water RT reactions were performed for 60 min at 48°C. The produced cDNA was subject to qRT-PCR in LightCycler 480 Real-Time PCR System (Roche Applied Sciences, Indianapolis, IN, USA) using SYBR Green Master Mix (LightCycler 480 SYBR Green master mix kit, Roche Applied Sciences) and gene-specific primers (Table S1). Data were analyzed as a ratio between gene of interest and a reference gene, TATA box binding protein (*Tbp*), and normalized to the mean of the LF samples.

Oral glucose tolerance test

After six weeks on experimental diets mice were subjected to 6 h fasted O-GTT. Early in the morning of the test day mice were placed in cages without feed and after six hours fasted blood glucose was measured in whole blood, taken from the tail vein by a Bayer Contour glucometer and glucose test strips (Bayer, Germany). Glucose was administered by oral gavage (2 mg glucose/g body mass) and blood glucose concentration was measured 15, 30, 60 and 120 minutes after glucose administration. Blood glucose incremental area under the curve (iAUC, mmol/L/h) was calculated as AUC above baseline value, i.e., 6 h fasted blood glucose, by applying the trapezoid rule to a plot of group mean blood glucose concentration versus time of measurements [33,34].

HOMA-IR and QUICKI

Based on 4 h fasted plasma glucose and insulin Homeostatic Model Assessment of Insulin Resistance (HOMA-IR) was calculated as follows: Glucose (mmol/l)×insulin (μU/ml)/22.5 [35] and Quantitative Insulin Sensitivity Check Index (QUICKI) was calculated as follows: 1/(log(insulin [mU/l])+log(glucose [mg/dl])) [36].

Indirect calorimetry and spontaneous locomotor activity

VO_2 and VCO_2 was measured in open-circuit indirect calorimetry cages as described previously [21]. In short, the mice were housed in CaloCages (Phenomaster, TSE Systems), equipped with infrared light-beam frames (ActiMot2). VO_2 and VCO_2 was measured for each cage, i.e., each mouse, for 1.9 min once every 30 min, while light-beam breaks were measured continuously. Measurements were performed for a total of 72 h while all groups were fed the LF diet and consecutively for 72 h on the respective HF diets. For each 72 h period of measurements the first 24 h were regarded as an adaptation period and only the subsequent 48 h were used for analyses; Based on two consecutive light (06.00–17.30 h) and dark (18.00–05.30 h) phases respiratory exchange ratio (RER) was calculated from VO_2 and VCO_2 and spontaneous locomotor activity was defined as total counts of light-beam breaks. Energy expenditure (EE) was calculated as follows; 16.3 kJ/L × L VO_2 +4.6 kJ/L × L VCO_2 [37].

Statistical analyses

The data represent group means ± SEM. After homogeneity of variances was established by Levene's test the data were subjected ANOVA analyses followed by Tukey's pair-wise comparisons and group means were considered statistically different at $P<0.05$. P values represent the overall outcome of the ANOVA and letters denote differences identified by post hoc tests. Repeatedly measured data, i.e., growth, energy intake, O-GTT, RER, activity and EE were analyzed by repeated measurements ANOVA followed by Tukey's post hoc.

Results

Reduced body mass gain, feed efficiency and adiposity in casein and cod/scallop fed mice

Casein fed mice gained significantly less body mass than LF and chicken fed mice, whereas cod/scallop fed mice gained less weight than chicken fed mice during six weeks of feeding (Fig. 1A-B). Energy intake was not significantly different between groups (Fig. 1C) and thus the feed efficiency was lower in casein and cod/scallop fed mice compared to LF and chicken fed mice (Fig. 1D). The dietary fat intake was similar in casein fed (769±29 mg fat/24 h), chicken fed (848±34) and cod/scallop fed (845±37), but significantly lower in LF fed mice (144.2±2.2) than in the three HF fed groups ($P<0.001$). However, significantly more fat was excreted in the feces in the casein fed mice (38.8±4.3 mg, $P<0.001$) than in LF fed (2.67±0.11), chicken fed (20.6±3.6) and cod/scallop fed mice (14.4±1.6). Moreover, chicken and cod/scallop fed mice excreted more fat in the feces than LF fed mice. Thus, apparent fat digestibility was lower in casein fed than in LF, chicken and cod/scallop fed mice (Fig. 1F). The nitrogen intake was lower in the casein fed (61.2±2.3 mg N/24 h, $P<0.001$) than in LF fed (90.7±1.4), chicken fed (72.2±2.9) and cod/scallop fed mice (77.6±3.3). Furthermore, nitrogen intake was higher in LF fed mice compared to all three HF fed groups. The fecal excretion

of nitrogen was lower in casein fed (27.1±2.9) and cod/scallop fed (22.1±1.1) than in LF fed mice (38.0±3.5). Moreover, nitrogen excretion was lower in cod/scallop fed than in chicken fed mice (33.8±2.6). Apparent nitrogen digestibility was thus higher in cod/scallop fed than in casein fed and chicken fed mice (Fig. 1H). Adiposity varied in the different groups after seven weeks of feeding (Fig. 1E); masses of eWAT and p/rWAT were lower in LF, casein and cod/scallop fed mice than in chicken fed mice. iWAT masses were lower in casein and cod/scallop fed mice than in chicken fed mice and lower in casein fed than in LF fed mice. iBAT masses were lower in casein and cod/scallop fed mice than in chicken fed mice. No differences were seen in soleus muscle and heart tissue masses between groups, but liver masses were increased in chicken fed compared to LF and casein fed mice, and kidney masses were increased in cod/scallop fed compared to LF and casein fed mice (Fig. 1G).

Elevated plasma and liver lipids in chicken fed mice

Obesity is associated with dysregulation of plasma lipids and ectopic fat accumulation, and thus, we measured plasma and liver lipids. Plasma metabolites and liver lipids measured in the 4 h fasted state are listed in Table 3. Chicken fed mice had increased plasma total cholesterol compared to LF and casein fed mice. Furthermore, plasma HDL cholesterol and LDL cholesterol levels were increased in chicken fed compared to LF and casein fed mice and increased in cod/scallop fed compared to LF fed mice. Casein fed mice had an increased HDL:total cholesterol ratio compared to LF, chicken and cod/scallop fed mice. Total bile acid levels were increased in LF fed compared to chicken fed mice. TG concentrations were increased in LF fed compared to casein and cod/scallop fed mice, and increased in chicken fed compared to casein fed mice. Glycerol was increased in LF fed compared to casein and cod/scallop fed mice. A tendency towards increased FFA was seen in LF fed mice ($P = 0.079$). No differences were seen in 4 h fasted plasma β-hydroxybutyrate or alanine aminotransferase between the groups (Table 3). Liver TG and total neutral lipid concentrations were higher in chicken fed than in casein and cod/scallop fed mice, while free cholesterol was increased in LF fed compared to casein fed mice and steryl ester was increased in LF fed compared to casein, chicken and cod/scallop fed mice. No differences were seen in hepatic diglyceride levels between groups (Table 3).

Hepatic expression of genes involved in de novo lipogenesis and gluconeogenesis is modulated by dietary protein source

Based on the differences in plasma and liver lipids we analyzed hepatic expression of mRNA encoding genes involved in lipogenesis and gluconeogenesis (Table 3). Expression of mRNA encoding stearoyl-CoA desaturase-1 (Scd-1), an enzyme catalyzing the conversion of SFA to MUFA, important for targeting FFA to either incorporation into lipoproteins (VLDL) or storage as TG in lipid-droplets, was higher in LF fed compared to casein, chicken and cod/scallop fed mice. Expression of mRNA encoding 3-Hydroxy-3-metylglutaryl-CoA reductase (Hmgcr) was increased in cod/scallop fed compared to LF fed mice. A strong tendency towards increased expression of mRNA encoding the lipogenic gene diacylglycerol acyltransferase 1 (Dgat1, $P = 0.054$) was seen in the LF and chicken fed mice compared to casein and cod/scallop fed mice. Expression of mRNA encoding phosphoenol pyruvate carboxykinase-1 (Pck-1), the rate limiting enzyme controlling gluconeogenesis by catalyzing the formation of phosphoenolpyruvate from oxaloacetate, was higher in LF and

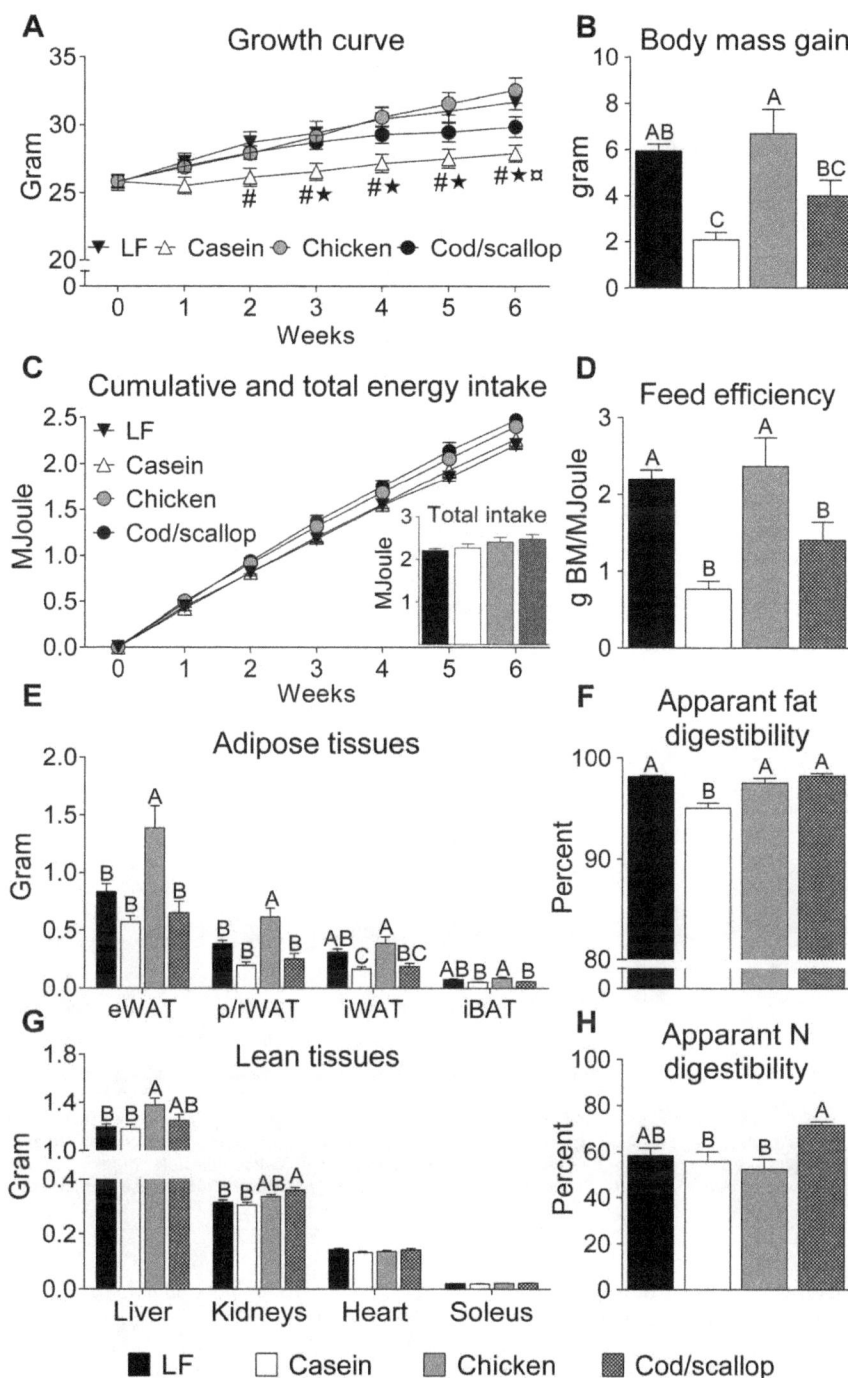

Figure 1. Casein and cod/scallop protein reduces body mass gain and adiposity despite equal energy intake. A: Growth curve during six weeks. B: Body mass gain. C: Cumulative and total energy intake. D: Feed efficiency. E: Adipose tissue masses. F: Apparent fat digestibility. G: Lean tissue masses. H: Apparent nitrogen digestibility. A-H in male C57BL/6J mice fed the experimental diets for six weeks. Data (Expt. 1) represent group means ($n = 7–8$) $\pm$ SEM analyzed by one-way ANOVA followed by Tukey's pair-wise comparisons. Body mass development and cumulative energy intake were analyzed by repeated measurements ANOVA followed by Tukey's post hoc. Means that do not share a letter are significantly different ($P<0.05$). # indicates significantly higher body mass in LF fed than in casein fed mice. * indicates significantly higher body mass in chicken fed than in casein fed mice. ¤ indicates significantly higher body mass in chicken fed than in cod/scallop fed mice.

cod/scallop fed than in casein fed mice. Expression of mRNA encoding hexokinase 2 *(Hk2)* was increased in LF fed compared to casein, chicken and cod/scallop fed mice. Expression of mRNA encoding phosphofructokinase, liver, B-type *(Pfkl)* was increased in LF fed compared to casein fed mice and expression of mRNA

encoding pyruvate kinase, liver and red blood cell *(Pklr)* was increased in LF fed compared to cod/scallop fed mice. No differences in expression of mRNA encoding the genes sterol regulatory element-binding transcription factor 1 *(Srebf1)*, acetyl-CoA carboxylase *(Acaca)*, fatty acid synthase *(Fasn)* and hexoki-

Table 3. 4 h fasted plasma metabolites, liver lipids and liver relative gene expression.

	LF	Casein	Chicken	Cod/scallop	*P* value
Plasma metabolites					
Total cholesterol (mmol/L)	3.24±0.18[b]	3.54±0.18[b]	4.53±0.19[a]	3.88±0.15[ab]	<0.001
HDL cholesterol (mmol/L)	2.68±0.13[c]	3.14±0.13[bc]	3.72±0.17[a]	3.23±0.13[ab]	<0.001
HDL:total cholesterol ratio	0.83±0.01[b]	0.89±0.01[a]	0.82±0.01[b]	0.83±0.01[b]	0.001
LDL cholesterol (mmol/L)	0.87±0.05[c]	0.97±0.08[bc]	1.42±0.08[a]	1.16±0.07[ab]	<0.001
Total bile acids (mmol/L)	3.0±0.29[a]	2.7±0.23[ab]	1.9±0.17[b]	2.4±0.19[ab]	0.009
TG (mmol/L)	0.75±0.07[a]	0.40±0.03[c]	0.65±0.05[ab]	0.50±0.06[bc]	<0.001
FFA (mmol/L)	0.46±0.04	0.32±0.04	0.27±0.03	0.33±0.07	0.079
Glycerol (mmol/L)	0.32±0.02[a]	0.24±0.02[b]	0.26±0.02[ab]	0.25±0.02[b]	0.025
β-hydroxybutyrate (mmol/L)	0.42±0.11	0.34±0.04	0.21±0.08	0.25±0.05	0.21
Alanine aminotransferase (U/L)	24±3	28±9	26±2	50±17	0.25
Liver lipids					
TG (mg/g)	29±7[ab]	26±3[b]	49±9[a]	25±3[b]	0.020
Total neutral lipid (mg/g)	35±7[ab]	30±3[b]	53±9[a]	29±3[b]	0.026
Cholesterol (mg/g)	2.9±0.09[a]	2.4±0.09[b]	2.5±0.06[ab]	2.7±0.15[ab]	0.037
Steryl ester (mg/g)	3.0±0.44[a]	1.6±0.16[b]	1.6±0.19[b]	1.2±0.10[b]	<0.001
Diglycerides (mg/g)	0.2±0.06	0.2±0.03	0.2±0.03	0.1±0.02	0.52
Liver relative mRNA expression					
Srebf1	1±0.28	0.82±0.27	1.22±0.25	1.52±0.34	0.36
Acaca	1±0.26	0.94±0.13	1.11±0.12	1.04±0.13	0.90
Fasn	1±0.39	0.81±0.14	0.76±0.10	0.89±0.15	0.88
Scd-1	1±0.25[a]	0.06±0.01[b]	0.03±0.01[b]	0.02±0.00[b]	<0.001
Dgat-1	1±0.06	0.70±0.06	0.98±0.13	0.73±0.09	0.054
Hmgcr	1±0.15[b]	1.87±0.44[ab]	1.95±0.34[ab]	2.64±0.49[a]	0.044
Pck-1	1±0.12[a]	0.32±0.05[b]	0.53±0.12[ab]	0.95±0.21[a]	0.006
Hk2	1±0.26[a]	0.43±0.12[b]	0.51±0.10[b]	0.48±0.04[b]	0.043
Hk4	1±0.19	0.81±0.06	1.21±0.25	0.91±0.11	0.42
Pfkl	1±0.13[a]	0.41±0.06[b]	0.85±0.15[ab]	0.75±0.10[ab]	0.014
Pklr	1±0.25[a]	0.55±0.08[ab]	0.71±0.13[ab]	0.41±0.05[b]	0.042

Data represent group means (*n*=6–8) ± SEM analyzed by one-way ANOVA followed by Tukey's post hoc. Means that do not share a superscript letter are significantly different (*P*<0.05). Abbreviations: TG, triglycerides; FFA, free fatty acids; *Srebf1*, sterol regulatory element-binding transcription factor 1; *Acaca*, acetyl-coenzyme A carboxylase alpha; *Fasn*, fatty acid synthase; *Scd-1*, stearoyl-CoA desaturase-1; *Dgat-1*, diacylglycerol acyltransferase-1; *Hmgcr*, 3-hydroxy-3-methylglutaryl-coenzyme A reductase; *Pck-1*, phosphoenol pyruvate carboxykinase-1; *Hk2*, hexokinase 2; *Hk4*, hexokinase 4; *Pfkl*, phosphofructokinase, liver, B-type; *Pklr*, pyruvate kinase liver and red blood cell.

nase 4 (*Hk4*) were observed, but expression of mRNA encoding *Srebf1* tended to be higher in cod/scallop fed than in casein fed mice (*P* = 0.067).

Decreased glucose tolerance in casein fed and increased insulin resistance-score in chicken fed mice

As obesity, visceral adiposity, and hepatic steatosis have been shown to associate with impaired glucose and insulin homeostasis, we subjected the mice to 6 h fasted O-GTT after six weeks of feeding. Chicken fed mice had increased 6 h fasted blood glucose levels compared to LF fed mice (Fig. 2A-B). Casein fed mice had higher blood glucose concentrations compared to LF and cod/scallop fed mice 30 and 60 minutes after glucose administration and chicken fed mice had higher blood glucose concentrations compared to LF fed mice 60 after administration (Fig. 2A). The glucose was administered according to body mass (2 mg glucose/g BM) and LF and chicken fed mice thus received a greater load of glucose than casein and cod/scallop fed mice (Fig. 2C). The calculated iAUC blood glucose (Fig. 2D) was higher in casein fed than in LF fed mice and furthermore tended to be higher in casein fed mice than in chicken and cod/scallop fed mice (*P* = 0.09). In 4 h fasted plasma collected at the termination of the mice after seven weeks of feeding, glucose concentrations were higher in chicken fed than in casein fed mice (Fig. 2F), while insulin concentrations tended to be increased in chicken fed mice (*P* = 0.09, Fig. 2G). HOMA-IR insulin-resistance-scores were higher in chicken fed than in casein fed animals and tended to be higher in chicken fed than in cod/scallop fed mice (*P* = 0.08, Fig. 2H). QUICKI insulin-sensitivity-scores were higher in casein fed than in chicken fed mice (Fig. 2I).

Difference in RER between light and dark phases abolished by HF feeding

To elucidate whether altered EE was an underlying mechanism behind differences in fat accretion, we utilized indirect calorimetry. During the 48 h of indirect calorimetry measurements that

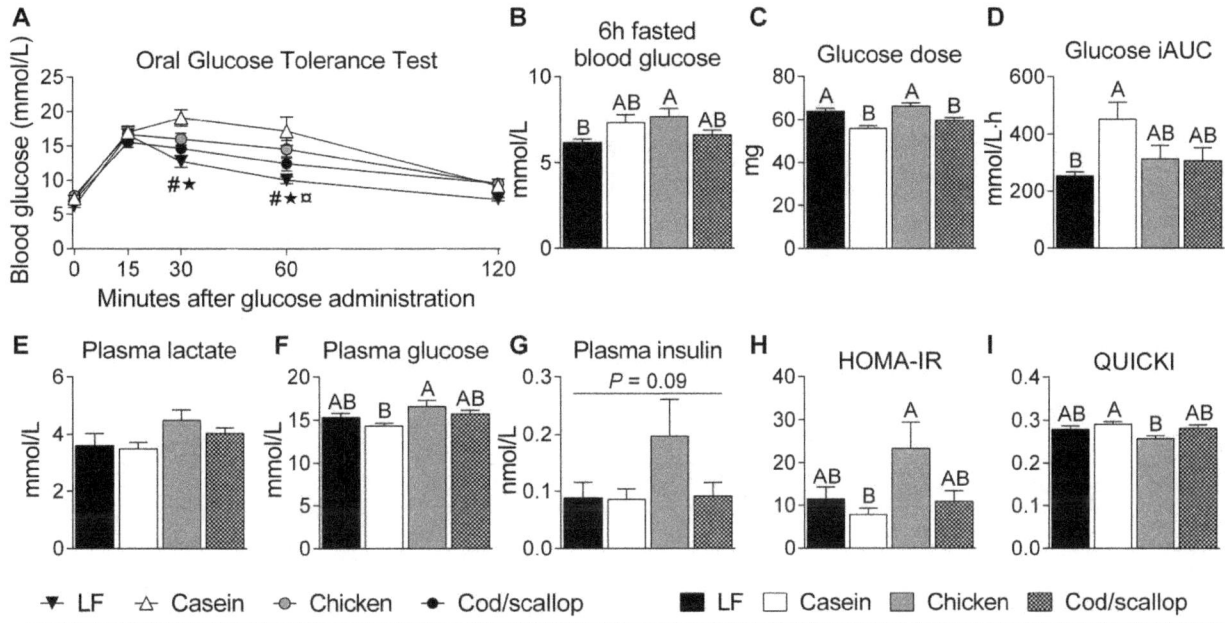

Figure 2. Casein tends to reduce glucose tolerance while chicken tends to cause increased plasma insulin concentration. A: Blood glucose measured before and at 15, 30, 60 and 120 minutes after oral administration of glucose (gavage, 2 mg/g body mass) during 6 h fasted oral glucose tolerance test in mice after six weeks on the experimental diets (O-GTT). B: 6 h fasted blood glucose. C: glucose dose administered by oral gavage. D: incremental blood glucose area under the curve (iAUC). E: Plasma lactate. F: Plasma glucose. G: Plasma insulin. E-G: concentrations measured in 4 h fasted plasma collected at the termination of the mice after seven weeks on the experimental diets. H: Homeostatic Model Assessment of Insulin Resistance (HOMA-IR) scores. I: Quantitative Insulin Sensitivity Check Index (QUICKI) scores. H-I: The scores were calculated based on 4 h fasted plasma glucose and insulin levels. Data (Expt. 1) represent group means ($n = 7$–8) $\pm$ SEM analyzed by one-way ANOVA followed by Tukey's pair-wise comparisons. O-GTT curve was analyzed by repeated measurements ANOVA followed by Tukey's post hoc. Means that do not share a letter are significantly different ($P<0.05$). # indicates significantly higher blood glucose in casein fed than in LF fed mice. * indicates significantly higher blood glucose in casein fed than in cod/scallop fed mice. ¤ indicates significantly higher blood glucose in chicken fed than in LF fed mice.

were analyzed, LF fed mice had higher RER in dark than in light phases ($P<0.0001$, Fig. 3A-B). After the shift to HF diets, RER decreased in both light and dark phases and the difference between light and dark phases was no longer evident (Fig. 3A-B). The different protein sources caused no differences in RER between the groups neither in light nor in dark phases (Fig. 3B).

Increased EE and a tendency towards increased activity in cod/scallop fed mice

Similarly to RER, activity level differed between light and dark phases in LF fed mice with higher activity levels during dark phases ($P<0.0001$, Fig. 3C-D). The initial activity levels measured while the mice were fed the LF diet were similar between groups and higher in dark than in light phases (Fig. 3D-E). Changing to HF diets did not change the activity during the dark phases (Fig 3D). However, HF feeding decreased activity level during light phases ($P = 0.018$, Fig. 3D). Total activity tended to decrease with the shift from LF diet to HF diets ($P = 0.07$, Fig. 3E). In dark phases cod/scallop fed mice tended to be more active than casein and chicken fed mice ($P = 0.09$, Fig. 3D) and a strong tendency towards higher total activity was seen in cod/scallop fed compared to casein and chicken fed mice ($P = 0.06$, Fig. 3E). Consistent with activity, EE was higher in dark than in light phases in LF fed mice. With the shift to HF diets, EE decreased in dark phases while no difference was seen between LF and HF feeding in light phases (Fig. 3F-G). No difference was seen between groups in light or dark phases while on LF diet, whereas EE tended to decrease ($P = 0.08$, Fig 3G) in casein fed compared to chicken and cod/

scallop fed mice in light phases and increased in cod/scallop fed compared to casein fed mice during dark phases (Fig. 3G).

Discussion

An increasing body of evidence supports a preventive role of HP diets against development of obesity. Less is known as to whether different protein sources consumed at normal dietary levels may differently affect energy balance. In the present study, we fed obesity-prone male C57BL/6J mice HF diets with either casein, chicken filet or a mixture of cod filet and scallop muscle as the protein source. Even though energy intake did not differ significantly, LF and chicken fed mice had higher feed efficiencies than the casein and cod/scallop fed mice, which after seven weeks of feeding translated into increased body masses and for the chicken fed mice also increased adipose tissue masses. Concomitantly, the chicken fed mice had deteriorated plasma lipid profile and enlarged liver mass with elevated hepatic TG levels. Thus, we demonstrate that different protein sources affect diet-induced obesity and associated co-morbidities in C57BL/6J mice when given at normal levels in a HF background diet.

Body fat accretion was reduced, evident as lower body mass gain, lower adipose tissue masses and reduced liver TG, in the casein and cod/scallop fed compared to the chicken fed mice. Interestingly, the apparent fat digestibility was reduced from an average of about 98% in LF, chicken and cod/scallop fed mice, to an average of about 95% in casein fed mice. Assuming that the apparent fat digestibility was constant for the entire seven week period, the casein fed mice absorbed approximately five g less fat

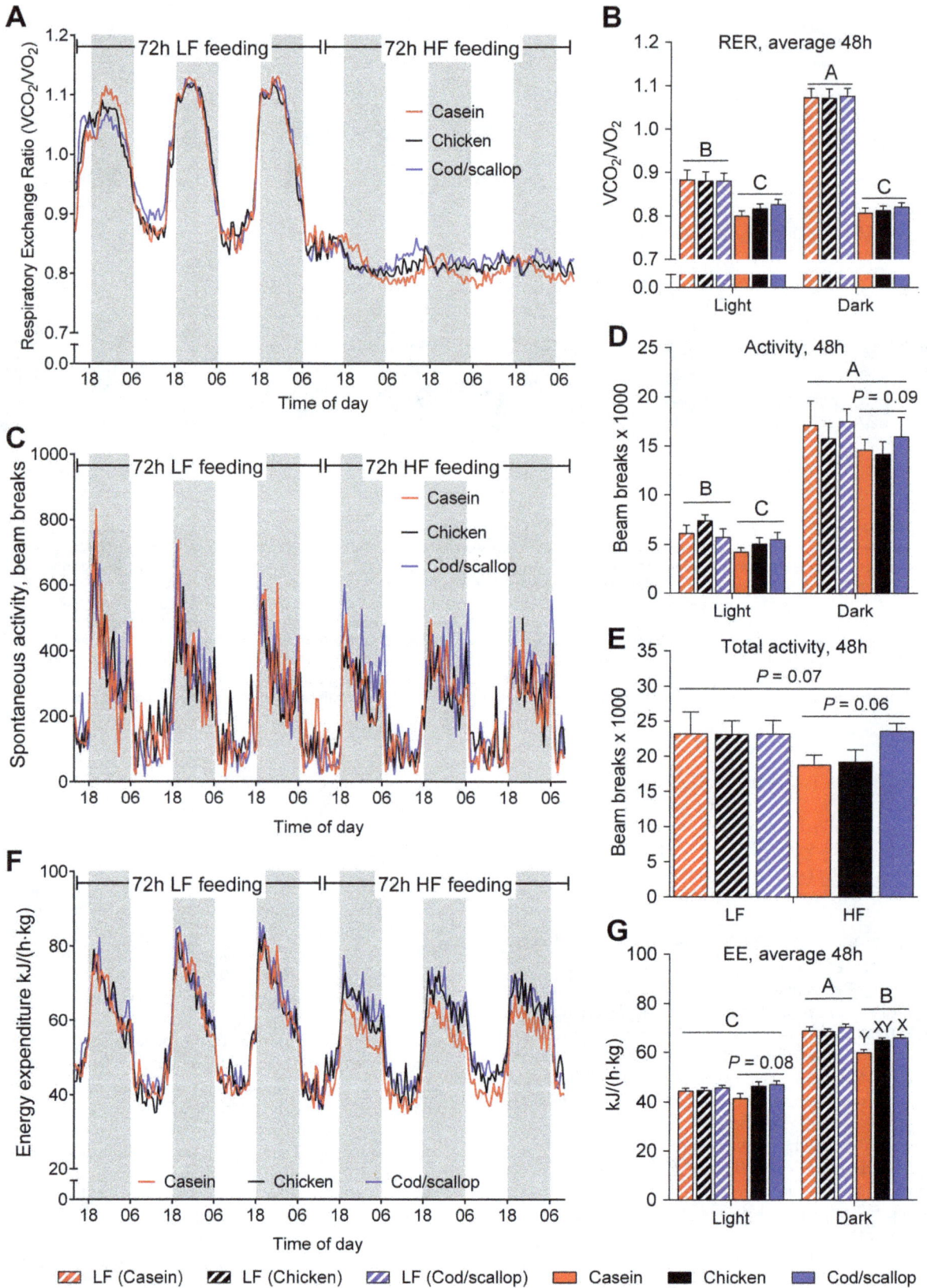

Figure 3. Protein sources affect energy expenditure and tend to affect spontaneous locomotor activity. A: RER in mice fed LF diet for 72 h and HF diets for 72 h in open-circuit indirect calorimetry cages. B: Average respiratory exchange ratio (RER) during 48 h on LF diet and HF diets in light and dark phases. C: Spontaneous locomotor activity during 72 h on LF diet and 72 h on HF diets. D: Spontaneous locomotor activity in light and dark phases during 48 h in mice fed LF diet and HF diets. E: Total spontaneous locomotor activity during 48 h in mice fed LF diet and HF diets. F:

Energy expenditure (EE) during 72 h on LF diet and 72 h on HF diets. G: Average EE during 48 h in light and dark phases in mice fed LF diet and HF diets. Data (Expt. 2) represent group means ($n = 9$–10) ± SEM analyzed by ANOVA followed by Tukey's pair-wise comparisons. RER, activity and EE data were analyzed by repeated measurements ANOVA followed by Tukey's post hoc. Means that do not share a letter are significantly different ($P < 0.05$).

than LF, chicken and cod/scallop fed mice. In mice, intake of a HF casein diet has previously been reported to cause higher fecal fat excretion and a leaner phenotype as compared to intake of a HF salmon diet [38]. Hence, it is likely that the reduced apparent fat absorption was a contributing factor to the reduced fat accretion in casein fed mice in the present study.

The cod/scallop fed mice maintained a lean phenotype, relative to chicken fed mice, without a reduction in fat absorption. To elucidate whether the protein sources modulated energy metabolism, we subjected the mice to indirect calorimetric measurements before onset of obesity at the transition from LF to HF feeding. HF diets have previously been shown to disturb feeding pattern and behavioral circadian rhythm in mice [39], such that the LF diet-induced fluctuations in RER between light and dark phases, reflecting different feed intake and substrate oxidations, are completely abolished after a switch to a casein based HF diet [40]. Accordingly, the RER was promptly reduced, and the differences in RER between light and dark phases disappeared after the switch to HF diets in the present study. There was no protein source effect on RER in the present study. However, following the transition from LF diet to HF diets EE decreased less in the cod/scallop fed than in casein fed mice, but we observed no significant difference in EE between chicken fed and cod/scallop fed mice that could explain the difference in adiposity. Our indirect calorimetry setup monitored gas exchange of each mouse for 1.9 minutes every 30 minutes, and it has been argued that the monitoring frequency has to be considerably higher in order to detect the 2–5% changes in diet-induced EE sufficient to elicit long term alterations on energy balance [41]. Decreased spontaneous locomotor activity has previously been demonstrated at the transition from LF to HF diets [39]. Accordingly, a decrease in spontaneous locomotor activity was observed in the light phases concurrent with a tendency towards decreased total activity ($P = 0.07$) after the shift from LF diet to HF diets in the present study. Importantly, cod/scallop feeding tended ($P = 0.06$) to avert this decrease in activity at the transition from LF to HF feeding. In line with this notion, we have previously observed an inverse correlation between locomotor activity and development of diet-induced obesity, without being able to detect differences in EE [21]. Indeed, whereas gas exchange was quantified at intervals (i.e. 1.9 min every 30 min), beam breaks were detected continuously, increasing the sensitivity of this measure as an indicator of EE. Therefore, differences in locomotor activity that nearly reached statistical significance ($P = 0.06$), are likely to reflect changes in EE that over time could explain the divergent fat accretion between the chicken and cod/scallop fed mice.

We have previously used another casein based HF diet (47 E% fat, 36 E% sucrose and 17 E% protein) to precipitate obesity and glucose intolerance in mice [12–14]. By increasing the fat content to 67 E% and reducing the sucrose content to 18 E%, the casein fed mice in the present study remained lean. Despite their lean phenotype, the casein fed mice were less glucose tolerant, when challenged in an O-GTT after six weeks of feeding. Cod protein intake has previously been associated with improved glucose metabolism in rats due to better peripheral insulin sensitivity as compared to casein feeding [30,42,43]. Moreover, in a randomized controlled intervention study with crossover design, insulin-resistant subjects exhibited improved insulin sensitivity [44] and reduced

levels of the inflammatory marker high-sensitivity C-reactive protein after intake of a cod based relative to a meat and dairy based diet for four weeks [45]. Therefore, both in the present study, as well as in studies with rats and humans, intake of cod as compared to casein is associated with improved glucose metabolism.

During HF feeding, metabolic adaptations to the elevated fat load occur by increasing mitochondrial content and oxidative capacity in liver [46,47] and skeletal muscle [40,48]. As a strong regulatory interaction exists between lipid and carbohydrates oxidation [49], HF feeding represses the use of glucose as an energy substrate (i.e. glycolysis) [40,46], a condition that could promote glucose intolerance. Based on the improved glucose clearance in the cod/scallop fed compared to casein fed mice in the present study as well as in HF cod fed rats reported by others [30,43], it is evident that dietary protein source affects glucose metabolism. However, our data did not indicate higher glycolysis or glucose utilization in the cod/scallop fed as compared to the casein fed mice and further studies are needed to elucidate the underlying mechanisms for the differences in glucose clearance.

The present study was not designed to identify underlying mechanisms, merely to elucidate whether diets with casein, chicken filet or a mixture of cod filet and scallop muscle modulate diet-induced obesity. As locomotor activity can be stimulated [50,51] and EE increased [52] by dietary taurine it is possible that the high taurine concentration of the cod/scallop diet contributed to the observed modulation of energy balance in the mice fed this diet. In addition, altered metabolism of branched-chain amino acid (BCAA) is likely associated with glucose dysregulation and the development of insulin-resistance [53]. In line with this notion, BCAA supplementation in a casein based HF diet impaired glucose tolerance in rats [54]. In the present study, the BCAA content was 39% higher in the casein diet than in the cod/scallop diet, which may have contributed to the observed differences in glucose tolerance. Elevated levels of BCAAs, leucine in particular, are associated with inhibition of insulin signaling through activation of the mammalian target of rapamycin pathway [55,56]. The altered glucose metabolism observed in the present study could therefore be related to differences in leucine intake. However, arguing against this is the recent findings that leucine supplementation in drinking water diminished HF feeding induced insulin resistance in mice [57,58]. Thus, further studies are needed to clarify whether varying amino acid content contributed to the observed differences in the present study.

In conclusion, protein sources are of importance for development of diet-induced obesity and associated co-morbidities in obesity-prone male C57Bl/6J mice fed HF diets. Whereas both casein and a mixture of cod and scallops prevented obesity, hepatic lipid accumulation and dyslipidemia, only cod/scallop fed mice remained glucose tolerant when challenged in an O-GTT. Further studies are needed to elucidate the underlying mechanisms for how the different protein sources induce these phenotypical alterations in HF fed mice.

Supporting Information

Table S1 Genes and corresponding primer sequences used for qRT-PCR.

Acknowledgments

The authors are grateful to Aase Heltveit and Øyvind Reinshol for technical assistance.

Author Contributions

Conceived and designed the experiments: BL HST LM KK. Performed the experiments: HST BL AKR KB. Analyzed the data: HST BL KK LM. Wrote the paper: HST BL. Read, revised and approved the final version of the manuscript: HST AKR KB LM KK BL.

References

1. Lowell BB, Spiegelman BM (2000) Towards a molecular understanding of adaptive thermogenesis. Nature 404: 652–660.

2. Hamilton TS (1939) The heat increments of diets balanced and unbalanced with respect to protein. Journal of Nutrition 17: 583–599.

3. Rothwell NJ, Stock MJ, Tyzbir RS (1983) Mechanisms of Thermogenesis Induced by Low Protein Diets. Metabolism-Clinical and Experimental 32: 257–261.

4. Kevonian AV, Vandertuig JG, Romsos DR (1984) Consumption of a Low Protein-Diet Increases Norepinephrine Turnover in Brown Adipose-Tissue of Adult-Rats. Journal of Nutrition 114: 543–549.

5. Miller DS, Payne PR (1962) Weight Maintenance and Food Intake. The Journal of Nutrition 78: 255–262.

6. Miller DS, Mumford P (1967) Gluttony. 1. An experimental study of overeating low- or high-protein diets. Am J Clin Nutr 20: 1212–1222.

7. Stock MJ (1999) Gluttony and thermogenesis revisited. Int J Obes Relat Metab Disord 23: 1105–1117.

8. Belza A, Ritz C, Sørensen MQ, Holst JJ, Rehfeld JF, et al. (2013) Contribution of gastroenteropancreatic appetite hormones to protein-induced satiety. The American Journal of Clinical Nutrition 97: 980–989.

9. Weigle DS, Breen PA, Matthys CC, Callahan HS, Meeuws KE, et al. (2005) A high-protein diet induces sustained reductions in appetite, ad libitum caloric intake, and body weight despite compensatory changes in diurnal plasma leptin and ghrelin concentrations. Am J Clin Nutr 82: 41–48.

10. Johnston CS, Day CS, Swan PD (2002) Postprandial thermogenesis is increased 100% on a high-protein, low-fat diet versus a high-carbohydrate, low-fat diet in healthy, young women. J Am Coll Nutr 21: 55–61.

11. Robinson SM, Jaccard C, Persaud C, Jackson AA, Jequier E, et al. (1990) Protein-Turnover and Thermogenesis in Response to High-Protein and High-Carbohydrate Feeding in Men. American Journal of Clinical Nutrition 52: 72–80.

12. Hao Q, Lillefosse HH, Fjaere E, Myrmel LS, Midtbo LK, et al. (2012) High-glycemic index carbohydrates abrogate the antiobesity effect of fish oil in mice. Am J Physiol Endocrinol Metab 302: E1097–1112.

13. Ma T, Liaset B, Hao Q, Petersen RK, Fjaere E, et al. (2011) Sucrose counteracts the anti-inflammatory effect of fish oil in adipose tissue and increases obesity development in mice. PLoS One 6: e21647.

14. Madsen L, Pedersen LM, Liaset B, Ma T, Petersen RK, et al. (2008) cAMP-dependent Signaling Regulates the Adipogenic Effect of n-6 Polyunsaturated Fatty Acids. Journal of Biological Chemistry 283: 7196–7205.

15. Klaus S (2005) Increasing the Protein: Carbohydrate Ratio in a High-Fat Diet Delays the Development of Adiposity and Improves Glucose Homeostasis in Mice. The Journal of Nutrition 135: 1854–1858.

16. Johnston CS, Tjonn SL, Swan PD (2004) High-protein, low-fat diets are effective for weight loss and favorably alter biomarkers in healthy adults. Journal of Nutrition 134: 586–591.

17. Larsen TM, Dalskov S-M, van Baak M, Jebb SA, Papadaki A, et al. (2010) Diets with High or Low Protein Content and Glycemic Index for Weight-Loss Maintenance. New England Journal of Medicine 363: 2102–2113.

18. Gilbert JA, Bendsen NT, Tremblay A, Astrup A (2011) Effect of proteins from different sources on body composition. Nutr Metab Cardiovasc Dis 21 Suppl 2: B16–31.

19. Liaset B, Hao Q, Jorgensen H, Hallenborg P, Du ZY, et al. (2011) Nutritional regulation of bile acid metabolism is associated with improved pathological characteristics of the metabolic syndrome. J Biol Chem 286: 28382–28395.

20. Liaset B, Madsen L, Hao Q, Criales G, Mellgren G, et al. (2009) Fish protein hydrolysate elevates plasma bile acids and reduces visceral adipose tissue mass in rats. Biochim Biophys Acta 1791: 254–262.

21. Lillefosse HH, Tastesen HS, Du ZY, Ditlev DB, Thorsen FA, et al. (2013) Hydrolyzed casein reduces diet-induced obesity in male C57BL/6J mice. J Nutr 143: 1367–1375.

22. Lillefosse HH, Clausen MR, Yde CC, Ditlev DB, Zhang X, et al. (2014) Urinary loss of tricarboxylic acid cycle intermediates as revealed by metabolomics studies: an underlying mechanism to reduce lipid accretion by whey protein ingestion? J Proteome Res 13: 2560–2570.

23. Shi J, Tauriainen E, Martonen E, Finckenberg P, Ahlroos-Lehmus A, et al. (2011) Whey protein isolate protects against diet-induced obesity and fatty liver formation. International Dairy Journal 21: 513–522.

24. Tranberg B, Hellgren LI, Lykkesfeldt J, Sejrsen K, Jeamet A, et al. (2013) Whey protein reduces early life weight gain in mice fed a high-fat diet. PLoS One 8: e71439.

25. Baer DJ, Stote KS, Paul DR, Harris GK, Rumpler WV, et al. (2011) Whey Protein but Not Soy Protein Supplementation Alters Body Weight and Composition in Free-Living Overweight and Obese Adults. Journal of Nutrition 141: 1489–1494.

26. Acheson KJ, Blondel-Lubrano A, Oguey-Araymon S, Beaumont M, Emady-Azar S, et al. (2011) Protein choices targeting thermogenesis and metabolism. Am J Clin Nutr 93: 525–534.

27. Fulgoni VL (2008) Current protein intake in America: analysis of the National Health and Nutrition Examination Survey, 2003–2004. The American Journal of Clinical Nutrition 87: 1554S–1557S.

28. Swan G (2004) Findings from the latest National Diet and Nutrition Survey. Proceedings of the Nutrition Society 63: 505–512.

29. Tastesen HS, Keenan AH, Madsen L, Kristiansen K, Liaset B (2014) Scallop protein with endogenous high taurine and glycine content prevents high-fat, high-sucrose-induced obesity and improves plasma lipid profile in male C57BL/6J mice. Amino Acids 46: 1659–1671.

30. Lavigne C, Tremblay F, Asselin G, Jacques H, Marette A (2001) Prevention of skeletal muscle insulin resistance by dietary cod protein in high fat-fed rats. Am J Physiol Endocrinol Metab 281: E62–71.

31. Mariotti F, Tome D, Mirand PP (2008) Converting nitrogen into protein–beyond 6.25 and Jones' factors. Crit Rev Food Sci Nutr 48: 177–184.

32. Liaset B, Julshamn K, Espe M (2003) Chemical composition and theoretical nutritional evaluation of the produced fractions from enzymic hydrolysis of salmon frames with Protamex(TM). Process Biochemistry 38: 1747–1759.

33. Allison DB, Paultre F, Maggio C, Mezzitis N, Pi-Sunyer FX (1995) The use of areas under curves in diabetes research. Diabetes Care 18: 245–250.

34. Floch J-PL, Escuyer P, Baudin E, Baudon D, Perlemuter L (1990) Blood Glucose Area Under the Curve: Methodological Aspects. Diabetes Care 13: 172–175.

35. Matthews DR, Hosker JP, Rudenski AS, Naylor BA, Treacher DF, et al. (1985) Homeostasis model assessment: insulin resistance and beta-cell function from fasting plasma glucose and insulin concentrations in man. Diabetologia 28: 412–419.

36. Katz A, Nambi SS, Mather K, Baron AD, Follmann DA, et al. (2000) Quantitative insulin sensitivity check index: a simple, accurate method for assessing insulin sensitivity in humans. J Clin Endocrinol Metab 85: 2402–2410.

37. Weir JBdV (1949) New methods for calculating metabolic rate with special reference to protein metabolism. The Journal of Physiology 109: 1–9.

38. Ibrahim MM, Fjaere E, Lock EJ, Naville D, Amlund H, et al. (2011) Chronic consumption of farmed salmon containing persistent organic pollutants causes insulin resistance and obesity in mice. PLoS One 6: e25170.

39. Kohsaka A, Laposky AD, Ramsey KM, Estrada C, Joshu C, et al. (2007) High-fat diet disrupts behavioral and molecular circadian rhythms in mice. Cell Metabolism 6: 414–421.

40. Koves TR, Ussher JR, Noland RC, Slentz D, Mosedale M, et al. (2008) Mitochondrial Overload and Incomplete Fatty Acid Oxidation Contribute to Skeletal Muscle Insulin Resistance. Cell Metabolism 7: 45–56.

41. Even PC, Nadkarni NA (2012) Indirect calorimetry in laboratory mice and rats: principles, practical considerations, interpretation and perspectives. American Journal of Physiology-Regulatory Integrative and Comparative Physiology 303: R459–R476.

42. Lavigne C, Marette A, Jacques H (2000) Cod and soy proteins compared with casein improve glucose tolerance and insulin sensitivity in rats. Am J Physiol Endocrinol Metab 278: E491–500.

43. Tremblay F, Lavigne C, Jacques H, Marette A (2003) Dietary cod protein restores insulin-induced activation of phosphatidylinositol 3-kinase/Akt and GLUT4 translocation to the T-tubules in skeletal muscle of high-fat-fed obese rats. Diabetes 52: 29–37.

44. Ouellet V, Marois J, Weisnagel SJ, Jacques H (2007) Dietary cod protein improves insulin sensitivity in insulin-resistant men and women: a randomized controlled trial. Diabetes Care 30: 2816–2821.

45. Ouellet V, Weisnagel SJ, Marois J, Bergeron J, Julien P, et al. (2008) Dietary cod protein reduces plasma C-reactive protein in insulin-resistant men and women. J Nutr 138: 2386–2391.

46. An Y, Xu W, Li H, Lei H, Zhang L, et al. (2013) High-Fat Diet Induces Dynamic Metabolic Alterations in Multiple Biological Matrices of Rats. Journal of Proteome Research 12: 3755–3768.

47. Guo Y, Darshi M, Ma Y, Perkins GA, Shen Z, et al. (2013) Quantitative Proteomic and Functioanl Analysis of Liver Mitochondria from High Fat Diet Diabetic Mice. Molecular & Cellular Proteomics.

48. van den Broek NMA, Ciapaite J, De Feyter H, Houten SM, Wanders RJA, et al. (2010) Increased mitochondrial content rescues in vivo muscle oxidative capacity in long-term high-fat-diet-fed rats. Faseb Journal 24: 1354–1364.

49. Randle PJ (1998) Regulatory interactions between lipids and carbohydrates: the glucose fatty acid cycle after 35 years. Diabetes/Metabolism Reviews 14: 263–283.

50. Idrissi A, Boukarrou L, Heany W, Malliaros G, Sangdee C, et al. (2009) Effects of Taurine on Anxiety-Like and Locomotor Behavior of Mice. In: Azuma J, Schaffer S, Ito T, editors. Taurine 7: Springer New York. pp. 207–215.

51. Murakami T, Furuse M (2010) The impact of taurine- and beta-alanine-supplemented diets on behavioral and neurochemical parameters in mice: antidepressant versus anxiolytic-like effects. Amino Acids 39: 427–434.

52. Tsuboyama-Kasaoka N, Shozawa C, Sano K, Kamei Y, Kasaoka S, et al. (2006) Taurine (2-aminoethanesulfonic acid) deficiency creates a vicious circle promoting obesity. Endocrinology 147: 3276–3284.

53. Newgard CB (2012) Interplay between lipids and branched-chain amino acids in development of insulin resistance. Cell Metab 15: 606–614.

54. Newgard CB, An J, Bain JR, Muehlbauer MJ, Stevens RD, et al. (2009) A branched-chain amino acid-related metabolic signature that differentiates obese and lean humans and contributes to insulin resistance. Cell Metab 9: 311–326.

55. Patti ME, Brambilla E, Luzi L, Landaker EJ, Kahn CR (1998) Bidirectional modulation of insulin action by amino acids. J Clin Invest 101: 1519–1529.

56. Tremblay F, Krebs M, Dombrowski L, Brehm A, Bernroider E, et al. (2005) Overactivation of S6 kinase 1 as a cause of human insulin resistance during increased amino acid availability. Diabetes 54: 2674–2684.

57. Macotela Y, Emanuelli B, Bång AM, Espinoza DO, Boucher J, et al. (2011) Dietary Leucine - An Environmental Modifier of Insulin Resistance Acting on Multiple Levels of Metabolism. PLoS ONE 6: e21187.

58. Binder E, Bermúdez-Silva FJ, André C, Elie M, Romero-Zerbo SY, et al. (2013) Leucine Supplementation Protects from Insulin Resistance by Regulating Adiposity Levels. PLoS ONE 8: e74705.

Searching for Novel Cdk5 Substrates in Brain by Comparative Phosphoproteomics of Wild Type and Cdk5$^{-/-}$ Mice

Erick Contreras-Vallejos[1], **Elías Utreras**[1], **Daniel A. Bórquez**[1¤], **Michaela Prochazkova**[2], **Anita Terse**[2], **Howard Jaffe**[3], **Andrea Toledo**[4], **Cristina Arruti**[4], **Harish C. Pant**[5], **Ashok B. Kulkarni**[2*], **Christian González-Billault**[1*]

1 Laboratory of Cellular and Neuronal Dynamics, Department of Biology, Faculty of Sciences, Universidad de Chile, Santiago, Chile, 2 Functional Genomics Section, National Institute of Dental and Craniofacial Research, National Institutes of Health, Bethesda MD, USA, 3 Protein and Peptide Facility, National Institute of Neurological Disorders and Stroke, National Institutes of Health, Bethesda MD, USA, 4 Laboratorio de Cultivo de Tejidos, Sección Biología Celular, Departamento de Biología Celular y Molecular, Facultad de Ciencias, Universidad de la República, Montevideo, Uruguay, 5 Laboratory of Neurochemistry, National Institute of Neurological Disorders and Stroke, National Institutes of Health, Bethesda MD, USA

Abstract

Protein phosphorylation is the most common post-translational modification that regulates several pivotal functions in cells. Cyclin-dependent kinase 5 (Cdk5) is a proline-directed serine/threonine kinase which is mostly active in the nervous system. It regulates several biological processes such as neuronal migration, cytoskeletal dynamics, axonal guidance and synaptic plasticity among others. In search for novel substrates of Cdk5 in the brain we performed quantitative phosphoproteomics analysis, isolating phosphoproteins from whole brain derived from E18.5 Cdk5$^{+/+}$ and Cdk5$^{-/-}$ embryos, using an Immobilized Metal-Ion Affinity Chromatography (IMAC), which specifically binds to phosphorylated proteins. The isolated phosphoproteins were eluted and isotopically labeled for relative and absolute quantitation (iTRAQ) and mass spectrometry identification. We found 40 proteins that showed decreased phosphorylation at Cdk5$^{-/-}$ brains. In addition, out of these 40 hypophosphorylated proteins we characterized two proteins, :MARCKS (Myristoylated Alanine-Rich protein Kinase C substrate) and Grin1 (G protein regulated inducer of neurite outgrowth 1). MARCKS is known to be phosphorylated by Cdk5 in chick neural cells while Grin1 has not been reported to be phosphorylated by Cdk5. When these proteins were overexpressed in N2A neuroblastoma cell line along with p35, serine phosphorylation in their Cdk5 motifs was found to be increased. In contrast, treatments with roscovitine, the Cdk5 inhibitor, resulted in an opposite effect on serine phosphorylation in N2A cells and primary hippocampal neurons transfected with MARCKS. In summary, the results presented here identify Grin 1 as novel Cdk5 substrate and confirm previously identified MARCKS as a a *bona fide* Cdk5 substrate.

Editor: Hemant K. Paudel, McGill University Department of Neurology and Neurosurgery, Canada

Funding: The work was supported by CONICYT-24120958 and MECESUP-UCH7013 (EC) Fondecyt 1095089 and ACT1114 (to CG-B), PAI 79100009 and Fondecyt 11110136 (EU), and the Divisions of Intramural Research, National Institute of Dental and Craniofacial Research (ABK) and the National Institute of Neurological Disorders and Stroke (HCP), National Institutes of Health, Bethesda, Maryland, United States of America. The funders had no role in study design, data collection and analysis, decision to publish, or preparation of the manuscript.

Competing Interests: The authors have declared that no competing interests exist.

* E-mail: chrgonza@uchile.cl (CGB); ak40m@nih.gov (ABK)

¤ Current address: Facultad de Medicina, Universidad Diego Portales, Chile.

Introduction

The complexity of cell functions derives not only from the particular content of individual proteins at a given time, but also from their post-translational modifications. The most common post-translational modification of proteins is the phosphorylation of serine, threonine, and tyrosine residues. This is particularly important to cellular mechanisms that integrate extracellular and intracellular signals through selective phosphorylation of proteins leading to specific functional outcomes. Protein phosphorylation plays a key role in many cellular functions such as cell signaling, apoptosis, cell migration, cytoskeletal dynamics and brain development. Cyclin-dependent kinase 5 (Cdk5) is an atypical member of the cyclin-dependent kinase family of which most of the

members are key regulators of the cell cycle [1,2]. Although Cdk5 is ubiquitously expressed, it is mainly active in post-mitotic neurons, where it is activated by neuron-specific activators p35 and p39 [3,4]. Cdk5 belongs to the proline-directed serine/threonine kinase group, and it phosphorylates many proteins possessing a canonical consensus sequence (S/T)PX(K/H/R) or at least a minimal consensus sequence (S/T)P [5]. Cdk5 is known to phosphorylate cytoskeletal proteins, signaling molecules, ion channels and regulatory proteins that participate in the normal function of the brain and also during neurodegenerative disorders [1,2,4,6]. A detailed analysis of Cdk5$^{-/-}$ mice, which display perinatal lethality and extensive neuronal migration defects, revealed that Cdk5 serves as a key regulator of neuronal migration, neurite outgrowth, and axonal path finding and dendrite development [7,8].

Given that a large number of the key cellular processes involve Cdk5 activity, suggesting that with the advent of new proteomic techniques many more Cdk5 substrates will be discovered. Interestingly, in the last few years many reports have indicated that Cdk5 also has extra-neuronal functions, such as regulating gene transcription, vesicular transport, apoptosis, cell adhesion, and migration in many cell types and tissues [2,9]. In the past, most of the Cdk5 substrates were discovered by classical biochemical and molecular approaches. However, there have been some efforts to identify Cdk5 substrates using high throughput screening (HTS). For example, Gillardon and colleagues used two-dimensional electrophoresis and mass spectrometry to study the protein phosphorylation patterns in cultured rat cerebellar granule neurons treated with Indolinone A, a Cdk5 inhibitor. Although these researchers did not find any specific substrates directly phosphorylated by Cdk5, their study demonstrated the changes in the phosphorylation status of certain proteins. Some results suggest that inhibiting Cdk5 activates stress proteins that may protect neurons against subsequent injurious stimuli [10]. Gillardon and colleagues also analyzed global changes in protein phosphorylation and gene expression in cultured cerebellar granule neurons by (^{32}P) orthophosphate labeling after the administration of a Cdk5 inhibitor. They found that Indolinone A treatment regulated protein phosphorylation and gene expression of candidates involved in neuronal survival, neurite outgrowth, and synaptic functions [11]. More recently, using high-density protein microarrays of Indolinone A-treated cerebellar granule neurons the same group identified two other potential Cdk5 substrates: Protein phosphatase 1 regulatory subunit 14A and Coiled-coil domain containing protein 97 [12]. Another phosphoproteomic study using a solid-phase capture-release-tag approach identified and suggested that Tau and MAP2 contain a Cdk5 phosphorylation motif [13].

In this study, we carried out systematic phosphoproteomic analysis of the Cdk5$^{+/+}$ and Cdk5$^{-/-}$ mice brains. We analyzed E18.5 mouse brain lysates using IMAC followed by iTRAQ and mass spectrometry. We discovered forty proteins with decreased phosphorylation at Cdk5 motifs in the Cdk5$^{-/-}$ brain. These proteins contained one or more phosphopeptides containing the canonical consensus phosphorylation motif for Cdk5 [5]. From the list of 40 protein candidates derived from HTS, we selected two proteins related to neurite outgrowth for further analysis. They were MARCKS (Myristoylated Alanine-Rich C Kinase Substrate) and Grin1 (G-protein-regulated inducer of neurite outgrowth). MARCKS was originally discovered as the main PKC substrate in the CNS [14,15], but it is also phosphorylatable by proline-directed kinases [16]. Interestingly for the purpose of this study is the fact that Cdk5 phosphorylates MARCKS at Ser25 in chick neuroblasts in vivo (Ser27 in mammals) [29]. Grin1 is highly expressed in the nervous system during development and it has been suggested that it has an important function in neuronal migration and brain formation [17]. The results presented here confirmed that chick MARCKS is phosphorylated by murine endogenous Cdk5 when it, was overexpressed in neuroblastoma N2a cells, altogether with p35. For the first time, we also show that Grin1 is a substrate for this kinase in analogous experimental conditions. Both proteins were not phosphorylated when Cdk5 was inhibited by roscovitine.

Materials and Methods

Animals

Cdk5 knockout mice (Cdk5$^{-/-}$) were generated as previously described [7]. Female Cdk5$^{+/-}$ mice pregnant were euthanized at E18.5 with a lethal injection of Ketamine/Xilacine mix. The genotype of each mouse was determined by PCR from DNA obtained from tail biopsies. For primary culture of rat hippocampal neurons, female rats Sprague-Dawley pregnant were euthanized at E18.5 with a lethal injection of Ketamine/Xilacine mix. These studies were performed in compliance with the National Institutes of Health's Guidelines on the Care and Use of Laboratory and Experimental Animals. All experimental procedures were approved by the Animal Care and Use Committee of the National Institute of Dental and Craniofacial Research, NIH, and the Bioethical Committee of the Faculty of Sciences, University of Chile, according to the ethical rules of the Biosafety Policy Manual of the National Council for Scientific and Technological Development (FONDECYT).

Phosphoprotein enrichment

Ten whole brains of E18.5 from Cdk5$^{+/+}$ and Cdk5$^{-/-}$ mice were homogenized using the buffer provided in the PhosphoProtein Purification Kit (QIAGEN #37101, Valencia, CA, USA), in accordance with the manufacturer's protocol. Briefly, protein extracts were loaded in columns, washed, eluted, and quantified by the Bradford method [18].

Protein digestion

One mg of phosphoprotein obtained from brain lysates was dried in the SpeedVac (ThermoSavant, Farmingham, NY, USA) and then dissolved in 8 M urea, 0.4 M NH$_4$HCO$_3$ for reduction by dithiothreitol and alkylation by indole acetic acid. After dilution into 2 M urea, 0.1 M NH$_4$HCO$_3$, tryptic digestion was performed as described before [19].

Quantitative proteomics (iTRAQ) of phosphoproteins

The identification and quantitation of phosphoproteins obtained from Cdk5$^{+/+}$ and Cdk5$^{-/-}$ mice were performed as previously described [20]. Briefly, we performed an iTRAQ procedure which is a non-gel based technique that incorporates isotope-coded covalent labeling of the N-terminus and side-chain amines of peptides with tags of varying mass (114.1; 115.1; 116.1; 117.1) (Absciex, Foster City, CA, USA). The samples were combined and analyzed by tandem mass spectrometry (MS/MS). In order to identify the labeled peptides and corresponding proteins, we used SEQUEST (http://fields.scripps.edu/sequest/) to quantify the low molecular mass reporter ion generated by the fragmentation of the attached tag and consequently the peptides and proteins from which they originated.

LC/MS/MS analysis

Samples were analyzed by LC/MS/MS on LTQ XL (linear trap quadraplole) with 2 Surveyor MS Pump plus HPLC pumps and Micro AS (Thermo Scientific, Waltham, MA, USA) and they were equipped with an Advance ESI (electrospray ionization) source (Michrom Bioresources Inc., Auburn, CA, USA). The equipment was used with an instrument configuration, columns, gradient, and source conditions as previously described [19]. The LTQ XL was set up to acquire a survey scan between m/z 400–1400 followed by a PQD MS/MS spectrum on each of the top 10 most abundant ions in the survey scan. Source conditions were as previously described and are listed here as follows: capillary temperature, 165°C; sheath gas flow, 2 U; spray voltage, 1.6 kV.

Key optimized PQD instrument parameters (32) were as follows: CE, 35%; isolation width, 3 m/z; activation Q, 0.700; activation time, 0.100 ms; minimum signal threshold, 10,000 cts; dynamic exclusion, repeat count 2, repeat duration 30 s, exclusion duration 60 s; MS/MS target, 4.0 X e^4; maximum fill time, 100 ms; 4 microscans.

Data analysis and iTRAQ quantitation

PQD MS/MS spectra were searched in the mouse database utilizing BioWorks 3.3.1 SP1 (Thermo Scientific) for site-specific phosphopeptide identification and iTRAQ quantification. The search parameters were set and data analysis was done as previously described [19]. Briefly, the search parameters were set as follows: static modifications: C = 57.0215, N-term = 144, and K = 144; differential modifications: S, T, Y = 79.9799 and M = 16. The search results were reported in descending order of the X correlation (XC) score subject to the default charge vs. XC filter: $1^+ = 1.50$, $2^+ = 2.00$, and $3^+ = 2.50$. Sequence assignments including the specific phosphorylated residue, were based on the selection of the phosphopeptide with the highest XC score, which is concurrent with the second ranked peptide displaying a ΔCn (the difference in the normalized XC score between the top scoring sequence and the next highest scoring sequence) of 0.1. MS/MS spectra were manually reviewed for spectral quality and the assignment of most major ions.

Transient transfection of N2A cell line

Mouse neuroblastoma N2A cells (ATCC#CCL-131) were transiently co-transfected with pBC12/pCMV-60k plasmid (chicken MARCKS) from Dr. Perry Blackshear [21] and pBI-p35/EGFP vector [22] and tTA vector in Opti-MEM with Lipofectamin 2000 (Invitrogen, Carlsbad, CA, USA) in DMEM. After 4 h of transfection, the solution was changed for DMEM plus 10% FBS, and the cells were treated with roscovitine (20 μM) during 20 h until protein extracts were made.

Transient transfection of primary hippocampal neurons in culture

Rat primary hippocampal neurons were performed as described before [23]. Briefly, hippocampal neurons from Sprague-Dawley rat embryos (E18.5) were dissociated after treatment for 20 min with 0.25% (w/v) trypsin (Gibco). 3×10^5 cells were plated in coverslips previously coated with poly-D-lysine 1 mg/ml (Sigma-Aldrich) in Neurobasal medium (Gibco) including 10% horse serum and Glutamine (Gibco). Then, neurons were transiently transfected with 0.6 μg of plasmid DNA (0.3 μg of CMV-GFP vector and 0.3 μg of chicken MARKCS vector) for 2 h and then the medium was replaced with Neurobasal medium supplemented with B27 and Glutamax (Gibco) in the absence of serum. Cells were kept in a humidified atmosphere of 5% CO2 at 37°C during 48 h and half of the covers were treated with roscovitine 10 μM. After 24 h neurons were fixed for posterior immunofluorescence against phosphor Ser25 MARCKS and DAPI.

Western Blot

Protein concentration was determined by the Bradford protein assay [18]. In brief, 30 μg of protein were loaded and run in 10% or 12% polyacrylamide gels at 100 V. The proteins were then transferred to nitrocellulose membranes in 120 mM glycine, 125 mM Tris, 0.1% SDS and 20% methanol, at 200 mA for 1 h. Then, the nitrocellulose membranes were blocked in 5% non-fat dry milk in PBS1X-0.1% Tween-20 for 1 h at room temperature and incubated overnight with primary antibodies diluted in 5% non-fat dry milk in PBS1X-0.1% Tween-20 at 4°C. The nitrocellulose membranes were incubated with the following antibodies: mAb 3C3 (mouse monoclonal anti-S25p-MARCKS [24]) diluted in blocking solution; Polo52 (rabbit polyclonal anti-MARCKS antibody, serum diluted 1:2000 [25]); anti-β-actin 1:2000 (sc-69879, Santa Cruz Biotechnology); anti-p35 (sc-820, Santa Cruz Biotechnology); α-tubulin (T9026) and phospho-serine (P5447) antibodies were obtained from Sigma (St. Louis, MO, USA); Phospho-MAPK/CDK Substrates Rabbit mAb #2325 (Cell Signaling Technology, Denver, MA, USA); anti-Grin1 116 was obtained from Dr. Tohru Kozasa from Department of Pharmacology, College of Medicine, University of Illinois, Chicago USA. Secondary antibodies horseradish peroxidase-conjugated goat anti-mouse and anti-rabbit antibodies were obtained from Jackson ImmunoResearch (West Grove, PA, USA). Secondary antibody goat anti-mouse IgG-HRP (31430, Thermo Scientific) used at a dilution of 1:10000 in blocking solution. Nitrocellulose membranes were washed three times with PBS containing 0.1% Tween-20 for 15 min and then the labeling was visualized with ECL reagent (32106, Thermo Scientific). All of Western blot data are representative of at least three independent experiments.

Immunoprecipitation assays

500 μg of protein from the whole mouse brain or the transfected N2A cells were immunoprecipitated in 500 μl of TPER buffer (Pierce) with 1–2 μg of Grin1, MARCKS, or Cdk5 antibodies overnight at 4°C in a shaker. Then, 30 μl of Protein A/G (Sigma) was added and incubated for 4 h at 4°C in a shaker. The immunocomplexes were centrifuged at 4°C during 5 min at 1000×g and washed 4 times in cold PBS1X. The loading buffer 5× (tris-HCl 62 mM, Glycerol 25%, SDS 2%, β-mercaptoethanol 10%, bromophenol blue 0.1%) was added into the beads and samples, and they were boiled at 95°C for 5 min and ran in SDS-PAGE gels.

Gene ontology analysis

The PANTHER (Protein analysis through evolutionary relationships) Classification System was used to classify proteins. The proteins were classified according to family and subfamily of molecular function, biological process and signaling pathway [26,27]. The Visualization and Integrated Discovery (DAVID) software was also used to analyze the data [28].

Statistical Analysis

All experiments were performed at a minimum of three times. A statistical evaluation was performed with GraphPad Prism software, version 5.0 (GraphPad, San Diego, CA). The significant differences between experiments were assessed by an unpaired Student test where α was set to 0.05.

Results

Phosphoproteomic analysis of Cdk5$^{-/-}$ brain

The proteomic analysis of iTRAQ-labeled tryptic peptides from control and Cdk5 null brains revealed 78 non-redundant phosphorylated sites that were decreased in Cdk5$^{-/-}$ brain. The majority of these phosphosites (55/78, 70%) present in 40 phosphoproteins have a minimal consensus Cdk5 motif of (S/T)P (Table 1). However, a decreased phosphorylation at these sites may also reflect the indirect effects of other proline-directed kinases regulated by Cdk5. Similar indirect effects have been reported for the kinases in the MEK-ERK [29] or JNK [30] pathways. In order to further refine identification of potential sites

Table 1. Putative Cdk5-dependent phosphorylation sites downregulated in Cdk5$^{-/-}$ brains.

Name	Gene	Site	% of decrease	PSSM score	NetphosK score
40S ribosomal protein S3	Rps3	T221[b,c]	97.4	1.1205	0.53
Abl interactor 1	Abi1	S183	76.7	0.9482	0.59
Alpha-adducin	Add1	S12	92.7	0.7623	0.57*
APC membrane recruitment protein 2	Amer2	S589	100	-	0.50
Bcl-2-associated transcription factor 1	Bclaf1	S510	100	0.8775	0.51
		S656	100	-	0.56*
Calmodulin-regulated spectrin-associated protein 1	Camsap1	S1069[c]	100	1.2446	0.54
Calmodulin-regulated spectrin-associated protein 3	Camsap3	S368	100	0.7122	0.66*
Clathrin coat assembly protein AP180	Snap91	S300	92.5	0.7577	0.69*
Collapsin response mediator protein 1	Dpysl1	T509[a,c]	57.2	0.5071	0.49
Collapsin response mediator protein 2	Dpysl2	T509[a]	78.1	0.7100	0.40
Collapsin response mediator protein 4	Dpysl3	T509[a,c]	100	0.7929	0.48
DNA ligase 1	Lig1	S51[c]	100	0.7076	0.60*
E3 ubiquitin-protein ligase TRIM2	Trim2	S428	91.1	0.7835	0.68*
G protein-regulated inducer of neurite outgrowth 1	*Gprin1*	*S369*	*74.3*	*0.8067*	*0.68**
		S691	100	0.6710	0.44
Growth-associated protein 43	Gap43	S96	34.5	0.9276	0.36
		T172	100	0.8226	0.43
Hepatoma-derived growth factor	Hdgf	S165	75	0.8228	0.44
Heterogeneous nuclear ribonucleoprotein D0	Hnrnpd	S83	37.1	-	0.52*
Kinesin light chain 1	Klc1	S459	100	1.0709	0.39
MARCKS-related protein	Marcksl1	S22	84.3	1.1036	0.43
Microtubule-associated protein 1B	Map1b	T527	100	1.0861	0.47
		S1307	46.6	1.0589	0.35
		S1438	47.4	1.1018	0.53*
Microtubule-associated protein 2	Map2	T1650	87.9	-	0.69*
Microtubule-associated protein tau	Mapt	T468	30	0.7591	0.51*
		S470	42.4	0.9090	0.56*
		T473[a]	30	0.7174	0.68*
		T504[a,b]	46.6	0.6513	0.70*
		T523[a]	93.2	0.7551	0.73*
		S527[a,b]	97.6	0.5220	0.68*
Myristoylated alanine-rich C-kinase substrate	*Marcks*	*S27[c]*	*80.7*	*0.9177*	*0.53**
		S138	32.9	0.8514	0.47
		T143[c]	41.1	0.6899	0.55
Na(+)/H(+) exchange regulatory cofactor NHE-RF1	Slc9a3r1	S275	51.1	0.5991	0.60*
Nestin	Nes	S1837	91	0.6777	0.64*
Neuronal migration protein doublecortin	Dcx	T336	67.8	0.7194	0.73*
		S339[a]	89.3	0.7502	0.74*
Phosphatidylinositol 4-kinase beta	Pi4kb	T517	100	0.9211	0.43
Plakophilin-4	Pkp4	S509	100	0.9879	0.48*
Programmed cell death protein 4	Pdcd4	S94	100	0.8618	0.58*
Protein SDE2 homolog	Sde2	S269	100	0.6656	0.54
Ras GTPase-activating protein-binding protein 1	G3bp1	S231	100	0.8606	0.38
Ras GTPase-activating protein-binding protein 2	G3bp2	T227	92.5	0.9730	0.37
Rho GTPase-activating protein 1	Arhgap1	S51	83.1	-	0.46
RNA-binding protein Raly	Raly	S135[c]	55.7	1.0370	0.48
SAFB-like transcription modulator	Sltm	S289	89.9	-	0.37
Serine/threonine-protein kinase DCLK1	Dclk1	S32	34.3	-	0.71*

Table 1. Cont.

Name	Gene	Site	% of decrease	PSSM score	NetphosK score
		S334	64.8	0.5840	0.66*
Stathmin	Stmn1	**S25**[a,b,c]	78.7	0.4295	0.36
		S38[a,b,c]	95.3	1.0112	0.72*
Stathmin-2	Scg10	**S62**	52.4	0.9175	0.48
T-box brain protein 1	Tbr1	**S84**	100	0.8802	0.38
U1 small nuclear ribonucleoprotein 70	Snrnp70	**S226**[c]	84.9	0.9905	0.57*

In **bold** type phosphorylated sites previously described in other mouse brain phosphoproteomic studies but lacking an assigned protein kinase responsible for such a phosphorylation.
[a]: phosphorylated by Cdk5,
[b]: phosphorylated by Cdk1;
[c]: phosphorylated by Cdk2.
PSSM scores determined as Borquez et al, 2013. NetphosK scores calculated using public protein prediction tool.
Asterisks (*) indicated Cdk5 is the best kinase for a given site.
Candidates validated in the present study are presented in italics type.

directly phosphorylated by Cdk5, we performed bioinformatics analysis using two sequence-based prediction tools for Cdk5 substrates: a position specific scoring matrix (PSSM) [5] and a web tool based in artificial neural networks (ANN), called NetPhosK [31]. The scores calculated by both tools are shown in Table 1. Although PSSM allows for better accuracy in predictions [5], NetPhosK allows a comparative study, with a parallel scoring for 16 other kinases. Interestingly, for 27 sites in 20 phosphoproteins (49%), Cdk5 is the most likely kinase that phosphorylates these sites (Table 1 asterisks). Two of these potential direct Cdk5 substrates, corresponding to Ser27 of MARCKS and Ser369 of Grin1, were selected for further analysis.

Gene ontology analysis of altered phosphoproteins in Cdk5$^{-/-}$ brain

In order to get an overview of the altered phosphoprotein profile in Cdk5$^{-/-}$ brains, we classified all of the phosphorylated proteins with the canonical Cdk5 motif which we identified in our experimental analysis in three different groups, namely signaling pathways (Figure 1A), molecular functions (Figure 1B) and biological processes (Figure 1C) according to their annotation in PANTHER. Thirteen of 40 phosphoproteins (30%) were classified in the category of the signaling pathways (Figure 1A). This classification assigned the phosphoproteins to eight different signaling pathways: 1) VEGF signaling (1/13); 2) axon guidance mediated by semaphorins (3/13); 3) angiogenesis (1/13); 4) cytoskeletal regulation by Rho GTPase (2/13); 5) Alzheimer disease-presenilin (1/13); 6) Alzheimer disease-amyloid secretase (1/13); 7) pyrimidine metabolism (3/13); and 8) PDGF signaling (1/13) (Figure 1A). In terms of molecular functions, we identified that 21 of the 40 phosphoproteins that were associated with binding category (GO: 0005488) which is defined as the selective, non-covalent, often stoichiometric, interaction of a molecule with one or more specific sites on another molecule (Figure 1B). These functions correspond to protein binding, receptor binding, cytoskeleton binding and nucleic acid binding. Some of these phosphoproteins belong to more than one binding subcategory. Finally, the 40 phosphoproteins identified in this study were classified into biological processes corresponding to metabolic processes, cell communications, developmental processes and cellular components organization, among others (Figure 1C). Alternatively, by using the bioinformatics tool DAVID, 11 phosphoproteins were clustered into the neuron projection process

(GO: 0043005) which is defined as a prolongation or process extending from nerve cells (axon or dendrites) (data not shown).

Cdk5 and MARCKS phosphorylation at Ser27

MARCKS has at least 7 sites that can be phosphorylated by MAPK and/or Cdks [32]. By using phosphoproteomics analysis, we found three phosphorylation sites on MARCKS that significantly decreased in the Cdk5$^{-/-}$ brain: Ser27 (80.7%), Ser138 (32.9%) and Thr143 (44.1%) (Figure 2A). A recent report indicated a decreased phosphorylation of Ser25 on MARCKS in chicken retinal neuroblasts treated with roscovitine or olomuicine (Cdk5 inhibitors), suggesting that Cdk5 was mainly responsible for the phosphorylation of this epitope [25]. Interestingly, MARCKS could be phosphorylated at Thr143 in vitro by Cdk2-cyclin A [33]. Since mouse MARCKS-Ser27 is a homologue residue to chicken MARCKS-Ser25, we expressed chicken MARCKS in mouse N2A neuroblastoma cells. This was to study the potential changes in MARCKS phosphorylation using a specific antibody that recognizes chicken MARCKS phosphorylated at Ser25 (pSer25) (mAb3C3) [24]. Interestingly, chicken MARCKS lacks the serine residue located -1 respect to phosphorylated the serine (Figure 2B). This residue seems to be critical for epitope recognition, because the pSer25 antibody did not recognize rat or mouse MARCKS. We overexpressed chicken MARCKS in neuroblastoma N2A cells along with a tetracycline inducible system for p35 over-expression [22]. We found that the phosphorylation of Ser25 on chicken MARCKS was detected only when we over-expressed p35 (Figure 2C). However, this phosphorylation was undetectable when N2A cells co-transfected with MARCKS and p35 were treated with roscovitine (Figure 2C). Additionally, we evaluated the pharmacological inhibition of Cdk5 in primary hippocampal neurons co-transfected with chicken MARCKS and CMV-GFP vectors. To determine which neurons were transfected we evaluated GFP expressing neurons. pSer25 MARCKS was detected on soma (white head arrow) and neurites (white arrow) of control hippocampal neurons (Figure 2D). However, the treatment with roscovitine (10 µM during 24 h) significantly decreased expression of this phospho-epitope in both compartment suggesting that phosphorylation of Ser25 in MARCKs was dependent on Cdk5 activation (Figure 2D). Our results are supported by an earlier report in which the phosphorylation of chicken MARCKS pSer25 was inhibited by treatment with roscovitine [25]. We then used NetPhosK tool to evaluate which of

A

GO Signaling Pathway

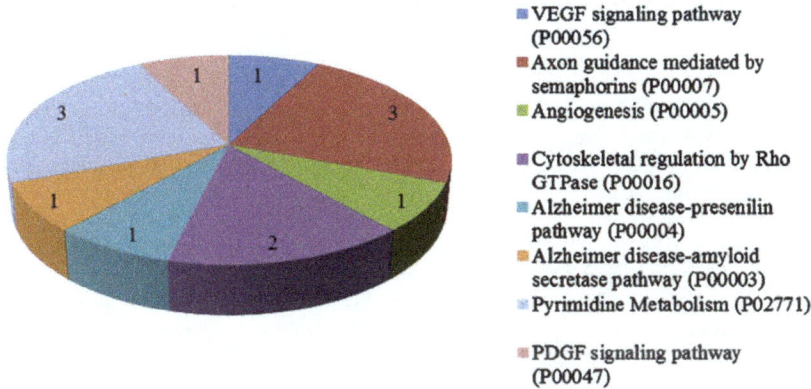

- VEGF signaling pathway (P00056)
- Axon guidance mediated by semaphorins (P00007)
- Angiogenesis (P00005)
- Cytoskeletal regulation by Rho GTPase (P00016)
- Alzheimer disease-presenilin pathway (P00004)
- Alzheimer disease-amyloid secretase pathway (P00003)
- Pyrimidine Metabolism (P02771)
- PDGF signaling pathway (P00047)

B

GO Molecular function

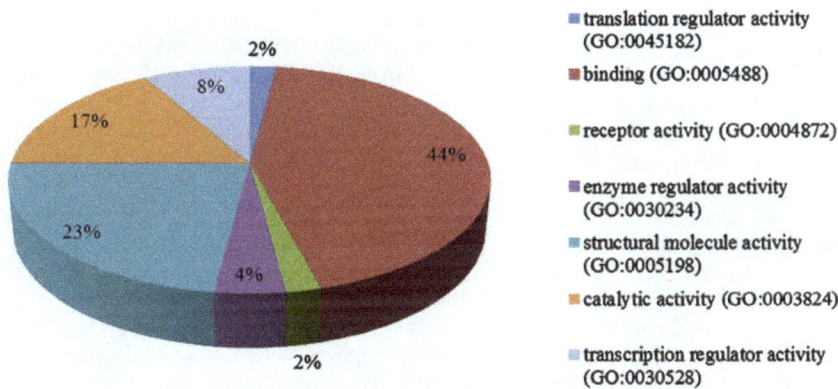

- translation regulator activity (GO:0045182)
- binding (GO:0005488)
- receptor activity (GO:0004872)
- enzyme regulator activity (GO:0030234)
- structural molecule activity (GO:0005198)
- catalytic activity (GO:0003824)
- transcription regulator activity (GO:0030528)

C

GO Biological Process

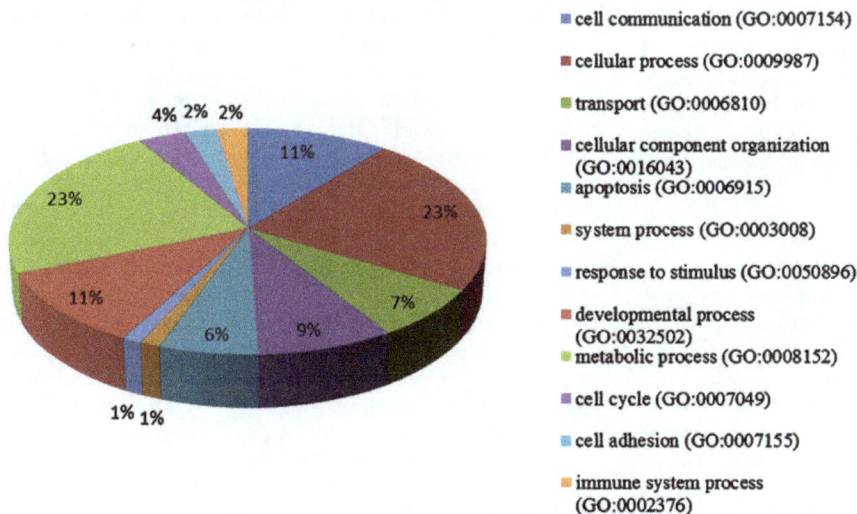

- cell communication (GO:0007154)
- cellular process (GO:0009987)
- transport (GO:0006810)
- cellular component organization (GO:0016043)
- apoptosis (GO:0006915)
- system process (GO:0003008)
- response to stimulus (GO:0050896)
- developmental process (GO:0032502)
- metabolic process (GO:0008152)
- cell cycle (GO:0007049)
- cell adhesion (GO:0007155)
- immune system process (GO:0002376)

Figure 1. Functional classification of the altered phosphoproteins in Cdk5$^{-/-}$ brain. The phosphorylated proteins identified in our analysis were classified into the following groups: A) Signaling pathways, B) Molecular functions and C) Biological processes. Percentages are expressed as gene hits against total number of process hits. The cellular process is defined as any process that is carried out at the cellular level, but not necessarily restricted to a single cell. The system process is defined as a multicellular organismal process carried out by any of the organs or tissues in an organ system. An organ system is a regularly interacting or interdependent group of organs or tissues that work together to carry out a biological objective. A response to stimulus is defined as any process that results in a change in state or activity of a cell or an organism (in terms of movement, secretion, enzyme production, gene expression, etc.) as a result of a stimulus. The process begins with detection of the stimulus and ends with a change in state or activity or the cell or organism.

the three MARCKS sites that were showing decreased phosphorylation in Cdk5 null brains was the best predicted Cdk5 site. As showed in Table 1, amongst Ser27, Ser138 and Thr143 sites, the bioinformatics approach identified Ser27 as a potential Cdk5 substrate the preferred Cdk5 site, confirming results from our phosphoproteomic analysis, heterologous expression and primary neuron assays (Table 1). MARCKS expression had been detected primarily in small dendrites, axons and axon terminal (Table S1). In summary, our results clearly indicate that Cdk5 phosphorylates MARCKS at Ser27 in the mouse brain.

Cdk5 and Grin1 phosphorylation at Ser369

Grin1 protein is highly expressed in the developing brain, whereas its expression is more restricted in adult stage [17]. It is highly enriched in growth cones, suggesting that it may be neuron-specific (Table S1). Here, by conducting phosphoproteomics analysis, we found that Grin1 displays two phosphorylation sites Ser369 (74.3% decrease) and Ser691 (100% decrease) that were significantly decreased in Cdk5$^{-/-}$ brain. Ser369 corresponded to a classical consensus sequence for Cdk5, which is conserved in mice and rats but not in humans (Figure 3A), while Ser691 is a KSP motif, which resembles the consensus site for Cdk5 phosphorylation in neurofilaments [34]. To confirm our phosphoproteomic analysis, we analyzed the expression of Grin1 in rat (B104) and mouse (N2A) neuroblastoma cells and the mouse brain. Grin1 antibody only recognized mouse protein, but it did not recognize rat protein (Figure 3B). In addition, we immunoprecipitated Grin1 from N2A cells and the mouse brain and we detected Cdk5 by Western blot (Figure 3C). Similarly we conducted the reciprocal immunoprecipitation with Cdk5 from N2A cells and the mouse brain and we detected Grin1 by Western blot (Figure 3D). These combined results suggest an interaction between Grin1 and Cdk5. Moreover, the levels of serine phosphorylation in Grin1 were considerably reduced in Cdk5$^{-/-}$ brain as detected with an antibody that recognizes phosphorylated SPXK motif (Figure 3E). This antibody preferentially recognizes Ser369. In addition, we found that the serine phosphorylation of Grin1 increased in N2A cells over-expressing p35, while roscovitine treatment of N2A cells over-expressing p35 had the opposite effect (Figure 3F). Moreover, by using bioinformatics tools NetPhosK we found that the best kinase that phosphorylated Ser369 on Grin1 is Cdk5 (Table 1). Our results confirm phosphoproteomic analysis and indicate that Cdk5 phosphorylates Grin1 at Ser369.

Discussion

Cdk5, a serine/threonine protein kinase, is involved in many important cellular processes associated with brain development and function. It is also implicated in disease processes associated with neurodegeneration. Cdk5 brings about its effect by phosphorylating a large number of target substrates, resulting in their activation or deactivation. These substrates play key roles in hippocampal neurogenesis, neuronal migration, cytoskeletal dynamics and synaptic plasticity [35]. It is therefore not surprising that novel substrates of Cdk5 that are being identified have

important functions in both CNS and PNS [6,36]. In the present study, we carried out comparative quantitative phosphoproteomic analysis of E18.5 Cdk5$^{+/+}$ and Cdk5$^{-/-}$ brains to search for new Cdk5 substrates. We carried out phosphoprotein enrichment using the IMAC procedure and utilized iTRAQ for peptide labeling and quantification. We found decreased phosphorylation in 55 Cdk5-consensus sites in 40 phosphoproteins from Cdk5$^{-/-}$ brain as compared with Cdk5$^{+/+}$ brains. The phosphoproteins found in our study are involved in different signaling pathways and regulate several cellular processes in the nervous system such as cytoskeletal regulation, axonal growth, axonal guidance and synapse formation. Cdk5 is known to regulate these biological processes through the phosphorylation of key substrates such as PAK1, β-catenin, Src, Nudel, synapsin, MUNC18, and amphphysin among others proteins [1,35].

MARCKS and Grin1 were two phosphoproteins among the total of 40 identified in our HTS phosphoproteomic analysis of Cdk5$^{-/-}$ brain. We further characterized the Cdk5-mediated phosphorylation of MARCKS and Grin1 in N2A cell line and mouse brain. MARCKS plays a pivotal role during neural development, since the inactivation of MARCKS gene results in abnormal brain development and perinatal death [37]. The functional role of MARCKS has been associated with regulation of growth cone adhesion, axon pathfinding [38] and dendritic spine morphology [39]. Interestingly, the defects found in mice lacking MARCKS were restored by the expression of certain mutated non-phosphorylated form of MARCKS, suggesting that some of the MARCKS functions were independent of its phosphorylation by PKC and possibly involved another phosphorylation sites phosphorylated by other kinase(s) [40]. In support of this suggestion, MARCKS phosphomutants which were used to restore the expression in null mice were indeed in a specific PKC region. MARCKS is also phosphorylated by MAPK and Cdks [32] and its Ser27 is phosphorylated in vitro by the Cdk2-cyclin E complex [41]. The phosphorylation of Ser27 (corresponding to Ser25 in chicken) is dependent on the integrity of actin cytoskeleton [42] and it has been found only in post mitotic neural cells during chicken embryo development [43,44] thus suggesting an important role in brain morphogenesis. In our study, we found decreased phosphorylation of MARCKS at three different sites: Ser27, Ser138 and Thr143 in Cdk5$^{-/-}$ brain. Interestingly, it was previously suggested that Ser27 (or Ser25 in chicken) could be phosphorylated by Cdk5 [25] and the bioinformatics analyses using a Cdk5 specific PSSM [5]. The NetphosK [31] tools confirmed that Ser27 had the higher probability to be phosphorylated by Cdk5. To further confirm this Cdk5-mediated phosphorylation at Ser27 site in MARCKS, we overexpressed chicken MARCKS and p35 into N2A cells and then evaluated Cdk5-dependent phosphorylation of Ser25 using an specific antibody [25]. Our results clearly indicated that phosphorylation of Ser25 was increased in N2A cells over-expressing p35 and it was decreased with the treatment with roscovitine of N2A cells and rat hippocampal neurons, suggesting that the phosphorylation of this residue is dependent of the Cdk5 activity. Additionally, MARCKS phosphorylation by PKC is

A

Ser27 Ser138 Thr143

145 169

Calmodulin-binding

1   309

B

Myristoylated alanine-rich protein kinase C substrate (S27)

Mus musculus	18	RPGEAAVAS**SPS**KANGQENG 37
Rattus novergicus	18	RPGEAAVAS**SPS**KANGQENG 37
Homo sapiens	18	RPGEAAVAS**SPS**KANGQENG 37
Bos taurus	18	RPGEAAVAS**SPS**KANGQENG 37
Gallus gallus	18	KPGEA-VAA**SPS**KANGQENG 36

C

MARCKS	-	+	+	+
BitetO-p35/tTA	-	-	+	+
Roscovitine (20 µM)	-	-	-	+

		kDa
pMARCKS (S25)		75
MARCKS		75
p35		35
β-actin		43

D

CONTROL

GFP   P-S25-MARCKS   TOPRO3   MERGE

20 µm

ROSCOVITINE

GFP   P-S25-MARCKS   TOPRO3   MERGE

20 µm

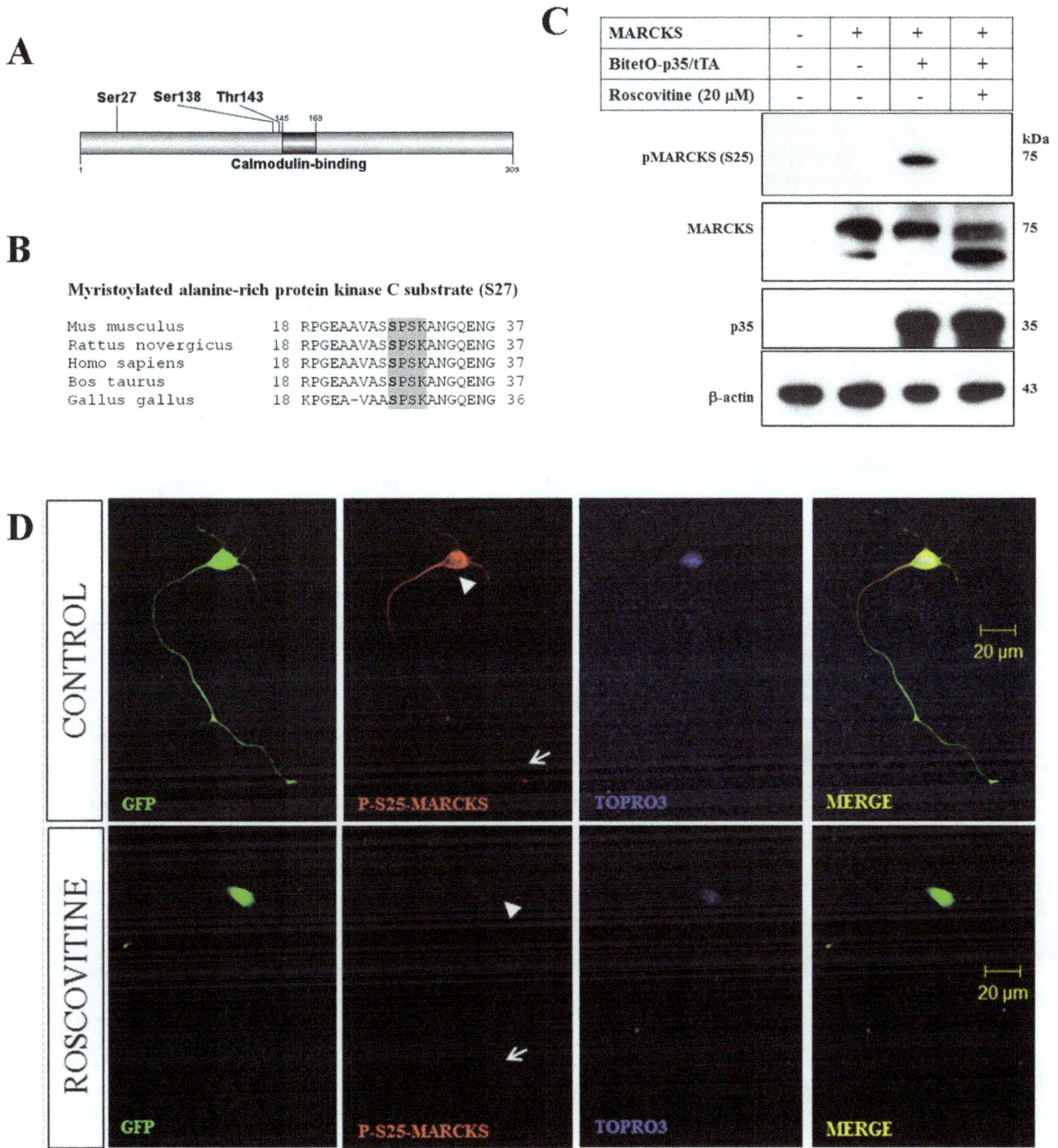

Figure 2. Cdk5 phosphorylation of MARCKS Ser27 in mouse. A) Schematic representation of MARCKS protein with the localization of phosphorylation sites that decreased in our phosphoproteomic study. B) MARCKS protein sequence alignment showing a conserved Cdk5 motif between species. Shaded boxes show conserved amino acids, bold amino acid is the phosphorylation site. C) Western blot analysis of pSer25, MARCKS and p35 in N2A cells co-transfected with MARCKS, MARCKS and pBI-p35/EGFP and MARCKS and pBI-p35/EGFP plus roscovitine. D) Immnunofluorescence of phosphor Ser25 MARCKS in hippocampal neurons transfected with MARCKS and CMV-GFP. Upper panel shows the control condition and lower panel shows roscovitine treated neurons. The expression of Phospho S25 MARCKS was detected on soma (white head arrow) and neurites (white arrow) of control cells. However, the treatment with roscovitine (10 µM during 24 h) decreased MARCKS phosphorylation at both compartments. Nuclei was stained with DAPI.

regulated by PP2A, PP1 and PP2B [25]. We hypothesize that the Cdk5-dependent phosphorylation of mouse MARCKS at Ser27 can positively regulate the binding of MARCKS to the plasma membrane and actin filaments. In this context, it was reported that the non-PKC phosphorylated MARCKS (membrane-bound) is involved in memory formation and contributes to stabilization of

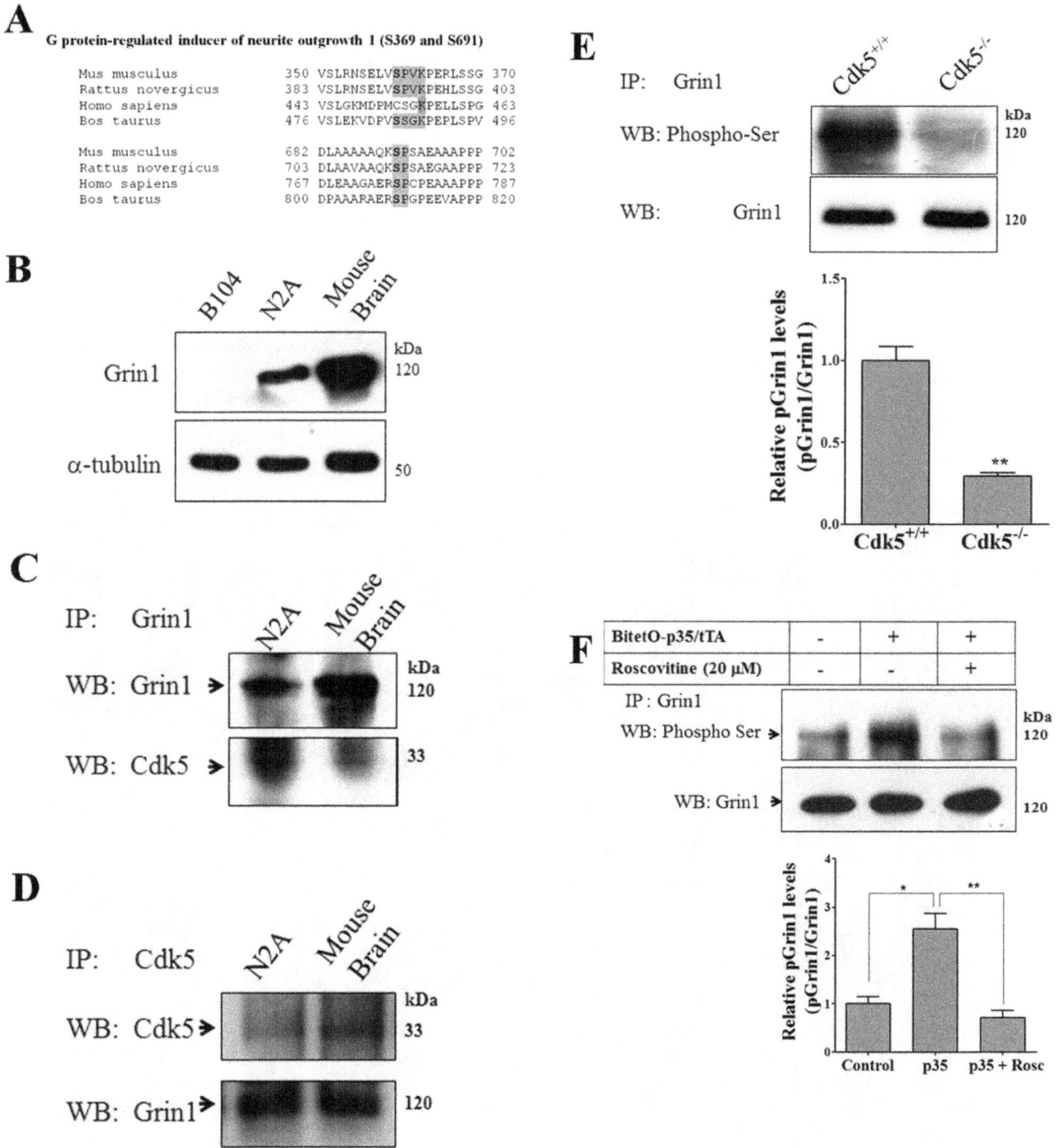

Figure 3. Cdk5 phosphorylates Grin1 at Ser369. A) Grin1 alignment showing Ser369 and Ser691 sequence as Cdk5 motifs. The shaded boxes show conserved amino acids, bold amino acid is the phosphorylation site. B) Detection of Grin1 in rat B104 and mouse N2A neuroblastoma cells and mouse brain. C) Immunoprecipitation of Grin1 and Western blot detection of Cdk5 and Grin1 in N2A cells and mouse brain. D) Immunoprecipitation of Cdk5 and Western blot detection of Grin1 and Cdk5 in N2A cells and mouse brain. E) Immunoprecipitation of Grin1 and Western blot detection of Grin1 and phospho Serine (using an antibody that recognize phosphorylation on SPXK) brain of $Cdk5^{+/+}$ and $Cdk5^{-/-}$ mice. F) Immunoprecipitation of Grin1 and Western blot detection of phospho serine and Grin1 in N2A cells transfected with pBI-p35/EGFP and tTA, and pBI-p35/EGFP and tTA plus roscovitine. All data are presented as mean and SEM ($n = 3$). * $p < 0.05$, ** $p < 0.01$ (t-Test).

synaptic morphology [45]. Because Cdk5 is also involved in learning and memory processes [46,47] we speculate that Cdk5 plays an important role in these processes through phosphorylation of MARCKS at Ser27 (Figure 4A).

We also found that Grin1 displayed decreased phosphorylation in two residues Ser369 and Ser691 in $Cdk5^{-/-}$ brain. Ser369 corresponds to Cdk5 SPXK motif [5] whereas Ser691 corresponds to the Cdk5 KSP motif [34]. Moreover, Grin1 has two more sites

A

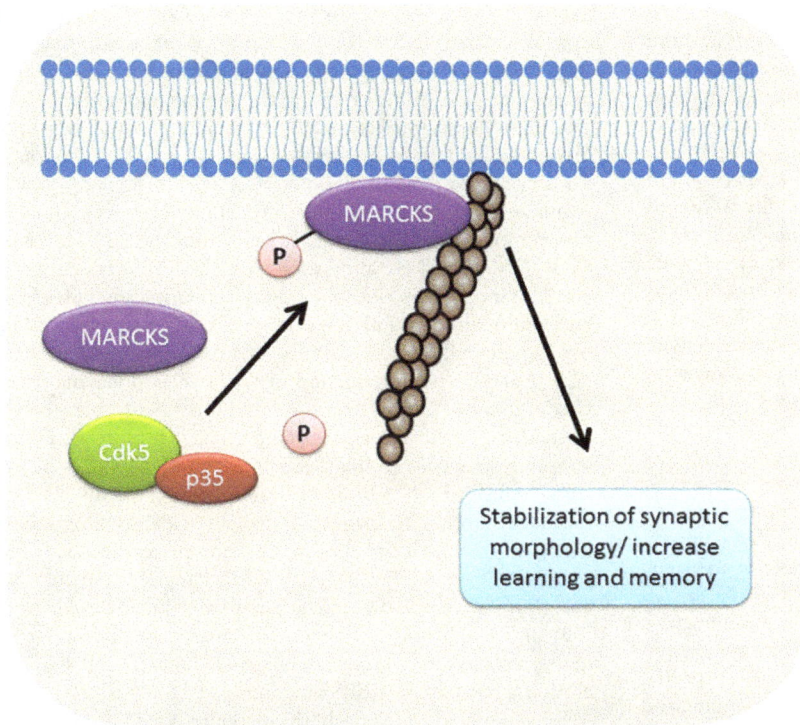

B

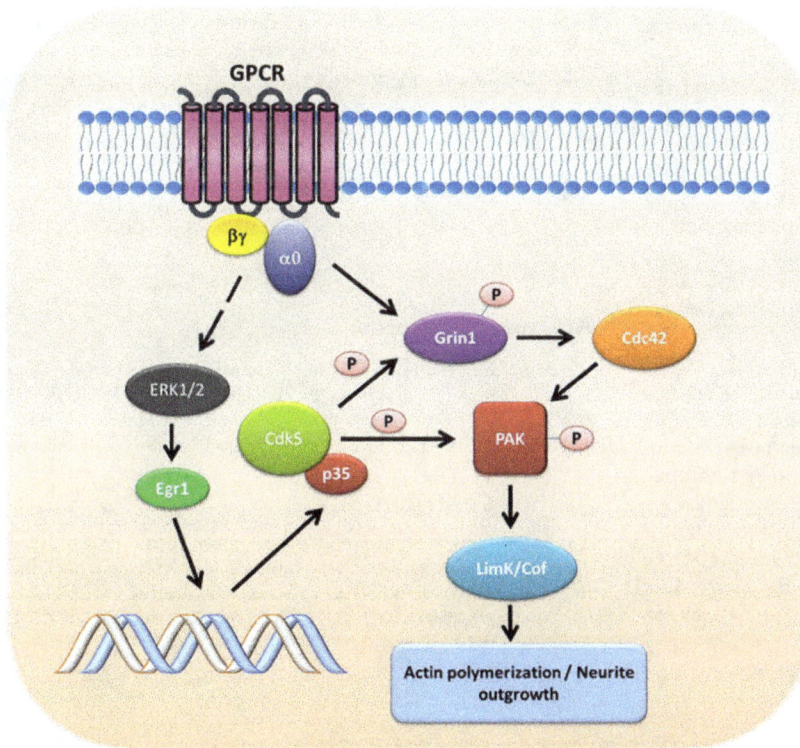

Figure 4. Proposed model illustrating potential roles of phosphorylated Grin1 by Cdk5. A) Phosphorylation of MARCKS by Cdk5 could modulate its interaction with actin filaments leading to stabilization of actin cytoskeleton. B) Involvement of Grin1 phosphorylation by Cdk5 in actin dynamics and neurite outgrowth. GPCR stimulation activates MAPK signaling pathway with increased of Egr1 and p35 expressions and subsequent increases in Cdk5 activity, which in turn phosphorylate Grin1. Additionally, GPCR stimulation promotes neurite outgrowth possibly mediated by the phosphorylation of Grin1 by Cdk5 and Cdc42-PAK-LimK-Cofilin pathway.

which include a minimal consensus motif for Cdk5 phosphorylation, Ser519 and Ser622. Although Ser519 and Ser622 sites in Grin1 were previously reported to be phosphorylated in brain [48,49], our phosphoproteomic analysis found significant decrease only in the phosphorylation of Ser369 and Ser691 sites. This suggests that the phosphorylation of Ser519 and Ser622 could be dependent on other kinases. Since we used an antibody that specifically detects phosphorylation by Cdk5 on the SPXK motif, Ser369 is the epitope in Grin1 phosphorylated by Cdk5. When we overexpressed p35 in N2a cells, we observed a significant increase in the serine phosphorylation of Grin1. This was recognized by the same antibody, while roscovitine treatment restored phosphorylation to the basal level. This indicated that phosphorylation of Ser369 on Grin1 is dependent on the Cdk5 kinase activity Grin1, Gap43 and $G\alpha_{i/o}$ protein are part of a G-couple receptor signaling pathway that regulates neurite growth in neural cells [50]. Interestingly, Gap43 is another protein which is differentially phosphorylated in Cdk5 null brains (Table 1). Grin1 does not contain conserved protein-protein interaction domains, however, it was reported its interaction with the activated subunits of G_z/G_i and G_o [51] which are the proteins associated with G protein coupled receptors (GPCRs). Grin1 is located mainly at neuronal growth cones and when it is co-expressed with G_o in N2A cells induces neurite elongation, suggesting that Grin1 is an effector of G_o [52]. Besides, the co-expression of constitutively active G_o and Grin1 are related to increase Cdc42 activity [17]. It was reported that Cdk5/p35 complex have been associated with motility and stabilization of growth cone during the axon elongation [53,54]. Our results suggest that the phosphorylation of Ser369 on Grin1 could be part of a network signaling controlled by Cdk5, regulating the elongation and maintenance of axons as well as the stability of growth cones. The stimulation of some GPCRs caused MAPK cascade activation [55]. Also, the signal transduction activated by second messenger-dependent kinases and the crosstalk between GPCRs and tyrosine kinases can induce ERK1/2 activation [56]. Interestingly, the ERK1/2 signaling pathway is a major regulator of Cdk5 activity through control of Egr1 and p35 expression [57–59]. Therefore, we suggest that extracellular signals that stimulate GCPRs with a subsequent activation of ERK1/2 can induce the expression of p35 increasing Cdk5 activity and maintaining a sustained response in time, reinforcing a potential signaling cascade through Grin1 necessary for axonal growth (Figure 4B).

Amongst the candidates identified to be potential Cdk5 substrates there is a group of proteins involved in the regulation of microtubule dynamics. This group encompasses Collapsin response mediator protein 1 (crmp1), Collapsin response mediator protein 2 (crmp2), Collapsin response mediator protein 4 (crmp4), microtubule-associated protein 1B (MAP1B), microtubule associated protein 2 (MAP2), tau, doublecortin (DCX) and stathmin. It has been previously shown that crmp1, crmp2 and crmp4 are phosphorylated by Cdk5 [60–62]. MAP1B is the first MAP expressed during nervous system development [63–64]. When phosphorylated by proline-directed protein kinases, such as gsk3β

[65], JNK [66] and Cdk5 [67], becomes highly enriched in the axonal compartment. Currently; antibodies directed against phosphorylated MAP1B are insensitive to Cdk5 inactivation. Therefore, it is likely that epitopes found differentially phosphorylated in this study may serve to identify novel MAP1B phosphorylation involved in axon formation. DCX is a microtubule-associated protein involved in neuronal migration [68]. Phosphorylation of Ser297 in DCX is mediated by Cdk5 and regulates neuronal migration [69]. However, other DCX phosphoepitopes had been described including Ser339 found in this study [70]. It will be interesting to address the consequences for Thr336 and Ser339 phosphorylation upon microtubule dynamics and neuronal migration, the canonical DCX functions. MAP2 is a novel potential Cdk5 substrate. Previously, it was shown that MAP2 can be phosphorylated in CAD cells displaying increased Cdk5 activity [71]. However, the functional role for Cdk5-dependent MAP2 phosphorylation still remains elusive. It is tempting to speculate that such phosphorylation may be related with changes in dendrite formation and plasticity. Tau, is an axonal microtubule associated protein widely expressed in the nervous system. It is abnormally phosphorylated in brain of patients with Alzheimer's disease [72]. Amyloid-β peptide induce tau phosphorylation by activating protein kinases such as gsk3β [73] and Cdk5 [74]. Therefore, decreased tau phosphoepitopes here presented may serve as molecular markers for neurodegeneration associated to Cdk5 functions.

In summary, our phosphoproteomics analysis of Cdk5 null brain identified decreased phosphorylation in several potential Cdk5 proteins that are involved in neuronal morphology, metabolism and signal transduction. These phosphoproteome data may provide a basis for identifying new Cdk5 substrates; however, further work is required to determine how the novel substrates or sites identified in our study regulates process such as cytoskeleton dynamics, neuronal migration and synapses formation and stability.

Supporting Information

Table S1

Acknowledgments

We thank Bradford Hall for helpful discussion. Lawrence Jones for expert editorial corrections, and Perry Blackshear for chicken MARCKS plasmid and Prof. Dr. Tohru Kozasa from Department of Pharmacology, College of Medicine, University of Illinois, Chicago USA for Grin1 reagents.

Author Contributions

Conceived and designed the experiments: EC-V ABK HJ CG-B. Performed the experiments: EC-V DAB EU MP. Analyzed the data: EC-V EU DAB A. Terse HJ A. Toledo CA HCP ABK CG-B. Contributed reagents/materials/analysis tools: CA. Wrote the paper: EC-V EU DB CA A. Toledo HCP HJ ABK CG-B.

References

1. Dhavan R, Tsai LH (2001) A decade of CDK5. Nat Rev Mol Cell Biol 2: 749–759.
2. Contreras-Vallejos E, Utreras E, Gonzalez-Billault C (2012) Going out of the brain: non-nervous system physiological and pathological functions of Cdk5. Cell Signal 24: 44–52.
3. Qin H, Chan MW, Liyanarachchi S, Balch C, Potter D, et al. (2009) An integrative ChIP-chip and gene expression profiling to model SMAD regulatory modules. BMC Syst Biol 3: 73.
4. Liu CC, Hu J, Kalakrishnan M, Huang H, Zhou XJ (2009) Integrative disease classification based on cross-platform microarray data. BMC Bioinformatics 10 Suppl 1: S25.
5. Borquez DA, Olmos C, Alvarez S, Di Genova A, Maass A, et al. (2013) Bioinformatic survey for new physiological substrates of Cyclin-dependent kinase 5. Genomics 101: 221–228.
6. Utreras E, Futatsugi A, Pareek TK, Kulkarni AB (2009) Molecular Roles of Cdk5 in Pain Signaling. Drug Discov Today Ther Strateg 6: 105–111.
7. Ohshima T, Ward JM, Huh CG, Longenecker G, Veeranna, et al. (1996) Targeted disruption of the cyclin-dependent kinase 5 gene results in abnormal

corticogenesis, neuronal pathology and perinatal death. Proc Natl Acad Sci U S A 93: 11173–11178.

8. Ohshima T, Hirasawa M, Tabata H, Mutoh T, Adachi T, et al. (2007) Cdk5 is required for multipolar-to-bipolar transition during radial neuronal migration and proper dendrite development of pyramidal neurons in the cerebral cortex. Development 134: 2273–2282.

9. Arif A (2012) Extraneuronal activities and regulatory mechanisms of the atypical cyclin-dependent kinase Cdk5. Biochem Pharmacol 84: 985–993.

10. Gillardon F, Schrattenholz A, Sommer B (2005) Investigating the neuroprotective mechanism of action of a CDK5 inhibitor by phosphoproteome analysis. J Cell Biochem 95: 817–826.

11. Gillardon F, Steinlein P, Burger E, Hildebrandt T, Gerner C (2005) Phosphoproteome and transcriptome analysis of the neuronal response to a CDK5 inhibitor. Proteomics 5: 1299–1307.

12. Schnack C, Hengerer B, Gillardon F (2008) Identification of novel substrates for Cdk5 and new targets for Cdk5 inhibitors using high-density protein microarrays. Proteomics 8: 1980–1986.

13. Tseng HC, Ovaa H, Wei NJ, Ploegh H, Tsai LH (2005) Phosphoproteomic analysis with a solid-phase capture-release-tag approach. Chem Biol 12: 769–777.

14. Wu WC, Walaas SI, Nairn AC, Greengard P (1982) Calcium/phospholipid regulates phosphorylation of a Mr "87k" substrate protein in brain synaptosomes. Proc Natl Acad Sci U S A 79: 5249–5253.

15. Palmer RH, Schonwasser DC, Rahman D, Pappin DJ, Herget T, et al. (1996) PRK1 phosphorylates MARCKS at the PKC sites: serine 152, serine 156 and serine 163. FEBS Lett 378: 281–285.

16. Mosevitsky MI (2005) Nerve ending "signal" proteins GAP-43, MARCKS, and BASP1. Int Rev Cytol 245: 245–325.

17. Nakata H, Kozasa T (2005) Functional characterization of Galphao signaling through G protein-regulated inducer of neurite outgrowth 1. Mol Pharmacol 67: 695–702.

18. Bradford MM (1976) A rapid and sensitive method for the quantitation of microgram quantities of protein utilizing the principle of protein-dye binding. Anal Biochem 72: 248–254.

19. Dosemeci A, Jaffe H (2010) Regulation of phosphorylation at the postsynaptic density during different activity states of Ca2+/calmodulin-dependent protein kinase II. Biochem Biophys Res Commun 391: 78–84.

20. Rudrabhatla P, Grant P, Jaffe H, Strong MJ, Pant HC (2010) Quantitative phosphoproteomic analysis of neuronal intermediate filament proteins (NF-M/H) in Alzheimer's disease by iTRAQ. FASEB J 24: 4396–4407.

21. Graff JM, Stumpo DJ, Blackshear PJ (1989) Molecular cloning, sequence, and expression of a cDNA encoding the chicken myristoylated alanine-rich C kinase substrate (MARCKS). Mol Endocrinol 3: 1903–1906.

22. Utreras E, Maccioni R, Gonzalez-Billault C (2009) Cyclin-dependent kinase 5 activator p35 over-expression and amyloid beta synergism increase apoptosis in cultured neuronal cells. Neuroscience 161: 978–987.

23. Henriquez DR, Bodaleo FJ, Montenegro-Venegas C, Gonzalez-Billault C (2012) The light chain 1 subunit of the microtubule-associated protein 1B (MAP1B) is responsible for Tiam1 binding and Rac1 activation in neuronal cells. PLoS One 7: e53123.

24. Zolessi FR, Duran R, Engstrom U, Cervenansky C, Hellman U, et al. (2004) Identification of the chicken MARCKS phosphorylation site specific for differentiating neurons as Ser 25 using a monoclonal antibody and mass spectrometry. J Proteome Res 3: 84–90.

25. Toledo A, Zolessi FR, Arruti C (2013) A novel effect of MARCKS phosphorylation by activated PKC: the dephosphorylation of its serine 25 in chick neuroblasts. PLoS One 8: e62863.

26. Thomas PD, Campbell MJ, Kejariwal A, Mi H, Karlak B, et al. (2003) PANTHER: a library of protein families and subfamilies indexed by function. Genome Res 13: 2129–2141.

27. Mi H, Lazareva-Ulitsky B, Loo R, Kejariwal A, Vandergriff J, et al. (2005) The PANTHER database of protein families, subfamilies, functions and pathways. Nucleic Acids Res 33: D284–288.

28. Huang da W, Sherman BT, Lempicki RA (2009) Systematic and integrative analysis of large gene lists using DAVID bioinformatics resources. Nat Protoc 4: 44–57.

29. Sharma P, Veeranna, Sharma M, Amin ND, Sihag RK, et al. (2002) Phosphorylation of MEK1 by cdk5/p35 down-regulates the mitogen-activated protein kinase pathway. J Biol Chem 277: 528–534.

30. Sun KH, Lee HG, Smith MA, Shah K (2009) Direct and indirect roles of cyclin-dependent kinase 5 as an upstream regulator in the c-Jun NH2-terminal kinase cascade: relevance to neurotoxic insults in Alzheimer's disease. Mol Biol Cell 20: 4611–4619.

31. Blom N, Sicheritz-Ponten T, Gupta R, Gammeltoft S, Brunak S (2004) Prediction of post-translational glycosylation and phosphorylation of proteins from the amino acid sequence. Proteomics 4: 1633–1649.

32. Taniguchi H, Manenti S, Suzuki M, Titani K (1994) Myristoylated alanine-rich C kinase substrate (MARCKS), a major protein kinase C substrate, is an in vivo substrate of proline-directed protein kinase(s). A mass spectroscopic analysis of the post-translational modifications. J Biol Chem 269: 18299–18302.

33. Chi Y, Welcker M, Hizli AA, Posakony JJ, Aebersold R, et al. (2008) Identification of CDK2 substrates in human cell lysates. Genome Biol 9: R149.

34. Pant AC, Veeranna, Pant HC, Amin N (1997) Phosphorylation of human high molecular weight neurofilament protein (hNF-H) by neuronal cyclin-dependent kinase 5 (cdk5). Brain Res 765: 259–266.

35. Su SC, Tsai LH (2011) Cyclin-dependent kinases in brain development and disease. Annu Rev Cell Dev Biol 27: 465–491.

36. Terada M, Yasuda H, Kogawa S, Maeda K, Haneda M, et al. (1998) Expression and activity of cyclin-dependent kinase 5/p35 in adult rat peripheral nervous system. J Neurochem 71: 2600–2606.

37. Stumpo DJ, Bock CB, Tuttle JS, Blackshear PJ (1995) MARCKS deficiency in mice leads to abnormal brain development and perinatal death. Proc Natl Acad Sci U S A 92: 944–948.

38. Gatlin JC, Estrada-Bernal A, Sanford SD, Pfenninger KH (2006) Myristoylated, alanine-rich C-kinase substrate phosphorylation regulates growth cone adhesion and pathfinding. Mol Biol Cell 17: 5115–5130.

39. Calabrese B, Halpain S (2005) Essential role for the PKC target MARCKS in maintaining dendritic spine morphology. Neuron 48: 77–90.

40. Scarlett CO, Blackshear PJ (2003) Neuroanatomical development in the absence of PKC phosphorylation of the myristoylated alanine-rich C-kinase substrate (MARCKS) protein. Brain Res Dev Brain Res 144: 25–42.

41. Manenti S, Yamauchi E, Sorokine O, Knibiehler M, Van Dorsselaer A, et al. (1999) Phosphorylation of the myristoylated protein kinase C substrate MARCKS by the cyclin E-cyclin-dependent kinase 2 complex in vitro. Biochem J 340 (Pt 3): 775–782.

42. Toledo A, Arruti C (2009) Actin modulation of a MARCKS phosphorylation site located outside the effector domain. Biochem Biophys Res Commun 383: 353–357.

43. Zolessi FR, Hellman U, Baz A, Arruti C (1999) Characterization of MARCKS (Myristoylated alanine-rich C kinase substrate) identified by a monoclonal antibody generated against chick embryo neural retina. Biochem Biophys Res Commun 257: 480–487.

44. Zolessi FR, Arruti C (2001) Sustained phosphorylation of MARCKS in differentiating neurogenic regions during chick embryo development. Brain Res Dev Brain Res 130: 257–267.

45. Solomonia RO, Apkhazava D, Nozadze M, Jackson AP, McCabe BJ, et al. (2008) Different forms of MARCKS protein are involved in memory formation in the learning process of imprinting. Exp Brain Res 188: 323–330.

46. Hawasli AH, Bibb JA (2007) Alternative roles for Cdk5 in learning and synaptic plasticity. Biotechnol J 2: 941–948.

47. Hawasli AH, Benavides DR, Nguyen C, Kansy JW, Hayashi K, et al. (2007) Cyclin-dependent kinase 5 governs learning and synaptic plasticity via control of NMDAR degradation. Nat Neurosci 10: 880–886.

48. Goswami T, Li X, Smith AM, Luderowski EM, Vincent JJ, et al. (2012) Comparative phosphoproteomic analysis of neonatal and adult murine brain. Proteomics 12: 2185–2189.

49. Huttlin EL, Jedrychowski MP, Elias JE, Goswami T, Rad R, et al. (2010) A tissue-specific atlas of mouse protein phosphorylation and expression. Cell 143: 1174–1189.

50. Nordman JC, Kabbani N (2012) An interaction between alpha7 nicotinic receptors and a G-protein pathway complex regulates neurite growth in neural cells. J Cell Sci 125: 5502–5513.

51. Iida N, Kozasa T (2004) Identification and biochemical analysis of GRIN1 and GRIN2. Methods Enzymol 390: 475–483.

52. Chen LT, Gilman AG, Kozasa T (1999) A candidate target for G protein action in brain. J Biol Chem 274: 26931–26938.

53. Connell-Crowley L, Le Gall M, Vo DJ, Giniger E (2000) The cyclin-dependent kinase Cdk5 controls multiple aspects of axon patterning in vivo. Curr Biol 10: 599–602.

54. Hahn CM, Kleinholz H, Koester MP, Grieser S, Thelen K, et al. (2005) Role of cyclin-dependent kinase 5 and its activator P35 in local axon and growth cone stabilization. Neuroscience 134: 449–465.

55. Bernasconi F, Malgaroli A, Vallar L (2006) Independent regulation of Rap1 and mitogen-activated protein kinase by the alpha chain of Go. Neurosignals 15: 180–189.

56. Luttrell DK, Luttrell LM (2003) Signaling in time and space: G protein-coupled receptors and mitogen-activated protein kinases. Assay Drug Dev Technol 1: 327–338.

57. Harada T, Morooka T, Ogawa S, Nishida E (2001) ERK induces p35, a neuron-specific activator of Cdk5, through induction of Egr1. Nat Cell Biol 3: 453–459.

58. Utreras E, Futatsugi A, Rudrabhatla P, Keller J, Iadarola MJ, et al. (2009) Tumor necrosis factor-alpha regulates cyclin-dependent kinase 5 activity during pain signaling through transcriptional activation of p35. J Biol Chem 284: 2275–2284.

59. Utreras E, Keller J, Terse A, Prochazkova M, Iadarola MJ, et al. (2012) Transforming growth factor-beta1 regulates Cdk5 activity in primary sensory neurons. J Biol Chem 287: 16917–16929.

60. Uchida Y, Ohshima T, Sasaki Y, Suzuki H, Yanai S, et al. (2005) Semaphorin3A signalling is mediated via sequential Cdk5 and GSK3beta phosphorylation of CRMP2: implication of common phosphorylating mechanism underlying axon guidance and Alzheimer's disease. Genes Cells 10: 165–179.

61. Horiuchi Y, Asada A, Hisanaga S, Toh-e A, Nishizawa M (2006) Identifying novel substrates for mouse Cdk5 kinase using the yeast Saccharomyces cerevisiae. Genes Cells 11: 1393–1404.

62. Tanaka H, Morimura R, Ohshima T (2012) Dpysl2 (CRMP2) and Dpysl3 (CRMP4) phosphorylation by Cdk5 and DYRK2 is required for proper positioning of Rohon-Beard neurons and neural crest cells during neurulation in zebrafish. Dev Biol 370: 223–236.

63. Gonzalez-Billault C, Avila J (2000) Molecular genetic approaches to microtubule-associated protein function. Histol Histopathol 15: 1177–1183.

64. Jimenez-Mateos EM, Paglini G, Gonzalez-Billault C, Caceres A, Avila J (2005) End binding protein-1 (EB1) complements microtubule-associated protein-1B during axonogenesis. J Neurosci Res 80:350–359.

65. Trivedi N, Marsh P, Goold RG, Wood-Kaczmar A, Gordon-Weeks PR (2005) Glycogen synthase kinase-3beta phosphorylation of MAP1B at Ser1260 and Thr1265 is spatially restricted to growing axons. J Cell Sci 118: 993–1005.

66. Kawauchi T, Chihama K, Nabeshima Y, Hoshino M (2003) The in vivo roles of STEF/Tiam1, Rac1 and JNK in cortical neuronal migration. EMBO J 22: 4190–4201.

67. Pigino G, Paglini G, Ulloa L, Avila J, Caceres A (1997) Analysis of the expression, distribution and function of cyclin dependent kinase 5 (cdk5) in developing cerebellar macroneurons. J Cell Sci 110 (Pt 2): 257–270.

68. Francis F, Koulakoff A, Boucher D, Chafey P, Schaar B, et al. (1999) Doublecortin is a developmentally regulated, microtubule-associated protein expressed in migrating and differentiating neurons. Neuron 23: 247–256.

69. Tanaka T, Serneo FF, Tseng HC, Kulkarni AB, Tsai LH, et al. (2004) Cdk5 phosphorylation of doublecortin ser297 regulates its effect on neuronal migration. Neuron 41: 215–227.

70. Graham ME, Ruma-Haynes P, Capes-Davis AG, Dunn JM, Tan TC, et al. (2004) Multisite phosphorylation of doublecortin by cyclin-dependent kinase 5. Biochem J 381: 471–481.

71. Tseng HC, Ovaa H, Wei NJ, Ploegh H, Tsai LH. (2005) Phosphoproteomic analysis with a solid-phase capture-release-tag approach. Chem Biol 12:769–777.

72. Hernández F, Lucas JJ, Avila J. (2013) GSK3 and tau: two convergence points in Alzheimer's disease. J Alzheimers Dis Suppl 1:S141–4.

73. Lucas JJ, Hernández F, Gómez-Ramos P, Morán MA, Hen R, Avila J. (2001) Decreased nuclear beta-catenin, tau hyperphosphorylation and neurodegeneration in GSK-3beta conditional transgenic mice. EMBO J 20:27–39.

74. Utreras E, Maccioni R, González-Billault C. (2009) Cyclin-dependent kinase 5 activator p35 over-expression and amyloid beta synergism increase apoptosis in cultured neuronal cells. Neuroscience 161:978–987.

Multiple Exposure and Effects Assessment of Heavy Metals in the Population near Mining Area in South China

Ping Zhuang[1], Huanping Lu[1], Zhian Li[1]*, Bi Zou[1], Murray B. McBride[2]

1 Key Laboratory of Vegetation Restoration and Management of Degraded Ecosystems, South China Botanical Garden, Chinese Academy of Sciences, Guangzhou, China, **2** Department of Crop and Soil Sciences, Cornell University, Ithaca, New York, United States of America

Abstract

The objective of this study was to investigate the levels of Cd, Pb, Cu and Zn in the environment and several important food sources grown and consumed in the vicinity of Dabaoshan mine in Southern China, and evaluate potential health risks among local residents. The Cd, Pb, Cu and Zn concentrations of arable soils and well water near the mines exceeded the quality standard values. The concentrations of Cd and Pb in some food crops (rice grain, vegetable and soybean) samples were significantly higher than the maximum permissible level. The Cd and Pb concentrations in half of the chicken and fish meat samples were higher than the national standard. The residents living near Dabaoshan mine had higher Cd and Pb levels in hair than those of a non-exposed population. The intake of rice was identified as a major contributor to the estimated daily intake of these metals by the residents. The hazard index values for adults and children were 10.25 and 11.11, respectively, with most of the estimated risks coming from the intake of home-grown rice and vegetables. This study highlights the importance of multiple pathways in studying health risk assessment of heavy metal exposure in China.

Editor: Aimin Chen, University of Cincinnati, United States of America

Funding: This research work was financially supported by the National Natural Science Foundation of China (No. 40871221 and No. 41301571) and the Research Fund Program of Guangdong Provincial Key Laboratory of Environmental Pollution Control and Remediation Technology (No. 2013K0008). The funders had no role in study design, data collection and analysis, decision to publish, or preparation of themanuscript.

Competing Interests: The authors have declared that no competing interests exist.

* E-mail: lizan@scbg.ac.cn

Introduction

Mining and smelting activities have had an important role for local and national economies; however, mining-related industries have commonly been performed in an uncontrolled way, giving rise to severe soil erosion and environmental problems, especially heavy metal pollution [1,2]. The discharge of acidic mine drainage (AMD), with elevated levels of heavy metals, can contaminate the downstream water, agricultural soils, food crops and biota and pose a health risk to residents near the mining areas [3,4]. Heavy metal contamination by mining is a major environmental concern on a global scale, particularly in developing countries. Within the global mineral resources industry, China has been one of the largest producers and consumers of several metalliferous and nonmetallic mineral commodities for many years. In China, there are about 8,000 state-owned mining enterprises and 230,000 collectively owned mines that produce hundreds of millions of tons of mining wastes annually [5]. As a result, health related incidents caused by heavy metal pollution in China have risen sharply since 2005, with major accidents attracting nationwide attention [6,7]. It was reported that high Cd in rice on the Chinese market was mainly the result of contaminated fields affected by AMD. Health risks in mine areas affect not only workers, but the whole population living around the areas, in particular children, so that millions of people in the world are estimated to be exposed to metals in mine areas [8–10].

There exist multiple exposure pathways for residents living close to mining or mineral-processing sites, including direct ingestion of soil and water, dermal contact by contaminated soil and water, inhalation of dusts, and consumption of food crops and animals. Various studies have been conducted to evaluate human health risks due to heavy metal exposure through soil [11], water [12], rice [13,14], vegetables [15,16], and even dust [17] from metalliferous mining areas throughout the world, such as Romania, Poland, Korea, China, and France. In general, dietary intake has been recognized as the main route of exposure for most populations, although inhalation can play an important role in very contaminated sites [18,19]. Simple media or pathway-specific approaches to risk assessment may fail to ensure public safety, so it is necessary to apply multi-pathway risk analysis involving all relevant environmental media to identify the dominant pathway of potential concern [9,20].

For metal toxicity monitoring and human health risk assessment, human hair has been widely used in biomonitoring of heavy metals in recent years to estimate environmental exposure levels and assess nutritional status [21]. As a metabolically inactive tissue, hair has become well established, especially for investigating levels of and changes in many heavy metals that accumulate in the body [22]. Human hair analysis has the advantages over other tissues of being less invasive to sample, more convenient to store and transport, and less hazardous to handle. Furthermore, hair levels of metals are less sensitive to immediate intake and could therefore also be a useful biological indicator in characterizing long-term exposure to the measured metal contaminant.

The aims of this study are: (1) to determine the levels of heavy metal concentrations in the environment and foodstuffs near the mining area, (2) to provide a better understanding of each exposure pathway and evaluate the potential health impacts of these metals on the general population. The results of this assessment will aid development of management options and health intervention policies for the affected areas near metal mining and smelting in China and around the world.

Materials and Methods

Ethics Statement

This research was approved by South China Botanical Garden, Chinese Academy of Sciences. The permit for each location was obtained by the authority of Shaxi town. The specific location of field study was provided in the text. The vegetables were purchased from the farmer. All necessary permits were obtained for the described field study. The field studies did not involve endangered or protected species. All participants were informed about the objectives and methods of the study before the investigation. And written consent was obtained from all participants. Data will be made available upon request.

Study Area

Dabaoshan mine (24°31′37″ N; 113°42′49″ E), the largest mine in South China, is located in eastern Shaoguan city, Guangdong, and has been in full-scale operation as a large-scale and integrative quarrying mine since the 1970s (Figure 1). The climate in this area is characterized by a humid subtropical climate, with an annual average temperature of 20.3°C and rainfall of 1762 mm. The Hengshi river (Southward) originates at the mine site and is the main drainage pathway for effluent from the Dabaoshan mine, with the Chuandu river (Northward) being another drainage

pathway. The rivers thus deliver significant quantities of heavy metals to numerous villages in this region. After about 40 years of exposure to several heavy metals, some local residents in the mining area have begun to acquire upper gastrointestinal diseases; specifically, oesophageal and stomach cancer are reported to be prevalent. Thus, certain villages around Dabaoshan mine have been termed endemic cancer villages, with a mortality rate approaching 56% in humans [23]. Previous studies in this area have reported that arable soil and foodstuffs in the vicinity of the mine were severely polluted by heavy metals [24] and that there was an increased risk of behavioral problems in school-aged children associated with metal exposures [25]. Wang et al [26] also reported elevated human cancer mortality rates among the metal-exposed populations in Dabaoshan mine area. Six sampling sites near the mine were selected as the study areas (See Supporting Information S1). The agricultural soil in this region was repeatedly irrigated with polluted water from the Hengshi river and Chuandu river. There are about 60–200 households in each village, and they have similar population structures, living conditions and lifestyle.

Sample Collection

Several different environmental media samples, including 122 soils, 54 rice grains, 320 vegetables, 30 soybeans, 12 sediments of fishponds, 150 fishes, 20 well waters, 48 chickens, and 64 hair samples were collected from different sites around Dabaoshan mine (Table S1). Rice grains and tens of species of home-grown vegetables (Table S2), and their corresponding soil samples were collected from SX, DS, FD, LQ, SB, and XJ villages during November 2007 – May 2008. There are 3–5 replications for each vegetable. Soil samples were collected from the top 0–10 cm layer. Soybean was sampled from DS and FD villages in August 2009. At each site of sampling, three to five subsamples were collected to form a composite sample which was stored in a polyethylene zip-bag, and immediately transported to the laboratory. During April - August 2009, a total of 24 wells were selected and sampled from

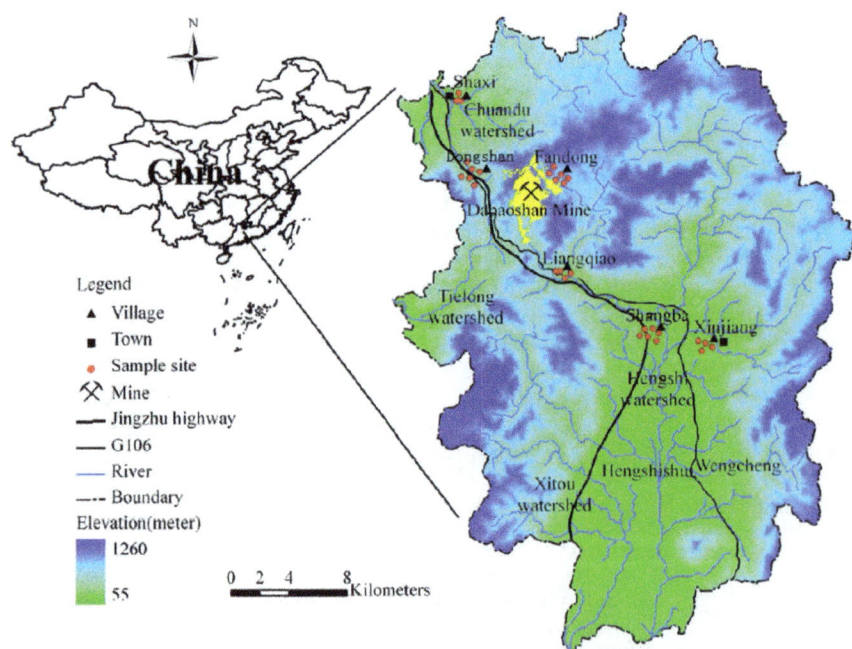

Figure 1. The location of sampling sites in the vicinity of Dabaoshan mine in Guangdong province in South China. The six sampled villages are Shaxi, Dongshan, Fandong, Liangqiao, Shangba and Xinjiang.

DS, FD and SB villages because these well waters were consumed by the local residents.

Sediment samples from 3 fishponds at FD village were collected from the surface down to a depth of 10 cm at five different locations, and these samples were pooled together. Six species of fish and sediment samples were sealed in polyethylene bags and kept cold on ice during transportation to the laboratory. The number of fish, total length, fresh weight (fw) and habitat of fish samples from three fishponds are shown in Table S3.

An experimental group of chicken was fed with rice grain (grown in FD contaminated soil), and the control group was fed with chickenfeed (bought from the market). The concentrations of Pb, Cd, Zn and Cu in the metal-enriched rice and chickenfeed fed with chickens are shown in Table S4. After feeding for 6 months, pectoral muscle of all chickens were separately dissected from the body of the specimens, frozen in liquid nitrogen, and freeze-dried. To characterize the exposure level and metal accumulation in human body, hair samples were taken from 24 inhabitants of both DS and FD village (n = 48). Sixteen further hair samples were collected from non-exposed populations. All the study participants gave their permission to be included in the study.

The soil samples were air-dried at room temperature, then pulverized and sieved through a 1 mm stainless-steel mesh. All home-grown food crops were washed thoroughly with Milli-Q water, and the fresh weight of the samples were recorded. All fish and sediment samples were sealed in polyethylene bags and kept cold on ice during transportation to the laboratory. Sediment samples were air-dried, crushed, sieved through a 2 mm screen, then pulverized and passed through a 0.2 mm mesh sieve.

Analysis of Samples

Four heavy metals (Cd, Pb, Cu and Zn) were analyzed in all the samples. Soil and sediment samples were digested in preparation for total metal analysis using a concentrated acid mixture (HNO_3, $HClO_4$ and HF). For vegetables and soybeans, dried samples were digested with HNO_3 and $HClO_4$ in a 5:1 ratio until a transparent solution was obtained [27]. Chicken muscle, fish muscle and hair samples were digested in HNO_3 (16 mol/L) and H_2O_2 (30%) by microwave. Each hair sample was cut into smaller sections, thoroughly mixed, and ample portions were washed using the method described by Altshul et al [28]. The concentrations of heavy metals in soil and foodstuff samples were analyzed using an atomic absorption spectrophotometer (AAS, GBC932AA), with the concentrations of Cd and Pb in rice grain, vegetables, fish and chicken being determined using graphite furnace atomic absorption spectrophotometer (GFAAS, GBC932AA). The hair digestion procedure was the same as the method described by Wang et al. [21]. The heavy metal concentrations of hair samples were analyzed with an inductively coupled plasma mass spectrometer (ICP-MS) (Agilent 7700x, Agilent Scientific Technology Ltd., USA).

Data Calculation

The translocation capability of heavy metals from the soil to the edible part of crops can be described using a bioaccumulation factor (BAF). The BAFs of Cd, Pb, Cu and Zn were calculated as follows:

$$BAF = \frac{Cplant}{Csoil}$$

where Cplant and Csoil on dry weight basis represent the heavy metal concentration in edible part of food crops and soils, respectively.

The average estimated daily intake (EDI) of heavy metals by the human subjects was calculated using the following equation, which is recommended by the US EPA [29].

$$EDI = \frac{C \times IR \times EF \times ED}{BW \times AT}$$

where EDI is the average daily intake or dose through ingestion (µg/kg bw/day); C is the heavy metal concentration in the exposure medium (mg/L or mg/kg); IR is the ingestion rate (L/day, or kg/day); EF is the exposure frequency (365 days/year); ED is the exposure duration (70 years, equivalent to the average lifespan); BW is the body weight (kg). Average adult and child body weights were considered to be 60 and 30 kg, respectively and AT is the time period over which the dose is averaged (365 days/year times number of exposure years, assumed to be 70 years in this study). A questionnaire-based survey was conducted in the studied villages to determine key risk factors such as dietary behaviors, daily activities and lifestyle of local people. We invited 50 local residents in each village to participate in the survey. In this study area, most foodstuffs are self-produced, whereas pork is purchased from the market. The average considered daily intakes of adults and children is shown in Table S5, according to the survey and reports by Ma et al. [30] and Zhai et al. [31].

The human health risk posed by heavy metal exposure are usually characterized by the target hazard quotient (THQ) [29], the ratio of the average estimated daily intake resulting from exposure to site media compared to the reference dose (RfD) for an individual pathway and chemical. Oral reference dose obtained from the Integrated Risk Information System [32], is an estimation of maximum permissible risk to a human population through daily exposure when taking into consideration a sensitive group during a lifetime. The applied RfD for Cd, Pb, Cu and Zn was 1.0, 4.0, 40, 300 µg/kg/d, respectively. The THQ based on non-cancer toxic risk is determined by

$$THQ = \frac{EDI}{RfD}$$

If the value of THQ is less than 1, the risk of non-carcinogenic toxic effects is assumed to be low. When it exceeds 1, there may be concerns for potential health risks associated with overexposure.

To assess the overall potential risk of adverse health effects posed by more than one metal, the THQs can be summed across contaminants to generate a hazard index (HI) to estimate the risk of a mixture of contaminants. The HI refers to the sum of more than one THQ for multiple substances and/or multiple exposure pathways. In the present study, the HI was used as a screening value to identify whether there is significant risk caused by heavy metals through average dietary consumption for the residents living near the Dabaoshan mine.

Quality Control

Appropriate quality assurance procedures and precautions were carried out to ensure reliability of the results for the different environmental media under investigation. Double distilled deionised water was used throughout the study. Glassware was properly cleaned, and the reagents were of analytical grade. Reagent blank determinations were used to correct the instrument readings. For validation of analytical procedures, a recovery study was carried out by spiking and homogenizing several already analyzed samples with varying amounts of standard solutions of the metals. Several standard reference materials (SRM) were obtained from the National Research Center for CRMs (Table S6) and used for

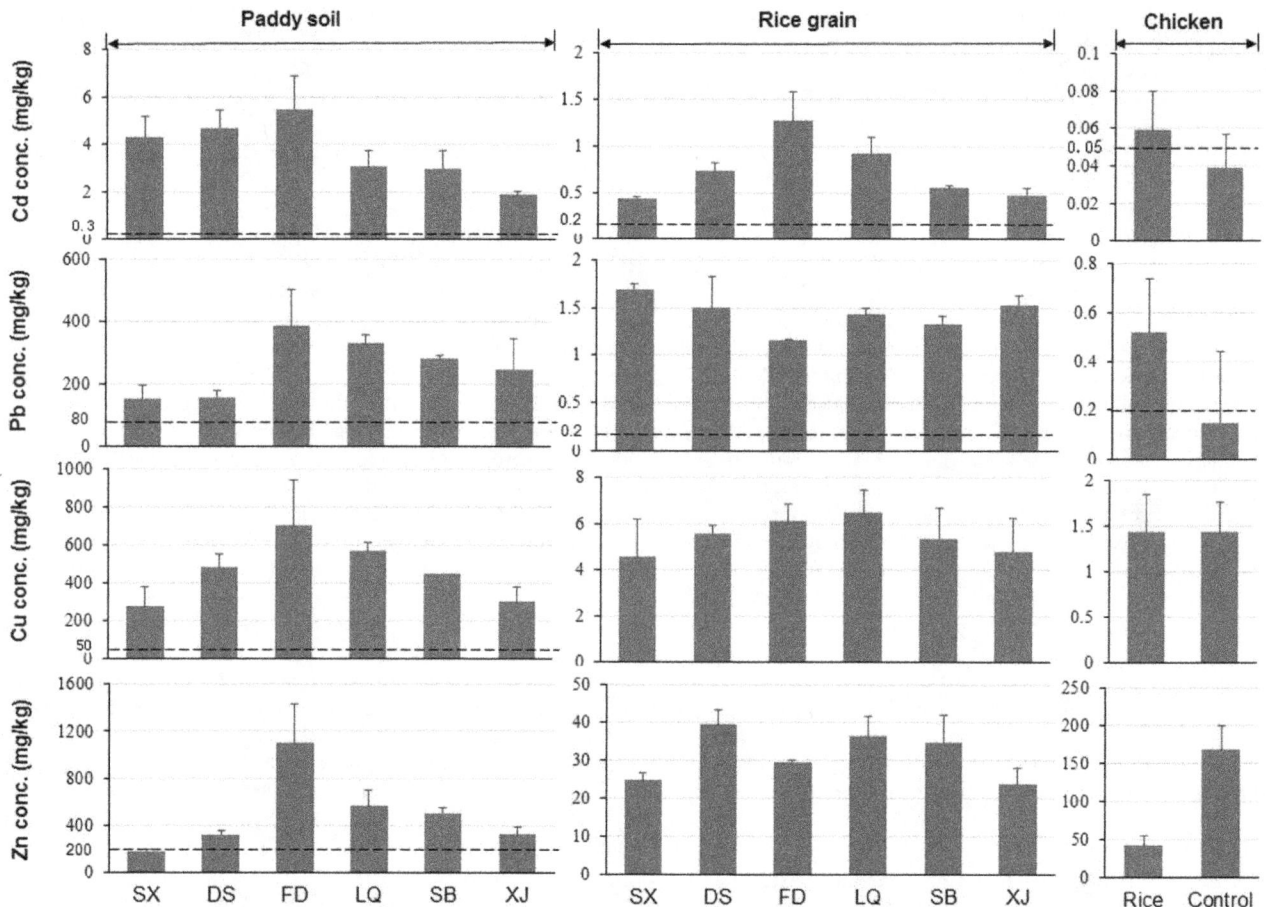

Figure 2. Comparison of concentrations (mg/kg, mean ± SD) of Cd, Pb, Cu and Zn and their respective Chinese national quality value (dotted line) in paddy soil, rice grain and chicken muscle samples collected at different localities in the vicinity of Dabaoshan mine. The number of samples is 9 for respective paddy soil, 8–10 for rice grain, 24 for each group chicken. For soil and rice samples, they were collected from six specific locations, including Shixi (SX), Dongshan (DS), Fandong (FD), Liangqiao (LQ), Shangba (SB) and Xinjiang (XJ). The chicken was fed with rice grain collected from Fandong village.

validation of the analytical procedure. Blank and drift standards were run after every twenty determinations to maintain instrument calibration. The coefficient of variation of replicate analyses was determined for the measurements to calculate analytical precision.

Results and Discussion

Heavy Metals in Different Environmental Media and Food Chain

The environmental sample analysis results showed widespread heavy metal (Cd, Pb, Cu and Zn) contamination in the different exposure media around the Dabaoshan mine (Figures 2–3). Among the six sites in the study area, the paddy soil samples collected from FD showed the highest concentrations of Cd (5.5 mg/kg), Pb (386 mg/kg), Cu (703 mg/kg), and Zn (1100 mg/kg), presumably because FD is located on the mountaintop of the Dabaoshan mine (Figure 2). In contrast, the soil samples collected from XJ, which is located far from Dabaoshan mines (>15 km), showed the lowest heavy metal concentrations. These results suggest that elevated heavy metal concentrations in soils are associated with the mining activities, and indicate that this area is unsuitable for agricultural use. The highest Cd concentration (4.9 mg/kg) in garden soils was found in DS village, whereas soil Pb (297 mg/kg) from LQ was significantly higher compared with

the other sites. Concerning co-located arable soils vegetated by various food crops, generally the paddy soil contained higher heavy metal concentrations than the garden soil. This may be due to the fact that the paddy soils were irrigated with highly heavy metal-contaminated stream water whereas the water sources of the garden soils were mainly derived from well water or rainfall.

In comparison to Chinese soil quality guideline values, the soil concentrations of Cd, Pb, Cu and Zn were frequently exceeded (Figures 2–3). Distance from the mine was a key determinant of soil pollution, as there are significant spatial differences among these six studied villages. This finding was consistent with the previous studies [3]. The heavy metal concentrations in soil samples around Dabaoshan mine were remarkably high in general, being comparable with those recorded near a Pb–Zn mine of Spain [33], and higher than those reported for the Songcheon Au–Ag mine in Korea [34], an old mining area in Romania [15] and an abandoned mine in Thailand [35]. From Figures 2–3, it is clear that the characteristics of soil significantly affect the soil level at which the heavy metals will exceed recommended limits in food crops and pose risks in terms of food safety and animal health.

The heavy metal concentrations in rice grain from the six sites decreased in the order Zn>Cu>Pb>Cd (Figure 2). The maximum concentrations of Cd (1.27 mg/kg) at FD and Pb

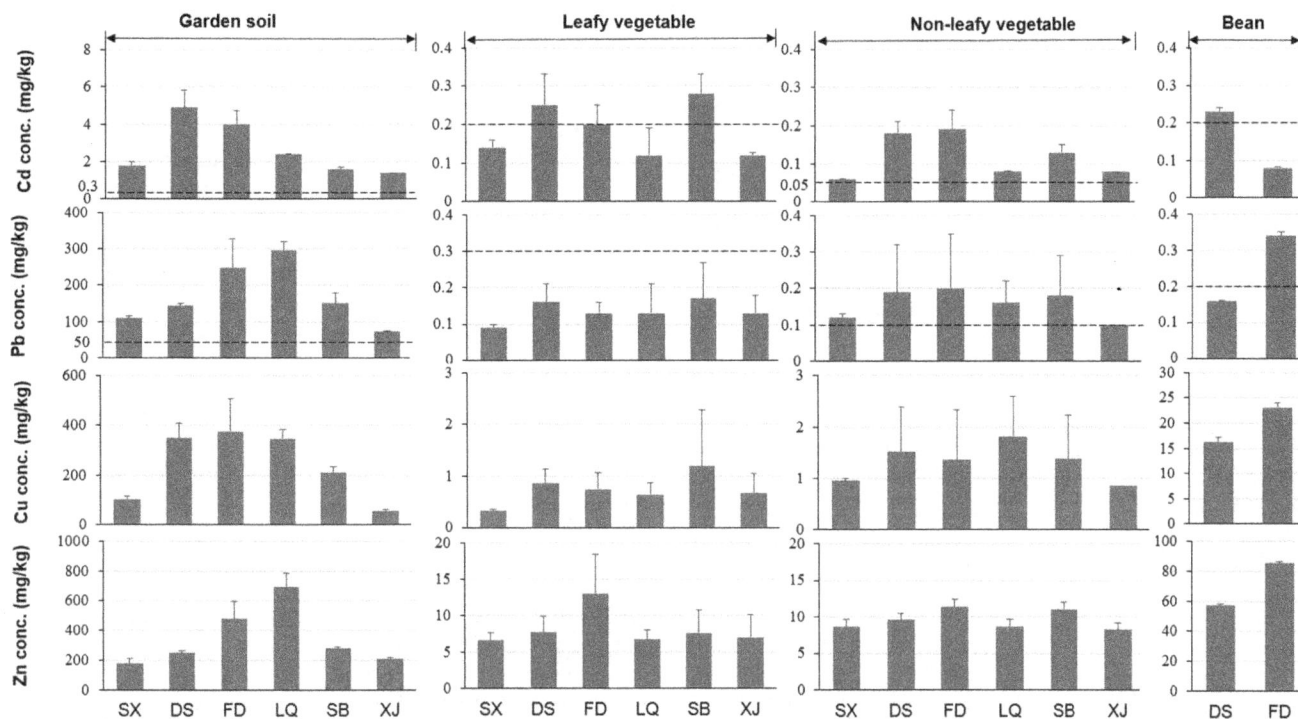

Figure 3. Comparison of concentrations (mg/kg, mean ± SD) of Cd, Pb, Cu and Zn and their respective Chinese national quality value (dotted line) in garden soil, leafy vegetables, non-leafy vegetables and bean samples from different locations. The number of samples is 9 for respective garden soil, 3–5 for each vegetable, 15 for respective soybean. For soil and vegetable samples, they were collected from six specific locations, including Shixi (SX), Dongshan (DS), Fandong (FD), Liangqiao (LQ), Shangba (SB) and Xinjiang (XJ). The bean samples were collected from DS and FD villages.

(1.69 mg/kg) at SX were approximately 6–8 folds greater than the maximum permissible level for both metals of 0.2 mg/kg of rice [36]. The relatively high heavy metal concentrations measured in the self-produced food crops were generally consistent with the elevated heavy metal concentrations found in the paddy and garden soils, although the consistently elevated Pb in rice grain did not follow the site-to-site variation of soil Pb concentration (Figure 2). In comparison with other mining areas, the highest Cd concentration (1.27 mg/kg) in rice grain collected from the present study area was 6–8 times higher than concentrations reported by Lee (0.15 mg/kg, in the Au-Ag-Pb-Zn mining area) [37] and by Ji (0.224 mg/kg, near abandoned metal mine) [38] from Korea, and about 5 times lower than those recorded in the Chenzhou Pb/Zn mining area from China [13], i.e. 6.99 mg/kg. The Pb concentrations in rice grain were higher than those reported for a Pb/Zn mining area from China (0.8 mg/kg) [13] and those reported for an abandoned metal mine from Korea [38].

The concentrations of heavy metals in garden soils and self-produced vegetables (e.g., mustard, Chinese cabbage, lettuce, spinach, garden pea, and tuber of sweet potato, etc. see Table S2) are presented in Figure 3. The leafy and non-leafy vegetables contained higher heavy metal concentrations, especially of Cd, in villages DS, FD and SB. Comparison of the results with the food quality guidelines as set by the Chinese government indicated that the Cd limit was exceeded in 60% of the vegetable samples. For soybean, the Cd concentration in DS and Pb concentration in FD (Figure 3) exceeded the maximum permissible level of 0.2 mg/kg [36]. The concentrations of heavy metals in homegrown vegetable samples around Dabaoshan mine were comparable with those reported in an old mining area of Romania [15], and higher than what was recorded in an abandoned metal mine area of Korea [38]. Of all the vegetables tested, leafy vegetables always contained higher Cd than the non-leafy vegetables (Figure 3).

Contamination of local wetlands, ponds, and rivers by sluicing waste and AMD could provide a pathway for heavy metals into

Table 1. Heavy metal concentrations (mg/L, mean ± SD) in well water sampled from three sites near Dabaoshan mine.

	SB	DS	FD	National standard[a]
Cd	0.0095±0.000	0.01±0.001	0.015±0.001	0.005
Pb	0.013±0.001	0.008±0.000	0.017±0.008	0.01
Cu	1.42±0.012	1.29±0.031	1.58±0.091	1.0
Zn	3.63±0.82	3.47±0.18	4.81±1.05	1.0

[a]Standards for drinking water quality (GB5749-2006) set by the Ministry of Health of the People's Republic of China.

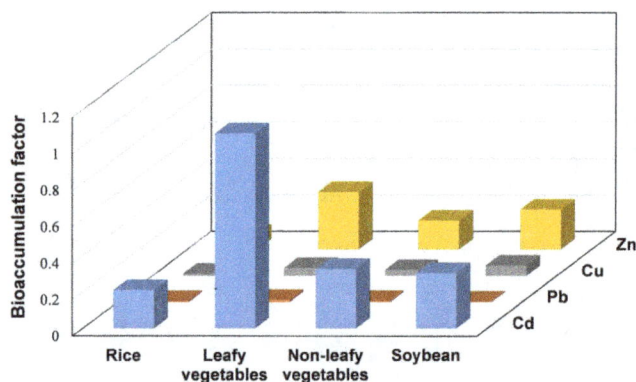

Figure 4. Bioaccumulation factor (BAF), a ratio of heavy metals concentration in the edible part of rice, leafy vegetables, non-leafy vegetables and soybean to that in the corresponding soil.

the aquatic food chain [39]. The Hengshi and Chuandu Rivers are the source of drinking and irrigation water for the residents near Dabaoshan mine, and groundwater also may be contaminated by the two rivers. The concentrations of heavy metals in the well water collected from DS, FD and SB villages exceeded the national standard (Table 1), indicating the drinking water in the local private wells were contaminated by acidic mine water from the Dabaoshan mine, which agreed with previous results [40]. In view of these results, a public water supply system was installed and the community was advised not to use private wells.

Heavy metal analyses in sediments taken from three different sites at FD village show that the levels of Cd (0.32–8.91 mg/kg), Pb (262–327 mg/kg), Cu (239–1477 mg/kg) and Zn (386–4524 mg/kg) were markedly higher than normal background levels (Figure S1). The heavy metal levels of fodder (Ryegrass) cultivated around the fishponds and fed to fish were 1.65, 3.95, 18.02 and 88.01 mg/kg fw for Cd, Pb, Cu and Zn, respectively. It is a positive result that average Cd, Cu and Zn concentrations in fish muscle tissue were within maximum permissible levels. However, the concentrations of Pb in more than 60% of the fish fresh samples from site 1 were above the maximum permissible level (Figure S1). These results suggested the high translocation of metals in site 1 were likely due to the low sediment pH values (5.22), which might cause the heavy metals to be more soluble and bioavailable [41]. Elevated concentrations of heavy metals in the water environment and its human health risk as a consequence of historical mining have been reported elsewhere all over the world [42,43]. Therefore, mining effluents increase metal levels in the aquatic system, and consumption by fish of contaminated sediments and water originating from mining operations is an important exposure route that can result in metal accumulation in fish.

Soil-to-plant transfer is one of the key components of human exposure to metals through the food chain. As shown in Figure 4, large variations in BAFs were observed among different food crops and metals. The results showed that BAFs of all food crops for these tested metals were in the order: Cd>Zn>Cu>Pb, consistent with a report by Li et al. [44]. The BAF values for Cd varied from 0.21 to 1.07, with the highest level in leafy vegetables. Leafy vegetables usually grow quickly and have high transpiration rate, which was in agreement with previous study [45]. The high BAFs of Cd and Zn for respective environmental media were similar to the results for some crops from Chenzhou Pb/Zn mine reported by Liu et al. [13]. Generally, the BAF of heavy metals is controlled

by the chemical speciation of heavy metals in soil, soil properties, such as pH and salinity, and plant physiological features [46,47]. In the present study, the lower pH in the sandy soil (Table S7) can increase the solubility of heavy metal and may transfer into the crop tissues.

In the third food chain exposure route investigated (soil-rice-chicken), it is notable that Cd and Pb concentrations in chicken muscle fed with metal-contaminated rice grain (grown in contaminated soil at FD) were higher than those of the control (Figure 2). The levels of Cd and Pb in muscle of chicken fed with contaminated rice exceeded the maximum permissible levels for Cd and Pb in meat prescribed by China. For livestock, the ingestion of rice or grass is the most relevant pathway for the intake of heavy metals [48]. From the data produced in the present study, it is clear that cultivation of food crops on contaminated soil for human or livestock consumption can potentially lead to the uptake and accumulation of trace metals in the edible plant parts, with a resulting risk to human and animal health in the Dabaoshan mining area.

Exposure Assessment

Human exposure to heavy metals in soils near the Dabaoshan mine might occur directly through the ingestion of soil or indirectly through consumption of locally grown vegetables and dairy and meat products from locally raised farm animals. Table 2 summarizes the results of the actual exposure assessment for adults and children exposed to heavy metals through five exposure pathways around Dabaoshan mine. For adults, the total EDI were 5.45, 10.59, 45.14 and 278.38 µg/kg bw/d for Cd, Pb, Cu and Zn, respectively, which were much higher than those reported for an abandoned metal mine from Korea [38]. The EDIs of Cd, Pb and Cu for adults through these five exposure pathways were 5.45, 2.64 and 1.13 fold higher than the RfD of these metals, respectively. It should be cautioned that the EDIs of heavy metals for children were slightly higher than those for adults. The biggest contribution to the intake of heavy metals came from rice, with tens of times as much as intake through others ingestion pathway, as shown in Table 2.

Health risks to residents in the study area through the consumption of agricultural products, and through inadvertent ingestion of soil were assessed by estimating target hazard quotients (THQ). A THQ value greater than 1 would indicate that a potential health risk may exist. The THQs of heavy metals from multiple consumption pathways is in decreasing order Cd>Pb>Cu>Zn (Table 2). The THQs of all heavy metal in rice were the highest among all the crops, which higher than those grown on reclaimed tidal flat soil in the Pearl River Estuary [44]. Considering that rice is a staple food in the diet of the people in the area, intake of heavy metals through rice consumption is likely to be a main source of heavy metal intake among residents in the area. The HI values (the sum of all THQs) through diet and soil for adults and children in the Dabaoshan mine area were 10.25 and 11.11, respectively; both are higher than reported in other areas [16,49]. These results suggest that the adults and children living around the Dabaoshan mine may experience adverse health effects. Considering THQ determination for the different metals and pathways for children, besides rice, the highest risk was for ingestion of vegetables for Cd (0.71), followed in descending order by ingestion of soil for Pb (0.32) and ingestion of bean for Cu (0.26). Concern has focused particularly on children who become exposed to heavy metals to a greater extent than adults, which may harm brain and nervous system development [50]. Despite Zn having the highest concentration among the metals, its THQ value

Table 2. Estimated daily intake (EDI, ug/kg bw/d) and target hazard quotient (THQ) for adult and children exposed to contaminants in the vicinity of Dabaoshan mine.

Metal	Pathways	EDI		THQ	
		Adult	Child	Adult	Child
Cd	Ingestion of rice	4.54	4.83	4.54	4.83
	Ingestion of vegetables				
	Leafy vegetables	0.54	0.42	0.54	0.42
	Non-leafy vegetables	0.26	0.21	0.26	0.21
	Bean	0.04	0.08	0.04	0.08
	Ingestion of meat			0.03	0.04
	Fish	0.03	0.04		
	Chicken	0.02	0.02	0.02	0.02
	Ingestion of soil	0.00	0.02	0.00	0.02
	Ingestion of water	0.02	0.02	0.04	0.04
	Total	5.45	5.63	5.47	5.65
Pb	Ingestion of rice	8.93	9.50	2.23	2.38
	Ingestion of vegetables				
	Leafy vegetables	0.39	0.31	0.10	0.08
	Non-leafy vegetables	0.24	0.19	0.06	0.05
	Bean	0.07	0.13	0.02	0.03
	Ingestion of meat				
	Fish	0.56	0.57	0.14	0.14
	Chicken	0.16	0.21	0.04	0.05
	Ingestion of soil	0.22	1.30	0.05	0.32
	Ingestion of water	0.02	0.02	0.01	0.00
	Total	10.59	12.23	2.65	3.06
Cu	Ingestion of rice	34.11	36.31	0.85	0.91
	Ingestion of vegetables				
	Leafy vegetables	2.13	1.67	0.05	0.04
	Non-leafy vegetables	1.67	1.34	0.04	0.03
	Bean	5.57	10.48	0.14	0.26
	Ingestion of meat				
	Fish	0.85	0.86	0.02	0.02
	Chicken	0.43	0.57	0.01	0.01
	Ingestion of soil	0.39	2.32	0.01	0.06
	Ingestion of water	2.69	2.15	0.07	0.05
	Total	45.14	53.55	1.20	1.39
Zn	Ingestion of rice	194.86	207.43	0.65	0.69
	Ingestion of vegetables				
	Leafy vegetables	23.45	18.33	0.08	0.06
	Non-leafy vegetables	14.62	11.70	0.05	0.04
	Bean	20.23	38.08	0.07	0.13
	Ingestion of meat				
	Fish	4.74	3.39	0.02	0.01
	Chicken	12.60	16.80	0.04	0.06
	Ingestion of soil	0.41	2.49	0.00	0.01
	Ingestion of water	7.46	5.96	0.02	0.02
	Total	278.38	304.17	0.93	1.02
Hazard index				10.25	11.11

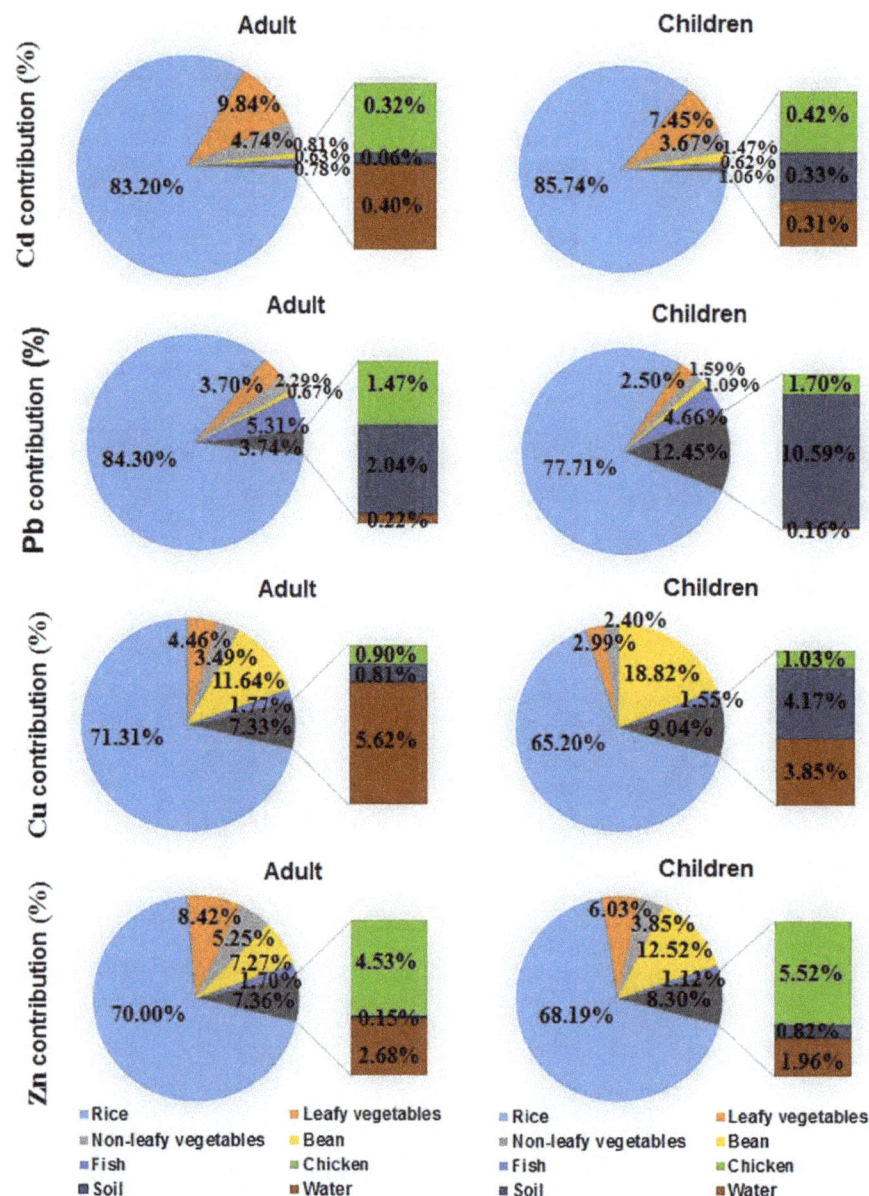

Figure 5. Relative contribution profile of daily human intake. Relative contribution was derived by dividing the daily intake estimate from a given source by the total daily intake estimate from all of the sources under investigation.

was below 1, indicating that Zn does not pose a health risk to the local residents.

Mining sites are usually characterized by soil contamination and a higher exposure to heavy metals for the population residing in the area [51,52]. Exposure level and risk assessment in the study area are compared with those of other studies [9,15,35,38,49,53–56] from mining and smelting sites around the world as shown in Table 3. The health risk for residents living in the vicinity of Dabaoshan mine associated with the consumption of local foodstuffs was markedly higher than that reported for other metal mines all over the world. Some evidences also revealed a large part of the population in the vicinity of Dabaoshan mine had contacted or ingested contaminated soil [24,57], contaminated water [23,40], consumed local agricultural products, such as foodcrops [24], fish [58] and chicken, over a long period of time. These concerns were further supported by the finding of behavioral

problems for school-aged children [25] and significantly elevated blood levels of metals (38.9 and 24.1 µg/L for Pb and Cd in high exposed area vs 4.46 and 1.87 µg/L for Pb and Cd in low exposed area) [26] in local residents from the same mining area. Liu et al. [23] and Wang et al. [26] reported that an increased risk of mortality from all cancer (e.g. enteron tumors) was probably associated with long-term environmental exposure to both Cd and Pb.

Source Allocation for Different Metals

When the profiles of relative contribution of soil and foodstuffs to daily heavy metal intake estimates from all the sources being investigated (e.g., rice grain, vegetables, beans, fish, chicken, water and soil) were compared between adults and children (Figure 5), intake of rice was the major source of Cd, Pb, Cu and Zn exposure, accounting for 65.2–85.7% of the estimated total daily

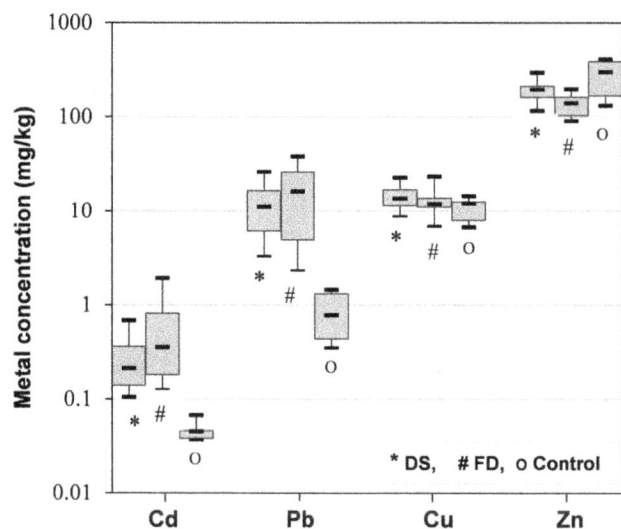

Figure 6. Box-plot diagram of four elements in hair samples from two villages (Dongshan, DS, Fandong, FD) in the vicinity of Dabaoshan mine and the control area (non-exposed site). Y-axis is presented in logarithmic scale. The central mark on each box is the median with the edges of the 25th and 75th percentiles.

intake in the study area. Interestingly, soil ingestion was a second important factor for daily Pb intake, accounting for 10.6% of children's exposure. Among all the main exposure pathways, the most important exposure pathway for heavy metals appeared to be the ingestion of locally grown rice, with about several times as much as intake through vegetables. The allocations of Cd and Pb from dietary exposure were comparable with that reported for Pb by Dong and Hu [20], but higher than the contribution from diet of 70% and 50% for Cd and Pb, respectively, estimated by Plant et al. [18].

The results of this study suggested that the contribution to total Cd and Pb exposure from drinking water is even lower than 1%,

which was consistent with the finding by Dong and Hu [20]. It is clear that the allocation of dietary exposure is the highest in the study area, and the contributions by the soil and water exposure pathways are relatively low. Specifically for children, a health risk from Pb was indicated, with the second most important exposure pathway being the direct ingestion of soil. The source allocation of Pb for soil accounted for 10.6% of children's exposure, which was similar to that reported for an abandoned metal mine area of Korea (11.1–16.8%) [38]. Staessen et al. [19] also found that 2–4% of the variance in the long-term body burden of Cd was directly related to consumption of vegetables, implying that the ingestion of locally grown vegetables is an important source for human Cd exposure.

Compared to other countries, the relatively high source contribution from the diet in China is due to the high concentrations of heavy metals in the food. The high heavy metal concentrations in grains and vegetables were mainly due to enrichment from the soil [59]. The substantial contribution of rice and soil to the intake of four heavy metals indicate that the status of metal contamination of soil and rice should receive further attention in the metal mine areas throughout the country. In fact, the potential health risk from heavy metals can be for some individuals much higher than our calculations based on average ingestion of water, food and soil. Some local inhabitants, who consume more contaminated locally grown food crops, breathe contaminated air, or smoke, might be exposed to a health risk from dietary heavy metals well above the calculated risk. As a result, great effort is required to control concentrations of heavy metals in soil to reduce dietary metal exposure, and ultimately to eliminate heavy metal exposure in China.

Heavy Metal Levels in Hair and Global Health Implications

Human hair can serve as a useful direct biomonitoring tool to assess the extent of heavy metal exposure to residents in metal-polluted areas. The heavy metal concentrations in the hair samples are shown in Figure 6. The hair levels of four heavy metals were in the order of Zn>Pb, Cu>Cd. The hair samples from the residents

Table 3. Comparison of estimated dietary intake (EDI) and hazard index (HI) of heavy metals in different areas around the world.

Sites	EDI (ug/kg bw/d) or HI	Cd	Pb	Cu	Zn	References
Dabaoshan mine, Shaoguan, China	EDI	5.45–5.63	10.6–12.2	50.4–59.1	292–320	In this study
	HI	5.45–5.63	2.65–3.06	1.26–1.48	0.97–1.07	
Abandoned metal mine, Kanchanaburi, Thailand	EDI	0.25–5.34	0.71–1.46			[35]
	HI	0.25–5.34	0.18–0.36			
Abandoned metal mine, Katowice, Poland	HI	2.6–2.7		0.025–0.041	0.11–0.18	[56]
Old mining area, Banat, Romania	HI	<1	>1	<1	<1	[15]
Lead-zinc mining area, Jiangsu, China	HI	1.81–3.32	3.01–16.2	0.164–0.189	0.28–0.30	[9]
Huludao Zn plant, Liaoning, China	EDI	0.70	1.36	45.5	204	[49]
	HI	0.749	0.364	1.22	0.731	
Pb/Sb smelter, Nanning, China	HI	7.42	1.35	0.146	0.323	[53]
Chatian mercury mining deposit Hunan, China	EDI	0.05–1.66	0.359–1.6			[55]
	HI	0.05–1.66	2.67–2.99			
Songcheon Au-Ag mine, Gangneung-si, Korea	HI	0.75		0.078	0.26	[54]
Abandoned metal mine, Korea	EDI	0–0.195	0.001–0.233	0.009–13.4	0.208–66.6	[38]
	HI	0.179	0.077	0.027	0.063	

living in the Dabaoshan mine area (DS and FD villages) contained markedly higher average Cd, Pb and Cu concentrations than those of non-exposed populations, with Cd around 5–11 times higher (0.27–0.57 mg/kg) and Pb 15–28 times higher (12.5–23.9 mg/kg) than the non-exposed levels, which might be a consequence of long term exposure of the local residents to the mining activities. Some subjects in FD showed even higher levels of Cd and Pb, with the maxima estimated at 1.94 and 38 mg/kg, respectively. The average levels of hair Pb, Cd, Cu and Zn in the study area were lower than the levels in the scalp hair of adults from S. Domingos mine, Portugal [60], but higher than those recorded in a lead-zinc mining area from China [9]. Interestingly, the Zn concentration in hair of a non-exposed population was higher than that of subjects living near the Dabaoshan mine, which might be due to the higher concentration of toxic metals like Cd and Pb in the vicinity, which can interact with Zn and replace it in the heme enzymes and metallothioneins [61]. Consistent with the higher THQ value of the residents in FD village, their hair samples exhibited higher than those in DS village. This finding agreed well to the conclusions of previous studies [9]. Therefore, the elevated hair Cd and Pb levels provide an independent indicator that residents in the Dabaoshan mine region might be at high risk of toxic metal exposure due to the elevated levels of heavy metals in food crops, fish or livestock.

Increasing evidence shows heavy metal pollution of mined areas to cause health damage to the local inhabitants [4,26]. For example, itai-itai disease was caused by Cd poisoning in Japan due to the Kamioka mine releasing this metal into river water that was then used to irrigate rice paddy soils [62]. Moreover, tens of thousands of Kabwe's residents (Zambia) suffered from severe lead poisoning, which resulted from artisanal re-mining of and exposures to wastes from historical lead-zinc mining and smelting [63]. In northern Nigeria, confirmed deaths of at least 400 children were the result of artisanal processing of lead-rich gold ores [64]. There is some evidence to suggest that local inhabitants may develop a natural tolerance to the high soil metal concentrations encountered in their environment, since non-local people who have supposedly suffered from heavy metal poisoning caused by the consumption of home produced vegetables grown in gardens reclaimed from former mine dumps [61,65]. Thus, by understanding human exposure pathways for toxic metals, the routes by which heavy metals may enter the body and cause health effects, scientists can help identify other mining areas that may pose the highest risk for heavy metal poisoning and need for medical surveillance and intervention.

Supporting Information

Figure S1 Comparison of concentrations (mg/kg) of Cd, Pb, Cu and Zn and their respective Chinese national quality value (dotted line) in muscle samples from three different fishponds at Fandong village. Y-axis of Pb, Cu and Zn concentrations are presented in logarithmic scale.

Table S1 Number of soil, rice, vegetable, soybean, water, sediment, fish and chicken samples collected from different sampling sites in the present study.

Table S2 The edible part and water content of vegetables.

Table S3 The number of fish, total length, fresh weight (fw) and habitat of fish samples.

Table S4 Heavy metal concentrations (mg/kg, mean ± SD) in the metal-enriched rice grain and chickenfeed for both groups of chicken.

Table S5 Average considered daily intakes of adults and children in the present study.

Table S6 Standard reference materials for respective sample under investigation.

Table S7 The pH and organic matter (OM) of paddy and garden soils from Dabaoshan mine.

Supporting Information S1 Description of study area.

Acknowledgments

We thank Prof. Aimin Chen (University of Cincinnati, USA), the editor and two reviewers (Prof. Iosif Gergen and an anonymous reviewer) for their comments on an early version of this paper.

Author Contributions

Conceived and designed the experiments: PZ ZAL. Performed the experiments: PZ HPL. Analyzed the data: PZ. Contributed reagents/materials/analysis tools: PZ BZ. Wrote the paper: PZ MBM.

References

1. Rybicka EH (1996) Impact of mining and metallurgical industries on the environment in Poland. Appl Geochem 11: 3–9.
2. Dudka S, Adriano DC (1997) Environmental impacts of metal ore mining and processing: a review. J Environ Qual 26: 590–602.
3. Benin AL, Sargent JD, Kalton M, Roda S (1999) High concentrations of heavy metals in neighborhoods near ore smelters in northern Mexico. Environ Health Perspect 107: 279–184.
4. Plumlee GS, Morman SA (2011) Mine wastes and human health. Element 7: 399–404.
5. Shu WS, Zhang ZQ, Lan CY (2000) Strategies for restoration of mining wastelands in China. Ecol Sci 19: 24–29.
6. Gao Y, Xia J (2011) Chromium contamination accident in China: viewing environment policy of China. Environ Sci Technol 45: 8065–8056.
7. Zhang XW, Yang LS, Li YH, Li HR, Wang WY, et al. (2012) Impacts of lead-zinc mining and smelting on the environment and human health in China. Environ Monit Assess 184: 2261–2273.
8. Plumlee GS, Durant JT, Morman SA, Neri A, Wolf RE, et al. (2013) Linking geological and health sciences to assess childhood lead poisoning from Artisanal gold mining in Nigeria. Environ Health Perspect 121: 744–750.
9. Qu CS, Ma ZW, Yang J, Liu Y, Bi J, et al. (2012) Human exposure pathways of heavy metals in a Lead-Zinc mining area, Jiangsu Province, China. PLoS ONE 7: e46793.
10. Wcisło E, Ioven D, Kucharski R, Szdzuj J (2002) Human health risk assessment case study: an abandoned metal smelter site in Poland. Chemosphere 47: 507–515.
11. Kapusta P, Szarek-Łukaszewska G, Stefanowicz AM (2011) Direct and indirect effects of metal contamination on soil biota in a Zn-Pb post-mining and smelting area (S Poland). Environ Pollut 159: 1516–1522.
12. Park JH, Choi KK (2013) Risk assessment of soil, water and crops in abandoned Geumryeong mine in South Korea. J Geochem Explor 128: 117–123.
13. Liu HY, Probst A, Liao BH (2005) Metal contamination of soils and crops affected by the Chenzhou lead/zinc mine spill (Hunan, China). Sci Total Environ 339: 153–166.
14. Zhu YG, Sun GX, Lei M, Teng M, Liu YX, et al. (2008) High percentage inorganic arsenic content of mining impacted and nonimpacted Chinese rice. Environ Sci Technol 42: 5008–5013.
15. Harmanescu M, Alda LM, Bordean DM, Gogoasa I, Gergen I (2011) Heavy metals health risk assessment for population via consumption of vegetables

grown in old mining area; a case study: Banat County, Romania. Chem Cent J 5: 64.

16. Wang X, Sato T, Xing B, Tao S (2005) Health risk of heavy metals to the general public in Tianjin, China via consumption of vegetables and fish. Sci Total Environ 350: 28–37.

17. Kerin EJ, Lin HK (2010) Fugitive dust and human exposure to heavy metals around the Red Dog Mine. Rev Environ Contam Toxicol 206: 49–63.

18. Plant J, Smith D, Smith B, Williams L (2001) Environmental geochemistry at the global scale. Appl Geochem 16: 1291–1308.

19. Staessen JA, Vyncke G, Lauwerys RR, Roels HA, Celis HG, et al. (1992) Transfer of cadmium from a sandy acidic soil to man: a population study. Environ Res 58: 24–34.

20. Dong ZM, Hu JY (2012) Development of lead source-specific exposure standards based on aggregate exposure assessment: Bayesian inversion from biomonitoring information to multipathway exposure. Environ Sci Technol 46: 1144–1152.

21. Wang T, Fu JJ, Wang YW, Liao CY, Tao YQ, et al. (2009) Use of scalp hair as indicator of human exposure to heavy metals in an electronic waste recycling area. Environ Pollut 157: 2445–2451.

22. Druyan ME, Bass D, Puchyr R (1998) Determination of reference ranges for elements in human scalp hair. Biol Trace Elem Res 62: 183–187.

23. Liu YS, Gao Y, Wang KW, Mai XH, Chen GD, et al. (2005) Etiologic study on alimentary tract malignant tumor in villages of high occurrence. China Trop med 5: 1139–1141. (in Chinese).

24. Zhuang P, McBride MB, Xia HP, Li NY, Li ZA (2009) Health risk from heavy metals via consumption of food crops in the vicinity of Dabaoshan mine, South China. Sci Total Environ 407: 1551–1561.

25. Bao QS, Lu CY, Song H, Wang M, Ling WH, et al. (2009) Behavioural development of school-aged children who live around a multi-metal sulphide mine in Guangdong province, China: a cross-sectional study. BMC Publ Health 9: 217.

26. Wang M, Song H, Chen WQ, Lu CY, Hu QS, et al. (2011) Cancer mortality in a Chinese population surrounding a multi-metal sulphide mine in Guangdong province: an ecologic study. BMC Publ Health 11: 319.

27. Allen SE, Grimshaw HM, Rowland AP (1986) Chemical analysis, in: Moore, P.D., Chapman, S.B. (Eds.), Methods in Plant Ecology. Oxford, London: Blackwell Scientific Publication.

28. Altshul L, Covaci A, Hauser R (2004) The relationship between levels of PCSs and pesticides in human hair and blood: preliminary results. Environ Health Perspect 112: 1193–1199.

29. US EPA (United States Environmental Protection Agency) (1997) Exposure factors handbook. EPA/600/P-95/002F. Washington, DC.

30. Ma WJ, Deng F, Xu YJ, Xu HF, Nie DP (2005) The study on dietary intake and nutritional status of residents in Guangdong, South China. J Prev Med 31: 1–5. (in Chinese).

31. Zhai FY, He YN, Ma GS, Li YP, Wang ZH, et al. (2005) Study on the current status and trend of food consumption among Chinese population. Chin J Epidemiol 26(7): 485–488. (in Chinese).

32. US EPA (United States Environmental Protection Agency) (2007) Integrated Risk Information System-database. Philadelphia PA; Washington, DC.

33. Rodríguez L, Ruiz E, Alonso-Azcárate J, Rincón J (2009) Heavy metal distribution and chemical speciation in tailings and soils around a Pb–Zn mine in Spain. J Environ Manage 90: 1106–1116.

34. Lee JS, Chon HT, Kim KW (2005) Human risk assessment of As, Cd, Cu and Zn in the abandoned metal mine site. Environ Geochem Health 27: 185–191.

35. Nobuntou W, Parkpian P, Kim Oanh NT, Noomhorm A, Delaune RD, et al. (2010) Lead distribution and its potential risk to the environment: Lesson learned from environmental monitoring of abandon mine. J Environ Sci Health A Tox/Hazard Subst Environ Eng 45: 1702–1714.

36. MHPRC (Ministry of Health of the People's Republic of China) (2012) The limits of pollutants in foods (GB 2762–2012) (in Chinese). Beijing, China: MHPRC.

37. Lee CG, Chon HT, Jung MC (2001) Heavy metal contamination in the vicinity of the Daduk Au-Ag-Pb-Zn mine in Korea. Appl Geochem 16: 1377–1386.

38. Ji K, Kim J, Lee M, Park S, Kwon HJ, et al. (2013) Assessment of exposure to heavy metals and health risks among residents near abandoned metal mines in Goseong, Korea. Environ Pollut 178: 322–328.

39. Moiseenko TI, Kudryavtseva LP (2001) Trace metal accumulation and fish pathologies in areas affected by mining and metallurgical enterprises in the Kola Region, Russia. Environ Pollut 114: 285–297.

40. Chen A, Lin C, Lu W, Wu Y, Ma Y, et al. (2007) Well water contaminated by acidic mine water from the Dabaoshan Mine, South China: Chemistry and toxicity. Chemosphere 70: 248–255.

41. Adhikari S, Ghosh L, Ayyappan S (2006) Combined effects of water pH and alkalinity on the accumulation of lead, cadmium and chromium to *Labeo rohita* (Hamilton). Int J Env Sci Tech 3: 289–296.

42. Borgmann U, Couillard Y, Grapentine LC (2007) Relative contribution of food and water to 27 metals and metalloids accumulated by caged *Hyalella azteca* in two rivers affected by metal mining. Environ Pollut 145: 753–765.

43. Proctor PD (1984) Heavy metal additions to the environment near mines, mills, and smelters, Southeast Missouri. In: Nriagu, J.O. (Ed.), Environmental Impacts of Smelters. Wiley, New York; 89–115.

44. Li QS, Chen Y, Fu HB, Cui ZH, Shi L, et al. (2012) Health risk of heavy metals in food crops grown on reclaimed tidal flat soil in the Pearl River Estuary, China. J Hazard Mat 227–228: 148–154.

45. Albering HJ, van Leusen SM, Moonen EJC, Hoogewerff JA, Kleinjans JCS (1999) Human health risk assessment: A case study involving heavy metal soil contamination after the flooding of the river Meuse during the winter of 1993–1994. Environ Health Perspect 107: 37–43.

46. Golia EE, Dimirkou A, Mitsios IK (2008) Influence of some parameters on heavy metals accumulation by vegetables grown in agricultural soils of different soil orders. B Environ Contam Toxicol 81: 80–84.

47. McLaughlin MJ, Whatmuff M, Warne M, Heemsbergen D, Barry G, et al. (2006) A field investigation of solubility and food chain accumulation of biosolid-cadmium across diverse soil types. Environ Chem 3: 428–432.

48. Rodrigues SM, Pereira ME, Duarte AC, Römkens PF (2012) Soil-plant-animal transfer models to improve soil protection guidelines: a case study from Portugal. Environ Int 39: 27–37.

49. Zheng N, Wang QC, Zhang XW, Zheng DM, Zhang ZS, et al. (2007) Population health risk due to dietary intake of heavy metals in the industrial area of Huludao city, China. Sci Total Environ 383: 81–89.

50. Rieuwert JS, Farago ME, Cikrt M, Bencko V (2000) Differences in lead bioavailability between a smelting and a mining area. Water Air Soil Pollut 122: 203–229.

51. Alonso E, Cambra K, Martinez T (2001) Lead and cadmium exposure from contaminated soil among residents of a farm area near an industrial site. Arch Environ Health 56: 278–282.

52. Paoliello M, De Cpitani E, Cunha F, Matsuno T, Carvalho M (2002) Exposure of children to lead and cadmium from a mining area of Brazil. Environ Res 88: 120–128.

53. Cui YJ, Zhu YG, Zhai RH, Chen DY, Huang ZH, et al. (2004) Transfer of metals from soil to vegetables in an area near a smelter in Nanning, China. Environ Int 30: 785–791.

54. Lim HS, Lee JS, Chon HT, Sager M. (2008) Heavy metal contamination and health risk assessment in the vicinity of the abandoned Songcheon Au-Ag mine in Korea. J Geochem Explor 96: 223–230.

55. Sun HF, Li YH, Ji YF, Yang LS, Wang WY, et al. (2010) Environmental contamination and health hazard of lead and cadmium around Chatian mercury mining deposit in western Hunan Province, China. Trans Nonferrous Met Soc China 20: 308–314.

56. Wcisło E, Ioven D, Kucharski R, Szdzuj J. (2002) Human health risk assessment case study: an abandoned metal smelter site in Poland. Chemosphere 47(5): 507–515.

57. Zhou JM, Dang Z, Cai MF, Liu CQ (2007) Soil heavy metal pollution around the Dabaoshan mine, Guangdong province, China. Pedosphere 17: 588–94.

58. Zhuang P, Li ZA, McBride MB, Zou B (2013) Health risk assessment for consumption of fish originating from ponds near Dabaoshan mine, South China. Environ Sci Pollut Res 20(8): 5844–5854.

59. Hough RL, Breward N, Young SD, Crout NM, Tye AM, et al. (2004) Assessing potential risk of heavy metal exposure from consumption of home-produced vegetables by urban populations. Environ Health Perspect 112: 215–221.

60. Pereira R, Ribeiro R, Goncalves F (2004) Scalp hair analysis as a tool in assessing human exposure to heavy metals (S. Domingos mine, Portugal). Sci Total Environ 327: 81–92.

61. Waalkes MP (2003) Cadmium carcinogenesis. Mutat Res 533: 107–120.

62. Abrahams PW (2002) Soils: their implications to human health. Sci Total Environ 291: 1–32.

63. Branan N (2008) Mining leaves nasty legacy in Zambia. Geotimes website. Available: http://www.geotimes.org/jan08/article.html?id = nn_zambia.html. Accessed 2008 Jan.

64. UNEP/OCHA (2010) Lead Pollution and Poisoning Crisis, Environmental Emergency Response Mission, Zamfara State, Nigeria, United Nations Environment Program, September/October.

65. Thomas R (1980) Arsenic pollution arising from mining activities in south-west England. Inorganic Pollution and Agriculture. London: HMSO, 142–158.

Getting More Out of Less – A Quantitative Serological Screening Tool for Simultaneous Detection of Multiple Influenza A Hemagglutinin-Types in Chickens

Gudrun S. Freidl[1,2]*, **Erwin de Bruin**[2], **Janko van Beek**[1,2], **Johan Reimerink**[2], **Sjaak de Wit**[3], **Guus Koch**[4], **Lonneke Vervelde**[5¤], **Henk-Jan van den Ham**[1], **Marion P. G. Koopmans**[1,2]

1 Department of Viroscience, Erasmus Medical Center, Rotterdam, The Netherlands, 2 Emerging Infectious Diseases, Division of Virology, Centre for Infectious Diseases Research, Diagnostics and Perinatal Screening, Centre for Infectious Disease Control, National Institute for Public Health and the Environment (RIVM), Bilthoven, The Netherlands, 3 GD Animal Health Service (AHS), Deventer, The Netherlands, 4 Central Veterinary Institute (CVI), Lelystad, The Netherlands, 5 Utrecht University, Utrecht, The Netherlands

Abstract

Current avian influenza surveillance in poultry primarily targets subtypes of interest for the veterinary sector (H5, H7). However, as virological and serological evidence suggest, surveillance of additional subtypes is important for public health as well as for the poultry industry. Therefore, we developed a protein microarray enabling simultaneous identification of antibodies directed against different HA-types of influenza A viruses in chickens. The assay successfully discriminated negative from experimentally and naturally infected, seropositive chickens. Sensitivity and specificity depended on the cut-off level used but ranged from 84.4% to 100% and 100%, respectively, for a cut off level of $\geq$1:40, showing minimal cross reactivity. As this testing platform is also validated for the use in humans, it constitutes a surveillance tool that can be applied in human-animal interface studies.

Editor: Hiroshi Nishiura, The University of Tokyo, Japan

Funding: This project was funded by the European Commission's FP7 program under the umbrella of the Antigone project - ANTIcipating the global onset of novel epidemics (project number 278976, http://antigonefp7.eu/). Part of the serum samples were generated within the FES program on Avian Influenza financed by the Ministry of Economic Affairs, the Netherlands. The funders had no role in study design, data collection and analysis, decision to publish, or preparation of the manuscript.

Competing Interests: The authors have declared that no competing interests exist.

* Email: gudrun.freidl@rivm.nl

¤ Current address: The Roslin Institute and R(D)SVS, University of Edinburgh, Edinburgh, United Kingdom

Introduction

Avian influenza A viruses (AIV) belong to the family *Orthomyxoviridae* and comprise eight gene segments consisting of negative sense single-stranded RNA. The classification of AIV into different subtypes is based on two surface structures, hemagglutinin (HA) and neuraminidase (NA). To date, 18 distinct HA-types and 11 NA-types are known of [1–3]. With the exception of subtypes H17N10 and H18N11 of which RNA was recently detected in bats, aquatic birds constitute reservoirs for AIV, usually without showing signs of disease [1,2]. To date, influenza A viruses have crossed the species barrier to humans, swine, aquatic mammals, domestic poultry, birds of prey, horses, mustelids, civets, felines and canines [4–6]. Several avian and swine influenza viruses have zoonotic potential. While AIV subtype A (H5N1) virus infections have had the largest economic and public health impact so far, AIV with HA types 6, 7, 9 and 10 have also caused virologically confirmed human infection with varying severity [4,6]. Until recently, human H7-infections have been associated with mild symptoms in humans. However, since early 2013, a newly emerging H7-subtype, A(H7N9), has formed an exception by causing a more severe clinical picture and death in

about 36% of the recorded patients, possibly related to specific host susceptibility factors [7,8]. Although the symptoms shown by patients largely resembled infection with highly pathogenic (HP) A(H5N1), the manifestation in poultry – the putative source of direct human infection – is different [9,10]. Unlike HP A(H5N1) viruses that cause severe illness and death in poultry, this novel influenza A(H7N9) strain causes subclinical infection in poultry, which allowed the virus to spread unnoticed over a large geographic region in China [10]. Consequently, the general population can be exposed to animals shedding this virus without warning signs. Indeed, serological investigations in poultry workers suggest more widespread infections in humans, possibly reflecting mild or unapparent illness [11].

This example and additional serological evidence for human infection with influenza viruses *other than* H5, H7, H9 and H10 – including H4, H6 and H11 [12–14] – highlight the importance of influenza monitoring at the human-animal interface, where humans are currently sentinels for circulation of zoonotic viruses [15,16]. Therefore, ideally, future serological studies evaluating influenza viruses at the human-animal interface would include these "neglected" subtypes.

Given the ability of AIV H5 and H7 to mutate into HP forms and the economic consequences associated with such infections, a compulsory European Union-wide surveillance system was implemented in 2005 [17]. In the Netherlands, serological monitoring is more intensive than required by EU-regulations [18] and includes screening of all poultry flocks at least once a year and high risk-groups, e.g. free-range flocks every three months. In practice, a representative number of farms and individuals per country are pre-screened with an indirect or competitive enzyme-linked immunosorbent assay (ELISA), identifying antibodies against conserved regions (matrix or nucleoprotein) that all influenza virus subtypes have in common [19]. Upon a positive pre-screening result, the presence of H5- or H7-antibodies is confirmed or ruled out by means of a hemagglutination inhibition (HI)-assay, and flocks are tested for active virus circulation. While this screening regimen meets the requirements for veterinary surveillance, the characterization of non-H5 and -H7 but ELISA-positive samples may be relevant for the poultry industry and for public health.

Here, we describe the development and use of a protein microarray (PA) that enables simultaneous screening for antibodies to multiple influenza HA-types in poultry, using minute quantities of serum (10 μl) that can be collected through routine veterinary surveillance.

Materials and Methods

Sera

Three different serum sets (hereafter referred to as group 1–3) were used to evaluate the performance of the PA for the use in chicken:

1. **Negative sera.** Negative sera were obtained from different sources. In total 38 chicken sera which tested negative by ELISA (Idexx FlockChek AI, MultiS-Screen Ab Test Kit, Hoofddorp, the Netherlands) were used:

1a) One serum pool of 52-week-old, specific pathogen free (SPF) white layers (flock from GD AHS)

1b) Ten sera from 3-week-old, non-infected, non-vaccinated, conventional Lohman Brown layers

1c) 27 sera from a commercial 6-week-old Ross broiler flock (hereafter named "negative field chickens")

2. **Consecutive serum samples from SPF chickens experimentally infected with live field strains.** Four groups of 15 white SPF laying hens (GD AHS) were intratracheally infected with live field strains (0.5 ml; ~10^6 EID$_{50}$) belonging to the subtypes H5N2, H6N2, H7N1 or H9N2 (Table 1) at 12 weeks of age. For the duration of the experiment infected chickens were kept in isolators with twelve hours light per day, 20–25°C and were given ad libitum access to food and water. Serum was collected from the wing vein at day 7, 14 and 22 post infection (p.i.) and seropositivity was confirmed by testing sera at one dilution (1:8) by standard HI-assay, as is done routinely in the animal health service. Therefore, data were available as positive/negative results only.

3. **Sera from outbreaks of avian influenza detected during routine surveillance in the Netherlands.** To evaluate applicability of the test in the field, we analyzed samples from four different laying hen flocks having undergone past infection with low pathogenic (LP) AIV subtypes, hereafter named "naturally infected field chickens". All flocks were identified as

AIV exposed by ELISA-testing (Idexx FlockChek AI, MultiS-Screen Ab Test Kit) of samples collected during routine surveillance performed by the AHS. HI typing of sera, and/or virus isolation and virus typing (CVI, Lelystad) confirmed initial diagnosis. Samples were derived from two outbreaks caused by subtype H6, both in flocks of 16-month-old, free-range brown laying hens (outbreak 1: n = 10; outbreak 2: n = 7). In addition, ten sera seropositive for LP H7N3 were obtained from 16-month-old, free-range brown layers, and eight sera from an H9N2-outbreak in 19-month-old, brown laying hens housed in cages were screened. Individual HI-titers were available for the H9- and one H6-outbreak (outbreak 2). Sera of the remaining two outbreaks were screened qualitatively at one dilution only (1:8).

Ethics statement

All experiments were approved by the Animal Experimental Committee of the Faculty of Veterinary Medicine of the Utrecht University or the Animal Welfare Committee (DEC) of the GD Animal Health Service, Deventer, the Netherlands, in accordance with the Dutch regulations on experimental animals.

Production of protein microarray-slides and sample analysis

We used a modification of the technique that has been described elsewhere [20]. In our study, 22 recombinant HA1-proteins comprising representatives of 13 different subtypes (Table 1) were printed onto 16-pad nitrocellulose slides as described before [20]. Antigens were produced in human embryonic kidney (HEK) cells, were purified by HIS-tag and were delivered at a protein concentration of 1 mg/ml (see manufacturer for details, Table 2). To determine the optimal working concentration for the recombinant HA1-proteins used in the PA, checkerboard titrations were performed for each protein using four different dilutions (2×, 4×, 8×, 16×). When necessary, proteins were concentrated using Amicon Ultra-0.5 mL Centrifugal Filters for Protein Purification and Concentration according to manufacturer's instructions (Merck Millipore, Massachusetts, USA) and checkerboard titrations were repeated thereafter.

Prior to testing, all sera were inactivated in a water bath at 56°C for one hour. For serum analysis, four slides fixed in a FAST frame slide holder (Whatman, Kent, UK) could be used simultaneously. Each holder accommodated up to seven sera and one in house-standard. Serum was titrated in two fold dilution series ranging from 1:20 (10 μl of serum) to 1:2560. Known negative sera were tested in two-fold dilutions ranging from 1:20 to 1:160. An in house-standard, comprising of a serum-pool of hyperimmunized chickens infected with strains of subtypes H5, H6, H7 and H9 was included in each test run. After serum incubation, bound antibodies were visualized using a Cy5 AffiniPure rabbit anti-chicken IgY Fc-fragment-specific conjugate (Jackson ImmunoResearch, West Grove, USA) diluted in Blotto Blocking Buffer (Thermo Fisher Scientific Inc., Rockford, MA, USA) and 0.1% Surfact-Amps (Thermo Fisher Scientific Inc.) at a concentration of 1:1300. IgY represents the avian equivalent of mammalian IgG [21].

Data analysis and statistics

Fluorescent signals were quantified and converted into titers as described before [20]. The PA spanned a detection range of titers from 1:20 to 1:2560. We calculated geometric mean titers (GMTs) including 95% confidence intervals (CI) as well as homologous versus heterologous GMT ratios of the validation data using GraphPad Prism for Windows (Version 6.03, GraphPad Software

Table 1. Recombinant HA1-proteins used on the PA and viruses used for infection of chickens of group 2.

Proteins	Subtype	Strain	Source
H1.18	H1N1	A/South Carolina/1/18	IT
H1.33	H1N1	A/WS/33	IT
H1.99	H1N1	A/New Caledonia/20/99	IT
H1.07	H1N1	A/Brisbane/59/2007	IT
H1.09	H1N1	A/California/6/2009	IT
H2.05	H2N2	A/Canada/720/05	IT
H3.68	H3N2	A/Aichi/2/1968	SB
H3.03	H3N2	A/Wyoming/3/03	IT
H4.02	H4N6	A/mallard/Ohio/657/2002	E
H5.97	H5N1	A/Hong Kong/156/97 (clade 0)	IT
H5.06	H5N1	A/Turkey/15/2006 (clade 2.2)	G
H5.02	H5N8	A/duck/NY/191255-59/2002 (LP)	SB
H5.07	H5N3	A/duck/Hokkaido/167/2007 (LP)	SB
H6.07	H6N1	A/northern shoveler/California/HKWF115/2007	SB
H7.03	H7N7	A/Chicken/Netherlands/1/03	IT
H8.79	H8N4	A/pintail duck/Alberta/114/1979	E
H9.99	H9N2	A/Guinea fowl/Hong Kong/WF10/99	IT
H9.07	H9N2	A/Chicken/Yunnan/YA114/2007	G
H11.02	H11N2	A/duck/Yangzhou/906/2002	IT
H12.91	H12N5	A/green-winged teal/ALB/199/1991	IT
H13.00	H13N8	A/black-headed gull/Netherlands/1/00	IT
H16.99	H16N3	A/black-headed gull/Sweden/5/99	IT

Infection	Subtype	Strain	GISAID accession number
	H5N2	A/chicken/Belgium/150/1999	EPI238402
	H6N2	A/turkey/Massachusetts/3740/1965	EPI3187
	H7N1	A/parrot/Northern Ireland/VF-73-67/73	EPI6514
	H9N2	A/chicken/Saudi Arabia/SP02525/3AAV/2000	AHS

LP, low pathogenic; IT, Immune Technology Corp.; SB, Sino Biological Inc.; E, e-enzyme; G, Genscript; AHS, from Animal Health Service, Deventer, the Netherlands.

Inc., California, USA). Log2-transformed median antibody titer ratios of field chickens were plotted in R (R Foundation for Statistical Computing, version 2.15). For consecutively collected samples, seroconversion or a significant rise was defined as a ≥4 fold increase in antibody titer [22]. Correlations between the PA and HI-test were calculated using a two-sided Spearman's rank correlation coefficient (ρ). A p-value of less than 0.05 was considered statistically significant.

The overall antibody reactivity for all seropositive individuals was visualized by means of a heat map, generated by applying hierarchical clustering (pairwise correlation distance and Ward's method) to log-transformed titers. No cut off titer was applied to the data. Bright red color indicates high titers whereas faint red and white corresponds to low titers and no reactivity, respectively. Amino acid (AA) sequence similarity of HA1s was determined using a fast algorithm with pairwise alignment in Bionumerics (version 6.6, Applied Maths).

Antigen stability and batch control

Antigen quality and stability between different batches was tested using an in-house serum pool comprising HA-specific polyclonal rabbit-antisera (Immune Technology Corp., New York,

USA) raised against all antigens included on the PA. Testing the last slide from each batch of 25 slides showed that all antigens were stable over time (data not shown). Prior experiments showed that spotted PA slides containing recombinant influenza HA1-proteins are stable for at least one year (unpublished data). Day-to-day variation was controlled for by correcting all titers according to the reactivity of the reference antigen H6.07 against the in house-standard, as previously described [20].

Results

Four out of 38 negative sera (1.5%) – all four belonging to group 1c (negative field chickens) – showed minor low-level reactivity with titers ranging between 21 and 30 against antigens H2.05 and H12.91, respectively. All other samples tested negative for all antigens (data not shown). These findings result in a specificity of the PA of 94.6% to 100% at a cutoff titer of >1:20 across all antigens, and of 100% when the cutoff was raised to ≥1:40 or higher.

In contrast, all experimentally infected chickens (group 2) seroconverted to the homologous antigens, although the kinetics of response differed slightly. H5- and H6-infected animals were the

Table 2. Sensitivities (%) for microarray antigens corresponding to subtype of virus strains used for infection of SPF chickens (group 2) according to time point of serum collection and different cut of levels. Bold font indicates 100% sensitivity.

	Virus subtype	H5				H6	H7	H9	
p.i.	**Cut off ≥**	**H5.97**	**H5.06**	**H5.02**	**H5.07**	**H6.07**	**H7.03**	**H9.99**	**H9.07**
Day 7	1:20	**100.0**	**100.0**	85.7	**100.0**	**100.0**	80.0	86.7	73.3
	1:40	92.9	92.9	78.6	**100.0**	**100.0**	73.3	73.3	66.7
	1:80	85.7	85.7	57.1	85.7	**100.0**	66.7	20.0	26.7
	1:160	78.6	78.6	28.6	57.1	**100.0**	53.3	6.7	13.3
	1:320	50.0	57.1	21.4	0.0	93.3	40.0	0.0	0.0
	1:640	14.3	28.6	0.0	0.0	86.7	6.7	0.0	0.0
	1:1280	0.0	0.0	0.0	0.0	53.3	0.0	0.0	0.0
	1:2560	0.0	0.0	0.0	0.0	13.3	0.0	0.0	0.0
Day 14	1:20	**100.0**	**100.0**	**100.0**	**100.0**	**100.0**	**100.0**	**100.0**	93.3
	1:40	**100.0**	**100.0**	85.7	**100.0**	**100.0**	**100.0**	93.3	86.7
	1:80	92.9	92.9	78.6	**100.0**	**100.0**	**100.0**	80.0	86.7
	1:160	85.7	78.6	57.1	**100.0**	**100.0**	93.3	20.0	46.7
	1:320	78.6	71.4	42.9	71.4	**100.0**	66.7	13.3	26.7
	1:640	71.4	50.0	14.3	28.6	**100.0**	60.0	6.7	20.0
	1:1280	21.4	14.3	0.0	14.3	80.0	0.0	0.0	6.7
	1:2560	7.1	7.1	0.0	0.0	40.0	0.0	0.0	0.0
Day 22	1:20	**100.0**	**100.0**	**100.0**	**100.0**	**100.0**	**100.0**	**100.0**	**100.0**
	1:40	**100.0**	**100.0**	**100.0**	**100.0**	**100.0**	**100.0**	**100.0**	**100.0**
	1:80	**100.0**	93.3	**100.0**	**100.0**	**100.0**	**100.0**	**100.0**	93.3
	1:160	**100.0**	86.7	**100.0**	**100.0**	**100.0**	78.6	46.7	73.3
	1:320	86.7	73.3	86.7	80.0	**100.0**	64.3	20.0	46.7
	1:640	80.0	53.3	66.7	53.3	**100.0**	21.4	6.7	20.0
	1:1280	73.3	40.0	26.7	26.7	73.3	0.0	0.0	6.7
	1:2560	13.3	0.0	6.7	13.3	40.0	0.0	0.0	0.0
Days combined	1:20	**100.0**	**100.0**	95.3	**100.0**	**100.0**	93.2	95.6	88.9
	1:40	97.7	97.7	88.4	**100.0**	**100.0**	90.9	88.9	84.4
	1:80	93.0	90.7	79.1	96.6	**100.0**	88.6	66.7	68.9
	1:160	88.4	81.4	62.8	89.7	**100.0**	75.0	24.4	44.4
	1:320	72.1	67.4	51.2	58.6	97.8	56.8	11.1	24.4
	1:640	55.8	44.2	27.9	34.5	95.6	29.5	4.4	13.3
	1:1280	32.6	18.6	9.3	17.2	68.9	0.0	0.0	4.4
	1:2560	7.0	2.3	2.3	6.9	31.1	0.0	0.0	0.0

p.i., post infection.

fastest to show 100% seroconversion at day 7 p.i. at a cut off of ≥ 1:40 for at least one PA-antigen used, whereas for H7- and H9- exposed animals complete seroconversion (100% of animals) occurred at a later time point (Figure 1, Table 2). At day 22 p.i. all animals showed a significant (≥ 4-fold) titer increase. With advancing antibody rise (at days 14 and 22 p.i.), sensitivities further increased for antigens matching the infecting subtype. In addition, we combined all serum collection time points to investigate the ability of the PA to identify positive individuals in different stages of antibody development and sensitivity remained high (Table 2).

Interestingly, although H5-infected SPF chickens were inoculated with a low-pathogenic H5-strain (Table 1), we observed the strongest antibody response against H5.97, an antigen representing HP AIV clade 0 (Table 1, Figure 1). For the H9-infection cohort, chickens showed mixed antibody reactivity against the two H9-antigens, with half the individuals reacting stronger against H9.99 and the other half displaying a higher titer against H9.07 at day 7 p.i. One individual had an equally high titer for both antigens at that time point. At day 14 and 22 p.i., reactivity profiles shifted towards H9.07 in the majority of chickens, ten and nine out

of 15, respectively, displaying a higher titer against H9.07 compared to H9.99 (data not shown).

Cross-reactivity against heterologous antigens of experimentally infected chickens (group 2)

In general, we observed some degree of heterogeneity in kinetics and cross reactivity of antibody responses within all infection groups (Figure 1). The ratio of homologous versus heterologous GMTs of all sampling days combined ranged from 1.8 to 57.9 in H5-, 19.1 to 161.1 in H6-, versus 12.8 to 27.4 for H7- and 4.6 to 13.3 in H9-infected individuals (Figures 1 and 2). The highest level of cross reactivity was observed in H5-infected animals reacting with the H2-antigen (GMT-ratio 1.8-4.2). Nevertheless, a clear distinction between homologous and heterologous reactivity was observed for the remaining antigens, with GMT ratios of >4 for all other antigen combinations (Figures 1 and 2). Therefore, the infecting strain could clearly be identified independent of the cutoff level chosen (Figure 1). To minimize or dismiss the "noise" caused by cross-reacting antibodies, the application of a cutoff level of ≥1:80 seems appropriate (Table 2, Figure 1).

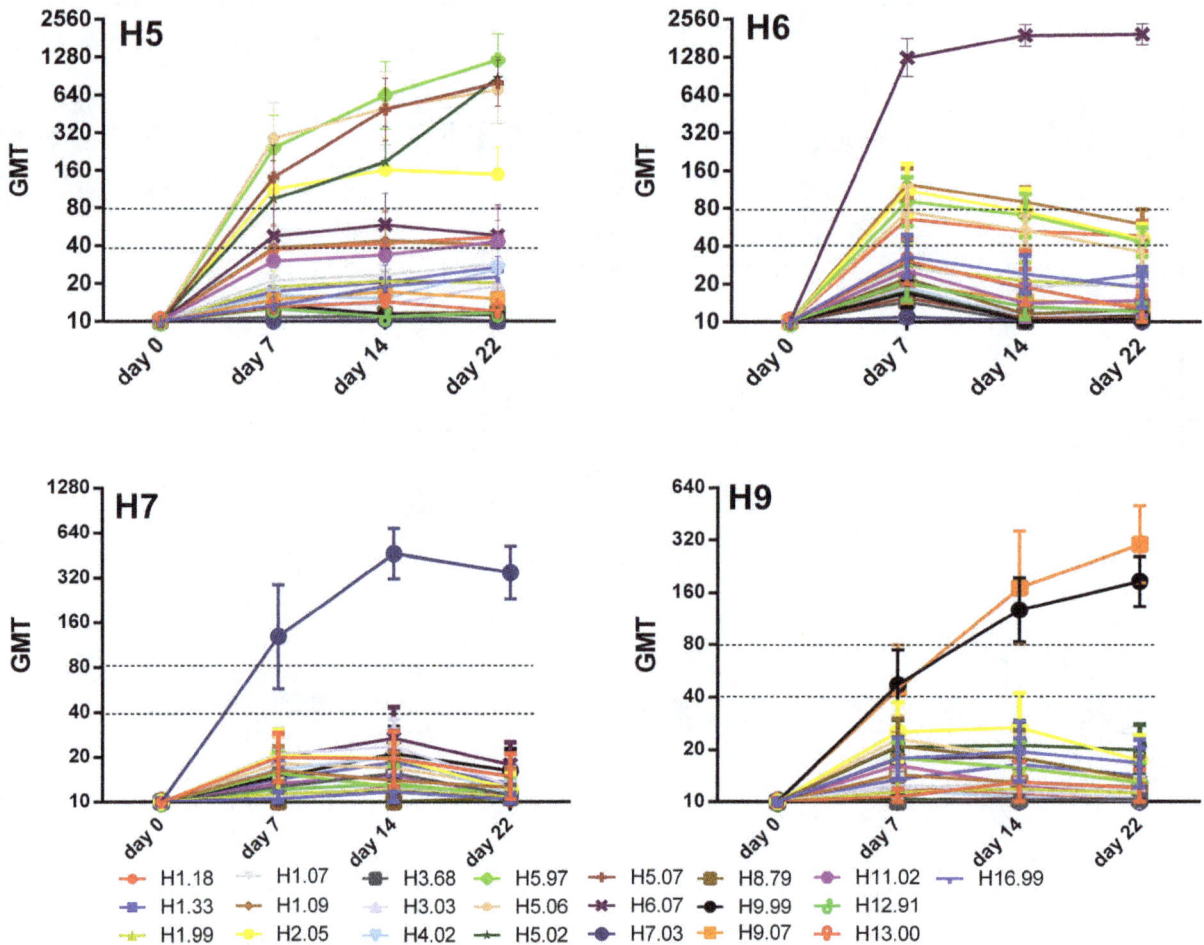

Figure 1. Kinetics of serological responses of SPF chickens after intratracheal infection with live virus (group 2). Titles of each graph indicate infection group. X-axes depict the day of serum collection post infection. Y-axes indicate geometric mean titers (GMT). Error bars represent 95% confidence intervals of the measurements. Note differences in log-scale. Heterologous reactions above the dotted line represent cross-reactive responses with a titer higher than 1:40 or 1:80, respectively.

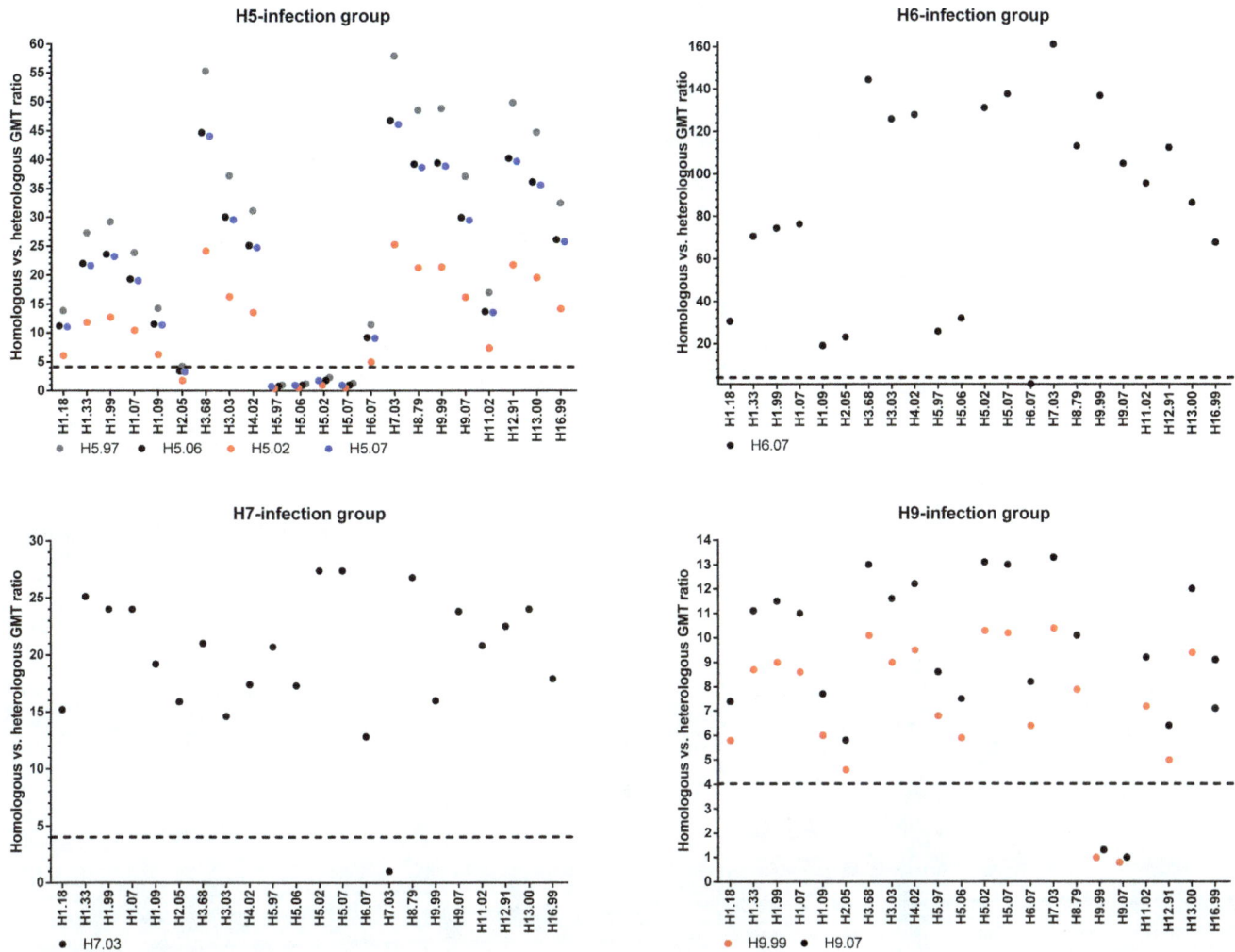

Figure 2. Homologous versus heterologous geometric mean titer (GMT) ratios for different groups of experimentally infected chickens (H5, H6, H7, H9) for all sampling days combined. A high homologous versus heterologous ratio in GMT indicates low cross-reactivity and vice versa. For instance, as for the H6-infection group the GMT against the homologous antigen H6.07 is 1668.8 and the GMT against the heterologous antigen H7.03 is 10.4, the homologous vs. heterologous ratio is the highest (~161), implying that the level of cross-reactivity is lowest for the H7-antigen in the H6-infection group. The dotted line demarkates a ratio of 4. Note differences in scale.

Serological profiles of naturally infected laying hens (group 3)

Serum samples from naturally infected field chickens showed similar discriminatory serological profiles compared with the data from the validation experiments (Figure 3). In the analysis, we combined the data of both H6-outbreaks. The PA correctly identified 100% of the tested field chickens as positive up to a cut-off titer of ≥1:80 (data not shown). Cross-reactivity was negligible for H6- and H7-infected individuals and generally matched the patterns observed in group 2 (Figure 3, light red, light green). Among the field chickens naturally infected with H9, we observed somewhat more cross reactivity (Figure 3 and 4). Nevertheless, the infecting subtype was still evident by resulting in the highest median log2-titer ratio (Figure 4).

Overall, the PA results showed good correlation with the HI-assay. Spearman's rank correlation coefficients showed strong, significant associations between the HI-assay and PA-antigens H9.07 ($\rho = 0.804$, $p = 0.021$) and H6.07 ($\rho = 0.850$, $p = 0.029$), whereas a relatively strong but not statistically significant

association could be demonstrated between HI-data and PA-antigen H9.99 ($\rho = 0.600$, $p = 0.121$).

Discussion

Here we present a highly sensitive and specific multiplex-screening tool to detect antibodies against different HA-types of AIV in chickens. We show that the PA discriminates between negative and experimentally infected, seropositive chickens. We further demonstrate that our test can serve as a surveillance tool in commercial field chicken flocks, by reliably identifying the infecting subtypes in laying hens from free-range- and indoor husbandry. An asset of the technique is that it requires a minute quantity of serum (5–10 µl) to simultaneously screen for *multiple* subtypes, whereas the HI-assay usually requires about the same amount to detect antibodies against only *one* subtype [23]. This characteristic is particularly advantageous when screening small animal species of which only small volumes of sera are available.

Analysis of consecutive sera of SPF chickens infected with live field strains of different AIV subtypes showed that the PA was able

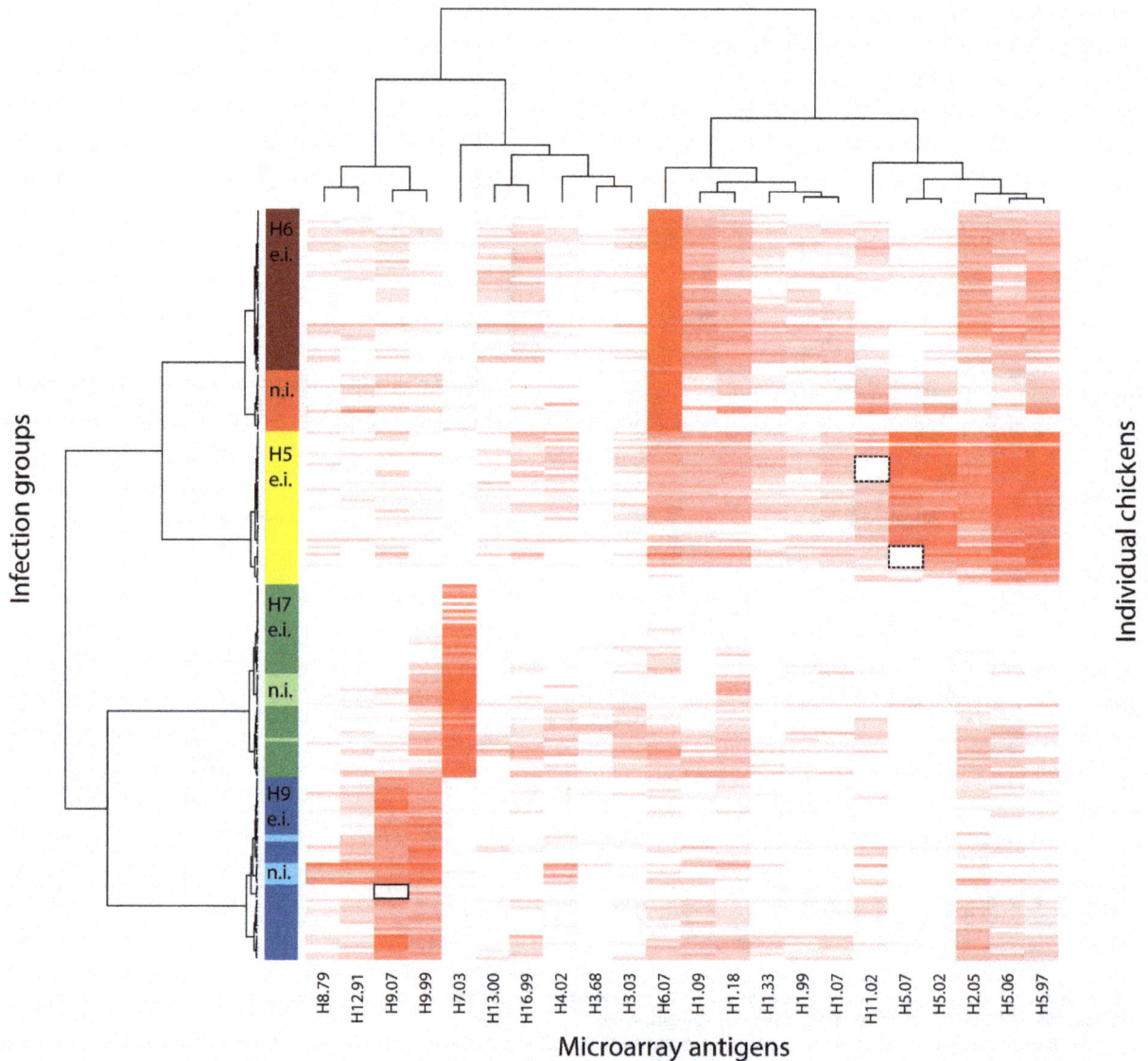

Figure 3. Heat map depicting serological patterns of naturally (n.i.) and experimentally infected (e.i.) chickens (H5, H6, H7, H9) spanning all samplings days. Dendrograms reflect clustering based on similarity of serological profiles. Microarray antigens are depicted on the X-axis. Different infection groups are color coded on the Y-axis according to the avian influenza virus subtype causing the infection. Rows represent reaction profiles of individual chickens across the entire antigen panel. Columns represent the reactivity of all individuals against a specific antigen as stated on the X-axis. Intensity of the red color is proportional to the log-titer height. Black dotted squares indicate missing antigens H5.07 (n = 6) and H11.02 (n = 7) due to spotting failure. Black square with solid line indicates no biological reactivity against the H9.07 antigen. The clustering algorithm automatically excluded negative sera.

to quantify varying titer heights per sampling time point and infection group. Such variation could either be due to differences in immunogenicity of strains used for infection [24], infectious dose [24,25], different chicken breeds or genetic lineages [24,26]. From a technical aspect, differing quality of antigens used on the PA and distant relatedness of strains used for infection and the assay antigen [27] could account for the differences in titer heights between infection groups. As infectious dose and breed were the same for all experimental infection cohorts and antigen quality was checked prior to testing and monitored throughout the experiment, these factors can be disregarded as a possible source of variability.

Lee et al. [28] speculated that immunogenicity, and therefore antibody titer heights, can depend on the protein itself and can

vary between strains of different subtypes in chickens immunized with different DNA-vaccines. Failure to regularly update antigens in HI-assay can result in a reduced ability to detect antibodies against more recent field isolates [27] and it is unclear if this also can be observed in our assay system. The strains used for the infections of group 2 animals were closely related to the strains from which antigens were produced, with the lowest level of AA-identity for antigen H9.07 (94.4%) (Table 3). This lower AA-identity in combination with individual variation could be a possible explanation why H9.07 did not yet react at day 7 p.i. for some experimentally infected chickens of group 2 (Figure 3, square with black solid line). On the other hand, the lower AA-identity of H9.07 did not seem to have a major influence, as this antigen showed a higher GMT at day 14 and 22 p.i. in H9-infected

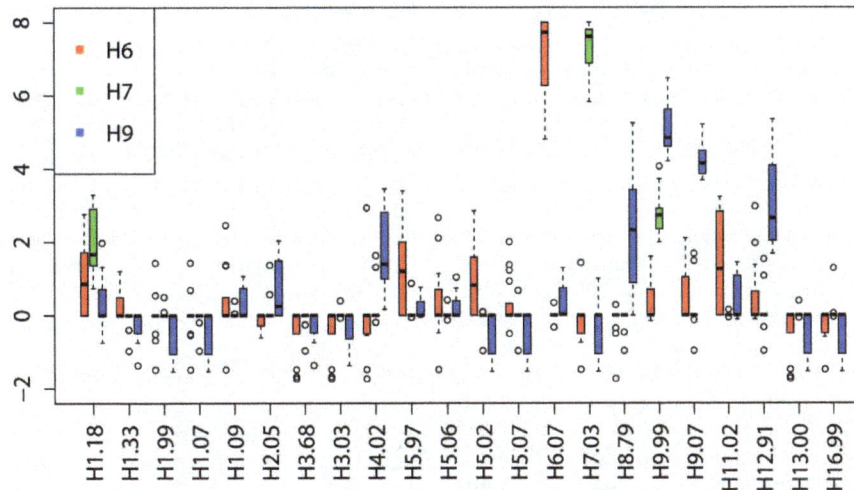

Figure 4. Antibody profiles of field chickens expressed as log2-transformed median antibody titer ratios plotted per outbreak. Antibody titer ratios were derived by log transforming the data, calculating the median antibody reactivity across all antigens included on the PA and subtracting it from the antibody reactivity against individual antigens. This was calculated for every chicken. By doing that, every individual's values are normalized according to its own background reactivity. Individual ratios were summarized in boxplots. Horizontal bars within each box represent log2-transformed median antibody titer ratios per antigen and outbreak. Chickens naturally infected with H6 are depicted in red (n = 17), H7 in green (n = 10) and H9 in blue (n = 8). The two H6-outbreaks were combined in this analysis.

chickens of group 2, compared to H9.99, which was 97.5% similar to the infecting strain (Table 3). It is not known how AA differences in HA1 translate to antigenic reactivity in the PA system. Cattoli et al. [29] examined serological responses of drift variants of H5 strains in chickens using HI- and microneutralization assay. Of the 11 AA substitutions found in the HA1, the researchers demonstrated that only five substitutions sufficed to cause antigenic drift. These findings stress that a high AA sequence similarity in the HA1 of two strains does not necessarily translate into similar serological reactivity, if critical substitutions occur in epitopes influencing antigenicity. Hence, AA sequence similarity is not a good indicator for antigenicity and cross reactivity, so no inferences about the compatibility between the viruses used for infection and PA-antigens can be made.

Overall, the observed cross-reactivities were negligible in comparison to the titer height of the antigens matching the subtype of infection. Interestingly, we noted that heterologous patterns largely reflected phylogenic relationships. The 16 currently known HA-types derived from birds divide into two phylogenetic groups which further segregate into 5 clades. Group 1 consists of 3 clades (H1, H2, H5 and H6; H8, H9 and H12; H11, H13 and H16) whereas group 2 comprises 2 clades (H3, H4 and H14; H7, H10 and H15) [1,30]. Heterosubtypic immunity has mainly been attributed to cytotoxic T-cells specific for internal proteins [31], but neutralizing antibodies also play an important role in protection [32,33]. To date, a number of broad reacting intra-subtype-, intra-clade-, intra-group- and inter-group-specific neutralizing monoclonal antibodies have been identified [34–37].

Table 3. Amino acid (AA) similarity matrix for strains of a particular subtype used to infect SPF chickens of group 2 (in bold) versus PA-antigens per HA-type (table 1).

H5N2	*100.00*				
H5.07	*97.37*	100.00			
H5.02	*96.15*	96.42	100.00		
H5.97	*95.75*	95.39	93.52	100.00	
H5.06	*95.29*	95.45	94.21	97.48	100.00
H6N2	*100.00*				
H6.07	*96.01*	100.00			
H7N1	*100*				
H7.03	*97.11*	100.00			
H9N2	*100.00*				
H9.99	*97.49*	100.00			
H9.07	*94.39*	93.77	100.00		

Similarity was calculated based on the HA1 part of the hemagglutinin (sequence length 318 AA). Percentages in bold and italics denote similarity between strains used for infection versus corresponding PA-antigen.

Of all vaccination cohorts, H5-vaccinated chickens displayed the highest level of cross-reactivity with antigen H2.05 (Figures 1, 2 and 3). This finding is not surprising due to the high sequence similarity of these two subtypes [28,38]. Likewise, H9-positive serum cross-reacted somewhat with members of the same clade, H8 and H12. Together with the calculation of the median-log2-titer ratios – as was performed for the field chickens in this study –, the knowledge of these patterns can be useful in distinguishing cross-reactivity from potential dual infections involving subtypes of different clades. Although we only tested one serum of a chicken simultaneously immunized with influenza virus strains belonging to two different subtypes (H7 and H9), the PA showed clear antibody titers against both HA-types (median log2-titer ratio for H7.03 = 8, H9.99 = 6 and H9.07 = 5.8, respectively) with no cross-reactivity to other antigens (median log2-titer ratio = 0). This capacity can be especially interesting for regions where AI surveillance is not implemented in poultry and where animals might experience multiple consecutive- or co-infections with different subtypes. To further investigate this potential, serum of experimentally infected chickens consecutively or simultaneously immunized with different subtypes would need to be analyzed, which were not available in this study.

Heterologous reaction was lowest in chickens experimentally and naturally infected with subtype H7 compared to other serum cohorts. This can possibly be explained by the fact that, apart from H3- and H4-antigens, no other representatives of phylogenetic group 2 (H10, H14, H15) were included in the PA setup. Similarly, Latorre-Margaleff et al. [39] found that after infection with a certain subtype, infection with the homologous- or subtypes within the same clade and group were uncommon, suggesting heterosubtypic immunity.

In this project, we showed that the PA can discriminate between different HA-types. Strain-discrimination was not possible yet with the PA, when more than one antigen per subtype was included, e.g., H5. This intra-subtype reactivity is not unexpected since a study found an intra-subtype similarity (based on AA-sequences of the HA1) of >92%, whereas inter-subtype identity based on AA-similarity was much lower (38.5%) [40]. Broad intra-subtype reactivity is exploited in diagnostics. Ducatez et al. [41] discovered that ancestral strain A of HP H5N1 as well as strains belonging to clade 2.2 (represented by H5.06 in our study) proved to be the most suitable antigen as they correctly identified most HP H5N1 antigens/-sera of other clades [41]. On the other hand, as genetic changes can lead to escape mutants eliciting different serological responses, it is important to monitor and regularly update the PA-antigen setup, as is done for other serological assays [27]. The extent to which strain discrimination can be achieved by means of the PA is currently focus of a follow up project.

It is important to stress that the PA does not give information on the presence or absence of neutralizing antibodies and can therefore not be used to determine the immune status, i.e. protection. In serological avian influenza surveillance the HI assay is currently the gold standard with a sensitivity and specificity of 98.8% and 99.5%, respectively [42]. Overall, the PA showed a good correlation with the HI test. Other currently known serological multiplex techniques for the use in poultry, e.g. bead-based Luminex assays, either target conserved regions of influenza virus (nucleoprotein, matrix protein, non-structural protein 1) [43], screen for antibodies against HA-types relevant for the poultry sector (H5 and H7) [44] or combine the two approaches, eg. nucleoprotein with H5 [45]. In addition, simultaneous serological screening for influenza virus in combination with other poultry diseases of economic importance (e.g., Newcastle Disease Virus, Infectious Bronchitis Virus, Infectious Bursal Disease Virus) are described in the literature [46,47]. To our knowledge, the PA technique is the first to allow simultaneous detection of influenza virus antibodies against more than two HA-types in chickens.

In this study, we aimed at including the full range of HA-types known to be present in birds at the time. The dependence on commercial availability lead to the random assembly of antigens of Eurasian as well as North American lineages and failure to cover all AIV subtypes. It is known that strains descending from Eurasian and North American lineages of H5 and H7 differ antigenically, as is reflected in differing titer heights in serological assays [28]. Therefore, to achieve optimal results, the PA should ideally comprise antigens relevant and topical for the region in which the test is to be deployed. A limitation that should be acknowledged is that the PA has only been tested with sera of subtypes H5, H6, H7 and H9. To evaluate the performance against other subtypes, additional serum cohorts would need to be analyzed. Furthermore, the PA is limited to the detection of HA-type specific antibodies and cannot identify antibodies against the neuraminidase. It is not known as to what extent NA-specific antibodies influence reactivities against the HA-proteins (due to steric hindrance) in this testing platform [28].

In conclusion, we present a sensitive and specific test for the simultaneous detection of HA-type specific antibodies against different AIVs in chicken that requires very low amounts of serum. In combination with a screening-ELISA targeting antibodies against a conserved region of AIV, the PA can provide a valuable epidemiological surveillance tool to monitor dispersal of different subtypes. As this testing platform is also validated for the use in humans [20,48] it lends itself for conducting exposure studies at the human-animal interface. Current research centers on the development of the PA for the use in swine.

Acknowledgments

We would like to thank Huub van der Sande from the Animal Health Service, Deventer who kindly organized serum panels for group 1, 2 and 3. Furthermore, we thank Dr. Maarten Schipper for advice on statistical analysis.

Author Contributions

Conceived and designed the experiments: GF EDB JVB JR MK. Performed the experiments: GF JVB JR EDB MK. Analyzed the data: GF JVB HJVDH MK. Contributed reagents/materials/analysis tools: JVB HJVDH SDW GK LV. Wrote the paper: GF EDB MK.

References

1. Fouchier RAM, Munster V, Wallensten A, Bestebroer TM, Herfst S, et al. (2005) Characterization of a Novel Influenza A Virus Hemagglutinin Subtype (H16) Obtained from Black-Headed Gulls. J Virol 79: 2814–2822. doi: 10.1128/JVI.79.5.2814-2822.2005.

2. Tong S, Li Y, Rivailler P, Conrardy C, Castillo DAA, et al. (2012) A distinct lineage of influenza A virus from bats. Proc Natl Acad Sci 109: 4269–4274. doi:10.1073/pnas.1116200109.

3. Tong S, Zhu X, Li Y, Shi M, Zhang J, et al. (2013) New World Bats Harbor Diverse Influenza A Viruses. PLoS Pathog 9: e1003657. doi:10.1371/journal.ppat.1003657.

4. Malik Peiris JS (2009) Avian influenza viruses in humans. Rev Sci Tech 28 (1): 161–173.

5. Reperant LA, Rimmelzwaan GF, Kuiken T (2009) Avian influenza viruses in mammals. Rev Sci Tech Int Off Epizoot 28: 137–159.

6. Freidl GS, Meijer A, de Bruin E, de Nardi M, Munoz O, et al. (2014) Influenza at the animal-human interface: a review of the literature for virological evidence of human infection with swine or avian influenza viruses other than A(H5N1). Euro Surveill.19(18).

7. WHO (2014) WHO risk assessment of human infection with avian influenza A(H7N9) virus. As of 27 June 2014; Available: http://www.who.int/influenza/human_animal_interface/influenza_h7n9/Risk_Assessment/en/.

8. Wang Z, Zhang A, Wan Y, Liu X, Qiu C, et al. (2014) Early hypercytokinemia is associated with interferon-induced transmembrane protein-3 dysfunction and predictive of fatal H7N9 infection. Proc Natl Acad Sci U S A 111: 769–774. doi:10.1073/pnas.1321748111.

9. Chen Y, Liang W, Yang S, Wu N, Gao H, et al. (2013) Human infections with the emerging avian influenza A H7N9 virus from wet market poultry: clinical analysis and characterisation of viral genome. Lancet 381: 1916–1925. doi:10.1016/S0140-6736(13)60903-4.

10. Lam TT-Y, Wang J, Shen Y, Zhou B, Duan L, Cheung C-L, et al. The genesis and source of the H7N9 influenza viruses causing human infections in China. Nature. 2013; 502(7470): 241–4.

11. Yang S, Chen Y, Cui D, Yao H, Lou J, et al. (2013) Avian-Origin Influenza A(H7N9) Infection in Influenza A(H7N9)-Affected Areas of China: A Serological Study. J Infect Dis 15: 265–269. doi:10.1093/infdis/jit430.

12. Kayali G, Ortiz EJ, Chorazy ML, Gray GC (2010) Evidence of Previous Avian Influenza Infection among US Turkey Workers. Zoonoses Public Health 57: 265–272. doi:10.1111/j.1863-2378.2009.01231.x.

13. Gill JS, Webby R, Gilchrist MJ, Gray GC (2006) Avian influenza among waterfowl hunters and wildlife professionals. Emerg Infect Dis 12: 1284.

14. Kayali G, Barbour E, Dbaibo G, Tabet C, Saade M, et al. (2011) Evidence of Infection with H4 and H11 Avian Influenza Viruses among Lebanese Chicken Growers. PLoS ONE 6. doi:10.1371/journal.pone.0026818.

15. Koopmans M, de Jong MD (2013) Avian influenza A H7N9 in Zhejiang, China. Lancet 381: 1882–1883. doi:10.1016/S0140-6736(13)60936-8.

16. Gray GC, Kayali G (2009) Facing pandemic influenza threats: the importance of including poultry and swine workers in preparedness plans. Poult Sci 88: 880.

17. European Commission (2006) Council Directive 2005/94/EC of 20 December 2005 on Community measures for the control of avian influenza and repealing Directive 92/40/EEC. Available: http://eur-lex.europa.eu/LexUriServ/LexUriServ.do?uri=CELEX:32005L0094:EN:NOT.

18. Gonzales JL, Stegeman JA, Koch G, de Wit SJ, Elbers ARW (2012) Rate of introduction of a low pathogenic avian influenza virus infection in different poultry production sectors in the Netherlands. Influenza Other Respir Viruses (1): 6–10.

19. World Organisation for Animal Health (OIE) (2012) Avian Influenza. OIE Terrestrial Manual. Version adopted by the World Assembly of Delegates of the OIE in May 2012. Vol. Chapter 2. 3. 4. Available: http://www.oie.int/fileadmin/Home/fr/Health_standards/tahm/2.03.04_AI.pdf.

20. Koopmans M, de Bruin E, Godeke G -J, Friesema I, van Gageldonk R, et al. (2012) Profiling of humoral immune responses to influenza viruses by using protein microarray. Clin Microbiol Infect 18: 797–807. doi:10.1111/j.1469-0691.2011.03701.x.

21. Sharma JM (1991) Overview of the avian immune system. Vet Immunol Immunopathol 30: 13–17. doi:10.1016/0165-2427(91)90004-V.

22. Wood JM, Gaines-Das RE, Taylor J, Chakraverty P (1994) Comparison of influenza serological techniques by international collaborative study. Vaccine 12: 167–174.

23. WHO Global Influenza Surveillance Network (2011) Manual for the laboratory diagnosis and virological surveillance of influenza. 153 p. Available: http://www.who.int/influenza/resources/documents/manual_diagnosis_surveillance_influenza/en/index.html.

24. Ladman BS, Rosenberger SC, Rosenberger JK, Pope CR, Gelb J Jr (2008) Virulence of low pathogenicity H7N2 avian influenza viruses from the Delmarva peninsula for broiler and leghorn chickens and turkeys. Avian Dis 52: 623–631.

25. Lu H, Castro AE (2004) Evaluation of the infectivity, length of infection, and immune response of a low-pathogenicity H7N2 avian influenza virus in specific-pathogen-free chickens. Avian Dis 48: 263–270.

26. Vervelde L, de Geus E, Jansen C, Heller DE (2011) Contribution of the genetic background to the immune response of broilers vaccinated or challenged with LPAI H9N2. BMC Proc 5: S5. doi:10.1186/1753-6561-5-S4-S5.

27. Escorcia M, Carrillo-Sanchez K, March-Mifsut S, Chapa J, Lucio E, et al. (2010) Impact of antigenic and genetic drift on the serologic surveillance of H5N2 avian influenza viruses. BMC Vet Res 6: 57.

28. Lee C-W, Senne DA, Suarez DL (2006) Development and Application of Reference Antisera against 15 Hemagglutinin Subtypes of Influenza Virus by DNA Vaccination of Chickens. Clin Vaccine Immunol 13: 395–402. doi:10.1128/CVI.13.3.395-402.2006.

29. Cattoli G, Milani A, Temperton N, Zecchin B, Buratin A, et al. (2011) Antigenic Drift in H5N1 Avian Influenza Virus in Poultry Is Driven by Mutations in Major Antigenic Sites of the Hemagglutinin Molecule Analogous to Those for Human Influenza Virus. J Virol 85: 8718–8724. doi:10.1128/JVI.02403-10.

30. WHO (1980) A revision of the system of nomenclature for influenza viruses: a WHO memorandum. Bull World Health Organ 58: 585–591.

31. Boon ACM, Mutsert G de, Baarle D van, Smith DJ, Lapedes AS, et al. (2004) Recognition of Homo- and Heterosubtypic Variants of Influenza A Viruses by Human CD8+ T Lymphocytes. J Immunol 172: 2453–2460.

32. Nguyen HH, van Ginkel FW, Vu HL, McGhee JR, Mestecky J (2001) Heterosubtypic immunity to influenza A virus infection requires B cells but not CD8+ cytotoxic T lymphocytes. J Infect Dis 183: 368–376. doi:10.1086/318084.

33. Tumpey TM, Renshaw M, Clements JD, Katz JM (2001) Mucosal Delivery of Inactivated Influenza Vaccine Induces B-Cell-Dependent Heterosubtypic Cross-Protection against Lethal Influenza A H5N1 Virus Infection. J Virol 75: 5141–5150. doi:10.1128/JVI.75.11.5141-5150.2001.

34. Chen Y, Qin K, Wu WL, Li G, Zhang J, et al. (2009) Broad Cross-Protection against H5N1 Avian Influenza Virus Infection by Means of Monoclonal Antibodies that Map to Conserved Viral Epitopes. J Infect Dis 199: 49–58. doi:10.1086/594374.

35. Smirnov YA, Lipatov AS, Gitelman AK, Okuno Y, Van Beek R, et al. (1999) An epitope shared by the hemagglutinins of H1, H2, H5, and H6 subtypes of influenza A virus. Acta Virol 43: 237–244.

36. Ekiert DC, Friesen RHE, Bhabha G, Kwaks T, Jongeneelen M, et al. (2011) A Highly Conserved Neutralizing Epitope on Group 2 Influenza A Viruses. Science 333: 843–850. doi:10.1126/science.1204839.

37. Mueller M, Renzullo S, Brooks R, Ruggli N, Hofmann MA (2010) Antigenic Characterization of Recombinant Hemagglutinin Proteins Derived from Different Avian Influenza Virus Subtypes. PLoS ONE 5: e9097. doi:10.1371/journal.pone.0009097.

38. Air GM (1981) Sequence relationships among the hemagglutinin genes of 12 subtypes of influenza A virus. Proc Natl Acad Sci U S A 78: 7639–7643.

39. Latorre-Margalef N, Grosbois V, Wahlgren J, Munster VJ, Tolf C, et al. (2013) Heterosubtypic Immunity to Influenza A Virus Infections in Mallards May Explain Existence of Multiple Virus Subtypes. PLoS Pathog 9: e1003443. doi:10.1371/journal.ppat.1003443.

40. Dugan VG, Chen R, Spiro DJ, Sengamalay N, Zaborsky J, et al. (2008) The Evolutionary Genetics and Emergence of Avian Influenza Viruses in Wild Birds. PLoS Pathog 4: e1000076. doi:10.1371/journal.ppat.1000076.

41. Ducatez MF, Cai Z, Peiris M, Guan Y, Ye Z, et al. (2011) Extent of antigenic cross-reactivity among highly pathogenic H5N1 influenza viruses. J Clin Microbiol: JCM.01279–11. doi: 10.1128/JCM.01279-11.

42. Comin A, Toft N, Stegeman A, Klinkenberg D, Marangon S (2012) Serological diagnosis of avian influenza in poultry: is the haemagglutination inhibition test really the "gold standard"? Influenza Other Respir Viruses. doi:10.1111/j.1750-2659.2012.00391.x.

43. Watson DS, Reddy SM, Brahmakshatriya V, Lupiani B (2009) A multiplexed immunoassay for detection of antibodies against avian influenza virus. J Immunol Methods 340: 123–131. doi:10.1016/j.jim.2008.10.007.

44. Molesti E, Wright E, Terregino C, Rahman R, Cattoli G, et al. (2014) Multiplex Evaluation of Influenza Neutralizing Antibodies with Potential Applicability to In-Field Serological Studies. J Immunol Res 2014: e457932. doi:10.1155/2014/457932.

45. Lupiani B, Mozisek B, Mason PW, Lamichhane C, Reddy SM (2010) Simultaneous detection of avian influenza virus NP and H5 antibodies in chicken sera using a fluorescence microsphere immunoassay. Avian Dis 54: 668–672.

46. Wang X, Shi L, Tao Q, Bao H, Wu J, et al. (2010) A protein chip designed to differentiate visually antibodies in chickens which were infected by four different viruses. J Virol Methods 167: 119–124. doi:10.1016/j.jviromet.2010.03.021.

47. Pinette MM, Rodriguez-Lecompte JC, Pasick J, Ojkic D, Leith M, et al. (2014) Development of a duplex Fluorescent Microsphere Immunoassay (FMIA) for the detection of antibody responses to influenza A and newcastle disease viruses. J Immunol Methods 405: 167–177. doi:10.1016/j.jim.2014.02.004.

48. Boni MF, Chau NVV, Dong N, Todd S, Nhat NTD, et al. (2013) Population-level antibody estimates to novel influenza A/H7N9. J Infect Dis 208(4): 554–8. doi:10.1093/infdis/jit224.

Experimentally Infected Domestic Ducks Show Efficient Transmission of Indonesian H5N1 Highly Pathogenic Avian Influenza Virus, but Lack Persistent Viral Shedding

Hendra Wibawa[1,2,3], John Bingham[1*], Harimurti Nuradji[1,2,4], Sue Lowther[1], Jean Payne[1], Jenni Harper[1], Akhmad Junaidi[3¤], Deborah Middleton[1], Joanne Meers[2]

1 Commonwealth Scientific and Industrial Research Organisation (CSIRO), Australian Animal Health Laboratory, Geelong, Australia, 2 School of Veterinary Science, The University of Queensland, Brisbane, Australia, 3 Disease Investigation Centre Regional IV Wates, Yogyakarta, Indonesia, 4 Indonesian Research Center for Veterinary Science, Bogor, West Java, Indonesia

Abstract

Ducks are important maintenance hosts for avian influenza, including H5N1 highly pathogenic avian influenza viruses. A previous study indicated that persistence of H5N1 viruses in ducks after the development of humoral immunity may drive viral evolution following immune selection. As H5N1 HPAI is endemic in Indonesia, this mechanism may be important in understanding H5N1 evolution in that region. To determine the capability of domestic ducks to maintain prolonged shedding of Indonesian clade 2.1 H5N1 virus, two groups of Pekin ducks were inoculated through the eyes, nostrils and oropharynx and viral shedding and transmission investigated. Inoculated ducks (n = 15), which were mostly asymptomatic, shed infectious virus from the oral route from 1 to 8 days post inoculation, and from the cloacal route from 2–8 dpi. Viral ribonucleic acid was detected from 1–15 days post inoculation from the oral route and 1–24 days post inoculation from the cloacal route (cycle threshold <40). Most ducks seroconverted in a range of serological tests by 15 days post inoculation. Virus was efficiently transmitted during acute infection (5 inoculation-infected to all 5 contact ducks). However, no evidence for transmission, as determined by seroconversion and viral shedding, was found between an inoculation-infected group (n = 10) and contact ducks (n = 9) when the two groups only had contact after 10 days post inoculation. Clinical disease was more frequent and more severe in contact-infected (2 of 5) than inoculation-infected ducks (1 of 15). We conclude that Indonesian clade 2.1 H5N1 highly pathogenic avian influenza virus does not persist in individual ducks after acute infection.

Editor: Todd Davis, Centers for Disease Control and Prevention, United States of America

Funding: The study was jointly funded by CSIRO and the Australian Centre for International Agriculture Research (ACIAR) grant number AH/2004/040. Hendra Wibawa was supported by an ACIAR John Allwright Fellowship. The funders had no role in study design, data collection and analysis, decision to publish, or preparation of the manuscript.

Competing Interests: The authors have declared that no competing interests exist.

* E-mail: John.Bingham@csiro.au

¤ Current address: Directorate General of Livestock and Animal Health Services, Ministry of Agriculture, Jakarta, Indonesia

Introduction

Although an outbreak of highly pathogenic avian influenza (HPAI) in poultry due to H5N1 virus was first reported in 1959 [1], only the A/goose/Guangdong/1/96 lineage H5N1 viruses have spread widely and have persisted over time. Since the first isolation of the progenitor virus in southern China in 1996 [2], this "Eurasian H5N1 HPAI" virus lineage has spread to over 60 countries throughout Asia and into Europe and Africa [3] and has continued to circulate for more than 16 years. These viruses continue to evolve via mutation and genetic reassortment with other avian influenza (AI) viruses, resulting in multiple virus genotypes and geographically related sublineages [4,5]. Most H5N1 HPAI virus outbreaks have occurred in domestic poultry, either in backyard or small commercial farms, indicative of the high incidence rate in these species and resulting in the death or forced culling of more than 400 million domestic poultry [6]. Although H5N1 HPAI viruses have not acquired efficient transmission among people, direct transmission of virus from

poultry to humans has caused severe disease and death of 375 people from 630 confirmed cases [7]. Thus, these viruses pose a major challenge for both veterinary and human public health.

The role of wild birds in the transmission and spread of the Eurasian lineage of H5N1 HPAI viruses remains controversial [8–10]. Both domestic and wild birds, including migratory waterfowl, free-range village poultry, poultry sold through live bird markets and fighting cocks are likely to be involved in the spread of H5N1 HPAI virus [3,9,11]. Difficulties in controlling local and regional movement of poultry and their products, problems in handling the trade (particularly illegal) of live birds, and limited participation of poultry farmers in control strategies are considered as significant factors contributing to the H5N1 HPAI virus epidemic [8,9,12]. In Asia, backyard farms are a common feature in villages, where biosecurity measures are inadequately employed, access to veterinary services is often limited [13,14] and chickens and waterfowl, including domestic ducks, are commonly raised together [15–17]. Ducks, particularly mallard-type breeds, are considered central to the maintenance and transmission of H5N1

HPAI viruses because they can replicate these viruses without suffering clinical disease [18–21]. Previous studies indicate that domestic ducks are a likely source of H5N1 HPAI viral infection to chickens in smallholder duck farms in Indonesia and husbandry practices of ducks within villages could increase the risk for H5N1 HPAI [15,22]. In addition, natural reassortment between different AI virus subtypes and endemic H5N1 HPAI viruses can occur in domestic ducks, leading to recurrent interspecies transmission and genetic drift [5]. Preventing transmission events of H5N1 HPAI virus from or into ducks is a key factor in minimizing HPAI virus spread. Therefore, gaining more knowledge on the patterns of H5N1 HPAI virus transmission in this species will assist efforts to control the disease.

Previous studies showed that experimentally infected ducks could shed low pathogenic avian influenza (LPAI) virus for up to 18–20 days post inoculation [23–25], while most H5N1 HPAI viruses have been reported to be shed by ducks for only 2–5 days post inoculation [20,26–29]. However, more persistent H5N1 HPAI virus shedding was shown by Hulse-Post *et al.* [21] indicating that mallard ducks can shed this virus up to 17 days post inoculation, despite seroconversion with a significant titer of hemagglutinating antibodies. They found that in sera taken from extended infections, the neutralizing titer against the inoculated virus was higher than against the virus isolated at the last day of shedding. These researchers hypothesized that because H5N1 HPAI viruses are less pathogenic to ducks, they are able to evolve rapidly in ducks through continuous selection of immunological escape mutants within a host. In this hypothesis, as the duck host develops neutralizing antibodies against the challenge virus, new antigenic mutant viruses arise that are not neutralized by the antibody response, and the host is therefore able to maintain the infection after the immune response develops. The implications of this concept are profound: if this occurs widely, then ducks may be the principal drivers for H5N1 HPAI virus evolution. They could remain inapparently infectious for longer than previously recognized, amplifying their infection risk for other susceptible populations. Such long-term infections of H5N1 HPAI virus in ducks have not been described elsewhere. In the study described here, we investigate the importance of this concept in domestic ducks for an H5N1 HPAI virus from Indonesia.

Our previous studies indicated that shedding of Indonesian H5N1 HPAI viruses in Pekin ducks occurred over only a short period, between 1 and 8 days post inoculation, with viral decline coinciding with the development of antibodies [28,30]. Shedding of virus after this period was not detected by the conventional methods (virus isolation of swab media). To our knowledge, there are no studies that have examined methodically the viral infection and shedding after the acute infection stage in ducks. Also, there have been few comparisons between the shedding patterns of infectious virus and viral RNA in H5N1 HPAI animal models. Therefore, in the present study we attempted to investigate further the importance of extended infection and shedding in Pekin ducks after recovery from H5N1 HPAI infection, by detecting not only presence of virus in swabs but also through detecting transmission to contact ducks. A comprehensive analysis was conducted comparing the shedding of infectious virus and viral RNA from the oral and cloacal routes through the acute infection and post-acute infection periods. The results showed that shedding of infectious virus occurred over a relatively short period for 1–8 days post inoculation (dpi) and viral transmission was not detected after 10 dpi. Shedding of viral RNA was detected over a much longer period. We discuss the implications of our findings for understanding the transmission biology and evolution of H5N1 HPAI virus in ducks and for the interpretation of viral RNA detection for diagnostic purposes.

Materials and Methods

Ethics Statement

All birds were handled and cared for in accordance with the animal welfare standard operating procedures of CSIRO-Australian Animal Health Laboratory (AAHL), Geelong, Australia, which are based on the recommendations in the National Health and Medical Research Council's Australian Code of Practice for the Care and Use of Animals for Scientific Purposes. The experimental procedures were conducted with the approval of AAHL's Animal Ethics Committee (number 1377).

Animals and Bio-containment

Twenty-nine 4 week-old Pekin ducks were used in this study. Before the inoculation date, each duck was given a unique numbered leg band. Ducks were divided into four groups by random assignment and housed in separate rooms at microbiological physical containment level 3 at AAHL with husbandry procedures as described previously [28].

Virus

An Indonesian H5N1 HPAI clade 2.1 virus, A/duck/BBVW-1003-34368/2007 (IDN 34368), was used in this study. This virus strain had been passaged, from the original duck specimen, three times in embryonated chicken eggs (specific-pathogen free) to obtain the working stock virus. The stock virus was diluted 1:10 in sterile phosphate buffer saline (PBS), then inoculated at a total volume of 0.5 ml into each duck in the inoculated groups through pathways considered closest to the natural route: 1–2 drops of inoculum into each eye and nostril, and the remainder instilled into the mouth. A back titration established that the inoculum contained $10^{7.7}$ EID$_{50}$/0.1 ml, which approximately equates to $10^{8.4}$ EID$_{50}$ in the 0.5 ml administered to each duck.

Study Design

Two groups of ducks (group 1, n = 5 and group 3, n = 10) were inoculated with the virus and kept in two separate rooms. Two other groups, group 2 (n = 5) and group 4 (n = 9) were contact ducks for inoculated groups 1 and 3, respectively. Our previous study demonstrated that Pekin ducks experimentally infected with this clade 2.1.1 virus showed subclinical acute infection that occurred between 1–8 dpi, as determined by detection of infectious virus in tissues and in oral swabs [30]. To investigate virus transmission during the acute infection stage, group 2 (contact ducks) was transferred into group 1 (inoculated ducks) at 1 dpi. To determine virus transmission during the period after acute infection (hereafter referred to as the post-acute infection stage) groups 3 (inoculated) and 4 (contact) were mixed at 10 dpi. To avoid the transmission from the contaminated room environment, the ten inoculated ducks of group 3 were securely transferred into the uncontaminated room containing group 4. Ducks in all groups were monitored after the virus inoculation or after they were mixed. Oral and cloacal swabs were collected once before the inoculation date and daily after the inoculation date; all the swab samples were stored at −80°C before testing. Sera were collected from each duck as 2–3 ml clotted blood from the wing vein with the following schedule: once before inoculation for all groups; at 8 and 15 dpi for groups 1 and 2 (equivalent to 7 and 14 days post-contact (dpc) for group 2); at 8, 15, 22, 29, and 34 dpi for group 3; and 22, 29 and 34 dpi (12, 19, 24 dpc) for group 4. For welfare reasons, clinically affected ducks were euthanized with intravenous pentobarbital sodium (Lethabarb, Virbac Animal Health, Australia) once they developed moderate clinical signs.

Serology

The hemagglutination inhibition (HI), virus neutralization (VN) and blocking enzyme- linked immunosorbent assay (bELISA) tests were used to detect the presence of antibodies prior to and after virus inoculation. Blood samples in EDTA were collected from each duck at the designated times. After overnight incubation at 4°C, blood was centrifuged at 1000 g for 10 minutes to separate the sera. Sera then were heat-treated at 56°C for 1 hour immediately after separation. Homologous antigen, either inactivated or live homologous virus (IDN34368), was used in the HI and the VN tests respectively, while recombinant influenza nucleoprotein (NP) antigen was used for bELISA. For HI and VN tests, sera were first diluted 1:2 before testing; sera for HI were diluted (adsorbed) with 10% chicken red blood cells (RBCs) or with 10% horse RBCs, while for VN sera were diluted with phosphate buffer saline (PBS). These sera were further diluted 1:2 with PBS at the start of the HI and VN tests; therefore, the minimal detectable titre given by the HI and VN tests was 4. For bELISA, 1:10 diluted sera in serum diluent (PBS with 0.05% Tween 20 and 1% skim milk) were used and the cut-off was set at 60% inhibition. All the tests were conducted using methods as previously described [31].

Virology

Virus isolation in eggs was performed for all oral and cloacal swab media collected from ducks in groups 1 and 2. For group 3, it was conducted on the swab media obtained from ducks at 1–15 dpi and from selected days afterward (17, 20, 24, and 29 dpi), whereas for group 4, it was performed for the swab media of ducks from 1, 5 and 7 dpc. For this, 0.2 ml undiluted swab medium was inoculated into the allantoic cavity of 9 to 11-day-old embryonated eggs (3 eggs inoculated per sample). Eggs were observed for the death of embryos each day for 3 days after inoculation. The allantoic fluids from all dead eggs were tested for the presence of influenza virus by the HA test using standard methods [32]. A swab was considered virus-positive if at least one of the inoculated eggs was HA-positive.

For three ducks that showed clinical signs during the experiment (ducks #85, #94, #97), 100–150 mg of brain, heart, lung, pharynx, trachea, air sacs, spleen, pancreas, small intestine and skeletal muscle were collected following euthanasia into tubes containing 1 ml of transport media and 0.4 to 0.5 μg of 1.0 mm silicon carbide beads (Daintree Scientific, St. Helens, Tasmania, Australia). Tissue homogenates were titred for virus in Vero cells grown in culture media (EMEM plus HEPES, glutamine, penicillin and streptomycin, fungizone, and 10% fetal calf serum/FCS), as described previously [28]. Because the dilutions started from 1:10, the lowest detectable titres, equivalent to CPE occurring in a single well at the lowest dilution, was $10^{0.7}$ 50% tissue culture infectious doses ($TCID_{50}$) per 0.1 ml. The $TCID_{50}$ was calculated by the Reed and Muench method [33].

RNA extraction and real-time reverse transcriptase-PCR (rRT-PCR)

RNA extraction and rRT-PCR were performed from the same oral and cloacal swab media as used for the virus isolation. Viral RNA was extracted from 50 μl of each swab medium. Viral RNA was also extracted from tissue homogenates of the two contact ducks from group 2, which were euthanized at 6 and 10 dpi. The MagMAX™-96 Viral RNA Isolation Kit (Ambion®, Applied Biosystem) with the Ambion® 96-well Magnetic-Ring stand were used for the RNA extraction as per the manufacturer's protocol. RNA was then tested for the presence of hemagglutinin (HA) gene of H5 subtype using a multiplex TaqMan assay with AgPath-ID One-Step RT-PCR Kit (Applied Biosystem). The assay consists of two sets of primers and probes targeting two different regions (C-terminus and N-terminus) of the hemagglutinin gene (sequences available upon request). The rRT-PCR assays were conducted using AAHL's standard conditions: 45°C for 10 min, 95°C 10 min, followed by 45 cycles of 95°C 15 sec and 60°C 45 sec. In this study, swab media with an undetectable Ct were assigned a value of 45 for calculation of the mean Ct scores for each group and for presentation of data (Figure 1 and 2).

Histopathology and immunohistochemistry

A range of tissues and structures, including heart, brain, spleen, liver, kidney, lung, trachea, air sacs, esophagus, proventiculus, gizzard, thymus, bursa, pancreas, small intestines, cecal tonsil, third eyelid and head samples, were obtained from three euthanized ducks which showed clinical signs of disease during the experiment (ducks #85, #94, #97). Tissues were placed into 10% neutral buffered formalin and fixed for 1–7 days, before they were processed for histology and immunohistochemistry (IHC) staining for detection of tissue lesions and influenza A nucleoprotein antigen, respectively. Procedures used to perform the histology and IHC followed the method as previously described [30].

Results

In this study, the trends of infectious virus and viral RNA shedding of ducks inoculated with Indonesian clade 2.1 H5N1 HPAI virus were determined by virus isolation and the monitoring of contact ducks and by rRT-PCR of oral and cloacal swabs, respectively. Prior to challenge, all the ducks were antibody negative by HI, VN and bELISA tests, swabs were all negative by virus isolation in embryonated eggs and H5 rRT-PCR Ct values were undetectable (≥45). These test results indicate that the ducks had not had prior exposure to influenza virus.

Serology

The antibody responses were measured by bELISA, HI using chicken RBCs (HI-C), HI using horse RBCs (HI-H), and VN tests and these responses were used to confirm that infection had occurred in individual ducks.

The antibody response in inoculated (group 1) and acute stage contact ducks (group 2) is shown in Table 1. At 8 dpi, 100% (5/5) of the inoculated ducks had seroconverted, as detectable antibodies shown by bELISA (>60% inhibition) and both HI tests, but no neutralizing antibodies were detected. At the same time (equivalent to 7 dpc), none of the contact ducks of group 2 had seroconverted by the HI and VN tests, but 60% (3/5, including duck #94 at 5 dpc) were positive by bELISA. The levels of detectable HI and VN antibodies increased about 1–3 $\log_2$ from 8 to 15 dpi, whereas the levels of bELISA appeared to remain steady over the same time. At 9 dpc, contact duck #85 had seroconverted by the bELISA and both HI tests, but not the VN test. At termination (15 dpi or 14 dpc, respectively), all 5 inoculated ducks and 2 of the 3 remaining contact ducks had significant antibodies by all tests. One contact duck (#71) developed only a marginal bELISA signal at 14 dpc, despite indication of infection since 2 dpc.

In group 3 (inoculated), the proportion of ducks that had seroconverted at 8 dpi varied among the different tests: 9/10

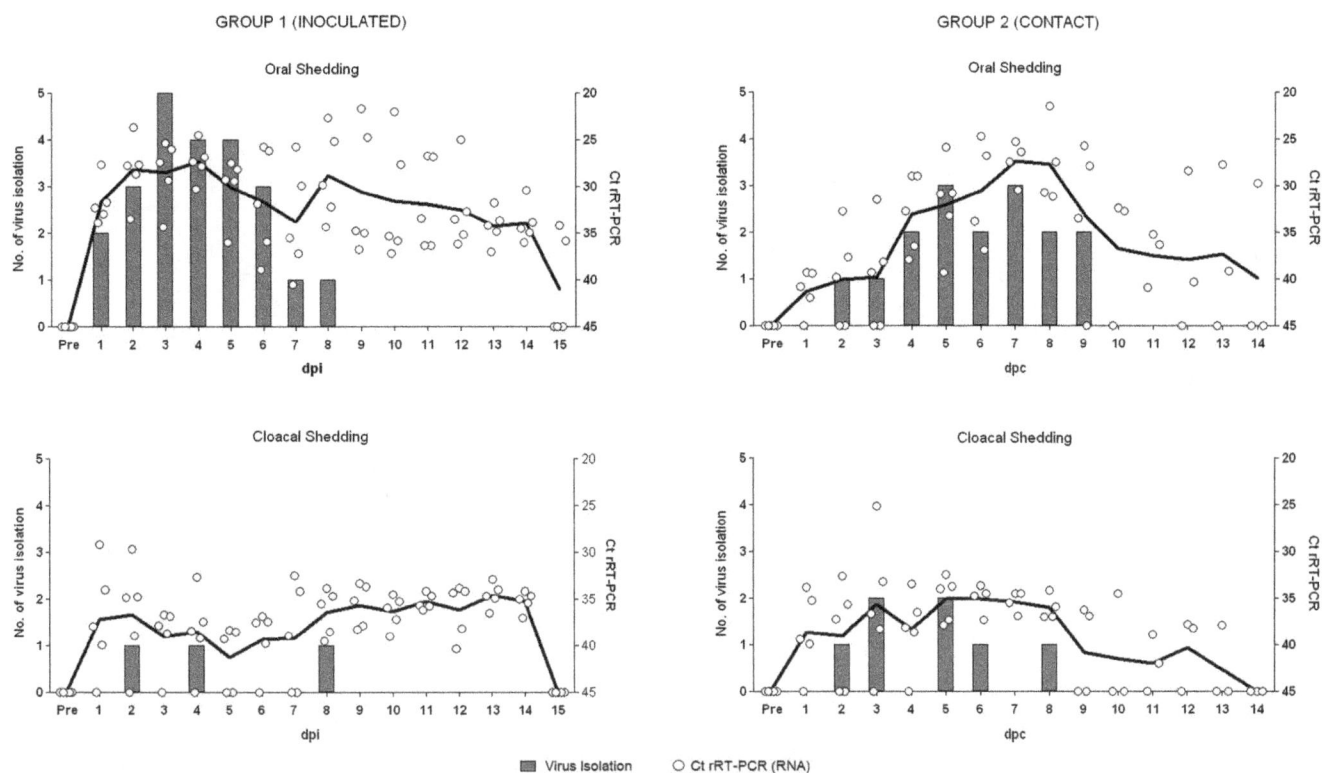

Figure 1. Virus isolation and H5 RNA Ct values for oral and cloacal swabs from the inoculated ducks (group 1) and the acute stage-contact ducks (group 2). Five contact ducks of group 2 (charts at right) were mixed with five inoculated ducks of group 1 (charts at left) at one day post inoculation (1 dpi). Swabs were obtained from all ducks in group 1 for 1–15 dpi, while those for group 2 were collected from 5 ducks from 1–5 day post-contact (dpc), 4 ducks from 6–9 dpc and 3 ducks from 10–15 dpc as 2 ducks were euthanized at 5 and 9 dpc. Vertical column bars indicate the number of ducks positive for virus isolation in embryonated chicken eggs, while white circles indicate the H5 RNA Ct values from rRT-PCR. Circles for Ct values have been staggered to avoid overlap. The means of Ct values from each time point are connected (solid black line); undetectable Ct values were given the value of 45 for calculation of the means.

(bELISA), 8/10 (HI-H), 7/10 (HI-C) and 5/10 (VN) (Table 2). At the later sampling times (15, 22, 29, 34 dpi), these proportions were more comparable, with all ducks, except one, having seroconverted by all four tests. Duck #80 remained seronegative by bELISA and without detectable HI and VN antibodies throughout the experiment. In contrast, influenza antibodies were not detected to the end of the study at 34 dpi (24 dpc) in any of the nine post-acute stage contact ducks in group 4 after they were mixed with the inoculated ducks of group 3.

Viral shedding

The patterns of viral shedding, including route and duration, were determined by detecting infectious virus or H5-RNA in oral and cloacal swabs of ducks from groups 1, 2 and 3. Because none of the contact ducks in group 4 seroconverted, oral and cloacal swabs sampled only from selected days (1, 5 and 7 dpc) were tested for virus isolation and rRT-PCR in order to support the evidence of lack of viral infection.

Groups 1 and 2: acute infection stage shedding and transmission. Virus isolation and H5 RNA Ct values from oral and cloacal swabs of ducks in groups 1 and 2 are shown in Figure 1. The means of Ct values were incorporated to demonstrate the kinetics of RNA shedding. On at least one occasion in the study, all the ducks in both groups were virus isolation positive and had rRT-PCR Ct values ≤30.9 (Table 1, virology results), confirming that all ducks in these groups became infected.

In group 1, Ct values <40 were detected in oral swabs from all the five ducks at 1 dpi and this was consistently found at the later sampling times (Figure 1 and Table S1). At 1–7 dpi the proportion of oral swabs with detectable H5-RNA was always higher than that of cloacal swabs, but at 8–14 dpi the relative proportions in oral and cloacal swabs were more similar. The relative quantity of RNA shedding from oral swabs was higher (Ct values: 27.3–34.9) than cloacal swabs (Ct values: 34.5–38.7). Infectious virus was isolated in oral swabs from 2 of 5 ducks at 1 dpi, but all cloacal swabs were virus isolation negative at this time. The peak of infectious virus and viral RNA shedding was seen around 3 and 4 dpi, with a high proportion of ducks being virus isolation positive and a high concentration of viral RNA found on those days, especially in the oral swabs. In agreement with the results of viral RNA shedding, the proportion of virus isolation positive oral swabs was consistently higher than that of cloacal swabs. Although H5-RNA was detected in both cloacal and oral swabs until 14 and 15 dpi, respectively, infectious virus was only detected up to 8 dpi.

In group 2, H5-RNA was detected in both oral and cloacal swabs of the contact ducks, starting at 1 dpc (Figure 1, Table S1). The relative amount of oral-RNA shedding (Ct values: 26.2–37.5) was always higher than cloacal-RNA shedding (Ct values: 33.4–40.5), with the exceptions at 1 to 3 dpc. Oral-RNA shedding peaked at 7 dpc and in one bird continued to be shed through the oral route until the end of the study at 14 dpc. Infectious virus was detected in cloacal and oral swabs from 2 dpc to 8 or 9 dpc, with the daily proportion of virus isolation positive cloacal swabs often

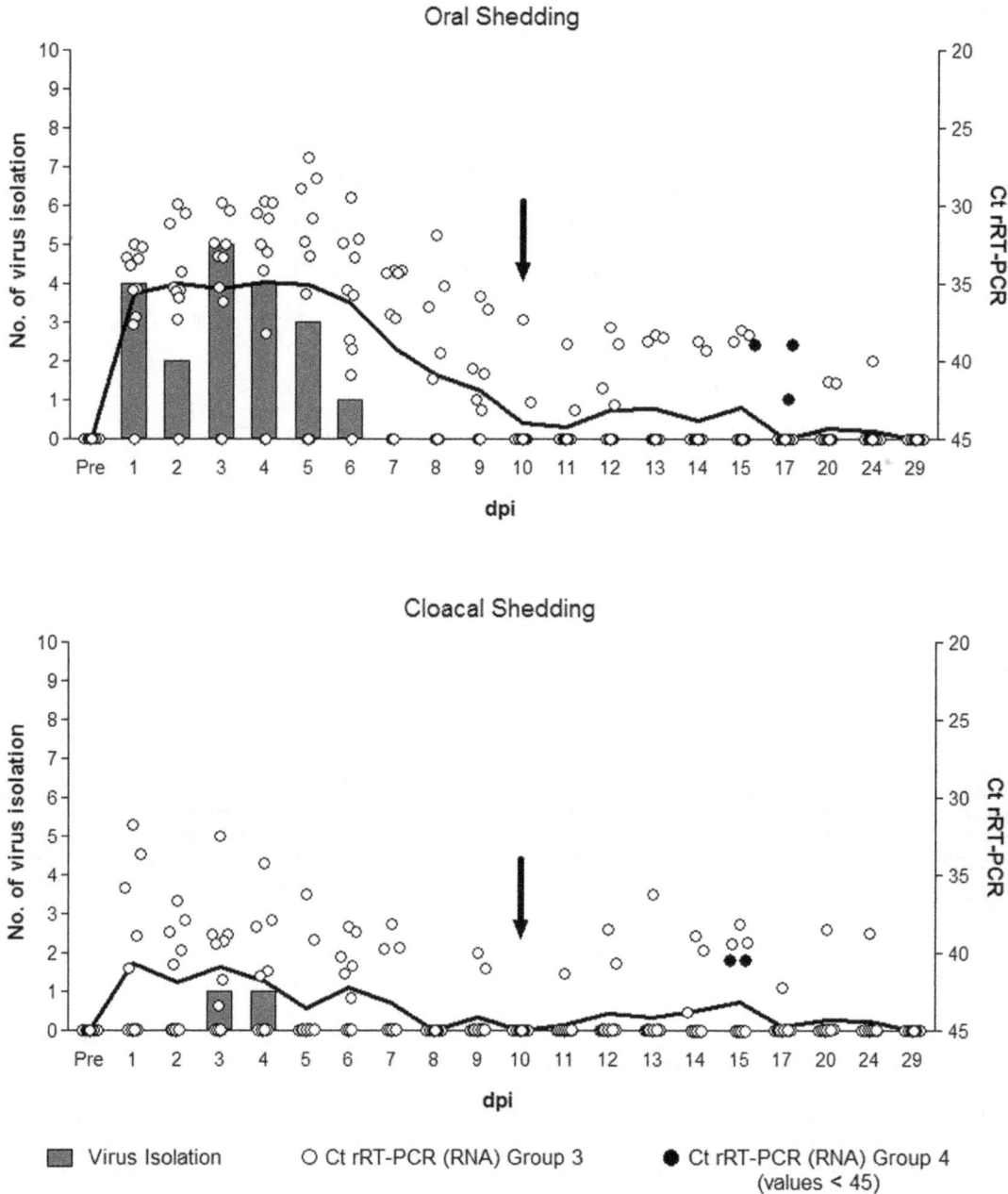

Figure 2. Virus isolation and H5 RNA Ct values for oral and cloacal swabs from the inoculated ducks (group 3) and the post-acute stage contact ducks (group 4). Swabs from ducks of group 3 (n = 10) were collected 1–15 day post inoculation (dpi) and at 17, 20, 24 and 29 dpi. Nine contact ducks (group 4) were mixed with the inoculated ducks of group 3 at 10 dpi (arrow). Virus isolation and rRT-PCR for group 4 was performed on oral and cloacal swabs collected from all the contact ducks at 1, 5 and 7 days post contact (dpc) (equivalent to 11, 15, 17 dpi); none of these swabs were virus isolation positive and only five swabs from different contact ducks had detectable Ct values (black circles). The mean Ct values for group 3 from each time point are connected (solid black line); undetectable Ct values were given the value of 45 for calculation of the means.

lower than oral swabs (Figure 1). Infectious virus was detected at least once in each bird. The peak of infectious virus shedding occurred around 5 and 7 dpc.

Groups 3 and 4 – post-acute infection stage shedding and transmission. In group 3, the proportion of oral swabs that contained detectable H5-RNA was consistently higher than with cloacal swabs (Figure 2, Table S2). The mean Ct value was also higher in oral swabs than in cloacal swabs over this period. H5-RNA was detected daily in oral swabs of four of 10 ducks from 1 to 9 or 10 dpi, but only intermittently detected afterwards. On the

other hand, in cloacal swabs H5-RNA was intermittent throughout most of the study period (Table S2). At 1 dpi, infectious virus was detected in oral swabs from 4 of 9 RNA-positive ducks and it was detected from the same swab type up to 6 dpi (Figure 2). In contrast, virus isolation positive cloacal swabs were found only at 3 and 4 dpi, from the same duck. Although one duck (#80) in this group was H5-antibody negative for 34 days observation, low H5 RNA levels (Ct value: 38.0–42.6) were detected intermittently in its oral (at 8, 10 and 15 dpi) and cloacal swabs (at 6 and 12 dpi), but no viable virus was detected.

Table 1. Serum antibody titers and viral detection in the H5N1 HPAI virus-inoculated (group 1) and acute stage contact (group 2) ducks.

| Bird ID | Pre-Inoculation | | | | 8 dpi | | | | 15 dpi | | | | VIª | rRT-PCRᵇ |
	ELISA	HI-C	HI-H	VN	ELI SA	HI-C	HI-H	VN	ELI SA	HI-C	HI-H	VN		
Group 1														
#73	36	<4	<4	<4	107	16	32	<4	104	128	128	128	+	27.8
#75	26	<4	<4	<4	103	8	32	<4	103	64	128	128	+	23.6
#77	34	<4	<4	<4	99	16	64	<4	105	128	128	128	+	26.8
#83	36	<4	<4	<4	101	32	64	<4	99	256	512	128	+	21.6
#92	30	<4	<4	<4	104	16	16	<4	104	128	256	128	+	24.5
Group 2														
#71	27	<4	<4	<4	55	<4	<4	<4	63	<4	<4	<4	+	30.5
#84	17	<4	<4	<4	54	<4	<4	<4	104	128	256	16	+	21.5
#85ᶜ	37	<4	<4	<4	102	<4	<4	<4	93	4	16	<4	+	24.7
#87	33	<4	<4	<4	68	<4	<4	<4	90	64	128	32	+	26.4
#94ᶜ	26	<4	<4	<4	64	<4	<4	<4	nd	nd	nd	nd	+	30.9

Five ducks inoculated with H5N1 HPAI virus (group 1) and five naive contact ducks (group 2) were mixed at one day post inoculation (dpi). Sera were collected at 4 days pre-inoculation for both groups and at 8 and 15 dpi (equivalent to 7 and 14 days post contact (dpc) respectively for group 2). Sera were tested with blocking ELISA (values are % inhibition; cut-off = 60%), HI using chicken RBCs (HI-C) and horse RBCs (HI-H), and virus neutralization (VN) tests. nd: not done.
ªSummary of virus isolation (VI): a+symbol indicates that virus was isolated at least once from either oral or cloacal swab from the duck during the duration of the trial.
ᵇSummary of Ct values for real-time reverse transcription polymerase chain reaction (rRT-PCR) results: the value is the lowest detected for the duck from either oral or cloacal swabs during the duration of the trial.
ᶜDucks 94 and 85 developed disease during the experiment; therefore the last sera were obtained when they were euthanized, for welfare reasons, at 5 dpc (shown under 8 dpi) and 9 dpc (shown under 15 dpi), respectively.

In post-acute stage contact ducks (Group 4), infectious virus was not found in any oral or cloacal swabs at the selected days: 1, 5 and 7 dpc (Figure 2, Table 2). However, low levels of H5-RNA (Ct values ≥38.9) were detected in swab media from 5 of the 9 ducks in this group: 3 oral and 2 cloacal swabs at 5 and 7 dpc (Figure 2).

Outcomes of H5N1 virus infection

There were no clinical signs in the majority of ducks in all groups after virus inoculation or contact, indicating that infection by the H5N1 HPAI virus used in this study was largely asymptomatic in ducks. However, three ducks developed signs of disease. Duck #97 in Group 3 developed mild head tilt which persisted until the termination of the trial at 34 dpi. Histological observation on this duck found mild chronic focal lymphocytic inflammation of the myocardium, Eustachian tube and a ganglion of the inner ear. The significance of these lesions is not clear. No viral antigen was detected in any of the tissues from this duck.

Two of five acute infection stage contact ducks of group 2 (duck #94 and #85) showed recumbency and depression for 1–2 days before they were euthanized at 5 and 9 dpc respectively. On necropsy examination a white membrane over the peritoneal organs and airsacs, and excessive peritoneal fluid, were noted in both ducks.

Histological examination of tissues of duck #94 found that disease consisted of inflammation of the respiratory system with systemic viral spread. There was severe acute localized necrotizing mononuclear cell airsacculitis, mild edema in the lungs and mononuclear cell inflammation in the pulp of occasional feathers. Abundant antigen was present in the epithelial tissues of the respiratory tract, and small to moderate amounts of antigen were found in a range of other tissues (Figure 3A, 3B and 3C; Table 3). In this duck, the tissues with the highest titers of virus and highest viral RNA levels were air sac, trachea and spleen (Table 3).

In duck #85, disease was characterized by generalized severe sub-acute inflammation of multiple organ systems, including heart, brain and airsacs. Lesions consisted of severe sub-acute diffuse myocarditis (Figure 3D), severe localized (cerebral) mononuclear cell encephalitis, severe sub-acute diffuse airsacculitis, moderate sub-acute diffuse mononuclear cell pectoral myositis, and mild mononuclear cell pterylitis in the pulp of occasional feathers. Small to moderate (although locally dense) amounts of antigen were found in the brain and other tissues (Figure 3; Table 3). In this duck, the greatest abundance of antigen was found in the feathers. Virus could not be isolated from any tissues except for trachea, which yielded a very low titre of virus. However, viral RNA was detected in all tissues tested, with the brain showing a relatively high viral RNA load (Ct = 19.9) (Table 3). It is probable that the peak of viral infection was passed at the time of euthanasia, as the levels of antigen and viral loads were relatively low despite severe histological lesions.

Discussion

In this study, we investigated the shedding of an Indonesian clade 2.1 H5N1 HPAI virus by domestic ducks after the acute infection stage, in an attempt to assess the potential impact of long-term infection of ducks in the evolution of H5N1 HPAI viruses in Indonesia. Although we found that ducks will shed and efficiently transmit H5N1 HPAI virus during acute infection, we found no evidence of shedding of viable virus or transmission after 10 dpi. None of the contact ducks (group 4) that were mixed with the inoculated ducks (group 3) after 10 dpi seroconverted or shed infectious virus. In contrast, in a separate contact group that was mixed during the acute stage of infection (at 1 dpi) virus was readily transmitted from infected ducks, which shed significant amounts of infectious virus and viral RNA. Although continued shedding of H5 viral RNA was shown in the inoculated ducks

Table 2. Serum antibodies and viral detection in the H5N1-infected (group 3) and post acute stage contact (group 4) ducks.

Bird ID	Pre-Inoculation				8 dpi				15 dpi				22 dpi				29 dpi				34 dpi				VI[a]	rRT-PCR[b]
	ELI SA	HI-C	HI-H	VN	ELI SA	HI-C	HI-H	VN	ELI SA	HI-C	HI-H	VN	ELI SA	HI-C	HI-H	VN	ELI SA	HI-C	HI-H	VN	ELI SA	HI-C	HI-H	VN		
Group 3																										
#72	29	<4	<4	<4	104	64	128	16	104	64	64	64	106	256	128	1024	108	128	128	512	97	64	64	1024	+	29.7
#76	35	<4	<4	<4	103	8	32	4	103	64	128	32	106	128	64	256	107	128	128	256	102	64	128	512	+	26.9
#80	28	<4	<4	<4	25	<4	<4	<4	24	<4	<4	<4	35	<4	<4	<4	30	<4	<4	<4	38	<4	<4	<4	–	38.0
#81	50	<4	<4	<4	95	64	64	8	104	256	256	128	102	256	128	1024	108	256	256	512	99	128	64	256	+	28.2
#82	23	<4	<4	<4	88	<4	<4	<4	88	8	8	8	96	8	8	16	84	8	8	16	77	4	8	16	+	33.8
#90	39	<4	<4	<4	88	4	8	<4	90	32	16	32	86	16	8	8	80	8	4	32	68	8	4	32	–	35.9
#93	31	<4	<4	<4	100	32	32	8	104	64	64	128	107	128	64	512	107	64	32	256	101	64	64	256	+	30.5
#97	33	<4	<4	<4	105	64	128	32	104	256	128	256	107	256	128	256	106	128	128	512	100	128	128	512	+	29.9
#98	36	<4	<4	<4	94	<4	32	<4	94	32	32	32	106	64	32	128	107	64	64	128	102	64	32	256	+	33.2
#99	18	<4	<4	<4	85	4	32	<4	101	64	64	128	105	128	32	512	105	128	64	512	104	128	64	512	+	28.9
Group 4																										
#74	36	<4	<4	<4	nd	nd	nd	nd	nd	nd	nd	nd	30	<4	<4	<4	22	<4	<4	<4	16	<4	<4	<4	–	38.9
#78	35	<4	<4	<4	nd	nd	nd	nd	nd	nd	nd	nd	31	<4	<4	<4	32	<4	<4	<4	18	<4	<4	<4	–	–
#79	22	<4	<4	<4	nd	nd	nd	nd	nd	nd	nd	nd	34	<4	<4	<4	25	<4	<4	<4	33	<4	<4	<4	–	40.8
#86	30	<4	<4	<4	nd	nd	nd	nd	nd	nd	nd	nd	27	<4	<4	<4	18	<4	<4	<4	34	<4	<4	<4	–	–
#89	28	<4	<4	<4	nd	nd	nd	nd	nd	nd	nd	nd	33	<4	<4	<4	30	<4	<4	<4	19	<4	<4	<4	–	40.7
#91	27	<4	<4	<4	nd	nd	nd	nd	nd	nd	nd	nd	27	<4	<4	<4	36	<4	<4	<4	24	<4	<4	<4	–	38.9
#95	33	<4	<4	<4	nd	nd	nd	nd	nd	nd	nd	nd	8	<4	<4	<4	26	<4	<4	<4	18	<4	<4	<4	–	–
#96	17	<4	<4	<4	nd	nd	nd	nd	nd	nd	nd	nd	9	<4	<4	<4	29	<4	<4	<4	26	<4	<4	<4	–	–
#100	18	<4	<4	<4	nd	nd	nd	nd	nd	nd	nd	nd	30	<4	<4	<4	23	<4	<4	<4	25	<4	<4	<4	–	42.4

Ten ducks (group 3) inoculated with H5N1 HPAI virus were mixed with nine naïve contact ducks (group 4) at 10 day post inoculation (dpi). Sera were collected at 4 days pre-inoculation for both groups and at 8, 15, 22, 29, and 34 dpi for both groups, which corresponded to 12, 19 and 24 days post contact (dpc) for group 4. Sera were tested using blocking ELISA (values are % inhibition; cut-off = 60%), HI with chicken RBCs (HI-C) and horse RBCs (HI-H), and virus neutralization (VN) tests. nd: not done.

aSummary of virus isolation (VI): a+symbol indicates that virus was isolated at least once from an oral or cloacal swab from the duck during the duration of the trial.

bSummary of Ct values for real-time reverse transcription polymerase chain reaction (rRT-PCR) results: the value is the lowest detected for the duck from either oral or cloacal swabs during the duration of the trial.

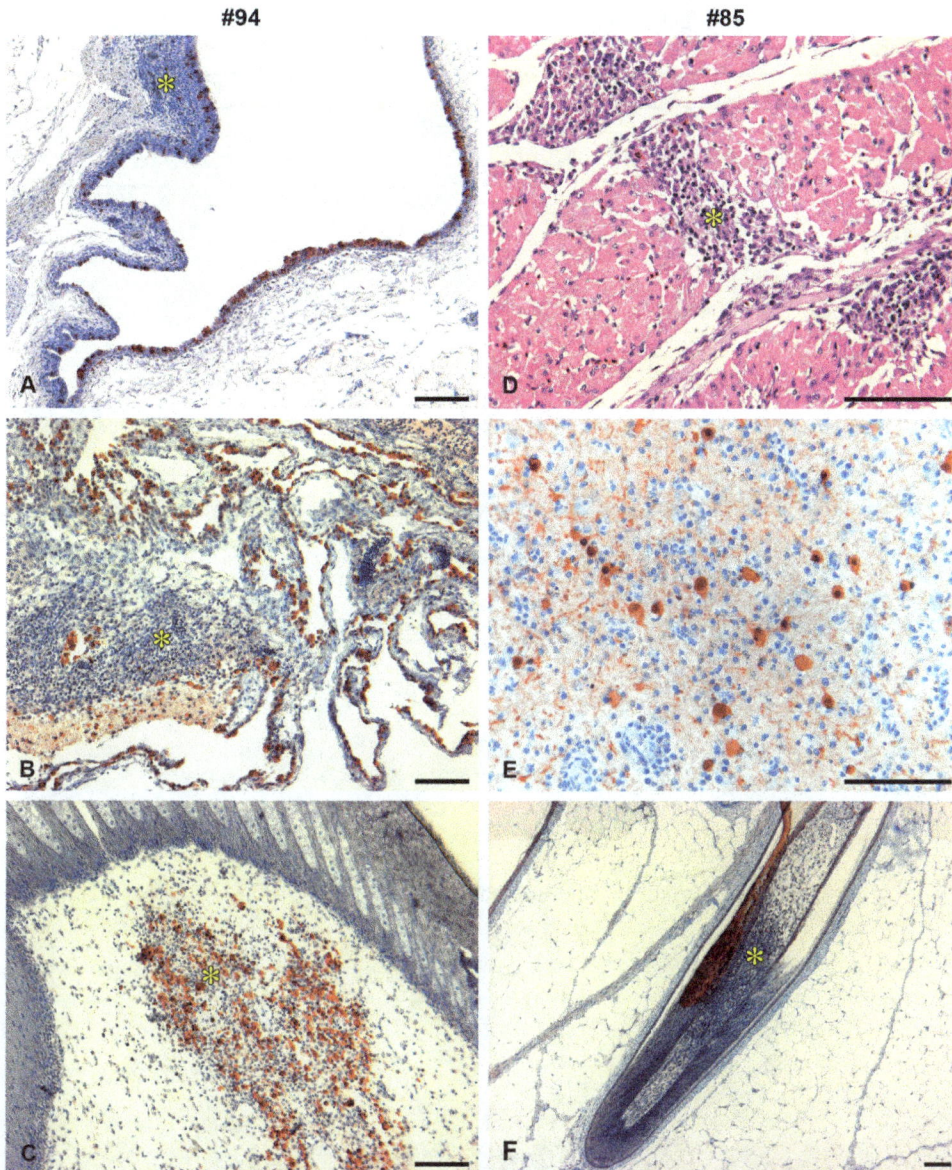

Figure 3. Histopathology and immunohistochemistry of H5N1 HPAI virus infection in acute stage contact-infected ducks #94 and #85. Ducks #94 (left, A–C) and #85 (right, D–F) were euthanized at 5 and 9 days post contact (dpc) following disease signs. A. Infraorbital sinus, showing influenza viral nucleoprotein antigen in epithelium and mononuclear cell inflammation (*). B. Air sac, showing abundant viral antigen in the epithelial membranes and in fibrinous exudates, and mononuclear cell infiltration (*); C. Feather, with abundant antigen staining associated with mixed cell inflammation (*) in the pulp; D. Heart, showing severe, sub-acute, diffuse mononuclear cell myocarditis (*); no antigen was detected in myocardium E. Brain, showing antigen in neuronal tissue associated with mononuclear cell encephalitis; F. Feather, with mononuclear cell pterilitis (*) observed in the pulp, and antigen staining in the epidermis. Haematoxylin and eosin stain (Fig. D) and immunohistochemistry stain (brown color) for influenza A virus nucleoprotein (Fig. A, B, C, E, F). All scale bars = 100 μm.

(group 3) up to 24 dpi, the negative virus isolation results from swabs and lack of evidence of infection in the post-acute infection stage contact ducks (group 4) indicated the absence of infectious virus shedding after 10 dpi. None of these contact ducks seroconverted although H5 RNA was detected in oral or cloacal swabs of a small number of ducks in group 4. It is likely that virus could not persist in the primary inoculated hosts because of the development of antibodies against H5N1 HPAI virus in those ducks by day 8 post-inoculation.

A previous study by Hulse-Post *et al.* [21] found that prolonged shedding of infectious H5N1 HPAI virus occurred in mallard ducks for up to 17 dpi despite seroconversion to the challenge virus. They hypothesize that neutralizing antibody does not completely eliminate all variants of the viral population, allowing repetitive infections of the host with the surviving variants. The extended, sub-clinical, infection period allows the host to transmit the immunologically novel variants to other hosts. However, our data indicate that, in domestic ducks infected with Indonesian H5N1 HPAI virus, the infectious period is short and its termination coincides with the rise of antibody. While our data do not preclude that viruses generated at the end of the infectious period may be antigenically different due to immunological

Table 3. Virus distribution in tissues of the two clinically affected and euthanized acute stage-contact ducks in group 2, as determined by immunohistochemistry, virus isolation and rRT-PCR.

	Duck #94	Duck #85
	(5 dpc)	(9 dpc)
Immunohistochemistry	**Antigen score[a]**	
Epithelia of respiratory tract:		
Infraorbital sinus	+++	−
Trachea	+	−
Bronchi	++	−
Lung air capillaries	+	−
Air sacs	+++	+
Brain[b]	−	++
Lymphoid tissue:		
Bursa (medulla)	+	−
Thymus (medulla)	+	−
Conjunctiva	+	−
Connective tissue, integument and associated tissues of head:		
Endosteum	+	+
Periosteum	+	+
Feather epidermis	−	+++
Feather pulp	+	−
Muscle:		
Myocardium	+	−
Pectoral skeletal muscle	−	+
Virus titration in Vero cells and rRT-PCR		
	Virus titre[c] (Ct value)	
Brain	<0.7 (31.4)	<0.7(19.6)
Heart	3.7 (20.7)	<0.7 (23.6)
Spleen	4.7 (19.2)	<0.7 (33.4)
Lung	2.7 (23.2)	<0.7 (30.1)
Pharynx	3.5 (23.9)	<0.7 (27.3)
Trachea	5.2 (19.9)	0.7 (27.8)
Air sacs	5.7 (15.5)	<0.7 (28.6)
Pancreas	3.7 (25.6)	<0.7 (32.2)
Small intestine	2.5 (27.0)	<0.7 (32.9)
Skeletal muscle	1.2 (26.3)	<0.7 (32.3)

[a]Influenza nucleoprotein antigen staining was present in single cells or small foci (+) or in localized clusters with moderate (++) to abundant (+++) quantities.
[b]Antigen found in neural tissues including neuron cell bodies, neuroglia and neuropil.
[c]Log_{10} $TCID_{50}$/0.1 ml of tissue homogenate.

selection, they indicate that the potential for this mechanism is considerably less than indicated by Hulse-Post *et al.* [21]. Given the short duration of the infectious period, the evolution of completely new immunological variants arising within a single duck host infected with this clade 2.1 virus, would appear to be an unlikely outcome.

Ducks were used as contacts in order to add an extra level of sensitivity for virus detection. Detection of transmission to a live host system would increase the probability of detection of viral shedding over the use of only *in vitro* methods such as swabbing. The use of ducks and not a different species, such as chickens, was considered appropriate as intra-species transmission is probably more efficient. Previous work at AAHL with H5N1 HPAI viruses indicated that Pekin ducks are a more sensitive recipient species

within contact-transmission trials than are chickens (J. Bingham, unpublished data). In this study, relatively large group sizes (10 and 9) of ducks were used to optimize the probability of transmission, given that shedding was expected to be low during the post-acute stage.

The results of this trial indicate that viral RNA is highly persistent. In both inoculated groups (groups 1 and 3) and in the acute infection stage-contact group (group 2), shedding of viral RNA was detected over longer periods than shedding of infectious virus. In the inoculated infected ducks (group 3), high level RNA shedding was only detected until 8–9 dpi. Viral RNA was shed intermittently at low levels at the later days of the trial in these ducks and in the post-acute infection stage of contact ducks (group 4), which otherwise showed no evidence of infection.

The rRT-PCR method is quick, safe and can handle larger numbers of samples more easily than virus isolation for H5N1 HPAI virus screening. At AAHL, the cycle threshold (Ct) for diagnostic positivity is normally set at <40, based on previously determined assays on positive control H5 RNA. However our data indicate that the stage of infection cannot easily be determined from rRT-PCR data alone, as H5 RNA continues to be detected weeks after infectious virus has disappeared. In addition, in the context of infectious risk, diagnosis of HPAI virus in ducks by detection of RNA is less meaningful than the level and duration of infectious virus shedding. The finding of Ct values below the threshold of positivity in ducks that otherwise have no evidence of infection indicates that contact with infectious birds may induce false positives by rRT-PCR. Although this molecular assay is an appropriate method for surveillance, its data must be interpreted carefully when making decisions on the infection status of individual birds, given that intra-flock cross contamination may occur during and shortly after an outbreak.

Consistent with some experimental studies using other Eurasian lineage H5N1 HPAI viruses [10,27–29], the present study demonstrated that the viral shedding in ducks, either infectious virus or H5 RNA, was more pronounced by the oral route than by the cloacal route. Our previous studies [28,30] indicated that the respiratory tract of ducks is the primary site for virus replication and the main source from which virus is shed from the oral cavity. This is contrary to LPAI viruses in which high viral concentration was found in fecal matter of ducks [10,34] and that LPAI virus transmission relies on fecal to oral transmission [35,36]. It would appear that with H5N1 HPAI, virus transmission among ducks is either through the air-borne route, where virus is expelled from the respiratory tract in droplets, or through dabbling, a behavioral feature of mallard-type ducks, which would facilitate the flushing of oral virus into water.

The inoculation of H5N1 HPAI virus into ducks is quickly followed within one to two days by oral and cloacal shedding; and the virus is efficiently transmitted at this time into naïve contact ducks. Interestingly, not all inoculated ducks developed evidence of infection: one duck (#80) did not seroconvert, no infectious virus was detected from it and Ct values for H5 RNA were intermittent and remained above 38.0. This duck appeared to have resisted not only the inoculation but also the contact challenge from its infected flockmates. Two other ducks (#82 and #90) seroconverted with low antibody titers, indicating infection had occurred, although only one of the ducks was positive on virus isolation. These data indicate that infection is variable at the dose administered in this study, even when followed by exposure to infected ducks.

In contrast with the inoculated ducks, of which only one of 15 showed mild neurological signs, two of five acute stage contact ducks developed clinical signs severe enough to warrant euthanasia. This suggests that contact-transmission may be more likely to induce severe disease or that more pathogenic mutants arose during the course of the study. Further investigation is important to understand the effects of transmission routes and to identify whether viruses isolated from these ducks harbor genetic mutations related to pathogenicity.

In conclusion, this study shows that domestic ducks efficiently acquired infection of Indonesian H5N1 HPAI virus through transmission during the acute infection stage. However, viral shedding occurred during a relatively short period and was eliminated coincident with the rise of influenza antibodies. Our studies indicated that there is no evidence of persistent shedding of infectious Indonesian H5N1 HPAI clade 2.1 virus in experimentally infected domestic ducks; they suggest that long-term viral infection and shedding in ducks is unlikely to be a significant factor in the ecological cycle and persistence of H5N1 HPAI virus. This supports the view that antigenic drift, which allows evolved viral types to re-infect a population of hosts, would take place over several transmission cycles in naive hosts, rather than in single hosts with extended infections. Models for assessing risk and for viral maintenance in duck populations should take this information in account.

Supporting Information

Table S1 Viral isolation and Ct rRT-PCR values for oral and cloacal swabs of the H5N1-inoculated (group 1) and acute infection stage-contact (group 2) ducks.

Table S2 Viral isolation and Ct rRT-PCR values for oral and cloacal swabs of the H5N1-inoculated ducks (group 3).

Acknowledgments

We thank the staff of various sections of AAHL: the Animal Services Unit for assistance in the animal trials, the Bio-Reagent Development and Protein Characterisation Group for their contribution of the rabbit antibody used for immunohistochemistry and the Serology Laboratory for assistance with serology tests.

Author Contributions

Conceived and designed the experiments: HW JB SL DM JM. Performed the experiments: HW JB HN JP JH. Analyzed the data: HW JB HN SL JM. Contributed reagents/materials/analysis tools: HW AJ. Wrote the paper: HW JB JM.

References

1. Alexander DJ (2007) An overview of the epidemiology of avian influenza. Vaccine 25: 5637–5644.
2. Guan Y, Smith GJ, Webby R, Webster RG (2009) Molecular epidemiology of H5N1 avian influenza. Rev Sci Tech 28: 39–47.
3. Alexander DJ, Brown IH (2009) History of highly pathogenic avian influenza. Rev Sci Tech 28: 19–38.
4. Chen H, Smith GJ, Li KS, Wang J, Fan XH, et al. (2006) Establishment of multiple sublineages of H5N1 influenza virus in Asia: implications for pandemic control. Proc Natl Acad Sci U S A 103: 2845–2850.
5. Vijaykrishna D, Bahl J, Riley S, Duan L, Zhang JX, et al. (2008) Evolutionary dynamics and emergence of panzootic H5N1 influenza viruses. PLoS Pathog 4: e1000161.
6. FAO (2011) H5N1 HPAI global overview: October–December 2011. FAO website. Available: [http://www.fao.org/docrep/015/an336e/an336e00.pdf]. Accessed 2013 September 20.
7. WHO (2013) Cumulative number of confirmed human cases for avian influenza A (H5N1) reported to WHO, 2003–2013. As reported on 29 August 2013. World Health Organization website. Available: http://www.who.int/influenza/human_animal_interface/EN_GIP_20130829CumulativeNumberH5N1cases.pdf. Accessed 2013 September 20.
8. Peiris JS, de Jong MD, Guan Y (2007) Avian influenza virus (H5N1): a threat to human health. Clin Microbiol Rev 20: 243–267.
9. Sims LD, Domenech J, Benigno C, Kahn S, Kamata A, et al. (2005) Origin and evolution of highly pathogenic H5N1 avian influenza in Asia. Vet Rec 157: 159–164.
10. Henaux V, Samuel MD (2011) Avian influenza shedding patterns in waterfowl: implications for surveillance, environmental transmission, and disease spread. J Wildl Dis 47: 566–578.
11. Kim JK, Negovetich NJ, Forrest HL, Webster RG (2009) Ducks: The "Trojan Horses" of H5N1 influenza. Influenza and Other Respiratory Viruses 3: 121–128.

12. CIVAS (2006) A review of free range duck farming systems in Indonesia and assessment of their implication in the spreading of the highly pathogenic (H5N1) strain of avian infl uenza (HPAI). Report from the Center of Indonesian Veterinary Analytical Studies (CIVAS) for the Food and Agricultural Organization of the United Nations. Available: http://civas.net/sites/default/files/report/FreeRangeDuck_report.pdf. Accessed 2013 September 20.

13. Webster RG, Peiris M, Chen H, Guan Y (2006) H5N1 outbreaks and enzootic influenza. Emerg Infect Dis 12: 3–8.

14. Forrest HL, Webster RG (2010) Perspectives on influenza evolution and the role of research. Anim Health Res Rev 11: 3–18.

15. Henning J, Wibawa H, Morton J, Usman TB, Junaidi A, et al. (2010) Scavenging ducks and transmission of highly pathogenic avian influenza, Java, Indonesia. Emerg Infect Dis 16: 1244–1250.

16. Achenbach JE, Bowen RA (2011) Transmission of avian influenza A viruses among species in an artificial barnyard. PLoS One 6: e17643.

17. FAO (2011) Approaches to controlling, preventing and eliminating H5N1 highly pathogenic avian influenza in endemic country. FAO website. Available: http://www.fao.org/docrep/014/i2150e/i2150e.pdf. Accessed 2013 September 20.

18. Keawcharoen J, van Riel D, van Amerongen G, Bestebroer T, Beyer WE, et al. (2008) Wild ducks as long-distance vectors of highly pathogenic avian influenza virus (H5N1). Emerg Infect Dis 14: 600–607.

19. Sturm-Ramirez KM, Hulse-Post DJ, Govorkova EA, Humberd J, Seiler P, et al. (2005) Are ducks contributing to the endemicity of highly pathogenic H5N1 influenza virus in Asia? J Virol 79: 11269–11279.

20. Tumpey TM, Suarez DL, Perkins LE, Senne DA, Lee JG, et al. (2002) Characterization of a highly pathogenic H5N1 avian influenza A virus isolated from duck meat. J Virol 76: 6344–6355.

21. Hulse-Post DJ, Sturm-Ramirez KM, Humberd J, Seiler P, Govorkova EA, et al. (2005) Role of domestic ducks in the propagation and biological evolution of highly pathogenic H5N1 influenza viruses in Asia. Proc Natl Acad Sci U S A 102: 10682–10687.

22. Henning J, Morton JM, Wibawa H, Yulianto D, Usman TB, et al. (2012) Incidence and risk factors for H5 highly pathogenic avian influenza infection in flocks of apparently clinically healthy ducks. Epidemiology and Infection: 1–12.

23. Higgins DA, Shortridge KF, Ng PL (1987) Bile immunoglobulin of the duck (Anas platyrhynchos). II. Antibody response in influenza A virus infections. Immunology 62: 499–504.

24. Kida H, Yanagawa R, Matsuoka Y (1980) Duck influenza lacking evidence of disease signs and immune response. Infect Immun 30: 547–553.

25. Hinshaw VS, Webster RG, Bean WJ, Sriram G (1980) The ecology of influenza viruses in ducks and analysis of influenza viruses with monoclonal antibodies. Comp Immunol Microbiol Infect Dis 3: 155–164.

26. Perkins LE, Swayne DE (2002) Pathogenicity of a Hong Kong-origin H5N1 highly pathogenic avian influenza virus for emus, geese, ducks, and pigeons. Avian Dis 46: 53–63.

27. Pantin-Jackwood MJ, Swayne DE (2007) Pathobiology of Asian highly pathogenic avian influenza H5N1 virus infections in ducks. Avian Dis 51: 250–259.

28. Bingham J, Green DJ, Lowther S, Klippel J, Burggraaf S, et al. (2009) Infection studies with two highly pathogenic avian influenza strains (Vietnamese and Indonesian) in Pekin ducks (Anas platyrhynchos), with particular reference to clinical disease, tissue tropism and viral shedding. Avian Pathol 38: 267–278.

29. Jeong OM, Kim MC, Kim MJ, Kang HM, Kim HR, et al. (2009) Experimental infection of chickens, ducks and quails with the highly pathogenic H5N1 avian influenza virus. J Vet Sci 10: 53–60.

30. Wibawa H, Bingham J, Nuradji H, Lowther S, Payne J, et al. (2013) The pathobiology of two Indonesian H5N1 avian influenza viruses representing different clade 2.1 sublineages in chickens and ducks. Comparative Immunology Microbiology and Infectious Diseases 36: 175–191.

31. Wibawa H, Henning J, Waluyati DE, Usman TB, Lowther S, et al. (2012) Comparison of serological assays for detecting antibodies in ducks exposed to H5 subtype avian influenza virus. BMC Vet Res 8: 117.

32. OIE (2012) Avian Influenza. In Manual of diagnostic tests and vaccines for terrestrial animals 2012. Website of the OIE Terrestrial Manual 2012. Available: http://www.oie.int/fileadmin/Home/eng/Health_standards/tahm/2.03.04_AI.pdf. Accessed 2013 September 20.

33. Reed IJ, Muench H (1938) A simple method for estimating fifty percent endpoints. Am J Hyg 27: 493–497.

34. Webster RG, Yakhno M, Hinshaw VS, Bean WJ, Murti KG (1978) Intestinal influenza: replication and characterization of influenza viruses in ducks. Virology 84: 268–278.

35. Hinshaw VS, Webster RG, Turner B (1979) Water-bone transmission of influenza A viruses? Intervirology 11: 66–68.

36. Hinshaw VS, Webster RG, Turner B (1980) The perpetuation of orthomyxo-viruses and paramyxoviruses in Canadian waterfowl. Can J Microbiol 26: 622–629.

A 20-Amino-Acid Deletion in the Neuraminidase Stalk and a Five-Amino-Acid Deletion in the NS1 Protein Both Contribute to the Pathogenicity of H5N1 Avian Influenza Viruses in Mallard Ducks

Yanfang Li[9], Sujuan Chen[9], Xiaojian Zhang, Qiang Fu, Zhiye Zhang, Shaohua Shi, Yinbiao Zhu, Min Gu, Daxin Peng*, Xiufan Liu*

College of Veterinary Medicine, Yangzhou University, Jiangsu Co-Innovation Center for the Prevention and Control of Important Animal Infectious Disease and Zoonoses, Yangzhou University, Yangzhou, Jiangsu, P.R. China

Abstract

Since 2003, H5N1-subtype avian influenza viruses (AIVs) with both a deletion of 20 amino acids in the stalk of the neuraminidase (NA) glycoprotein (A−) and a deletion of five amino acids at positions 80 to 84 in the non-structural protein NS1 (S−) have become predominant. To understand the influence of these double deletions in the NA and NS1 proteins on the pathogenicity of H5N1-subtype AIVs, we selected A/mallard/Huadong/S/2005 as a parental strain to generate rescued wild-type A−S− and three variants (A−S+ with a five-amino-acid insertion in the NS1 protein, A+S− with a 20-amino-acid insertion in the NA stalk, and A+S+ with insertions in both NA and NS1 proteins) and evaluated their biological characteristics and virulence. The titers of the AIVs with A− and/or S− replicated in DEF cells were higher than that of A+S+, and the A−S− virus exhibited a replication predominance when co-infected with the other variants in DEF cells. In addition, A−S− induced a more significant increase in the expression of immune-related genes in peripheral blood mononuclear cells of mallard ducks *in vitro* compared with the other variants. Furthermore, an insertion in the NA and/or NS1 proteins of AIVs resulted in a notable decrease in virulence in ducks, as determined by intravenous pathogenicity index, and the two insertions exerted a synergistic effect on the attenuation of pathogenicity in ducks. In addition, compared with A+S+ and A+S−, the A−S+ and A−S− viruses that were introduced via the intranasal inoculation route exhibited a faster replication ability in the lungs of ducks. These data indicate that both the deletions in the NA stalk and the NS1 protein contribute to the high pathogenicity of H5N1 AIVs in ducks.

Editor: Hui-Ling Yen, The University of Hong Kong, Hong Kong

Funding: This work was supported by the Major National Basic Research Development Program (973 Program) (grant number 2011CB505003), the Jiangsu High School Natural Science Foundation (10KJA230055), the Important National Science & Technology Specific Projects (2012ZX10004214001002), the Special Fund for Agro-scientific Research in the Public Interest (20100312), the National High-Tech Research and Development Program of China (2011AA10A209), the Qing Lan Project, and a Project Funded by the Priority Academic Program Development of Jiangsu Higher Education Institutions and the Yangzhou University Funding for Scientific Research (2012CXJ081). The funders had no role in study design, data collection and analysis, decision to publish, or preparation of the manuscript.

Competing Interests: The author have declared that no competing interests exist.

* E-mail: daxinpeng@yahoo.com (DP); xfliu@yzu.edu.cn (XL)

⊚ These authors contributed equally to this work.

Introduction

Avian influenza virus has a wide geographical distribution in poultry and wild birds and certain genotypes/subtypes exhibit continuous cross-species transmission to humans and other mammals, which has resulted in the global concern of a potential pandemic threat [1]. The viral surface glycoproteins hemagglutinin (HA) and neuraminidase (NA) are major determinants in the interspecies transmission and adaptation of influenza A viruses to a new host [2]. The sialidase activity of NA not only facilitates the release and diffusion of progeny virions but also initiates the viral infection process [3–5]. A deletion in the stalk region of the NA (A−) decreases the ability of NA to release the virus from cells [6–9] and alters the virulence of the virus [10,11]. In addition, a deletion in the stalk of the NA gene may be required for the adaptation of H5N1 influenza viruses from wild aquatic birds to poultry [12–19].

The non-structural (NS) gene of influenza A virus encodes two proteins, namely NS1 and NEP, which share ten amino acids from the first residues at the N-terminal of the ORF [20]. The NS1 protein is a multifunctional protein involved in various protein-protein and protein-RNA interactions. In addition, NS1 is responsible for the inhibition of host immune responses by regulating the production of interferons (IFN) in the infected cells [21–23], the downregulation of host apoptosis, the post-transcriptional block of cellular mRNA maturation [24], and the regulation of the pathogenicity of influenza A viruses [25,26]. A five-amino-acid deletion at positions 80 to 84 in the NS1 protein of H5N1-subtype AIVs (S−) appeared in 2000 [16,27–29], which has resulted in an increase in the virulence of H5N1 viruses in chicken and mice [30].

Table 1. Primers for the mutagenesis of the NA and NS genes of the H5N1 AIV SY strain.

Primer name		Primer sequences (5'→3')
mNA	Ba-NA-1[a]	TATTGGTCTCAGGGAGCAAAAGCAGGAGT
	NA-1d	TATGTCTGATTTACCCAGGTGTTGTTTTCATAAGTAATAATGCTT
		TGATTGCATGGTTCAACTTGGTGTTGATTCCCTGTCTGAATT
	NA-2u	ACTTATGAAAACAACACCTGGGTAAATCAGACATATGTC
		AACATCAGCAATACTAATTTTCTTACTGAGAAAGCTGTGGCTT
	Ba-NA-2	ATATGGTCTCGTATTAGTAGAAACAAGGAGTTTTTT
mNS	Bm-NS-1[b]	TATTCGTCTCAGGGAGCAAAAGCAGGGTG
	Bm-NS-1d	TTACGTCTCAATTGCCATTTTAAGTGCCTC
	Bm-NS-2u	TTACGTCTCGCAATTGCATCCAGCCCGACTTCAC
	Bm-NS-2	ATATCGTCTCGTATTAGTAGAAACAAGGGTGTTTT

Ba-NA-1[a], the restriction endonucleases site for BsaI is underlined.
Bm-NS-1[b], the restriction endonucleases site for BsmBI is underlined.

H5N1 influenza viruses with both a short NA stalk and a five-amino-acid deletion in the NS1 protein were first found in 2002 and were the prevailing strains by 2003. However, the role of the double deletions in the NA and NS1 proteins in the pathogenicity of H5N1 subtype AIVs remains unknown. In this study, four rescue viruses with or without deletions in the NA and NS1 proteins were obtained using a reverse genetics technique based on the wild-type H5N1-subtype AIV strain A/mallard/Huadong/S/2005, and their biological characteristics and virulence were determined.

Materials and Methods

Ethics Statement

All of the animal studies were approved by the Jiangsu Administrative Committee for Laboratory Animals (Permission number: SYXKSU-2007-0005) and complied with the guidelines for laboratory animal welfare and ethics of the Jiangsu Administrative Committee for Laboratory Animals.

Viruses and Cells

A/mallard/Huadong/S/2005(SY), which has a 20-amino-acid deletion in the NA stalk and a five-amino-acid deletion at residues 80–84 in the NS1 protein, was isolated from mallard ducks and identified as an H5N1-subtype highly pathogenic AIV by our lab [31]. MDCK, 293T, and Vero cells were purchased from the Shanghai Institute of Biological Science, CAS, and cultured in DMEM (Invitrogen, CA, USA) containing 10% fetal calf serum (FCS, HyClone, UT, USA). Primary duck embryo fibroblasts (DEF) or primary chick embryo fibroblast (CEF) cells were prepared from embryonated unvaccinated duck eggs or SPF chicken eggs and cultured in M199 (Invitrogen, CA, USA) containing 4% FCS.

Virus Mutagenesis

Based on the sequences of the NA and NS genes of Gs/GD/96, which possessed intact NA and NS genes, a 60-nucleotide fragment (TGC AAT CAA AGC ATT ATT ACT TAT GAA AAC AAC ACC TGG GTA AAT CAA ACA TAT GTC AAC, which is conserved in all H5N1 isolates) and a 15-nucleotide fragment (GCC ATT GCT TCC AGT, which is varied in different species-based isolates, the inserted 15 nucleotides can be found in chicken-, duck-, and goose- origin H5N1 viruses at 3.7%,

7.7%, and 62.5%, respectively) were inserted into the NA stalk and the NS1 genes of the AIV SY strain, respectively, through overlap PCR [30,32,33]. The primers used for the mutations are listed in Table 1. The modified NA and NS genes were cloned into the PHW2000 vector, verified through sequence analysis, and named pHW256-NA+ and pHW258-NS+, respectively. Virus rescue was performed as described previously [34,35]. Briefly, eight rescue plasmids (pHW251-PB2, pHW252-PB1, pHW253-PA, pHW254-HA, pHW255-NP, pHW256-NA, pHW257-M, and pHW258-NS) [31] with or without the substitution plasmids pHW256-NA+ and/or pHW258-NS+ were cotransfected into a mixture of 293T and MDCK cells. After 48 h, the culture mixtures were inoculated into 10-day-old SPF eggs to amplify the rescued viruses at 35°C. The allantoic fluids were tested individually for the presence of infectious virus through a standard hemagglutination assay using chicken red blood cells (CRBCs) [36]. The RNAs of the propagated rescue viruses were extracted and amplified, and each viral gene segment was sequenced to ensure the absence of unwanted mutations. The rescue viruses were named A−S− if the virus exhibited both deletions in the NA and NS1 proteins, A+S− if the virus exhibited the 20-amino-acid insertion in the NA stalk, A−S+ if the virus exhibited the five-amino-acid insertion in the NS1 protein, and A+S+ if the virus exhibited both insertions in the NA and NS1 proteins.

Growth Curve

Confluent MDCK, Vero, CEF, and DEF cells in 35-mm dishes were infected in duplicate with each rescue virus at a multiplicity of infection (MOI) of 0.01 and incubated at 37°C in the appropriate medium containing 1% FCS. The virus titers of the supernatants, which were collected at different time points, were determined as the number of 50% tissue culture infectious doses ($TCID_{50}$) per 1 ml of CEF cell culture using the method described by Reed and Muench [37].

NA Activity Assays

For the enzymatic assays, virus dilutions in U-bottomed microtiter plates were incubated with increasing concentrations (5 to 100 μM) of the fluorogenic substrate 4-methylumbelliferyl N-acetylneuraminic acid (4-MUNANA; Sigma, MO, USA), and the fluorescence of the released 4-methylumbelliferone was monitored using a Safire2 microplate reader (Tecan, Mannedorg, Switzerland). The kinetic parameters K_m and V_{max} were calculated by fitting the data to the appropriate Michaelis–Menten equations using KaleidaGraph software (Synergy Software) [11,15,33].

To determine the rate of virus elution from CRBCs, 50 μl of serial twofold dilutions of the viral stocks in phosphate-buffered saline (PBS) was incubated with 50 μl of 1% CRBC suspension in U-bottom microtiter plates. The plates were left on ice for 1 h to allow virus adsorption to the CRBCs and then transferred to a water bath at 37°C. The decrease in HA titer, which reflects the NA-mediated virus elution from CRBCs, was monitored for 24 h [15,33].

Antiviral Activity Assay of IFN-β

The antiviral activity of IFN-β was assayed as previously described [1,38]. Briefly, Vero cells plated at a density of 2×10^5 cells per well in 6-well plates were treated with recombinant human IFN-β (R&D systems, MN, USA) at different concentrations (100 U, 200 U, 400 U, 800 U, 1600 U, 2000 U, and 10,000 U) for 24 h in serum-free DMEM, and the cells were then inoculated with the viruses at an MOI of 0.0001. The culture supernatants were collected 72 h after inoculation for subsequent determination of the $TCID_{50}$ per 0.1 ml of the CEF cell culture.

Competition Inhibition Assay in vitro

Vero, MDCK, DEF, and CEF cells at a density of 2×10^5 cells per well in 6-well plates were used for the serial passaging. The A−S− virus, which was mixed equivalently with the A−S+, A+ S−, or A+S+ viruses (1×10^3 TCID50 per 0.1 ml of each), was inoculated into the monolayer cells at an MOI of 0.01. After adsorption for 1 h at 37°C, the inoculum was removed, and fresh medium containing 1% FCS was added to the wells. The inoculated cells were incubated at 37°C for 24 h or 48 h according to the viral growth rate in the different cells. When approximately 80% of the cytopathic effect was obtained, the medium was collected and centrifuged at 800×g for 5 min at 4°C to remove the debris, and the supernatant was named the P1 stock. Each virus mixture stock was diluted 1000 (Hemagglutination titers ≤5log2) or 10,000-fold (Hemagglutination titer ≥ 6log2) with medium containing 1% FCS and passaged continually with the same cells up to the 10th passage. All of the supernatants were collected and stored at −70°C until use.

The total RNAs of the P1, P5, and P10 mixture samples from the different cells were prepared through treatment with the Trizol reagent (Invitrogen, CA, USA), and the full-length cDNAs of the viruses were synthesized using a 12-bp random primer [39]. The total viral RNA copies were quantified by quantitative real-time PCR (qRT-PCR) using the primers for the matrix gene, and the viral RNA copies of the A+S−, A−S+, and A+S+ viruses were quantified using the primers for NA and/or NS genes (one of the primer pairs was located in the insertion regions) (Table 2). The percentages of the A+S−, A−S+, or A+S+ viruses in the mixture of viruses were counted by comparing the copies of the single-mutant virus with that of the total viruses. All of the real-time PCR reactions were performed under the following conditions: 95°C for 30 s and 40 consecutive cycles of 95°C for 5 s and 60°C for 30 s. For all reactions, melting curve analysis was performed to verify the product specificity.

Expression Levels of Immune-related Genes in Peripheral Blood Mononuclear Cells of Mallard Ducks

The whole blood was collected from six-week-old ducks, and the peripheral blood mononuclear cells (PBMCs) were purified by treatment with lymphocyte separation media (Mediatech Inc., Herndon, VA, USA). The PBMCs in RPMI-1640 (Invitrogen, CA, USA) with 2% FCS were plated in six-well plates at 2×10^6 cells per well and infected with each virus at an MOI of 1. The culture plates were gently rocked every 15 min for 1 h, and the media was then replaced with fresh media. The cells were harvested at 8 h postinfection, and the total RNA of these samples was extracted.

The quantification of the cytokine mRNA levels was performed according to the protocol described by Kuo et al. (2010) [40]. The primers for the IFN-α, MX1, IL-1β, IL-8, IL-10, IL-18, MHC-I, MHC-II, and TLR-7 genes of ducks were designed based on published sequences or previously reported primers [41,42]. All of the primers are listed in Table 3. The expression level of each gene relative to that of GAPDH was calculated using the threshold cycle $2^{-\triangle\triangle CT}$ method [43].

To determine the replication of the viruses in the duck PBMCs, the duck PBMCs were infected with each virus at an MOI of 1. The supernatant and cells were harvested at 4 h, 8 h, and 24 h. The numbers of the viruses were determined by quantifying the M gene copy numbers according to the above methods.

Virulence in Chickens and Mallard Ducks

To determine the effect of A− and S− on the virulence of the rescue viruses, ten six-week-old SPF chickens were inoculated intravenously with 0.1 ml of a 1:10 dilution of allantoic fluid and observed clinically over a period of 10 days. The intravenous pathogenicity index (IVPI) was determined according to the OIE standard [44]. The IVPIs of these viruses in six-week-old mallard ducks without AIV antibody were similarly determined.

To further determine the virulence of the viruses in ducks, six-week-old mallard ducks were randomly divided into six groups with 12 ducks per group. The ducks in groups 1 through 5 were inoculated intranasally with 0.1 ml of SY or one of the four rescue viruses at a dose of 1×10^6 EID$_{50}$, and the ducks in group 6 were challenged with sterile PBS as a negative control. On days 1, 3, 5, and 7 post-challenge, three ducks from each group were euthanized, and their heart, liver, spleen, lungs, kidneys, and brain were collected. The tissues samples were homogenized in PBS with antibiotics and titrated through inoculation in 10-day-embryonated chicken eggs. Oropharyngeal and cloacal swabs were collected from each group on days 3, 5, and 7 post-challenge. The swabs were placed immediately in PBS, and an aliquot was titrated through inoculation in 10-day-embryonated chicken eggs for the examination of virus shedding. All of the animals were housed in animal biosafety level 3 facilities at Yangzhou University.

Statistical Analysis

The viral titers and viral loads are expressed as the mean ± standard deviation (SD). The expression levels of the immune-related genes are presented as the mean fold change ± SD. The statistical analyses were performed using an independent-sample t test. Differences with a P value of less than 0.05 were regarded as statistically significant.

Results

Prevalence of H5N1 Viruses with Double Deletions in NA and NS1 Proteins

All available sequences of the NA and NS1 genes from H5N1 viruses isolated between 1996 and 2012 were downloaded from GenBank, and the frequency of H5N1 viruses with double deletions in the NA and NS1 proteins was calculated. The results of the statistical analysis revealed that double deletions in the NA and NS1 proteins of H5N1 viruses were first found in 2002, and

Table 2. SYBR green real-time PCR primers for the identification and quantification of the NA and NS genes containing or not containing amino-acid deletions in the viral cDNAs.

Target genes	Primer name	Primer sequences	Target sequence number
M gene	M-F	AAGTGGCTTTTGGCCTAGTGTG	EU195395 (422–443)
	M-R	TGATTAGTGGGTTGGTGATGGTT	EU195395 (498–520)
NA gene	NA-F[a]	GAAAACAACACCTGGGTAAATCAG	EU195394 (−)
	NA-R	CCATCCTCTAATGGGGCAAA	EU195394 (212–231)
NS gene	NS-F[b]	AATGGCAATTGCATCCAGC	EU195396 (−)
	NS-R	AACCTGCCACTTTCTGCTTGG	EU195396 (305–325)

NA-F[a], the primer was targeted to the nucleotide sequence of the 20-amino-acid insertion in the NA stalk.
NS-F[b], the primer was targeted to the nucleotide sequence of the five-amino-acid insertion in the NS1 protein.

the numbers of these viruses were markedly increased in 2003. The ratio of H5N1viruses with double deletions in the NA and NS1 proteins was increased up to 90% in 2004 and thereafter (Table 4), which indicates that this type of virus has become predominant worldwide. In addition, the ratio of H5N1 viruses with double deletions in the NA and NS1 proteins isolated from land-based poultry was higher than that from domestic waterfowl in the early stage.

Virus Rescue and Viral Replication in Different Cells

Four rescue viruses were generated. All of these viruses shared the same PB2, PB1, PA, HA, NP, and M genes derived from SY and carried different modified NA and/or NS genes. The $TCID_{50}$ assay using CEF cells was performed to determine the replication kinetics of the viruses in Vero, MDCK, CEF, and DEF cells. As shown in Figure 1, the titers of the four viruses were similar to each other in Vero or CEF cells. However, the titers of A+S− and A+S+ in MDCK cells were approximately 1.5 $\log_{10}$ $TCID_{50}$/ml higher than those of A−S− and A−S+, and the titers of A+S+ in DEF cells were approximately 0.5 $\log_{10}$ $TCID_{50}$/ml lower than those of the other three viruses at 12 and 24 h postinfection. The wild-type strain SY displayed a similar growth pattern as the rescue A−S− in the four types of cells. These results suggest that both A− and S− can improve the viral replication in DEF cells at the early stages of AIV infection.

Enzymatic Activity of the Neuraminidase

To evaluate the possible effects of A− and S− on the neuraminidase activity, the enzymatic parameters of SY and the four rescue viruses were determined using the MUNANA fluorogenic substrate. As shown in Table 5, the K_m values, which reflect the affinity for the substrate, for the viruses with A− were very similar ($P>0.05$) but approximately 1.4- to 2.9-fold lower (reflecting a higher affinity) than those obtained for the viruses with long-stalk NA ($P<0.05$). In addition, the V_{max} values, which depend on both the specific activity and the amount of enzyme in the reaction, for the viruses with A− were 1.56- to 2.01-fold lower than those obtained for the viruses with long-stalk NA ($P<0.05$), which indicates that A− decreases the enzymatic activity of the neuraminidase toward small MUNANA substrates.

The elutions of the viruses from CRBCs were also determined. The complete elution of the viruses with long-stalk NA occurred within 6 h, whereas the viruses with short-stalk NA were completely eluted from the CRBCs after a 12-h incubation at 37°C. This finding indicates that A− reduced the rate of viral elution from CRBCs.

IFN Resistance of the Rescue Viruses

To evaluate the resistance of the viruses to IFN, Vero cells were pretreated with different concentrations of IFN and infected with the viruses. The replication of A+S+ in Vero cells was completely inhibited in the presence of IFN-β at a concentration of 400 U, and the replication of A+S− was fully inhibited at an IFN-β

Table 3. Real-time PCR primers for detection of the expression levels of immune-related genes in the PBMCs of mallard ducks.

Target genes	Forward primers	Reverse primers
GAPDH	ATGTTCGTGATGGGTGTGAA	CTGTCTTCGTGTGTGGCTGT
DMX1	TCACACGAAGGCCTATTTTACTGG	GTCGCCGAAGTCATGAAGGA
DIL-10	GGGGAGAGGAAACTGAGAGATG	TCACTGGAGGGTAAAATGCAGA
DIL-1β	GAGATTTTCGAACCCGTCACC	AGGACTGGGAGCGGGTGTA
DIL-8	AGGACAACAGAGAGGTGTGCTTG	GCCTTTACGATCCGCTGTACC
DIL-18	AGGTGAAATCTGGCAGTGGAAT	ACCTGGACGCTGAATGCAA
DIFN-α	TTGCTCCTTCCCGGACA	GCTGAGGGTGTCGAAGAGGT
DTLR-7	GTGGCAGCTTCAAGACAACA	TTAGTTGGCCATTCCAGGAC
DMHC-I	GAAGGAAGAGACTTCATTGCCTTGG	CTCTCCTCTCCAGTACGTCCTTCC
DMHC-II	CCACCTTTACCAGCTTCGAG	CCGTTCTTCATCCAGGTGAT

Table 4. Frequency of H5N1 viruses with double deletions in the NA and NS proteins from 1996 to 2012[a].

Year	Ratios[b]			Viruses with A− and S−			
	Viruses with A−	Viruses with S−	Viruses with A+ and S+	Total	Land-based poultry	Domestic waterfowl	Other sources[c]
1996-1999	10/16	0/16	6/16	0/16 (0[d])	0/7 (0)	0/4 (0)	0/5 (0)
2000	3/21	5/21	13/21	0/21 (0)	0/0 (0)	0/11 (0)	0/10 (0)
2001	5/22	15/22	2/22	0/22 (0)	0/7 (0)	0/14 (0)	0/1 (0)
2002	20/29	21/29	1/29	13/29 (44.8%)	5/8 (62.5%)	4/12 (33.3%)	4/9 (44.4%)
2003	38/46	42/46	1/46	35/46 (76.1%)	13/15 (86.7%)	11/14 (78.6%)	11/17 (64.7%)
2004	104/110	102/110	3/110	99/110 (90%)	48/49 (98.0%)	19/23 (82.6%)	32/38 (84.2%)
2005	114/121	115/121	6/121	114/121 (94.2%)	46/46 (100%)	31/36 (86.1%)	37/39 (94.9%)
2006	147/153	148/153	5/153	147/153 (96.1%)	49/49 (100%)	25/27 (92.6%)	73/77 (94.8%)
2007	134/138	133/138	4/138	133/138 (96.4%)	53/54 (98.1%)	39/41 (95.1%)	41/43 (95.3%)
2008	75/76	73/76	1/76	73/76 (96.1%)	42/42 (100%)	12/13 (92.3%)	19/21 (90.5%)
2009	52/55	51/55	3/55	51/55 (92.7%)	30/30 (100%)	2/4 (50.0%)	19/21 (90.5%)
2010	48/48	43/48	0/48	43/48 (89.6%)	20/22 (90.9%)	10/11 (90.9%)	13/15 (86.7%)
2011	68/69	68/69	1/69	68/69 (98.6%)	15/15 (100%)	13/14 (92.9%)	40/40 (100%)
2012	10/10	10/10	0/10	10/10 (100%)	0/0 (0)	9/9 (100%)	1/1 (100%)

[a]All available sequences of both NA and NS1genes from H5N1 viruses deposited in Genbank were selected.
[b]The numbers indicate the ratio of the viruses to the total H5N1 viruses or the sources-based isolates in the indicated year.
[c]Other sources: The viruses from wild birds, mammals (including humans), and environmental samples.
[d]Percentage of H5N1 viruses with A− and S− isolated in the indicated year.

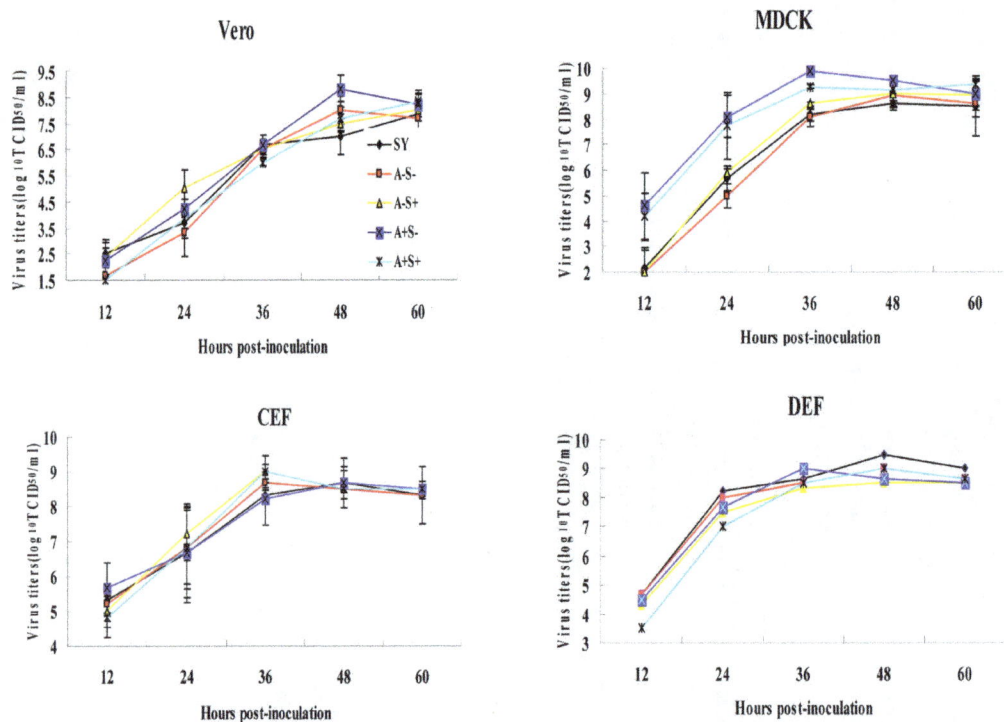

Figure 1. Growth kinetics of the viruses in Vero, MDCK, CEF, and DEF cells. The cells were infected with the wild-type strain and the four rescue viruses at an MOI of 0.01 TCID$_{50}$/cell, and the culture media were harvested at the indicated times after infection. The virus titers at each time point are presented as the mean $\pm$ SD of duplicate experiments.

concentration of 1600 U. However, the titers of A−S+ and A−S− were still detectable in the presence of IFN-β at a concentration of 10,000 U (Table 6), which indicates that A− and S− both enhance the interferon resistance of the viruses and that A− plays a more important role.

SYBR Green Real-time PCR Assay

A mixture of cDNAs of the A−S− and A+S+ viruses at the same concentration of approximately 4.00×10^4 copies/μl was used to test the specificity and accuracy of the SYBR green real-time PCR assay. The results indicated that the average amount of the NA gene without deletion (from the A+S+ virus) was 1.98×10^4 copies/μl. The average amount of the NS gene without deletion (from the A+S+ virus) was 1.97×10^4 copies/μl, and the average amount of the M gene (from the A−S− and A+S+ viruses) was 3.97×10^4 copies/μl. The percentage of the A+S+ virus was approximately 49.75%, and the percentage of the A−S− virus

was approximately 50.25% ($P>0.05$). In addition, the plasmids pHW256-NA+, pHW258-NS+, and pHW257-M at the concentrations of 4.5×10^6 copies/μl, 6.0×10^5 copies/μl, and 3.3×10^6 copies/μl, respectively, were evaluated by assays for five replicate tests, and the average concentrations were 4.504×10^6 copies/μl, 5.982×10^5 copies/μl, and 3.302×10^6 copies/μl for each plasmid, respectively. Furthermore, the plasmids mixtures (pHW256-NA+ and pHW256-NA, or pHW258-NS+ and pHW258-NS) were tested using the assays, and only the copy numbers of plasmids pHW256-NA+ or pHW258-NS+ were detected (data not shown). These data indicate that the SYBR green real-time PCR method can efficiently detect the proportion of the viruses with intact NA or NS genes in the virus mixtures of interest.

Competitive Growth on Different Cells

The A−S− virus, which was mixed with A−S+, A+S−, or A+S+, was serially passaged in Vero, MDCK, CEF, and DEF cells for

Table 5. Enzymatic properties of the NA protein of H5N1 viruses.

Virus	K$_m$ (μM)[a]	V$_{max}$ (fluorescence U/S)[a]	V$_{max}$ ratio[b]	Elution time (h)
SY	275.57±8.62	13.02.33±0.27	1.00	12
A−S−	192.7±3.35	15.69±0.56	1.20	12
A−S+	244.9±3.37	10.49±0.09	0.81	12
A+S−	551.2±19.8	26.18±0.58	2.01	6
A+S+	381.97±2.9	20.36±0.12	1.56	6

[a]The results are presented as the mean $\pm$ SD from three independent determinations on duplicate samples using dilutions of the H5N1 viruses.
[b]V$_{max}$ ratio of the rescue viruses to the wild-type SY virus.

Table 6. IFN-β resistance of H5N1 viruses.

Viruses	Titers after pretreatment with different concentrations of IFN-β[a]							No treatment
	100 U	200 U	400 U	800 U	1600 U	2000 U	10,000 U	
SY	7.67	7.5	6.67	6.67	4.5	4.5	2.67	7.5
A−S−	7.33	7	7	6.33	4.67	3.67	1.33	7
A−S+	7	6.5	6.5	6	4.33	4.33	1.83	7.5
A+S−	6.67	7	6.67	5.33	<	<	<	6.5
A+S+	6.5	6.33	<	<	<	<	<	6.67

[a]Vero cells were pretreated with different concentrations of recombinant human IFN-β at 37°C. After 24 h, the cells were infected with the viruses at an MOI of 0.0001. The virus titers ($\log_{10}$TCID$_{50}$/0.1 ml) were measured 72 h after infection. The values indicate the means of three experiments. <: titer <0.5.

ten generations. The cDNAs of the P1, P5, and P10 samples from different cells were detected using the above-described SYBR green real-time PCR assay. The results indicated that the viral percentage of A−S− in the P1, P5, and P10 mixture of A−S− and either A+S+ or A+S− were not significantly different from each other in Vero and CEF cells, whereas the percentages of A+S+ and A+S− in the P10 samples obtained from the MDCK cells were 1.5% and 17.4%, respectively, and the percentages of A+S+ and A+S− in the P10 samples from the DEF cells were 5.8% and 0.5%, respectively (Fig. 2). The percentage of A−S− in the mixture of A−S− and A−S+ was increased slightly after serial passage in DEF and CEF cells. These data indicate that the A−S− virus replicates predominantly in DEF cells that are co-infected with other variants.

Expression Levels of Immune-related Genes in the PBMCs of Mallard Ducks *in vitro*

To investigate the effect of the four rescue viruses on the host response, the PBMCs of mallard ducks were challenged with the viruses, and the induced expression levels of immune-related genes were determined at 8 h postinfection. There was no significant difference in the expression level of the anti-inflammatory cytokine IL-10 among the PBMCs infected with the different viruses. In contrast, the expression levels of the IFN-α, MX1, IL-1β, IL-8, IL-18, MHC-I, MHC-II, and TLR-7 genes in the PBMCs infected with SY and A−S− were significantly increased. There was slight upregulation or downregulation of the expression of immune-related genes in the PBMCs infected with the other variants (Fig. 3A, 3B). These results indicate that the virus with both A− and S− induced higher expression of the immune-related genes in PBMCs. When the growth curves of the viruses were determined in the PBMCs, only A+S+ displayed a significant delay in growth rate at 4 h postinfection, and there were no significant difference in growth rate among the other three viruses (Fig. 3C), which displayed similar trends in DEF cells.

Virulence in Chickens and Mallard Ducks

To determine the effect of A− and S− on viral virulence, chickens or mallard ducks were challenged intravenously with the four rescue viruses. The IVPIs of the viruses in chickens ranged from 2.96 to 3.00, which indicates that all of these viruses are highly pathogenic to chickens. However, the IVPIs of A+S+, A+S−, A−S+, and A−S− in ducks were 0.054, 1.336, 1.307, and 2.314, respectively, which indicates that A+S+ is slightly pathogenic to mallard ducks, whereas A−S+, A+S−, and A−S− are all highly pathogenic to mallard ducks (the A−S− virus was the most virulent strain to mallard ducks).

Because the virulence of the four viruses exhibited differences in mallard ducks through the intravenous route, mallard ducks were challenged intranasally with the four viruses, and the viral loads in the main organs of the infected ducks were determined by viral culture. The mean viral titers in the lungs, livers, kidneys, and spleens of A+S+ or A+S–infected ducks was lower than the mean viral titers detected in the groups infected with A−S+, A−S−, or SY on day 3 postinfection ($P<0.05$). The mean viral titer in the hearts of the A+S+-infected mallard ducks was lower than that obtained in the hearts of the ducks infected with A−S+, A−S−, or SY on day 3 postinfection ($P<0.05$). The mean viral titers in the brains of A+S+-infected ducks was lower than the mean viral titers detected in the groups infected with A−S+, or A−S− on day 5 postinfection ($P<0.05$) (Fig. 4). In addition, in both A+S+ and A+S− groups, virus replication appears to be delayed in most organs, compared with that measured in both the A−S+ and A−S− groups.

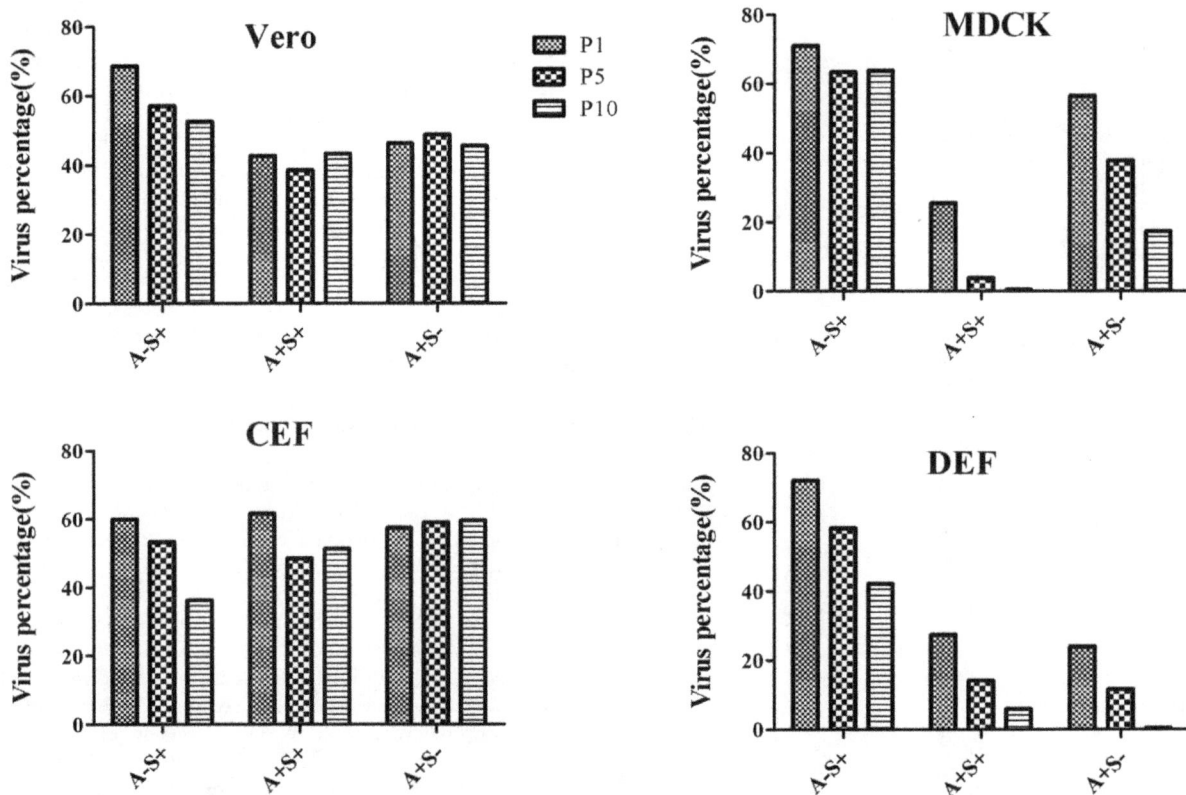

Figure 2. Serial passage of the mixture of two rescue viruses in Vero, MDCK, CEF, and DEF cells. The A−S− virus was mixed with A−S+, A+S−, or A+S+ (approximately 1×10^3 TCID50 per 0.1 ml of each virus), and the mixture was inoculated into different cells and then serially passaged for 10 generations. The percentages of A−S+, A+S−, and A+S+ in the P1, P5, and P10 samples of the different cells were detected by SYBR green real-time PCR assay.

Viral shedding was detected in both oropharyngeal and cloacal swabs from infected ducks on days 3, 5, and 7 postinfection. On day 3 postinfection, the viral shedding ratios obtained from the oropharyngeal swabs from the groups infected with A+S+, A+S−, A−S+, and A−S− were 22.2% (2/9), 44.4% (4/9), 100% (9/9), and 88.9% (8/9), respectively (Table 7), which was correlated with the viral titers in the lungs at the same time point (Fig. 4). On day 7 postinfection, the viral shedding ratios obtained from the oropharyngeal swabs from group infected with A+S+ and A+S− were 100% and for both, which was also correlated with the higher viral titers in lungs at same time point. However, at this time point (on day 7 postinfection), there was no detectable viral shedding in both the A−S+− and the A−S− infected groups. Compared with the ducks infected with A+S+ and A+S−, the viral shedding of ducks infected with A−S+ and A−S− reached a peak two days earlier. On day 3 postinfection, the viral shedding ratios in the cloacal swabs samples from ducks infected with A+S−, A−S+, or A−S− were 11.1% (1/9), 66.7% (6/9), 33.7% (3/9), respectively. However, there was no detectable viral shedding in the cloacal swabs samples from ducks infected with A+S+. These results indicate that mallard ducks infected with A+S+ shed the progeny virus only through the oropharynx, whereas mallard ducks infected with A+S−, A−S+, and A−S− shed the progeny virus through both the oropharynx and cloaca.

Discussion

According to the deletion length and location in the NA stalk, H5N1 viruses isolated in 1997 were divided into four groups: a long NA stalk, a short NA stalk with a 20-amino-acid deletion at positions 49 to 68, a short NA stalk with a 20-amino-acid deletion at positions 55 to 74, and a short NA stalk with a 19-amino-acid deletion at positions 55 to 73. Since that time, only long NA stalks and short NA stalks with a 20-amino-acid deletion at positions 49 to 68 have been observed in H5N1 viruses, which indicates that the other two types of viruses had a selective evolutionary disadvantage. H5N1 viruses with a short NA stalk and a five-amino-acid deletion from position 80 to 84 in the NS1 protein were first observed in 2002, became predominant in 2003, and have continued to exhibit a very high ratio (approximately 90%) in subsequent isolates. In addition, the deletion in both NA and NS1 proteins of H5N1 viruses was biased for land-based poultry in the early stage. However, there were few isolates of other subtypes of influenza virus that have contained the double deletions in the NA and NS1 proteins. It is possible that the H5N1 viruses with double deletions in the NA and NS1 proteins have a prevailing advantage and are stably maintained in poultry. It is worthwhile to note that H5N1 viruses have been found to be highly pathogenic to ducks since 2002 [45,46]. To investigate the role of double deletions in the NA and NS1 proteins in the pathogenicity of H5N1-subtype AIVs, a series of rescue viruses, which were derived from a H5N1 AIV with double deletions in the NA and NS1 proteins, was obtained by reverse genetics. We found that these rescue viruses all

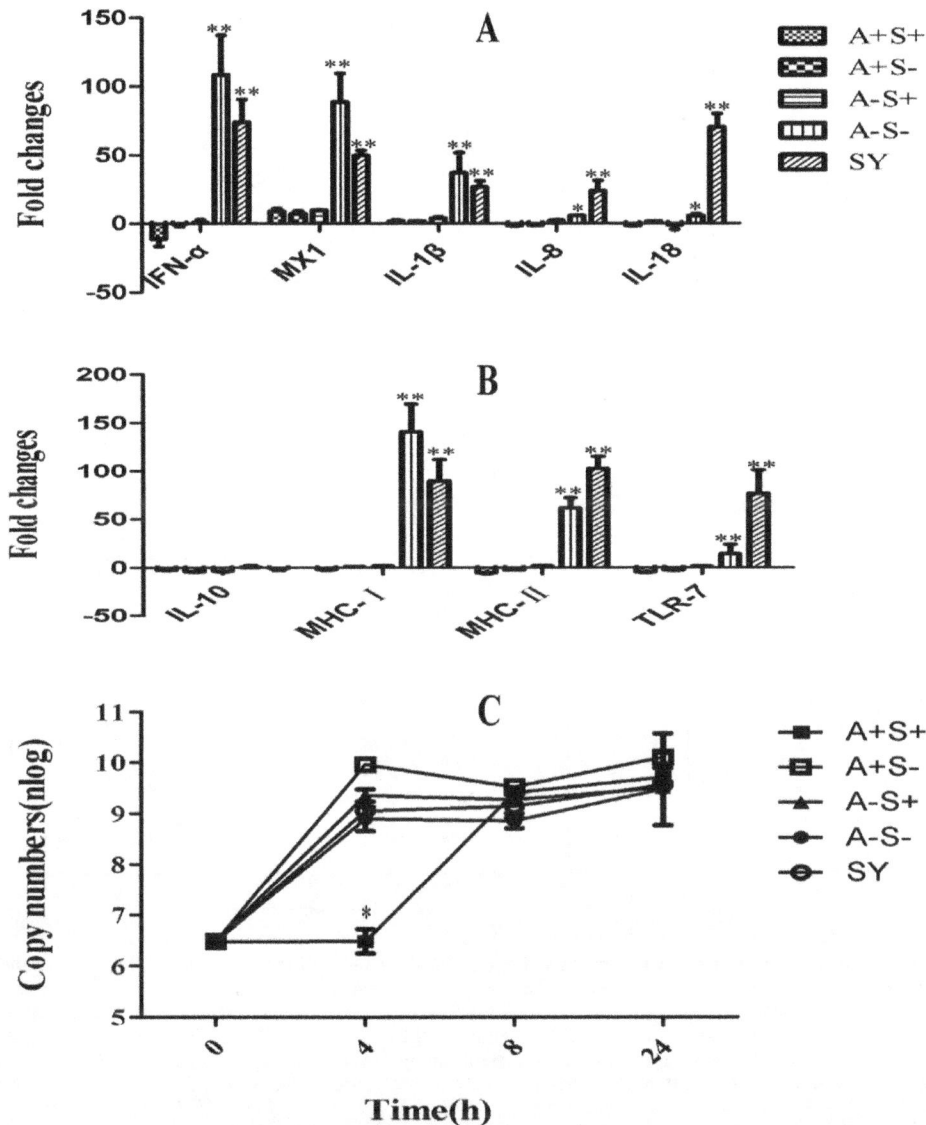

Figure 3. Real-time RT-PCR quantification of the expression of immune-related genes in mallard PBMCs and growth kinetics of the viruses in mallard PBMCs. Mallard PBMCs in a six-well plate were inoculated with SY and the four rescue viruses at an MOI of 1 TCID$_{50}$/cell. The total RNA was extracted from the PBMCs at 8 h postinfection, and equal amounts of RNA (1 µg) from each sample were used for RT-PCR. The gene expression was normalized to the expression level of the GAPDH gene and is presented as the fold increase relative to the results observed with mock-treated cells. The data represent the mean fold changes ± SD (A and B). Mallard PBMCs were also infected with these viruses at an MOI of 1 copy/cell, and the supernatant and cells were harvested at 4 h, 8 h, and 24 h post-infection. These samples were determined by quantitative real-time PCR (qRT-PCR) using the primers for the matrix gene. The numbers of the viruses are presented as the mean ± SD of duplicate experiments (C).

replicated efficiently in embryonated chicken eggs, which indicates that the presence or absence of the deletion in the NA stalk and the NS1 protein of H5N1 viruses did not significantly change their viral replication ability in embryonated chicken eggs.

In accordance with previous reports [33,47], at the early stage of viral infection, the titers of A+S− and A+S+ in MDCK cells were approximately 1.5 log$_{10}$ TCID$_{50}$/ml higher than those of A−S− and A−S+, which indicates that the replication ability of the viruses with a long-stalk NA in MDCK cells was better than that of the viruses with a short-stalk NA. The enzymatic activities of the neuraminidase of viruses with a long-stalk NA were higher than those of viruses with a short-stalk NA, as judged by their higher rates of elution from CRBCs. Thereafter, higher NA activity facilitated the release and diffusion of progeny virions, which

resulted in a higher replication ability of the viruses with a long-stalk NA in MDCK cells. MDCK cells express high amounts of both α2,3 and α2,6-gal sialyl glycoconjugates [48,49]. Vero cells express a high amount of α2,3-linked receptors and a relatively low amount of α2,6-gal-linked receptors, and avian cell lines (QT-6 and DF-1) express a high amount of α2,3-gal-linked receptors [48]. Thus, the match of the viral NA activity and the viral binding ability to cellular receptors contributes to the replication ability of H5N1 viruses in different cell lines. It is thus reasonable that the growth pattern of the viruses in Vero, CEF, and DEF cells are different from that in MDCK cells.

Vero cells are IFN-α/β-deficient, and the replication abilities of the four rescue viruses in Vero cells were similar. However, there was a significant difference in viral growth in the IFN-β-pretreated

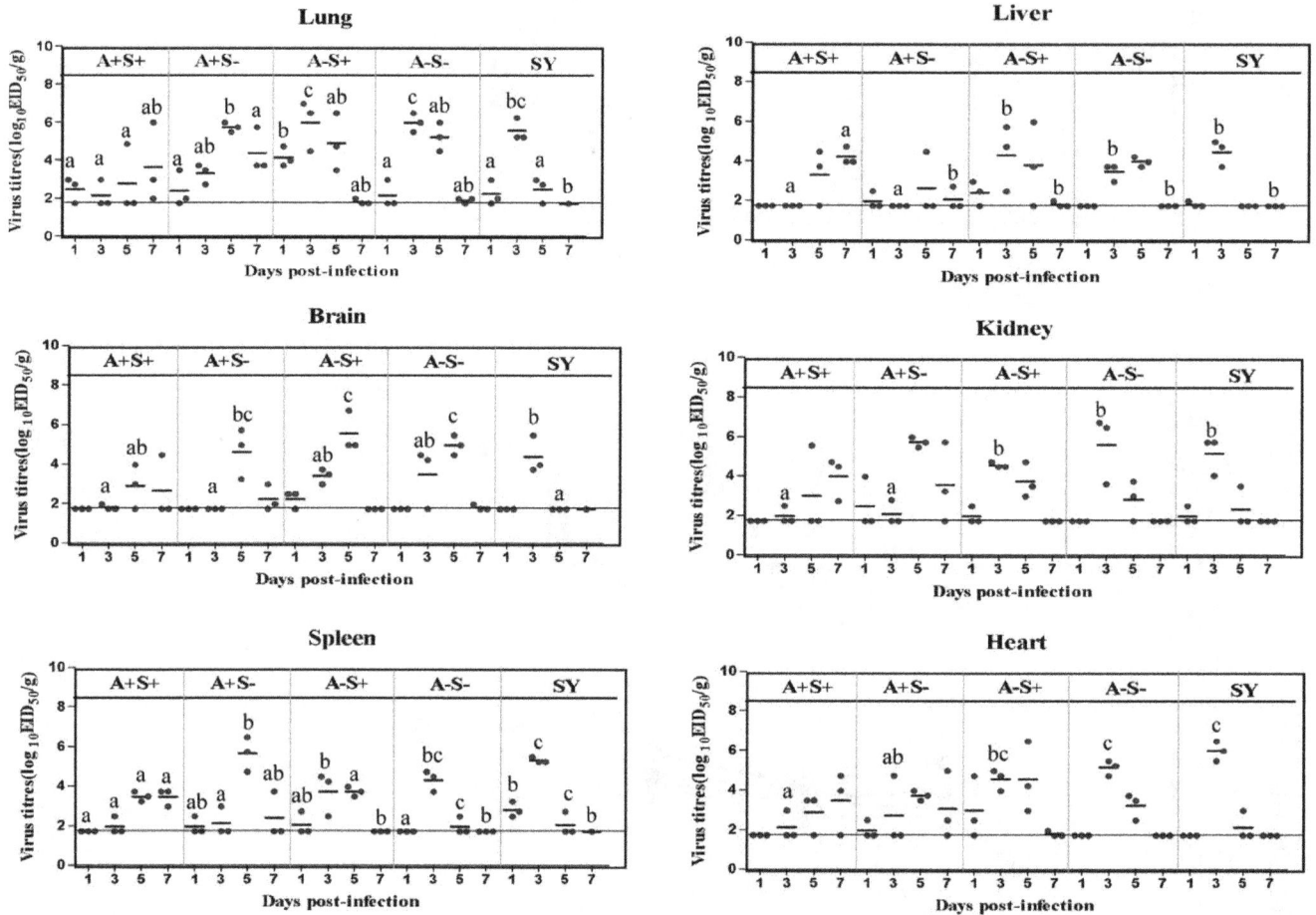

Figure 4. Replication kinetics of SY and the four rescue viruses in mallard ducks. The virus titers in the lungs, livers, hearts, spleens, kidneys, and brains of mallard ducks inoculated intranasally with $10^6 EID_{50}/0.1$ ml of SY and the four rescue viruses were determined. Each horizontal bar represents the mean virus titer in $log_{10} EID_{50}/g$ of tissue. The horizontal line indicates the lower limit of detection. Different lowercase letters indicate significant differences among SY and the four viruses infected groups on the same day postinfection ($P<0.05$). For example, on day 3 postinfection, the mean viral titers in the lungs of A+S+ (a) and A+S− (ab) groups were significantly lower than that of A−S+ (c) and A−S− (c) groups, and there were no significant differences between the A+S− (ab) and SY (bc) groups. There were also no significant differences among the A−S− (c), A−S− (c) and SY (bc) groups.

Table 7. Virus shedding of oropharyngeal and cloacal swabs of mallard ducks inoculated with 0.1 ml of 10^6 EID_{50} H5N1 viruses.

Virus	NO. of positive swabs/NO. of total swabs ($log_{10}EID_{50}\pm$ SD)					
	Oropharyngeal swabs			**Cloacal swabs**		
	3 dpi[A]	**5 dpi**	**7 dpi**	**3 dpi**	**5 dpi**	**7 dpi**
A+S+	2/9 (3.00±0.71)	4/6 (3.63±0.97)	3/3 (3.00±0.50)	0/9(<0.5)	0/6(<0.5)	0/3(<0.5)–
A+S−	4/9 (2.38±0.92)	6/6 (2.79±0.86)	3/3 (1.58±1.23)	1/9(1.5)	2/6 (1.75±0.01)	2/3 (1.63±1.24)
A−S+	9/9 (3.22±1.37)	6/6 (2.50±1.22)	0/3 (<0.5)	6/9 (1.33±0.79)	3/6 (1.50±0.66)	0/3(<0.5)
A−S−	8/9 (3.16±0.88)	4/6 (2.13±0.92)	0/3(<0.5)	3/9 (1.67±0.14)	0/6(<0.5)	0/3(<0.5)
SY	8/9 (2.81±0.40)	4/6 (2.81±1.20)	0/1 [B] (<0.5)	7/9 (1.32±0.45)	1/6(1.25)	0/1 (<0.5)

The oropharyngeal and cloacal swabs were collected for virus isolation from each group at 3, 5, and 7 days postinfection.
[A]dpi: days postinfection.
[B]Two out of three ducks were dead on day 6 postinfection.

Vero cells, which indicates that the interferon-resistance abilities of these viruses were different. The order of the interferon resistance ability from high to low of these viruses was A−S+ = A−S−>A+ S−>A+S+. These results suggest that both A− and S− enhance the interferon-resistance ability of H5N1 AIVs. It has been reported that the NS1 protein is critical for the influenza virus to antagonize the host cell IFN response [1,23,50]. However, there is no report on the NA protein of influenza virus participation in the viral resistance to the host IFN response. Therefore, the mechanism through which the short-stalk NA protein counteracts the anti-viral activity of IFN-β needs to be further investigated.

It has been reported that RIG-I expression is an intracellular RNA sensor that detects the presence of vRNA, leading to induced expression of IFN-β [51]. CEF cells are derived from chickens, which lack RIG-I [52], and may, therefore, fail to induce the expression of IFN-β via this pathway. Further study confirmed that no IFN expression was observed in CEF cells infected with avian influenza viruses [53]. This may explain that the viral percentage of A−S− in the mixture of A−S−, and A+S+ or A+ S− were not significantly different from each other in CEF cells and IFN-deficient Vero cells in the competition assay. However, RIG-I expression is detected in both MDCK cells [54] and duck cells [52]. Although both A+S+ and A+S− displayed higher replication ability in MDCK cells when compared with A−S−, A−S− replicated dominantly when co-infected with A+S− or A+ S+ in MDCK and DEF cells. The interferon-resistance ability of A−S− virus might contribute to its replication predominance to some extent, and the precise mechanism of the replication advantage of the A−S−virus over the A+S+ virus in the competition assay needs to be further studied.

To evaluate the effect of A− and S− on the viral pathogenicity in poultry, the IVPIs of these viruses in chickens and mallard ducks were measured. It has been reported that a deletion in the NA stalk of H1N1 AIV results in increased virulence for chickens [11], and a deletion in the NA stalk of H5N1 AIV results in no significant difference in the virulence for mice via the intranasal route [9]. Because of the high pathogenicity of the parental virus SY in SPF chickens, the IVPIs of the A−S− virus and its variants for chickens were all similar and thus did not reflect the effect of A− and S− on the viral pathogenicity of these viruses in chickens. However, the IVPIs of A−S+ and A+S− for mallard ducks were significantly higher than that of A+S+ and lower than that of A− S−, which indicates that both A− and S− result in a marked increase in the virulence of the viruses for mallard ducks. In addition, this finding demonstrated that A− and S− exert a synergistic effect on the virulence of H5N1 viruses for mallard ducks. We also found that the PBMCs of mallard ducks infected with A−S− displayed a significant cytokine response, although the growth rate of A−S− was similar to that of A−S+ or A+S−. It

was hypothesized that the high expression levels of IFNs and proinflammatory cytokine genes in PBMCs may play an important role in the high pathogenicity of A−S− to mallard ducks through the intravenous route.

We also monitored the viral pathogenicity in mallard ducks after intranasal inoculation. Different from the intravenous inoculation, the viruses (except for SY with two deaths out of nine ducks) at dosages of 1×10^6 EID$_{50}$ caused serious clinical signs but no death in mallard ducks within the observation period. Compared to the A+S+-inoculated mallard ducks, the mallard ducks infected with A−S+, A+S−, and A−S− presented higher virus titers in the lungs and brain. Furthermore, compared with A+S+ or A+S−, A−S+ and A−S− displayed faster replication ability in the lungs of mallard ducks. In addition, the viral shedding results demonstrated that mallard ducks infected with A+S+ shed the progeny virus only through the larynx, whereas mallard ducks infected with the other viruses shed the progeny virus through not only the larynx but also the cloaca. It is worth determining the viral replication ability in the intestines of ducks in a future study. These data suggest that with stronger ability to resist the interferon inhibition (as shown in Table 6), the A−S− and A−S+ viruses possessed stronger ability to overcome or suppress the host immune system and achieved increased viral replication ability. In addition, both A− and S− enhanced the viral replication ability and shedding of H5N1-subtype AIVs in mallard ducks and the S− of H5N1 viruses had less of an effect on the virulence than A− when the virus infection occurred via the intranasal route.

In summary, H5N1 AIVs with double deletions in the NA and NS1 genes have been the prevailing strains in recent years. The rescue virus with both a short-stalk NA and a deletion in the NS1 protein exhibited increased interferon resistance, competitive inhibition in DEF cells, increased expression of immune-related genes in the PBMCs of mallard ducks, and increased virulence in mallard ducks compared with the rescue virus with intact NA and NS1 protein. Although H5N1 AIVs with double deletion in the NA and NS1 genes most likely occur in chicken in nature but not in waterfowl, the genotype is maintained in waterfowl once the virus is re-introduced [9,30]. Our data indicate that both deletions in the NA stalk and the NS1 protein contribute to the high pathogenicity of H5N1 AIVs in ducks and that these deletions may play an important role in the maintenance and circulation of these viruses in poultry.

Author Contributions

Conceived and designed the experiments: YL DP XL. Performed the experiments: YL SC XZ QF ZZ YZ MG SS. Analyzed the data: YL SC DP. Contributed reagents/materials/analysis tools: YL SC. Wrote the paper: YL SC DP XL.

References

1. Seo SH, Hoffmann E, Webster RG (2004) The NS1 gene of H5N1 influenza viruses circumvents the host anti-viral cytokine responses. Virus Res 103: 107–113.
2. Neumann G, Kawaoka Y (2006) Host range restriction and pathogenicity in the context of influenza pandemic. Emerg Infect Dis 12: 881–886.
3. Matrosovich MN, Matrosovich TY, Gray T, Roberts NA, Klenk HD (2004) Neuraminidase is important for the initiation of influenza virus infection in human airway epithelium. Journal of virology 78: 12665–12667.
4. Ohuchi M, Asaoka N, Sakai T, Ohuchi R (2006) Roles of neuraminidase in the initial stage of influenza virus infection. Microbes Infect 8: 1287–1293.
5. Suzuki T, Takahashi T, Guo CT, Hidari KI, Miyamoto D, et al. (2005) Sialidase activity of influenza A virus in an endocytic pathway enhances viral replication. Journal of virology 79: 11705–11715.
6. Baigent SJ, McCauley JW (2001) Glycosylation of haemagglutinin and stalk-length of neuraminidase combine to regulate the growth of avian influenza viruses in tissue culture. Virus Res 79: 177–185.

7. Castrucci MR, Kawaoka Y (1993) Biologic importance of neuraminidase stalk length in influenza A virus. J Virol 67: 759–764.
8. Els MC, Air GM, Murti KG, Webster RG, Laver WG (1985) An 18-amino acid deletion in an influenza neuraminidase. Virology 142: 241–247.
9. Matsuoka Y, Swayne DE, Thomas C, Rameix-Welti MA, Naffakh N, et al. (2009) Neuraminidase stalk length and additional glycosylation of the hemagglutinin influence the virulence of influenza H5N1 viruses for mice. Journal of virology 83: 4704–4708.
10. Zhou H, Yu Z, Hu Y, Tu J, Zou W, et al. (2009) The special neuraminidase stalk-motif responsible for increased virulence and pathogenesis of H5N1 influenza A virus. PLoS One 4: e6277.
11. Munier S, Larcher T, Cormier-Aline F, Soubieux D, Su B, et al. (2010) A genetically engineered waterfowl influenza virus with a deletion in the stalk of the neuraminidase has increased virulence for chickens. Journal of virology 84: 940–952.

12. Banks J, Speidel ES, Moore E, Plowright L, Piccirillo A, et al. (2001) Changes in the haemagglutinin and the neuraminidase genes prior to the emergence of highly pathogenic H7N1 avian influenza viruses in Italy. Arch Virol 146: 963–973.

13. Campitelli L, Mogavero E, De Marco MA, Delogu M, Puzelli S, et al. (2004) Interspecies transmission of an H7N3 influenza virus from wild birds to intensively reared domestic poultry in Italy. Virology 323: 24–36.

14. Giannecchini S, Campitelli L, Calzoletti L, De Marco MA, Azzi A, et al. (2006) Comparison of in vitro replication features of H7N3 influenza viruses from wild ducks and turkeys: potential implications for interspecies transmission. J Gen Virol 87: 171–175.

15. Matrosovich M, Zhou N, Kawaoka Y, Webster R (1999) The surface glycoproteins of H5 influenza viruses isolated from humans, chickens, and wild aquatic birds have distinguishable properties. J Virol 73: 1146–1155.

16. Guan Y, Poon LL, Cheung CY, Ellis TM, Lim W, et al. (2004) H5N1 influenza: a protean pandemic threat. Proc Natl Acad Sci U S A 101: 8156–8161.

17. Croville G, Soubies SM, Barbieri J, Klopp C, Mariette J, et al. (2012) Field monitoring of avian influenza viruses: whole-genome sequencing and tracking of neuraminidase evolution using 454 pyrosequencing. Journal of clinical microbiology 50: 2881–2887.

18. Li J, Zu Dohna H, Cardona CJ, Miller J, Carpenter TE (2011) Emergence and genetic variation of neuraminidase stalk deletions in avian influenza viruses. PloS one 6: e14722.

19. Yamada S, Shinya K, Takada A, Ito T, Suzuki T, et al. (2012) Adaptation of a duck influenza A virus in quail. Journal of virology 86: 1411–1420.

20. Inglis SC, Barrett T, Brown CM, Almond JW (1979) The smallest genome RNA segment of influenza virus contains two genes that may overlap. Proc Natl Acad Sci U S A 76: 3790–3794.

21. Kochs G, Garcia-Sastre A, Martinez-Sobrido L (2007) Multiple anti-interferon actions of the influenza A virus NS1 protein. J Virol 81: 7011–7021.

22. Egorov A, Brandt S, Sereinig S, Romanova J, Ferko B, et al. (1998) Transfectant influenza A viruses with long deletions in the NS1 protein grow efficiently in Vero cells. J Virol 72: 6437–6441.

23. Garcia-Sastre A, Egorov A, Matassov D, Brandt S, Levy DE, et al. (1998) Influenza A virus lacking the NS1 gene replicates in interferon-deficient systems. Virology 252: 324–330.

24. Hale BG, Randall RE, Ortin J, Jackson D (2008) The multifunctional NS1 protein of influenza A viruses. J Gen Virol 89: 2359–2376.

25. Basler CF, Reid AH, Dybing JK, Janczewski TA, Fanning TG, et al. (2001) Sequence of the 1918 pandemic influenza virus nonstructural gene (NS) segment and characterization of recombinant viruses bearing the 1918 NS genes. Proc Natl Acad Sci U S A 98: 2746–2751.

26. Zhu Q, Yang H, Chen W, Cao W, Zhong G, et al. (2008) A naturally occurring deletion in its NS gene contributes to the attenuation of an H5N1 swine influenza virus in chickens. J Virol 82: 220–228.

27. Lipatov AS, Andreansky S, Webby RJ, Hulse DJ, Rehg JE, et al. (2005) Pathogenesis of Hong Kong H5N1 influenza virus NS gene reassortants in mice: the role of cytokines and B- and T-cell responses. J Gen Virol 86: 1121–1130.

28. Li KS, Guan Y, Wang J, Smith GJ, Xu KM, et al. (2004) Genesis of a highly pathogenic and potentially pandemic H5N1 influenza virus in eastern Asia. Nature 430: 209–213.

29. Zhou H, Jin M, Chen H, Huag Q, Yu Z (2006) Genome-sequence analysis of the pathogenic H5N1 avian influenza A virus isolated in China in 2004. Virus Genes 32: 85–95.

30. Long JX, Peng DX, Liu YL, Wu YT, Liu XF (2008) Virulence of H5N1 avian influenza virus enhanced by a 15-nucleotide deletion in the viral nonstructural gene. Virus Genes 36: 471–478.

31. Tang YH, Wu PP, Sun Q, Peng DX, Zhang WJ, et al. (2008) [Role of amino acid residues at positions 322 and 329 of hemagglutinin in virulence of H5N1 avian influenza virus]. Bing Du Xue Bao 24: 340–344.

32. Long JX, Xue F, Peng Y, Gu M, Liu XF (2006) [The deletion of nucleotides of NS gene from 263 to 277 of H5N1 increases viral virulence in chicken]. Wei Sheng Wu Xue Bao 46: 301–305.

33. Zhang W, Xue T, Wu X, Zhang P, Zhao G, et al. (2011) Increase in viral yield in eggs and MDCK cells of reassortant H5N1 vaccine candidate viruses caused by insertion of 38 amino acids into the NA stalk. Vaccine 29: 8032–8041.

34. Shi H, Liu XF, Zhang X, Chen S, Sun L, et al. (2007) Generation of an attenuated H5N1 avian influenza virus vaccine with all eight genes from avian viruses. Vaccine 25: 7379–7384.

35. Hoffmann E, Krauss S, Perez D, Webby R, Webster RG (2002) Eight-plasmid system for rapid generation of influenza virus vaccines. Vaccine 20: 3165–3170.

36. Killian ML (2008) Hemagglutination assay for the avian influenza virus. Methods Mol Biol 436: 47–52.

37. Reed LJ, Muench H (1938) A simple method of estimating fifty per cent endpoints. American Journal of Epidemiology 27: 493–497.

38. Seo SH, Hoffmann E, Webster RG (2002) Lethal H5N1 influenza viruses escape host anti-viral cytokine responses. Nat Med 8: 950–954.

39. Hoffmann E, Stech J, Guan Y, Webster RG, Perez DR (2001) Universal primer set for the full-length amplification of all influenza A viruses. Arch Virol 146: 2275–2289.

40. Kuo RL, Zhao C, Malur M, Krug RM (2010) Influenza A virus strains that circulate in humans differ in the ability of their NS1 proteins to block the activation of IRF3 and interferon-beta transcription. Virology 408: 146–158.

41. Adams SC, Xing Z, Li J, Cardona CJ (2009) Immune-related gene expression in response to H11N9 low pathogenic avian influenza virus infection in chicken and Pekin duck peripheral blood mononuclear cells. Mol Immunol 46: 1744–1749.

42. Liang QL, Luo J, Zhou K, Dong JX, He HX (2011) Immune-related gene expression in response to H5N1 avian influenza virus infection in chicken and duck embryonic fibroblasts. Mol Immunol 48: 924–930.

43. Livak KJ, Schmittgen TD (2001) Analysis of relative gene expression data using real-time quantitative PCR and the 2(-Delta Delta C(T)) Method. Methods 25: 402–408.

44. Edwards S (2006) OIE laboratory standards for avian influenza. Dev Biol (Basel) 124: 159–162.

45. Zhou J, Sun W, Wang J, Guo J, Yin W, et al. (2009) Characterization of the H5N1 highly pathogenic avian influenza virus derived from wild pikas in China. Journal of virology 83: 8957–8964.

46. Li Y, Shi J, Zhong G, Deng G, Tian G, et al. (2010) Continued evolution of H5N1 influenza viruses in wild birds, domestic poultry, and humans in China from 2004 to 2009. Journal of virology 84: 8389–8397.

47. Luo G, Chung J, Palese P (1993) Alterations of the stalk of the influenza virus neuraminidase: deletions and insertions. Virus Res 29: 321.

48. Lee CW, Jung K, Jadhao SJ, Suarez DL (2008) Evaluation of chicken-origin (DF-1) and quail-origin (QT-6) fibroblast cell lines for replication of avian influenza viruses. J Virol Methods 153: 22–28.

49. Seo SH, Goloubeva O, Webby R, Webster RG (2001) Characterization of a porcine lung epithelial cell line suitable for influenza virus studies. J Virol 75: 9517–9525.

50. Geiss GK, Salvatore M, Tumpey TM, Carter VS, Wang X, et al. (2002) Cellular transcriptional profiling in influenza A virus-infected lung epithelial cells: the role of the nonstructural NS1 protein in the evasion of the host innate defense and its potential contribution to pandemic influenza. Proc Natl Acad Sci U S A 99: 10736–10741.

51. Yoneyama M, Kikuchi M, Natsukawa T, Shinobu N, Imaizumi T, et al. (2004) The RNA helicase RIG-I has an essential function in double-stranded RNA-induced innate antiviral responses. Nature immunology 5: 730–737.

52. Barber MR, Aldridge JR Jr, Webster RG, Magor KE (2010) Association of RIG-I with innate immunity of ducks to influenza. Proceedings of the National Academy of Sciences of the United States of America 107: 5913–5918.

53. Sutejo R, Yeo DS, Myaing MZ, Hui C, Xia J, et al. (2012) Activation of type I and III interferon signalling pathways occurs in lung epithelial cells infected with low pathogenic avian influenza viruses. PloS one 7: e33732.

54. Chen S, Sheng C, Liu D, Yao C, Gao S, et al. (2013) Enhancer of zeste homolog 2 is a negative regulator of mitochondria-mediated innate immune responses. J Immunol 191: 2614–2623.

A Comparison of Next Generation Sequencing Technologies for Transcriptome Assembly and Utility for RNA-Seq in a Non-Model Bird

Findley R. Finseth*¤, Richard G. Harrison

Department of Ecology and Evolutionary Biology, Cornell University, Ithaca, New York, United States of America

Abstract

De novo assembled transcriptomes, in combination with RNA-Seq, are powerful tools to explore gene sequence and expression level in organisms without reference genomes. Investigators must first choose which high throughput sequencing platforms will provide data most suitable for their experimental goals. In this study, we explore the utility of 454 and Illumina sequences for *de novo* transcriptome assembly and downstream RNA-Seq applications in a reproductive gland from a non-model bird species, the Japanese quail (*Coturnix japonica*). Four transcriptomes composed of either pure 454 or Illumina reads or mixtures of read types were assembled and evaluated for the same cost. Illumina assemblies performed best for *de novo* transcriptome characterization in terms of contig length, transcriptome coverage, and complete assembly of gene transcripts. Improvements over the Hybrid assembly were marginal, with the exception that the addition of 454 data significantly increased the number of genes annotated. The Illumina assembly provided the best reference to align an independent set of RNA-Seq data as ~84% of reads mapped to single genes in the transcriptome. Contigs constructed solely from 454 data may impose problems for RNA-Seq as our 454 transcriptome revealed a high number of indels and many ambiguously mapped reads. Correcting the 454 transcriptome with Illumina reads was an effective strategy to deal with indel and frameshift errors inherent to the 454 transcriptome, but at the cost of transcriptome coverage. In the absence of a reference genome, we find that Illumina reads alone produced a high quality transcriptome appropriate for RNA-Seq gene expression analyses.

Editor: Cynthia Gibas, University of North Carolina at Charlotte, United States of America

Funding: This research was funded by an NSF DDIG DEB-1010757 to FRF and RGH, P.E.O. Scholar, Cornell Sigma Xi, Andrew W. Mellon, and Paul. F. Feeny awards to FRF. The funders had no role in study design, data collection and analysis, decision to publish, or preparation of the manuscript.

Competing Interests: The authors have declared that no competing interests exist.

* Email: far25@cornell.edu

¤ Current address: Division of Biological Sciences, University of Montana, Missoula, Montana, United States of America

Introduction

Until recently, evolutionary and population-genomic research was restricted to the small number of taxa considered model organisms. Modern next-generation sequencing technologies offer the opportunity to generate massive (and increasing) amounts of sequence data easily and affordably. Today, the potential for large-scale genomic investigations exists for virtually any study system [1–4]. One approach adopted by the non-model research community is shotgun-sequencing of transcriptomes [1–4]. With the advent of deep, parallel sequencing of cDNA ("RNA-Seq") researchers can quantify expression variation in a high-throughput and cost-effective manner [5,6]. Given options in terms of sequencing platform and bioinformatics workflow, a pressing question is what is the optimal strategy to harness both the static (sequence-level) and dynamic (expression-level) nature of transcriptomes of non-model species.

Until recently, investigators predominantly utilized long sequencing reads generated by the 454 GS-FLX (Roche Diagnostics Corporation; hereafter "454") sequencing platform to facilitate *de novo* transcriptome assembly [2,3], e.g., [7–11]. Although 454 is appropriate for assembly, the millions of short reads produced by

Illumina (Illumina, Inc.) are preferred for RNA-Seq as detection of differential expression is sensitive to sequencing depth [3,5,12,13]. One approach to RNA-Seq has been to map Illumina short reads onto a reference constructed from longer 454 reads [14,15]. With increasing read lengths produced by Illumina HiSeq technology (currently 125–150 bp), studies assembling *de novo* transcriptomes directly from Illumina data are emerging [16–23]. This approach is attractive, as data for transcriptome characterization and quantification are collected simultaneously. Recent work comparing technologies using real and simulated data suggest that hybrid assemblies combining 454 and Illumina reads yield the highest quality transcriptomes [24–26]. However, collecting both types of data may be cost-prohibitive. Here, we sequence a transcriptome of a non-model organism with both 454 and Illumina technologies, perform *de novo* assembly with each data type separately and in combination, and compare the various transcriptomes in terms of quality and utility for RNA-Seq. Our objective was to model approaches taken by those studying genomics of non-model organisms, considering cost as a potential limiting factor. Thus, we sequenced our transcriptome with both 454 and Illumina technologies at depths that were approximately the same cost (~$5000, Table S1).

Our transcriptome data derive from a reproductive tissue of a non-model species. Male Japanese quail (*Coturnix japonica*) possess a well-developed foam gland that produces a viscous secretion that is whipped into a stiff foam by contractions of the cloacal sphincter muscle [27,28]. During copulation, a male introduces semen and a large quantity of foam to a female's reproductive tract [29]. The foam gland is of interest to evolutionary biologists because it is an example of a novel trait [27,30], and it is likely involved in sexual selection [31,32]. Foam is also a key mediator of male fitness, influencing the outcome of sperm competition and improving several aspects of fertility and sperm performance [31–36].

We sequenced cDNA from the foam gland with both 454 and Illumina technologies and assembled transcriptomes following four schemes previously applied to species without genomic resources [3,14,24–26,37]. The first two assemblies were composed solely of reads from one or the other technology ("454" and "Illumina" transcriptomes).

The remaining assembly strategies utilized both types of reads initially subsampled to 50% of the raw data in order to keep costs comparable to the pure assemblies. The third assembly attempted to address known issues with systematic errors inherent to 454 sequencing (e.g., homopolymer errors; [4,38]). For this approach, we mapped Illumina reads onto a 454 assembly, identified points of discrepancy between the 454 contigs and the majority of Illumina reads, and created a corrected consensus sequence ("Corrected 454"). Finally, we constructed a hybrid assembly ("Hybrid") by merging contigs made by 454 and Illumina data, and performing an additional round of assembly on those. We chose to merge contigs, rather than assemble from raw reads, because recent work in non-model systems suggests that this method performs better than a merge-reads hybrid approach in terms of contig length, total transcriptome coverage, and number of genes identified [25].

Transcriptomes made of Illumina data are often assembled with de-Bruijn graph based strategies, but these tend to work poorly for 454 data due to indel-errors and low coverage [39]. As our objective was to construct four high quality assemblies economically (thereby mimicking the approach adopted by non-model researchers) we chose to use assemblers optimized for each data type. Many prior studies compared different transcriptome assemblers and there is some consensus regarding which assemblers perform optimally on different sequence types [16,39–43]. For 454 data, combining output from multiple assemblers produces the best transcriptomes [39]. Therefore, we chose the assembly pipeline iAssembler, which performs iterative assemblies with MIRA (4 cycles) and CAP3 (1 cycle), followed by automated error detection and correction [41]. For Illumina data we initially used the Trinity assembler, which has effectively reconstructed many transcriptomes from Illumina data [18,40,43]. All transcriptomes were subjected to an additional round of assembly with iAssembler, to reduce variation due to differences in assemblers.

We initially evaluated the transcriptome assemblies with standard metrics based on transcript length (e.g., N50, median contig length, etc.). *De novo* assembled transcriptomes can retain errors not captured by standard metrics, such as sequencing errors, insertions/deletions ("indels"), misassembled paralogs, chimeras, and/or partial transcripts [25,44]. Annotation-based metrics can be more informative of transcriptome quality than the popular length-based metrics [45]. Although *Coturnix* quail do not have a well-annotated genome available, Japanese quail diverged ~34 million years ago from the chicken (*Gallus gallus*) and exhibit conserved synteny and chromosomal structure with the chicken

genome [46–48]. Functional annotation using a related species' genome as a proxy reference is robust for species pairs diverged less than 100 million years [24]. Thus, we annotated our quail transcriptomes with the high-quality chicken transcriptome and assessed how well assembled contigs reproduced orthologous genes. Finally, we evaluated each transcriptome's utility for RNA-Seq by mapping data from an independent sample of foam glands to each assembly and comparing the alignments.

Methods

Subjects and RNA extraction

Japanese quail were lab-reared and housed on a 16:8 light:dark cycle. All study males were sexually mature, had phenotypically normal foam glands, and produced normal foam complements. A foam gland from a Japanese quail male approximately one year old was used to generate the 454 data. Foam glands from six Japanese quail males (two were one-year old and four were five months old), were used to generate the Illumina data for transcriptome assembly. For the independent RNA-Seq assessment, we sampled foam glands from six different Japanese quail males on winter light conditions (8:16 light:dark cycle, with lights on at 8:00) with testosterone replacement. Testosterone-replaced males have phenotypically normal foam glands and produce foam [49,50]. After euthanizing with CO_2, we immediately dissected out foam glands and froze samples on liquid nitrogen. We extracted RNA with the Agencourt RNAdvance Tissue Kit (Beckman Coulter) following the manufacturer's instructions with the exception that we performed half-reactions. RNA quality and concentration was assessed by agarose gel electrophoresis and NanoDrop spectrophotometry. We checked for RNA purity and integrity using an Agilent 2100 BioAnalyzer.

Library construction

454. We isolated mRNA from one µg total RNA, synthesized first-strand cDNA and generated ds cDNA following the manufacturer's instructions for the SMART Polymerase Chain Reaction (PCR) cDNA Synthesis Kit (Clontech Laboratories, Inc.), with the exception that we used SuperScript III Reverse Transcriptase (Invitrogen) as the reverse transcriptase and made adjustments accordingly. We amplified the cDNA, confirmed successful amplification via agarose gel electrophoresis, and cleaned the PCR products with the QIAquick PCR Purification Kit (Qiagen). We partially normalized our library subjecting amplified cDNA to hybridization and double-stranded nuclease (DSN) digestion following instructions from the TRIMMER cDNA Normalization Kit (Evrogen) except using only 1/8 and 1/16 concentrations of DSN. Size selection was performed with the QIAquick Gel Extraction Kit (Qiagen) according to manufacturer's instructions. We enzymatically fragmented the dsDNA with NEBNext dsDNA Fragmentase (New England BioLabs, Inc.), end polished using T4 polymerase (New England BioLabs, Inc.), phosphorylated 5′ ends with T4 kinase (New England BioLabs, Inc.), added an adenine to 3′ ends with NEB Taq (New England BioLabs, Inc.), and ligated Multiplex Identifier (MID) Adaptor #1 for GS FLX Titanium chemistry (Roche/454 Life Sciences) to ds cDNA using T4 ligase (New England BioLabs, Inc.). Throughout the normalization, end polishing, and ligation procedure, the ds cDNA was cleaned with the QIAquick PCR Purification Kit (Qiagen) when necessary. Cornell University's Genomics Facility at the Institute of Biotechnology performed ½ plate of 454 GS FLX sequencing with Titanium chemistry on the resulting library (Roche/454 Life Sciences) in April 2010.

Illumina. In January 2012, six Illumina libraries for the transcriptome assembly were prepared from approximately 1.2 µg total RNA using the TruSeq RNA Sample Preparation Kit (Illumina) following the manufacturer's instructions. We also prepared six samples from testosterone-replaced males for the independent RNA-Seq evaluations. All twelve samples were tagged with a unique adapter index, pooled, and single-end sequenced on one lane of an Illumina HiSeq 2000, with a target read length of 100 bp. Sequencing was performed by Cornell University's Genomics Facility at the Institute of Biotechnology in April 2012. Raw data for the sequencing runs is reported in Table S1.

Transcriptome assembly

454. Initial quality filtering of reads was performed by the Cornell University's Genomics Facility at the Institute of Biotechnology. SeqClean (http://sourceforge.net/projects/seqclean/) was used to trim low complexity sequences and short sequences (<90 bp). MID-1 and SMART adaptors were trimmed using both SeqClean and NextGENE (Softgenetics). We assembled the reads into unigenes using two rounds of iAssembler [41]. In all instances where iAssembler was applied, we used iAssembler version v1.2.2 with default parameters except that minimum overlap was set to 30 and 95% identity was used for sequence clustering and assembly [41]. Contigs and singletons from the first round of iAssembler served as input for the second round to produce 68,678 unigenes (42,484 of which were represented by singletons). We retained all unigenes over 200 bp for further analysis (47,859 unigenes).

Illumina. Initial quality filtering and barcode removal were performed by Cornell University's Genomics Facility at the Institute of Biotechnology. We used fastq-mcf version 1.04.636 (http://code.google.com/p/ea-utils/wiki/FastqMcf) to remove Illumina adaptors, trim low-quality terminal ends, discard short sequences, and filter reads. Fastq-mcf scans a sequence file for adapters and, based on a log-scaled threshold, determines a set of clipping parameters by initially evaluating a subsampled portion of the data. We used fastq-mcf with default parameters, except that we subsampled one million reads for threshold estimation, quality filtered for mean phred scores <20, and set the percentage of bad reads causing cycle removal to 1. We merged the six libraries into a single file and assembled a transcriptome using Trinity release 2012-06-08 with default parameters [40]. Contigs produced by Trinity were then clustered into 37,166 unigenes with iAssembler.

Hybrid. As one of our goals was to assemble libraries that represent approximately the same cost, prior to the Hybrid transcriptome assembly, we randomly subsampled 50% of the 454 and Illumina filtered reads using custom awk scripts. As a result, the Hybrid transcriptome represents roughly the same amount of sequencing cost as the 454 and Illumina transcriptomes. We then assembled the subsampled 454 and Illumina data with iAssembler and Trinity, respectively, as above. The output of these two preliminary assemblies were merged into a single file and assembled with iAssembler as before.

Corrected 454. We used Illumina data to correct errors with a 454 transcriptome using the Nesoni pipeline, version 0.85 (http://www.bioinformatics.net.au/software.nesoni.shtml). Nesoni utilizes the SHRiMP short read mapper to align short reads to an assigned reference [51]. Positions where disparity exists between the majority of reads and the reference are identified, corrected, and the consensus sequences forms a corrected sequence set. We input the 454 transcriptome as reference, Illumina data as reads, and created consensus sequences using default parameters, with the exception that we allowed reads to be mapped to multiple places. Because we wanted to maintain comparable sequencing costs, both the 454 reference transcriptome and Illumina reads reflect data initially sampled to 50% as generated during the Hybrid assembly (with one additional round of iAssembler for the 454 reference, for a total of two rounds of iAssembler). Only those transcripts with at least one aligned read were retained. Terminal N's were trimmed from the sequence and only sequences greater than 200 bp were retained. The cleaned consensus sequence set represents the Corrected 454 assembly.

Transcriptome evaluation

Unless specified, analyses were conducted in R version 2.15.1 and RStudio version 0.96.330. Figures were made in ggplot2 [52].

Standard metrics. For each assembly, we calculated standard metrics of quality including number of contigs, average contig length, median contig length, N50 (median contig size weighted by length), the distribution of contig lengths, and summed contig length [24,39]. We downloaded all chicken coding sequences from Ensembl version 68 (*G. gallus* assembly: WASHUC2) via the BioMart tool [53] and calculated the same standard metrics for comparison. Prior to computation of basic metrics, we removed contigs ≤200 bp in each dataset, as Trinity assemblies do not report contigs ≤200 bp. We were also interested in how well each assembly predicted open reading frames and identified open reading frames with OrfPredictor [54]. OrfPredictor outputs the 'best' open reading frame, which is the longest among the six possible reading frames for a putative transcript. For each assembly, we computed the frequency of contigs with no open reading frames and the distribution of the lengths of open reading frames.

Ortholog comparisons. We used data from the chicken to identify orthologs. All chicken protein sequences from Ensembl version 68 (*G. gallus* assembly: WASHUC2) were downloaded via the BioMart tool [53]. We filtered the protein set to remove redundant entries (*i.e.*, duplicates, alternative splice variants) by self-BLAST following Hornett and Wheat [24]. Briefly, for any pairwise BLASTp hit with an e-value $\leq 1 * 10^{-6}$, >90% similarity, and >33 amino acids in length, we removed the shorter of the two proteins. All BLAST steps were performed in parallel via Cornell University's Computational Biology Application Suite for High Performance Computing. The reciprocal best blast method was used to determine orthologs with a cutoff e-value of $1 * 10^{-6}$ [55–57]. We report the number of orthologs identified for each transcriptome assembly and present their distributions in a Venn diagram made with the VennDiagram package v1.6.5 in R [58].

For each contig from the various transcriptome assemblies, we computed the "ortholog hit ratio" as described by O'Neil *et al.* [62]. This ratio represents the length of a putative coding region of a contig divided by the length of the coding region of its orthologous transcript. The hit region of the best BLASTx result between a contig and its ortholog was used as a conservative estimate of the "putative coding region" of a contig. Only reciprocal best hits were used for ortholog hit ratio determination. Lengths are in amino acids. An ortholog completely represented by a contig would have a ratio of "1". Ratios less than 1 indicate instances where contigs only partially covered orthologs, while ratios greater than 1 usually indicate insertions in contigs.

Independent RNA-Seq assessment. We produced an independent Illumina RNA-Seq dataset from foam glands of six different foam-producing Japanese quail males to evaluate the utility of our various assemblies for gene expression analyses. The RNA-Seq data were merged into a single file and aligned using the Burrow-Wheeler transform as implemented in the aln algorithm of BWA with default parameters except that -q was set to 20 [59].

For the chicken, we relaxed an additional criterion given expected divergence between chicken and quail, setting -n to 0.1. The Nesoni pipeline (http://vicbioinformatics.com/nesoni.shtml) was used to generate statistics about the quality of the alignments to the various assemblies including the number of mapped/unmapped reads and the number of indels per 100,000 bp identified between each assembly and the majority of the RNA-Seq data. We calculated the number of uniquely mapped reads with samtools [60].

Data accessibility

Raw data have been deposited on the Short Read Archive under accession numbers SRR1346108 and SRR1352724.

Ethics statement

All animal procedures were approved by Cornell University's Institutional Animal Care and Use Committee under permit 2002–0117.

Results and Discussion

Standard transcriptome quality assessment

We sequenced foam glands with the 454 and Illumina platforms and assembled the raw data into four transcriptomes, each having approximately the same sequencing cost: two made solely from each type of data (454, Illumina), one that used Illumina data to correct errors in the 454 transcriptome (Corrected 454), and one that used both kinds of data as input (Hybrid). High quality assemblies possess near full-length contigs representing most of the actual transcriptome. We first evaluated each transcriptome assembly and the Chicken coding sequence set using a suite of standard metrics [24,39]. We use the Chicken coding sequence set as a tool for comparing the relative performance of the various transcriptomes, as gene length is highly conserved within eukaryotes [61], but recognize that the chicken transcriptome comprises a much more diverse collection of sequences as they derive from multiple tissues, life history stages, and sexes. Thus, our expectation is that the foam gland transcriptome should only contain a portion of the genes transcribed in the Chicken sequence set.

The Illumina assembly displayed the highest values across most standard metrics of transcriptome quality, followed by the Hybrid assembly (Table 1). The distribution of contig lengths are quite different between the 454- and Illumina-based datasets, although similar to patterns described previously (Figure 1a) [18,25]. Both the Illumina and Hybrid assemblies generated many contigs that were long (Figure 1; Table 1) and covered a large portion of the transcriptome (summed contig length in Table 1, which has been used previously as a proxy for transcriptome coverage [24]). In contrast, the 454 assembly tended to have short contigs (e.g., N50, mean, longest contig) and a low summed length (Figure 1; Table 1). Although the Hybrid transcriptome generated many long contigs, it also had proportionally more short contigs (Figure 1), deflating several standard metrics relative to the Illumina transcriptome (N50, mean in Table 1). We also find that the absolute longest contigs derive from the Illumina-only assembly (Table 1). Interestingly, the Hybrid transcriptome is a composite of both the 454 and Illumina transcriptomes in terms of contig length; at shorter length ranges, 454-like contigs dominate the Hybrid assembly, whereas long contigs from the Hybrid assembly more closely resemble the Illumina transcriptome (Figure 1). Previous work showed that hybrid, rather than Illumina-only, assemblies produced the largest summed contig lengths, although results were mixed regarding which technology

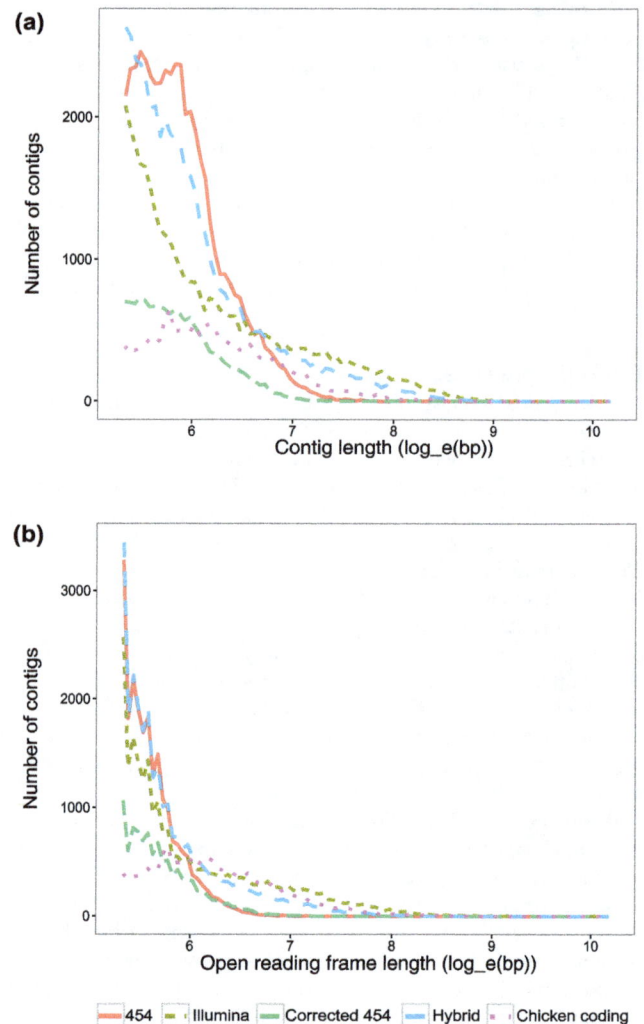

Figure 1. Distribution of contig lengths for each transcriptome assembly. a) Histogram of contig lengths (natural-log transformed) in nucleotide base pairs of each of the transcriptome assemblies and the Chicken coding sequence set. **b)** Histogram of open reading frame lengths (natural-log transformed) in base pairs predicted for each of the transcriptome assemblies and the Chicken coding sequence set. Legend applies to both graphs.

used singly yielded the next best outcome [24,25]. One study also found that transcriptome assemblies composed solely of Illumina reads had longer contigs than those composed only of 454 reads [24], but other studies did not get this result [25,37].

We note that the median contig lengths are very similar across all four transcriptomes, but that the means of the Illumina and Hybrid assemblies are much higher (Table 1). This is likely because all of the *de novo* assembled transcriptomes produced an excess of very short contigs (~200 bp) relative to the Chicken coding sequences, but the Hybrid and Illumina assemblies had many more long contigs (Figure 1a). This result highlights that all four assembly strategies have problems because they produce many short contigs. Further, though the Illumina and Hybrid assemblies generated many long contigs, it is worth noting that the much more diverse Chicken sequence set has fewer long sequences than either the Hybrid or Illumina transcriptomes (Figure 1). Thus, many of the long transcripts in the Hybrid and Illumina

Table 1. Standard metrics of transcriptome assembly (lengths and N50 in base pairs).

Assembly	Number of contigs	N50	Median contig length	Mean contig length	Maximum contig length	Summed contig length
454	47,859	410	336	395	3,387	18,888,486
Illumina	37,166	1297	389	749	12,391	27,823,843
Corrected 454	13,643	398	331	380	1,906	5,189,883
Hybrid	47,003	646	343	537	8,691	25,231,308
Chicken coding	23,392	2136	1068	1507	26,362	32,322,198

data sets may be isoforms (we removed isoforms from the Chicken sequence set for these analyses) or simply false.

The Corrected 454 assembly performed poorly across almost all basic metrics revealing short contigs representing a small portion of the expected transcriptome size (Table 1). This pattern arises in part because the preliminary 454 transcriptome constructed from 50% of the 454 data constitutes an upper limit in terms of number and length of contigs for the Corrected 454 assembly. For example, the dramatic decrease in the number of contigs from the 454 assembly is because, in addition to using only half the 454 data, the Corrected 454 transcriptome is also limited to consensus sequences between the preliminary 454 transcriptome and RNA-Seq data. Hence, only the subset of the preliminary 454 assembly with at least some mapped Illumina reads was retained. Improvements in the Corrected 454 versus the 454 transcriptomes are seen as revised errors within the assembly and would not be captured by standard metrics (Tables 2, 3).

Some errors in transcriptome assembly (e.g., homopolymer errors) can produce frameshifts, in which case downstream analyses reliant on properly called open reading frames would be difficult. Frameshifts can create premature stop codons, resulting in shorter open reading frames. We predicted open reading frames *in silico* for each transcriptome assembly and computed the frequency of contigs with no open reading frames, as well as the distribution of the lengths of open reading frames (Table 2, Figure 1b). All of the assemblies again produced an excess of short open reading frames compared to the chicken coding sequence (Figure 1b). The 454, Illumina, and Hybrid assemblies all produced a high number of open reading frames, but the 454 transcriptome did so at the cost of long contigs and a relatively high frequency of contigs with no reading frames (Table 2, Figure 1b). Only the Illumina and Hybrid transcriptomes produced a high number of contigs with long open reading frames, with the Illumina performing slightly better than the Hybrid assembly (Figure 1b; Table 2). Again, comparisons with the Chicken sequence set suggest many of the *de novo* assembled contigs may be isoforms or false transcripts. Correcting the 454

data with Illumina sequences decreased the proportion of contigs without open reading frames, suggesting this may be an effective strategy to remove nonsense errors in 454-based transcriptomes (Table 2).

Previously, approaches combining both 454 and Illumina data revealed significant improvements over either single technology using similar metrics [25,26]. Here, we find that Illumina data alone produces transcriptomes that are better in quality than assemblies incorporating both types of data. The discrepancy between our results and previous work may be due to aspects of our experimental design. Since our sequencing efforts, 454 has introduced GS FLX+ chemistry (Roche Diagnostics Corporation), which promises more reads that are longer (up to 1000 bp) than the GS FLX Titanium chemistry we used. Longer reads can improve transcriptome contiguity and reduce mis-assembly of short reads [26]. We chose to sequence one-half lane of Illumina and one-half plate of 454 for the transcriptome assemblies because these strategies had approximately the same cost. However, for this cost Illumina sequencing generated significantly more data (Table S1). The discrepancies in coverage could, therefore, explain many of the differences in transcriptome quality. Nevertheless, in construction of our Hybrid transcriptome, our merge-contigs approach started with more contigs from the 454 assembly (~45K) than the Illumina assembly (~32K), yet the Hybrid assembly performed much better than the 454-only transcriptome (Table 1, Figure 1). Another possibility is that differences in the levels of polymorphism in the input samples could influence transcriptome quality. The 454 data were produced from a single individual, whereas the Illumina data were generated from six males. Other studies using Trinity for *de novo* transcriptome assembly have found that contig length or gene recovery (but not accuracy) are negatively influenced by increased polymorphism [43,44]. Given that we find improved performance with our sampling that has increased polymorphism (i.e., Illumina), polymorphism differences likely do not explain our main results. Additionally, we sequenced a single tissue that expresses fewer genes than would be expressed across all tissues. Thus, assemblies generated with short Illumina

Table 2. Number and frequency of contigs with no open reading frames.

Assembly	# contigs with ORF	# contigs with no ORF	Frequency (%)
454	47,342	517	1.09
Illumina	36,961	206	0.55
Corrected 454	13,621	22	0.16
Hybrid	46,639	364	0.78
Chicken coding	17,031	2	0.01

Table 3. The number of deletions, and insertions per 100,000 bp identified between RNA-Seq and an assembly.

Assembly	Deletions	Insertions
454	307.36	61.47
Illumina	0.98	0.46
Corrected 454	17.04	0.93
Hybrid	64.93	79.73
Chicken coding	0.33	0.19

reads may be appropriate for sequencing a smaller number of genes, but hybrid assemblies may exhibit improvements as the number and diversity of expressed genes increase.

Comparisons with orthologs

De novo assembled transcriptomes from non-model species rely on BLAST-based annotations to provide information about gene identity and function. We exploited the fact that quail and chicken are closely related and determined quail-chicken orthologs via reciprocal best BLAST [55,56]. We find that assemblies that include some Illumina sequences outperform those built solely from 454 reads in terms of the number of orthologs identified, providing significantly more annotations. We identified at least 1,100 more orthologs from the Hybrid (8,547) and Illumina assemblies (7,918) than the 454 transcriptome (6,789) (Figure 2). Again, the Corrected 454 transcriptome was limited by its consensus-based assembly pipeline (3,367 orthologs). However, by aligning RNA-Seq data to the 454 dataset in the construction of the Corrected 454 assembly, we retained a higher proportion of contigs with orthologs (0.24) compared with the 454 transcriptome (0.14). Our results contrast with previous work that annotated similar numbers of [24,25] or more [37] contigs in assemblies from 454 data than Illumina data. Compared to the previous studies, we either generated more Illumina and less 454 sequence data, or implemented the newer Illumina HiSeq 2000 sequencing

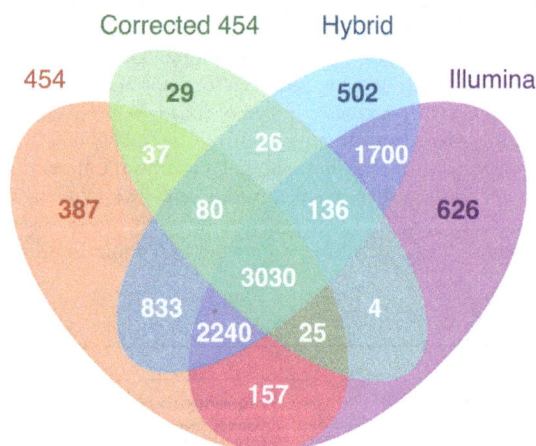

Figure 2. Venn diagram of the number of orthologs for each *de novo* assembled transcriptome. Orthologs were identified via reciprocal best BLAST with chicken and each transcriptome assembly. Non-white numbers indicate orthologs that were unique to one assembly.

technology (Table S1; [25,37]). Hybrid assemblies performed well across studies (present study, [24,25]).

If a research goal is to annotate the maximum number of genes, combining annotations from hybrid and single-data assemblies is the preferred method (Figure 2). The 454 reads contributed an additional 1,865 annotations over the Illumina assembly, whereas the Illumina data added 2,994 annotations over the 454 transcriptome. These are significant contributions, as only 9,812 contigs were annotated in total. Therefore, utilizing both types of data can substantially improve the number of gene annotations, although improvement is greater with Illumina. It should noted, however, that the quality of the annotations added by the 454 assembly may be low, as these annotations are likely represented by low coverage contigs that incompletely recover gene sequences (Figure 3).

Contigs from optimal assemblies represent full, not partial, gene sequences. We assessed how well contigs from each assembly reproduced ortholog length by calculating the "ortholog hit ratio" (Figure 3; [18,62]). This ratio is the length of the assembled contig length relative to the length of its chicken ortholog as defined by reciprocal best blast hits. Contigs representing fully assembled transcripts have ortholog hit ratios close to one. Values less than one represent partial contigs, whereas values greater than one generally (but not always) indicate an insertion in the assembled contig. Because we only examined the ortholog hit ratios from the reciprocal best BLAST hit, this metric is conservative (i.e., parts of orthologs may be represented by contigs that are not the reciprocal best hit). Still, Hornett and Wheat [24] found that the longest assembled contig per ortholog (which was often also the reciprocal best hit in our datasets) is the single best metric for assessing transcriptome performance.

All assemblies displayed many ratios less than one, suggesting that partial transcripts are a challenge for *de novo* assembled transcriptomes (Figure 3). The Illumina and Hybrid assemblies had many more fully assembled transcripts than either the 454 or Corrected 454 assemblies, with the Illumina assembly in particular revealing a high number of transcripts with ortholog hit ratios equal to one (Figure 3). The greater depth of coverage provided by Illumina sequencing may be partly responsible for the increase in the number of full-length or nearly full-length assembled transcripts [62]. The Illumina and Hybrid assemblies performed well at constructing complete transcripts across both small and large orthologous genes (Figure 4), whereas the ability of the 454 and Corrected 454 assemblies to build full transcripts degraded quickly with ortholog length (Figure 4). High completeness of transcripts across many ortholog sizes has been demonstrated previously for Illumina-only transcriptome assemblies [18,62].

Independent RNA-Seq assessment

A general challenge for RNA-Seq analyses is dealing with ambiguity in read mapping, and one proposed solution is to retain only uniquely mapped reads for detection of differential expression [6,63]. Therefore, for *de novo* transcriptome assemblies to be useful for many gene expression analysis, a large proportion of high quality RNA-Seq reads need to map unambiguously to a single contig with few errors. We generated Illumina sequences from foam glands of an independent set of Japanese quail males and aligned reads to each of the four transcriptome assemblies and the Chicken coding sequence set. To assess each assembly's utility for RNA-Seq, we calculated the total number of aligned reads and the number that mapped uniquely or ambiguously (Figure 5). Our results suggest that assemblies built from Illumina data alone offer the best combination of quantity (total number) and quality (proportion unique) of mapped reads for RNA-Seq.

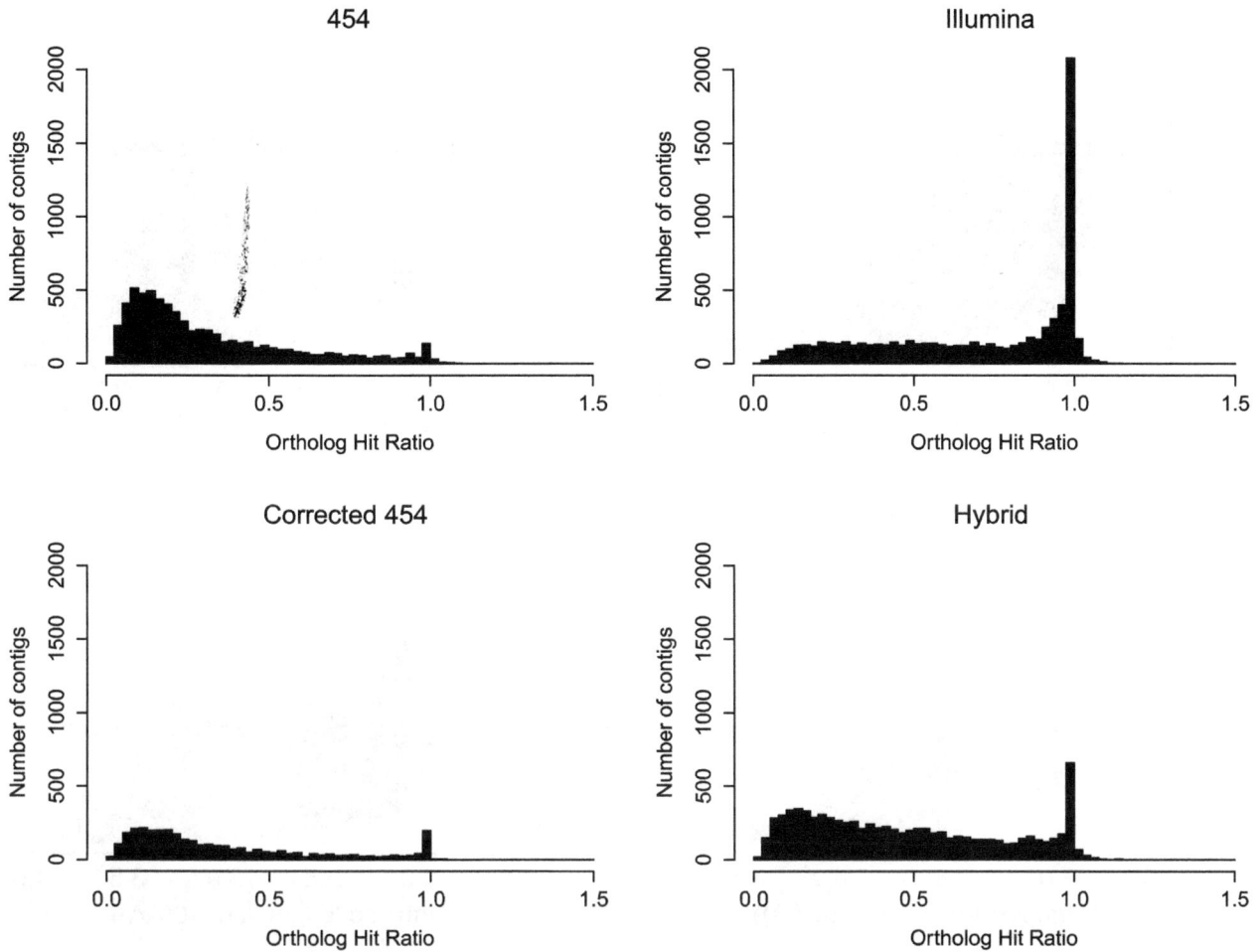

Figure 3. Ortholog hit ratios for each transcriptome assembly. Histograms of ortholog hit ratios (*i.e.*, contig lengths relative to ortholog length) for contigs generated from the 454, Illumina, Corrected 454, and Hybrid transcriptome assemblies. Ratios equal to 1 indicate fully assembled transcripts. Values <1 signify partial transcripts and values >1 than represent contigs with insertions relative to orthologs. Orthologs were determined by 1:1 reciprocal best blast hits with chicken.

The Illumina transcriptome allowed for the largest number of mapped reads, but at least half of the total reads mapped when aligned to any *de novo* assembled quail transcriptome (Figure 5). In contrast, a large proportion of reads remained unmapped when the chicken transcriptome was used as reference. Strikingly, the Illumina assembly allowed for a very high proportion of uniquely mapped reads (Figure 5), whereas any transcriptome built with 454 data resulted in a significant portion of ambiguously mapped reads (Figure 5).

Either something particular to the 454 reads or the assembly pipeline could be responsible for the high levels of ambiguity in the Hybrid and 454 transcriptomes. One option is that erroneous indels in the 454 transcriptome produce ambiguous mappings. Because correcting 454 transcriptomes reduces indels (Table 3) but does not reduce the frequency of ambiguously mapped reads (Figure 5), this is likely not the issue. A more promising explanation is that the 454 reads produced transcriptomes with a high number of contigs representing portions of the same genes causing Illumina reads to map to multiple contigs in the transcriptome. This is consistent with the observed excess of short reads and low ortholog hit ratios found in 454-based libraries (Figures 1a, 3). Additionally, though the final assembler used for the all four transcriptomes was the same, differences in the initial

assembler could have introduced biases that would make reads more or less likely to map uniquely. For example, the first round of assembly in all transcriptomes explicitly attempts to retain isoforms (Trinity, the MIRA cycles of iAssembler use the EST mode which keeps isoforms), but differences in how the isoforms are called may influence the frequency of shared exons between contigs, producing ambiguity [40,41]. Finally, even though the RNA-Seq data derived from an independent set of birds, the sampling and raw sequence data were nearly identical to strategies used for the Illumina transcriptome, and it may be unsurprising that it produces a higher proportion of uniquely mapped reads. Nevertheless, Illumina or similar short-read data are currently the standard for RNA-Seq projects, and our results suggest that Illumina-based assemblies will indeed be most appropriate for RNA-Seq experiments mapping to *de novo* assembled transcriptomes.

Biases inherent to next-generation sequencing can compromise accurate quantification of gene expression [64]. False indels are one type of bias that may be problematic for RNA-Seq. They result in fewer high quality mapped reads or more mis-assigned reads, both of which would negatively affect the detection of true differences in expression. Downstream applications with transcriptomes that rely on properly called open reading frames (e.g.,

454

Illumina

Corrected 454

Hybrid

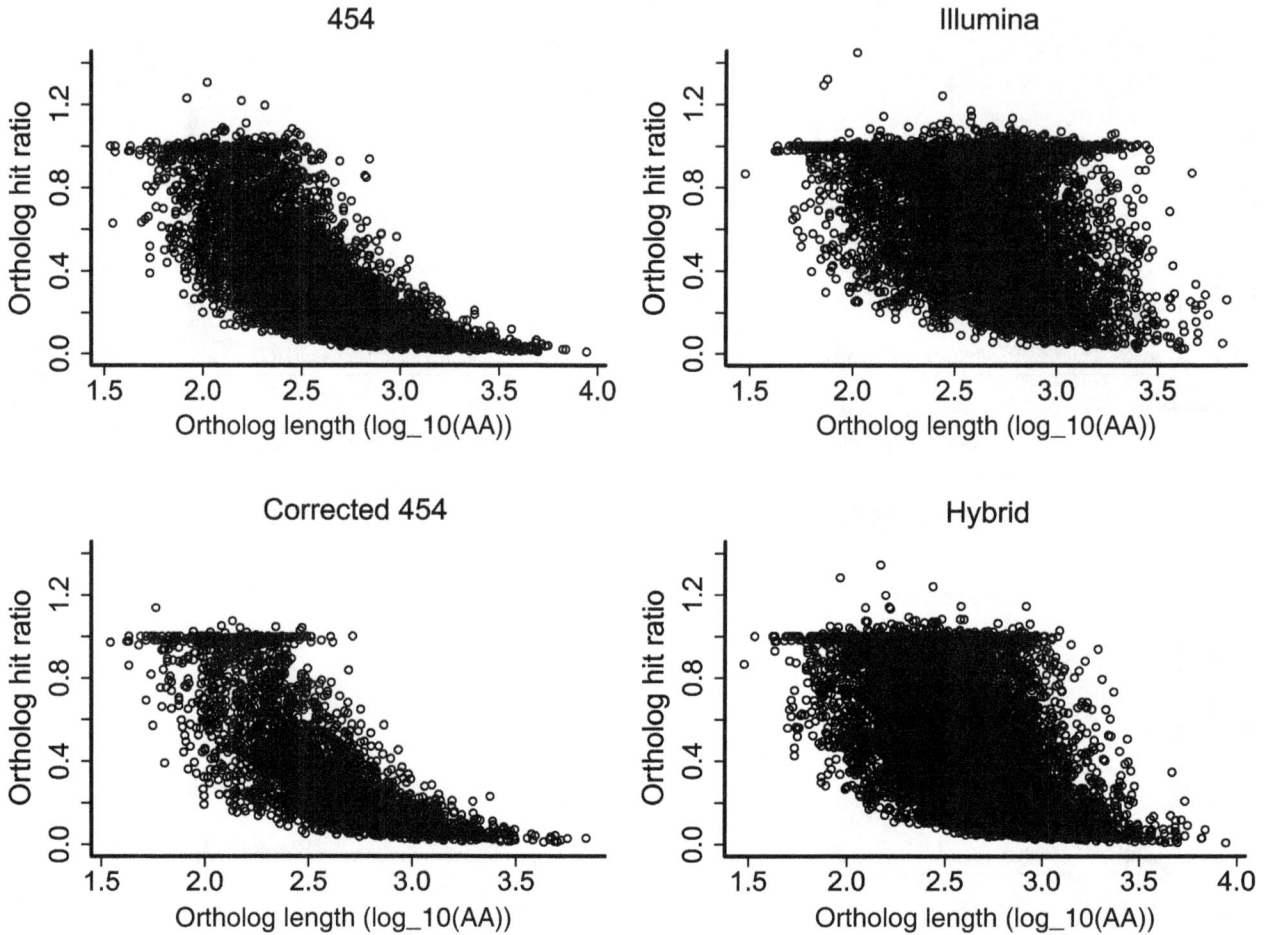

Figure 4. Relationship between ortholog hit ratio and ortholog length for each transcriptome assembly. The ortholog hit ratio standardizes contig lengths relative to ortholog length. Contigs representing complete transcripts will have ratios equal to 1. Ortholog lengths are in amino acids and were log_{10} transformed. Orthologs were determined by 1:1 reciprocal best blast hits with chicken.

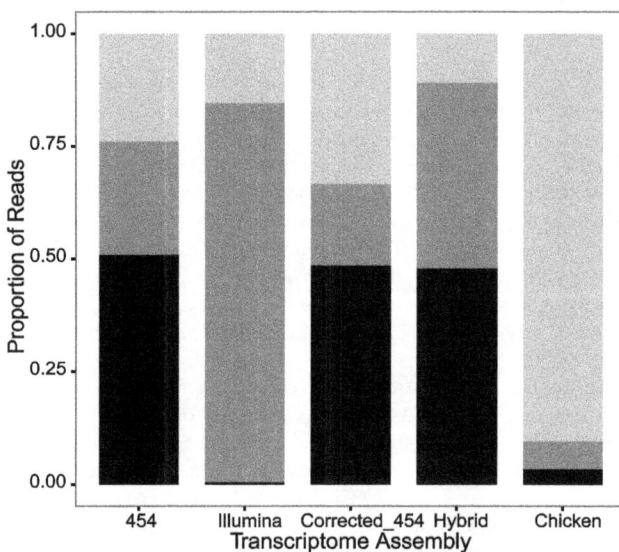

Figure 5. Performance of each assembly for RNA-Seq read mapping. Proportion of 88,446,213 RNA-Seq reads that mapped uniquely (grey), ambiguously (black), or were unmapped (light grey) to each of the transcriptomes.

calculation of evolutionary rates) would be further complicated by erroneous frameshifts produced by false indels. To assess potentially confounding errors in our various transcriptomes, we computed the number of indels per 100,000 bp identified between the consensus alignments of RNA-Seq reads and each assembly (Table 3). The Illumina assembly formed similar numbers of indels as the Chicken transcriptome (Table 3), which is reassuring given that the Chicken coding sequences are certainly in frame (Table 2). Strikingly, mapping RNA-Seq reads to the 454 or Hybrid assemblies produced two-three orders of magnitude more indels than alignments with the Illumina or Chicken transcriptome (Table 3). Indels could reflect errors in the assembled transcriptome, mistakes in the RNA-Seq data, or true polymorphisms. We believe our 454 data is at fault given a large reduction in indels after correcting the 454 assembly with Illumina data (Table 3), a low incidence of indels in the transcriptomes without 454 data including the high quality Chicken sequence set, and previously described homopolymer issues with 454 technology [4,38]. Since correcting the 454 transcriptome with Illumina data significantly reduces the frequency of seemingly erroneous indels, researchers doing RNA-Seq analyses with 454-based transcriptomes should consider performing a consensus-based correction step prior to detection of differential expression. It is worth noting that the quantification of differential gene expression can be robust to sequencing errors, though this finding was based on the errors and error rates specific to Illumina, not 454, sequencing [44].

Recent work suggests that directly mapping RNA-Seq reads to a related species' transcriptome (up to 15% divergent) outperforms mapping to *de novo* assembled transcriptomes in terms of accurately quantifying gene expression [44]. We aligned our RNA-Seq data to the Chicken sequence set. Comparisons of Japanese quail and chicken reveal on average 14% sequence divergence at protein-coding mitochondrial loci [65]. Despite levels of divergence within the recommended range, we found that directly mapping Japanese quail RNA-Seq reads to the Chicken transcriptome performed poorly, as few reads aligned, many of which had ambiguous assignments (Figure 5). In fact, all transcriptomes constructed *de novo* from Japanese quail data allowed for many more uniquely mapped reads than the chicken sequence set (Figure 5). Thus, we find that decent transcriptomes from a focal species serve as a better reference for RNA-Seq than excellent transcriptomes from a distant relative.

Differences in the nature of the data examined may explain the disparity between our results and previous work. To mimic reference-based mapping, Vijay *et al.* [44] introduced varying levels of divergence (5–30%) *in silico* to the zebra finch transcriptome and mapped simulated RNA-Seq reads, also from zebra finch, back to the various transcriptomes. Their datasets accounted for simple differences due to nucleotide polymorphisms and indels, but did not incorporate more complex forms of variation that could affect the ability to map RNA-Seq data (*e.g.*, inversions, gene rearrangements, duplications, exon shuffling). As we utilized non-simulated data, our reference-based mapping approach encompassed both simple and complex forms of sequence divergence that occurred after the Japanese quail and chicken lineages split. Increasing transcriptome complexity (size, paralogs, alternatively spliced isoforms) negatively affects both *de novo* transcriptome assembly and the ability to quantify gene expression [44]. Therefore, we caution directly mapping to a reference transcriptome from a model species, unless sequence differences between the target and reference are known to be simple.

Conclusion

We compared assemblies generated from mixtures of 454 and Illumina reads for *de novo* transcriptome assembly and utility for RNA-Seq analyses in a non-model species. The Illumina assembly often performed the absolute best in standard assays of transcriptome quality, though both the Hybrid and Illumina assemblies produced longer contigs covering more of the transcriptome than 454-based assemblies. Hybrid and Illumina assemblies also afforded more gene annotations that better reproduced ortholog lengths. However, if a goal is to identify the maximum number of annotations, utilizing both 454 and Illumina is preferred, as each contributes a significant number of annotations. Correcting the 454 library with Illumina data drastically reduced the error rate in terms of indels and premature stop codons, but at the cost of contig length and gene annotation.

The Illumina assembly offered the best reference for RNA-Seq data, delivering the highest number of uniquely mapped reads by far. Our results may be unsurprising given the vast differences in the number of reads generated by the two technologies. However, cost is often a limiting factor when working with non-model species and we spent approximately the same amount of money to generate both types of data.

A current challenge facing the non-model community is how to navigate the landscape of next-generation sequencing efficiently and economically. In the past, researchers considered a two-step approach, first building a transcriptome (often from 454 reads) that later served as a reference for mapping RNA-Seq reads, generally generated from a separate Illumina run (e.g., [14,15]). From sequencing one-half of an Illumina lane, we assembled a high quality transcriptome that consistently outperformed a 454 and mixed data transcriptome for less money. *De novo* assemblies made from paired-end Illumina sequences are likely to be even better than the results obtained here. Moreover, our Illumina data averaged 20 million reads per sample, which is well within the range the suggested number for robust detection of differential gene expression (10-30 million reads; [13], but see [66]). To be fair, our study represents a single snapshot in time and is conservative. Indeed, both sequencing platforms currently produce more data with increasing read lengths and fewer errors, at less cost. Although Roche has recently announced that they will be taking 454 technology off the market, our results are likely applicable to users of the popular Ion Torrent Personal Genome Machine sequencing platform, as the high rate of homopolymer-associated indel errors and mean read length are comparable to our 454 data [67–69]. In summary, for researchers on limited budgets with few genomic resources, the present study shows that sequencing transcriptomes with Illumina technology provides sufficient data for *de novo* assembly and RNA-Seq analysis in a single step.

Supporting Information

Table S1   Summary statistics of raw data generated for assemblies.

Acknowledgments

We would like to thank Steve Bogdanowicz and Jennifer Mosher for guidance and assistance with next generation sequencing; Elizabeth Adkins-Regan for providing and housing quail; Stephanie Iacovelli, Nicole Baran, and Dave Cerasale for quail handling; Tim van Deusen, Percy Smith, Linda Vann, and Stephanie Martin for quail care; the RNA-Seq reading group for advice; and the Harrison lab plus lab 'regulars' for feedback on an earlier version of the manuscript.

Author Contributions

Conceived and designed the experiments: FRF RGH. Performed the experiments: FRF. Analyzed the data: FRF. Contributed reagents/materials/analysis tools: FRF RGH. Wrote the paper: FRF RGH.

References

1. Ellegren H (2008) Sequencing goes 454 and takes large-scale genomics into the wild. Mol Ecol 17: 1629–1631. doi:10.1111/j.1365-294X.2008.03699.x.

2. Wheat CW (2010) Rapidly developing functional genomics in ecological model systems via 454 transcriptome sequencing. Genetica 138: 433–451. doi:10.1007/s10709-008-9326-y.

3. Ekblom R, Galindo J (2010) Applications of next generation sequencing in molecular ecology of non-model organisms. Heredity 107: 1–15. doi:10.1038/hdy.2010.152.

4. Hudson ME (2008) Sequencing breakthroughs for genomic ecology and evolutionary biology. Mol Ecol Resour 8: 3–17. doi:10.1111/j.1471-8286.2007.02019.x.

5. Wang Z, Gerstein M, Snyder M (2009) RNA-Seq: a revolutionary tool for transcriptomics. Nat Rev Genet 10: 57–63. doi:10.1038/nrg2484.

6. Wolf JBW (2013) Principles of transcriptome analysis and gene expression quantification: an RNA-seq tutorial. Mol Ecol Resour 13: 559–572. doi:10.1111/1755-0998.12109.

7. Vera JC, Wheat CW, Fescemyer HW, Frilander MJ, Crawford DL, et al. (2008) Rapid transcriptome characterization for a nonmodel organism using 454 pyrosequencing. Mol Ecol 17: 1636–1647. doi:10.1111/j.1365-294X.2008. 03666.x.

8. Reading BJ, Chapman RW, Schaff JE, Scholl EH, Opperman CH, et al. (2012) An ovary transcriptome for all maturational stages of the striped bass (*Morone saxatilis*), a highly advanced perciform fish. BMC Res Notes 5: 111. doi:10.1186/1756-0500-5-111.

9. Schwartz TS, Tae H, Yang Y, Mockaitis K, Van Hemert JL, et al. (2010) A garter snake transcriptome: pyrosequencing, de novo assembly, and sex-specific differences. BMC Genomics 11: 694. doi:10.1186/1471-2164-11-694.

10. Meyer E, Aglyamova GV, Wang S, Buchanan-Carter J, Abrego D, et al. (2009) Sequencing and de novo analysis of a coral larval transcriptome using 454 GSFlx. BMC Genomics 10: 219. doi:10.1186/1471-2164-10-219.

11. Ekblom R, Farrell LL, Lank DB, Burke T (2012) Gene expression divergence and nucleotide differentiation between males of different color morphs and mating strategies in the ruff. Ecol and Evol 2: 2485–2505. doi:10.1002/ece3.370.

12. Tarazona S, Garcia-Alcalde F, Dopazo J, Ferrer A, Conesa A (2011) Differential expression in RNA-seq: A matter of depth. Genome Res 21: 2213–2223. doi:10.1101/gr.124321.111.

13. Wang Y, Ghaffari N, Johnson CD, Braga-Neto UM, Wang H, et al. (2011) Evaluation of the coverage and depth of transcriptome by RNA-Seq in chickens. BMC Bioinformatics 12: S5. doi:10.1186/1471-2105-12-S10-S5.

14. Su CL, Chao YT, Alex Chang YC, Chen WC, Chen CY, et al. (2011) De novo assembly of expressed transcripts and global analysis of the *Phalaenopsis aphrodite* transcriptome. Plant Cell Phy 52: 1501–1514. doi:10.1093/pcp/pcr097.

15. Jensen JK, Schultink A, Keegstra K, Wilkerson CG, Pauly M (2012) RNA-Seq analysis of developing nasturtium seeds (*Tropaeolum majus*): identification and characterization of an additional galactosyltransferase involved in xyloglucan biosynthesis. Molecular Plant 5: 984–992. doi:10.1093/mp/sss032.

16. Feldmeyer B, Wheat CW, Krezdorn N, Rotter B, Pfenninger M (2011) Short read Illumina data for the de novo assembly of a non-model snail species transcriptome (Radix balthica, Basommatophora, Pulmonata), and a comparison of assembler performance. BMC Genomics 12: 317. doi:10.1186/1471-2164-12-317.

17. Crawford JE, Guelbeogo WM, Sanou A, Traoré A, Vernick KD, et al. (2010) De novo transcriptome sequencing in *Anopheles funestus* using Illumina RNA-Seq technology. PLoS ONE 5: e14202. doi:10.1371/journal.pone.0014202.g005.

18. Van Belleghem SM, Roelofs D, Van Houdt J, Hendrickx F (2012) De novo transcriptome assembly and SNP discovery in the wing polymorphic salt marsh beetle *Pogonus chalceus* (Coleoptera, Carabidae). PLoS ONE 7: e42605. doi:10.1371/journal.pone.0042605.t004.

19. Chen S, Yang P, Jiang F, Wei Y, Ma Z, et al. (2010) *De novo* analysis of transcriptome dynamics in the migratory locust during the development of phase traits. PLoS ONE 5: e15633. doi:10.1371/journal.pone.0015633.t002.

20. Xia Z, Xu H, Zhai J, Li D, Luo H, et al. (2011) RNA-Seq analysis and de novo transcriptome assembly of *Hevea brasiliensis*. Plant Mol Biol 77: 299–308. doi:10.1007/s11103-011-9811-z.

21. Etebari K, Palfreyman RW, Schlipalius D, Nielsen LK, Glatz RV, et al. (2011) Deep sequencing-based transcriptome analysis of *Plutella xylostella* larvae parasitized by *Diadegma semiclausum*. BMC Genomics 12: 446. doi:10.1186/1471-2164-12-446.

22. Birzele F, Schaub J, Rust W, Clemens C, Baum P, et al. (2010) Into the unknown: expression profiling without genome sequence information in CHO by next generation sequencing. Nucleic Acids Res 38: 3999–4010. doi:10.1093/nar/gkq116.

23. Moghadam HK, Harrison PW, Zachar G, Székely T, Mank JE (2013) The plover neurotranscriptome assembly: transcriptomic analysis in an ecological model species without a reference genome. Mol Ecol Resour 13: 696–705. doi:10.1111/1755-0998.12096.

24. Hornett EA, Wheat CW (2012) Quantitative RNA-Seq analysis in non-model species: assessing transcriptome assemblies as a scaffold and the utility of evolutionary divergent genomic reference species. BMC Genomics 13: 361. doi:10.1186/1471-2164-13-361.

25. Cahais V, Gayral P, Tsagkogeorga G, Melo-Ferreira J, Ballenghien M, et al. (2012) Reference-free transcriptome assembly in non-model animals from next-generation sequencing data. Mol Ecol Resour 12: 834–845. doi:10.1111/j.1755-0998.2012.03148.x.

26. Wall PK, Leebens-Mack J, Chanderbali AS, Barakat A, Wolcott E, et al. (2009) Comparison of next generation sequencing technologies for transcriptome characterization. BMC Genomics 10: 347. doi:10.1186/1471-2164-10-347.

27. Klemm R, Knight CE, Stein S (1973) Gross and microscopic morphology of glandula-proctodealis (foam gland) of *Coturnix c. japonica* (aves). J Morphol 141: 171–184.

28. Seiwert C, Adkins-Regan E (1998) The foam production system of the male Japanese quail: Characterization of structure and function. Brain Behav Evol 52: 61–80.

29. Coil WH, Wetherbee DK (1959) Observations on the cloacal gland of the Eurasian quail *Coturnix coturnix*. Ohio J Sci 59: 268–270.

30. Fujihara N (1992) Accessory reproductive fluids and organs in male domestic birds. Worlds Poult Sci J 48: 39–56.

31. Finseth FR, Iacovelli SR, Harrison RG, Adkins-Regan EK (2013) A nonsemen copulatory fluid influences the outcome of sperm competition in Japanese quail. J Evol Biol 26: 1875–1889. doi:10.1111/jeb.12189.

32. Cheng KM, McIntyre RF, Hickman AR (1989) Proctodeal gland foam enhances competitive fertilization in domestic Japanese quail. Auk 106: 286–291.

33. Adkins-Regan E (1999) Foam produced by male *Coturnix* quail: What is its function? Auk 116: 184–193.

34. Singh RP, Sastry KVH, Pandey NK, Singh KB, Malecki IA, et al. (2012) The role of the male cloacal gland in reproductive success in Japanese quail (*Coturnix japonica*). Reprod Fertil Dev 24: 405. doi:10.1071/RD11057.

35. Cheng KM, Hickman AR, Nichols CR (1989) Role of the proctodeal gland foam of male Japanese quail in natural copulations. Auk 106: 279–285.

36. Singh RP, H Sastry von K, Shit N, Pandey NK, Singh KB, et al. (2011) Cloacal gland foam enhances motility and disaggregation of spermatozoa in Japanese quail (*Coturnix japonica*). Theriogenology 75: 563–569. doi:10.1016/j.theriogenology.2010.09.028.

37. Milano I, Babbucci M, Panitz F, Ogden R, Nielsen RO, et al. (2011) Novel tools for conservation genomics: comparing two high-throughput approaches for SNP discovery in the transcriptome of the European hake. PLoS ONE 6: e28008. doi:10.1371/journal.pone.0028008.t007.

38. Gilles A, Meglécz E, Pech N, Ferreira S, Malausa T, et al. (2011) Accuracy and quality assessment of 454 GS-FLX Titanium pyrosequencing. BMC Genomics 12: 245. doi:10.1186/1471-2164-12-245.

39. Kumar S, Blaxter ML (2010) Comparing de novo assemblers for 454 transcriptome data. BMC Genomics 11: 571. doi:10.1186/1471-2164-11-571.

40. Grabherr MG, Haas BJ, Yassour M, Levin JZ, Thompson DA, et al. (2011) Full-length transcriptome assembly from RNA-Seq data without a reference genome. Nat Biotechnol 29: 644–652. doi:10.1038/nbt.1883.

41. Zheng Y, Zhao L, Gao J, Fei Z (2011) iAssembler: a package for de novo assembly of Roche-454/Sanger transcriptome sequences. BMC Bioinformatics 12: 453. doi:10.1186/1471-2105-12-453.

42. Rawat A, Elasri MO, Gust KA, George G, Pham D, et al. (2012) CAPRG: Sequence assembling pipeline for next generation sequencing of non-model organisms. PLoS ONE 7: e30370. doi:10.1371/journal.pone.0030370.g007.

43. Singhal S (2013) De novotranscriptomic analyses for non-model organisms: an evaluation of methods across a multiple-species data set. Mol Ecol Resour 13: 403–416. doi:10.1111/1755-0998.12077.

44. Vijay N, Poelstra JW, Künstner A, Wolf JBW (2013) Challenges and strategies in transcriptome assembly and differential gene expression quantification. A comprehensive in silico assessment of RNA-seq experiments. Mol Ecol 22: 620–634. doi:10.1111/mec.12014.

45. O'Neil ST, Emrich SJ (2013) Assessing De Novo transcriptome assembly metrics for consistency and utility. BMC Genomics 14: 465. doi:10.1186/1471-2164-14-465.

46. Kayang BB, Fillon V, Inoue-Murayama M, Miwa M, Leroux S, et al. (2006) Integrated maps in quail (Coturnix japonica) confirm the high degree of synteny conservation with chicken (Gallus gallus) despite 35 million years of divergence. BMC Genomics 7: 101. doi:10.1186/1471-2164-7-101.

47. Sasazaki S, Hinenoya T, Lin B, Fujiwara A, Mannen H (2006) A comparative map of macrochromosomes between chicken and Japanese quail based on orthologous genes. Anim Genet 37: 316–320.

48. Sasazaki S, Hinenoya T, Fujima D, Kikuchi S, Fujiwara A, et al. (2006) Mapping of expressed sequence tag markers with a cDNA-amplified fragment length polymorphism method in Japanese quail (*Coturnix japonica*). Animal Sci J 77: 42–46.

49. Schumacher M, Balthazart J (1983) The effects of testosterone and its metabolites on sexual-behavior and morphology in male and female Japanese quail. Physiol Behav 30: 335–339.

50. Adkins E (1977) Effects of diverse androgens on sexual-behavior and morphology of castrated male quail. Horm Behav 8: 201–207.

51. Rumble SM, Lacroute P, Dalca AV, Fiume M, Sidow A, et al. (2009) SHRiMP: Accurate Mapping of Short Color-space Reads. PLoS Comput Biol 5: e1000386. doi:10.1371/journal.pcbi.1000386.t004.

52. Wickham H (2009) ggplot2: elegant graphics for data analysis. New York: Springer Publishing Company, Incorporated.

53. Kasprzyk A (2011) BioMart: driving a paradigm change in biological data management. Database (Oxford) 2011: bar049. doi:10.1093/database/bar049.

54. Min XJ, Butler G, Storms R, Tsang A (2005) OrfPredictor: predicting protein-coding regions in EST-derived sequences. Nucleic Acids Res 33: W677–W680. doi:10.1093/nar/gki394.

55. Koonin EV (2005) Orthologs, paralogs, and evolutionary genomics. Annual Rev Genet 39: 309–338. doi:10.1146/annurev.genet.39.073003.114725.

56. Bork P, Koonin EV (1998) Predicting functions from protein sequences—where are the bottlenecks? Nat Genet 18: 313–318. doi:10.1038/ng0498-313.

57. Tatusov RL (1997) A genomic perspective on protein families. Science 278: 631–637. doi:10.1126/science.278.5338.631.

58. Chen H, Boutros PC (2011) VennDiagram: a package for the generation of highly-customizable Venn and Euler diagrams in R. BMC Bioinformatics 12: 35. doi:10.1186/1471-2105-12-35.

59. Li H, Durbin R (2009) Fast and accurate short read alignment with Burrows-Wheeler transform. Bioinformatics 25: 1754–1760. doi:10.1093/bioinformatics/btp324.

60. Li H, Handsaker B, Wysoker A, Fennell T, Ruan J, et al. (2009) The Sequence Alignment/Map format and SAMtools. Bioinformatics 25: 2078–2079. doi:10.1093/bioinformatics/btp352.

61. Xu L (2006) Average gene length is highly conserved in prokaryotes and eukaryotes and diverges only between the two kingdoms. Mol Biol Evol 23: 1107–1108. doi:10.1093/molbev/msk019.

62. O'Neil ST, Dzurisin JD, Carmichael RD, Lobo NF, Emrich SJ, et al. (2010) Population-level transcriptome sequencing of nonmodel organisms *Erynnis propertius* and *Papilio zelicaon*. BMC Genomics 11: 310. doi:10.1186/1471-2164-11-310.

63. Treangen TJ, Salzberg SL (2012) Repetitive DNA and next-generation sequencing: computational challenges and solutions. Nat Rev Genet 13: 36–46. doi:10.1038/nrg3117.

64. Fang Z, Cui X (2011) Design and validation issues in RNA-seq experiments. Brief Bioinform 12: 280–287. doi:10.1093/bib/bbr004.

65. Desjardins P, Morais R (1991) Nucleotide sequence and evolution of coding and noncoding regions of a quail mitochondrial genome. J Mol Evol 32: 153–161.

66. Toung JM, Morley M, Li M, Cheung VG (2011) RNA-sequence analysis of human B-cells. Genome Res 21: 991–998. doi:10.1101/gr.116335.110.

67. Glenn TC (2011) Field guide to next-generation sequencers. Mol Ecol Resour 11: 759–769.

68. Loman NJ, Constantinidou C, Chan JZM, Halachev M, Sergeant M, et al. (2012) High-throughput bacterial genome sequencing: an embarassment of choice, a world of opportunity. Nat Rev Microbiol 10: 598–606.

69. Quail MA, Smith M, Coupland P, Otto TD, Harris SR, et al. (2012) A tale of three next generation sequencing platforms: comparison of Ion Torrent, Pacific Biosciences and Illumina MiSeq sequencers. BMC Genomics 13: 341.

14

Origin and Loss of Nested LRRTM/α-Catenin Genes during Vertebrate Evolution

Pavel Uvarov[1], Tommi Kajander[2], Matti S. Airaksinen[1]*

1 Institute of Biomedicine, Anatomy, University of Helsinki, Helsinki, Finland, 2 Institute of Biotechnology, University of Helsinki, Helsinki, Finland

Abstract

Leucine-rich repeat transmembrane neuronal proteins (LRRTMs) form in mammals a family of four postsynaptic adhesion proteins, which have been shown to bind neurexins and heparan sulphate proteoglycan (HSPG) glypican on the presynaptic side. Mutations in the genes encoding LRRTMs and neurexins are implicated in human cognitive disorders such as schizophrenia and autism. Our analysis shows that in most jawed vertebrates, *lrrtm1*, *lrrtm2*, and *lrrtm3* genes are nested on opposite strands of large conserved intron of α-catenin genes *ctnna2*, *ctnna1*, and *ctnna3*, respectively. No *lrrtm* genes could be found in tunicates or lancelets, while two *lrrtm* genes are found in the lamprey genome, one of which is adjacent to a single *ctnna* homolog. Based on similar highly positive net charge of lamprey LRRTMs and the HSPG-binding LRRTM3 and LRRTM4 proteins, we speculate that the ancestral LRRTM might have bound HSPG before acquiring neurexins as binding partners. Our model suggests that *lrrtm* gene translocated into the large *ctnna* intron in early vertebrates, and that subsequent duplications resulted in three *lrrtm/ctnna* gene pairs present in most jawed vertebrates. However, we detected three prominent exceptions: (1) the *lrrtm3/ctnna3* gene structure is absent in the ray-finned fish genomes, (2) the genomes of clawed frogs contain *ctnna1* but lack the corresponding nested (*lrrtm2*) gene, and (3) contain *lrrtm3* gene in the syntenic position but lack the corresponding host (*ctnna3*) gene. We identified several other protein-coding nested gene structures of which either the host or the nested gene has presumably been lost in the frog or chicken lineages. Interestingly, majority of these nested genes comprise LRR domains.

Editor: Stephan C. F. Neuhauss, University Zürich, Switzerland

Funding: This study was supported by grants from the Academy of Finland and the Sigrid Juselius Foundation. The funders had no role in study design, data collection and analysis, decision to publish, or preparation of the manuscript.

Competing Interests: The authors have declared that no competing interests exist.

* E-mail: matti.airaksinen@helsinki.fi

Introduction

Members of the leucine-rich repeat transmembrane (LRRTM) family of neuronal proteins contain ten N-terminal LRR repeats, a single pass transmembrane domain, and a C-terminal cytoplasmic tail comprising a PDZ domain binding motif [1]. LRRTMs function as postsynaptic adhesion proteins in excitatory synapses [2] by interacting with presynaptic neurexins, similar to the neuroligins [3–7]. *LRRTM1* gene is associated with schizophrenia and handedness [8]. In rodents, LRRTM1 and LRRTM2 proteins have been shown to interact with neurexins, but there are also indications that all the four LRRTMs can bind to neurexins [3–6]. Recently, heparan sulfate proteoglycan (HSPG) glypican was identified as an alternative receptor for LRRTM4 and possibly for LRRTM3 [9,10].

In human and mouse genomes LRRTM1 is encoded by a single exon, whereas the first four coding nucleotides (ATGG) of other LRRTM genes (*Lrrtm2*, *Lrrtm3*, and *Lrrtm4*) are located in a separate first exon [1]. Three of the four LRRTM genes (*Lrrtm1* to *Lrrtm3*) are nested in a large conserved intron of α-catenin genes (*Ctnna1* to *Ctnna3*) [1]. Each mammalian *Ctnna* gene has 17 coding exons (encoding a protein of about 900 amino acids) and hosts one *Lrrtm* nested in the opposite orientation in a large (~50–450 kb in human) intron between coding exons 6 and 7: *Lrrtm1* is nested in *Ctnna2*, *Lrrtm2* in *Ctnna1*, and *Lrrtm3* in *Ctnna3*. *Lrrtm4* gene is not nested but is located within a few genes away from the *Lrrtm1*/

Ctnna2 gene pair in mammals [1]. Genes encoding for α-catenins exist in all metazoan animals analyzed [11], whereas LRRTM genes have only been found in vertebrate genomes [1].

Nested genes represent a subgroup of overlapping genes [12]: one gene ("nested") is situated totally inside another gene ("host"). Nearly all protein-coding nested genes are thought to have emerged by insertion of a corresponding DNA sequence into an intron of a pre-existing gene [13]. Most commonly, the internal/nested gene lies inside an intron of the larger host gene in the opposite orientation [12]. Nested genes that have a single coding exon presumably emerged by retrotransposition [13]. A gene may also become nested by fusion of two flanking genes or by acquisition of new exons. Alternatively, nested genes may originate *de novo* through accumulation of mutations inside a preexisting gene [12]. Once formed, a nested gene structure can be duplicated or lost during evolution. However, no loss of a nested gene structure encoding conserved proteins was reported in vertebrates in a previous study [13].

Here, we have studied the evolution of the LRRTM family. Our analysis suggests that in early vertebrates an ancestral *lrrtm* gene had become incorporated into a pre-existing *ctnna* intron that was followed by two duplications of the nested *lrrtm/ctnna* structure. We found that the nested *lrrtm/ctnna* gene structure is conserved in jawed vertebrates. However, the clawed frog (*Xenopus*) genome contains two notable exceptions. First, the genome contains *ctnna1* but lacks the corresponding nested (*lrrtm2*) gene. Second, the genome contains a clear *lrrtm3* ortholog in syntenic position but lacks the corresponding host (*ctnna3*) gene. A database analysis identified several other phylogenetically old nested gene structures comprising LRR-domain encoding genes that have apparently been lost in amphibian or avian lineages.

Although invertebrates, such as fruit fly and nematode have a neurexin (*nrxn*) gene ortholog [14,15], the evolution of the alternatively spliced *nrxn* AS4 exon, which encodes a loop sequence required for LRRTM binding in mammals [3–6], has not been investigated. Therefore, we also studied whether the alternative splicing of *nrxn* AS4 exon would have co-evolved with the appearance of *lrrtm*. We show that the AS4 exon emerged *de novo* in chordates, and that the mechanism of its alternative splicing may have evolved in the early vertebrates. Based on analysis of net charge of the extracellular LRR domains, we speculate that the first LRRTMs may have bound HPSGs before acquiring neurexins as binding partners.

Materials and Methods

Identification of Sequences

We searched the Ensembl genome database (release 72, Jun 2013) for the genomic location and structure of the annotated LRRTM and α-catenin gene homologs (by searching for their names/gene symbols) from the following species: human, chicken (*Gallus gallus*), Western (tropical) clawed frog (*Xenopus tropicalis*), coelacanth (*Latimeria chalumnae*), zebrafish (*Danio rerio*), and sea lamprey (*Petromyzon marinus*). *Lrtm* orthologs were also retrieved from other ray-finned fish genomes (*Gasterosteus aculeatus*, *Oryzias latipes*, *Takifugu rubripes*, and *Tetraodon nigroviridis*). In addition, we searched the tunicates (*Ciona intestinalis* and *Ciona savignyi*), amphioxus (*Branchiostoma floridae*, genome.jgi-psf.org/Brafl1), elephant shark (*Callorhinchus milii*, esharkgenome.imcb.a-star.edu.sg), spotted gar (*Lepisosteus oculatus*, pre.ensembl.org/Lepisosteus_oculatus), and the African clawed frog (*Xenopus laevis*, xenopus.lab.nig.ac.jp/assembly v7.1) genomes. We also searched the transcriptomes of clawed frogs (*X. laevis* and *X. tropicalis*, www.xenbase.org) and salamander (axolotl, *Ambystoma mexicanum*, www.ambystoma.org, assembly V4.0) for *lrtm* and *ctnna* homologs. If some LRRTM or α-catenin homologs seemed to be missing or incompletely annotated, we searched the corresponding genomes by using TBLASTN (blast.ncbi.nlm.nih.gov/) using the corresponding mouse and chicken protein sequences as a query and verified the hits by reciprocal BLAST searches (using default parameters). The N-terminal part of some LRRTM transcripts was curated manually to conform to the splice site consensus sequences. Identified shark and coelacanth CTNNA fragments were aligned and assembled manually. Isoelectric point (pI) values were calculated using Geneious 6.1.7 (Biomatters Ltd.) for the extracellular LRR-domains of LRRTMs (excluding the signal sequence and hinge domain). These pI values and accession numbers for the identified LRRTM and α-catenin sequences are provided in Table S1.

Analysis of Synteny

We identified human orthologs for genes surrounding the *lrrtm3* gene within *X. tropicalis* scaffold_7:33-34M (www.xenbase.org) and their chromosomal position in human genome using Ensembl. Presence of regions of conserved synteny (paralogous pairwise clusters) between the *CTNNA* gene regions within the human genome were analyzed using the Synteny Database (syntenydb.uoregon.edu/synteny_db/) using a sliding window size of 50 or 100 genes and *C. intestinalis* as outgroup [16]. Possible conserved synteny between vertebrate genomes (e.g. in regions containing the *lrrtm4* gene) was analyzed using the Genomicus database (v73 www.genomicus.biologie.ens.fr).

Alignment and Phylogenetic Analysis

The predicted LRRTM and CTNNA amino acid sequences were aligned using MAFFT v.7 (http://mafft.cbrc.jp/alignment/software/) [17] with default parameters. For LRRTM3 and LRRTM4 orthologs that have alternative C-terminal splice forms, only the shorter isoform (ending to -ECEV) was used. The alignment was edited using Geneious in order to remove positions (amino acid residues) of the LRRTM signal sequence and the extracellular juxtamembrane domain where more than half of the sequences had gaps. The LRRTM alignment is shown in Fig. S1. Phylogenetic trees were inferred using PhyML3.0 under the following model parameters (LG substitution model, empirical equilibrium frequencies, four gamma-distributed substitution rate categories and five random starting trees) with confidence estimates derived from 1000 bootstrap replicates [18]. Trees were rearranged with Geneious and visualized using the MEGA5 software [19].

Analysis of Selected Nested Gene Structures in Vertebrates

We also searched the Ensembl database for vertebrate orthologs for a subset of previously identified human different strand nested gene pairs [12]. We included for the search only those different strand nested gene pairs that were reported to be shared between human and mouse [12], and in which a protein-coding nested gene is flanked by protein-coding exons of the host gene. This selection resulted in 91 protein-coding different strand nested gene pairs for our analysis (Table S3). If annotated orthologs for the nested gene pair were identified (by searching for their names/gene symbols) in coelacanth or zebrafish (or in both), as well as in chicken and clawed frog (*X. tropicalis*) genomes, the nested structure was designated as conserved. If the nested gene structure was present in coelacanth or zebrafish but either the host or the nested gene, or both, were not annotated in either chicken or clawed frog genomes, the nested gene structure was designated as potentially lost (not conserved). The absence of these nested gene structures in chicken or in *X. tropicalis* genomes was verified by BLAST searches and by synteny analysis of adjacent genes.

Evolution of *neurexin* AS4 Exon and Alternative Splicing

To study when the *nrxn* AS4 exon emerged during evolution, we searched selected invertebrate and vertebrate genomes with BLASTP using a 160 amino acid residue fragment of mouse neurexin-1 protein (ENSMUSP00000125407, Refseq NP_064648.3) that is encoded by the AS4 and flanking exons (see Fig. S6). To estimate the relative percentage of *nrxn* transcripts in which the AS4 exon is skipped or retained in selected species (that contain the AS4 exon), we searched the NCBI expressed sequence tag database (dbEST) with TBLASTN with default parameters (BLOSUM62 matrix) using the above 160 amino acid fragment of mouse neurexin-1 as a query (Fig. S6). Hits that were considered relevant for the analysis were at least 80 amino acid long, aligned at least partially with the AS4 exon of the query, and had over 30% sequence identity (Table S6). This ruled out short fragments and distant (non-neurexin) sequences. The location of AU-rich sequence motifs in the introns flanking the *nrxn* AS4 exon (within 200 bp upstream and 200 bp downstream of the exon) was analyzed by text search.

Genomic PCR with Degenerate Primers

We purified *X. tropicalis* (obtained from the European Xenopus Resource Centre, www.port.ac.uk/research/exrc/) and chicken genomic DNA using the Wizard SV Genomic DNA Purification System (Promega, Madison, WI). Degenerate *ctnna3* primers were designed to conform to three conditions. (1) The primers efficiently amplify a corresponding genomic fragment of *ctnna3* from other vertebrate species. (2) The primers also amplify a corresponding genomic fragment of *ctnna1* and/or *ctnna2* from *X. tropicalis*, as well as from other vertebrate genomes, although with a lower efficiency compared to the corresponding fragment of *ctnna3*. This would serve as an internal positive control for the quality of genomic DNA and for the PCR amplification process itself. (3) The PCR product is at least 100 bp and the primer pairs belong to a single (conserved) *ctnna* exonic region.

iCODEHOP (COnsensus-DEgenerate Hybrid Oligonucleotide Primers) software [20] was used to design degenerate PCR primers from protein multiple alignments. One pair of degenerate primers that conformed to all the conditions was identified inside the last (and the longest) coding exon of the *ctnna3* gene: (a3-F) 5'-GGC TGC CAA RAA YYT NAT GAA YGC-3' and (a3-R) 5'- GGC TTC TTT KCN GGN GCY TTC AT-3'. Both primers recognize *ctnna3* sequences, which are highly conserved in different vertebrates (Fig. S3). Moreover, the primers amplify the corresponding genomic fragments of *ctnna1* and *ctnna2* from *X. tropicalis* genomic DNA (Fig. S4). The predicted size of the PCR products obtained with these primers for all known *ctnna* genes is 144 bp. Both primers have degeneracy (number of different nucleotide sequences in the primer pool) of 64.

We used a two-step PCR protocol and a PCR machine with a gradient temperature block option. Annealing temperature was kept 45°C for all samples for the first 5 cycles and then was increased up to 54–65°C for 8 different samples (gradient block) for the last 35 cycles. The PCR reactions were run on a 2% agarose gel and an expected product about 150 bp was observed in the reactions with annealing temperatures during the second step kept from 54.1°C up to 56.3°C. These PCR products were extracted from gel, pooled, and sequenced using the a3-F and a3-R primers.

Results

Phylogenetic Analysis of LRRTM and α-catenin Genes in Vertebrates

The LRRTM family is thought to be vertebrate-specific since clear LRRTM gene homologs were originally identified in several mammalian and teleost fish genomes but not in the fruit fly or nematode genomes [1]. To study the evolution of the LRRTM family in vertebrates, we collected all annotated *lrrtm* and *ctnna* genes, and noted their corresponding genomic structures and locations, from representative model organisms (human, chicken, African clawed frog, coelacanth, zebrafish, and sea lamprey), for which whole genome sequences are available (Fig. 1). Partial *lrrtm* and *ctnna* sequences were also obtained from the elephant shark [21] and spotted gar draft genomes (Tables S1 and S2). No *lrrtm* homologs could be found from the sea squirt (*Ciona intestinalis* and *Ciona savignyi*) or from the lancelet (*Branchiostoma floridae*) genomes. The best hits from these species correspond to Slit-like and other LRR-domain containing proteins as confirmed by reciprocal BLAST search (Table S4).

To correctly identify the subtypes of the new LRRTM protein sequences in the novel species and to provide a relative time point for the divergence of the different subtypes within each family, we aligned the predicted LRRTM sequences (Fig. S1) and generated phylogenetic trees using PhyML (Fig. 2A) and MrBayes (Fig. S2). Orthologs of each LRRTM family member (LRRTM1 to LRRTM4) from different jawed vertebrate species group together forming a clade. Individual family members in the tree are located in general as expected from the known vertebrate phylogeny. Among the four LRRTMs, the highest amino acid sequence identity is seen between LRRTM3 and LRRTM4 proteins in all the analyzed jawed vertebrate species (with average pairwise sequence identities of ~60%, Table S5). Consistent with this, the LRRTM3 and LRRTM4 clades cluster together in the phylogenetic trees. LRRTM2 proteins show higher (47–49%) pairwise sequence identity to LRRTM1 than to LRRTM3 or LRRTM4 proteins (~40%) in all the analyzed species (Table S5). Consistent with this, the LRRTM1 and LRRTM2 clades branch together (Fig. 2A).

In a similar way, we aligned CTNNA proteins and inferred phylogenetic trees (Fig. 2B and Fig. S2). The resulting tree topology has high bootstrap support and, in agreement with a previous study [11], shows that orthologs of CTNNA1 and CTNNA2 from different jawed vertebrates form separate clades that apparently originated by duplication from a common ancestor. The CTNNA3 orthologs from different jawed vertebrates also form a distinct clade that originated before the split of the CTNNA1 and CTNNA2 proteins. However, the CTNNA3 clade has diverged clearly more from the common ancestor than CTNNA1 and CTNNA2 clades.

Structure of *lrrtm/ctnna* Genes in Jawed Vertebrates

In all analyzed jawed vertebrate genomes (except the amphibians, see below), *lrrtm1* and *lrrtm2* are nested in a large intron between conserved coding exons 6 and 7 of α-catenin genes *ctnna2* and *ctnna1*, respectively. Similarly, *lrrtm3* gene resides in a homologous position (inside the large intron between coding exons 6 and 7) of the *ctnna3* gene in all annotated genomes of amniotes (mammals, reptiles, and birds), as well as in the lobe-finned fish coelacanth (*Latimeria chalumnae*) and the ray-finned fish spotted gar (*Lepisosteus oculatus*) genomes (Figs. 1 and 3, and Tables S1 and S2). Clear orthologs of all four *lrrtm* and three *ctnna* genes have also been found in the elephant shark genome. Nested gene structures of *lrrtm2/ctnna1* and *lrrtm3/ctnna3* are annotated, while the expected *lrrtm1/ctnna2* gene structure could not be verified because of the short size of the scaffold_422 which contains *lrrtm1* (esharkgenome.imcb.a-star.edu.sg). In contrast, the genomes of ray-finned fishes (other than the spotted gar, which diverged before the teleost fish-specific whole genome duplication [22]) lack both *lrrtm3* and *ctnna3*. Clear *lrrtm4* orthologs were found in all jawed vertebrate species analyzed. In mammals, *Lrrtm4* is located near the nested *Lrrtm1/Ctnna2* gene structure, whereas in other vertebrates, *lrrtm4* is located in a different chromosome than the *lrrtm1/ctnna2*. In contrast to other jawed vertebrates (shark, coelacanth, and tetrapods), which have a single *lrrtm4* ortholog, the analyzed genomes of ray-finned fishes (other than the spotted gar) contain four *lrrtm4* orthologs located as two closely situated genes in two chromosomes, each pair on a single chromosome being phylogenetically closer to each other (Figs. 1 and 2A, and data not shown).

Analysis of paralogous clusters of genes using the Synteny Database (syntenydb.uoregon.edu/synteny_db/) found suggestive evidence of conserved synteny between human *CTNNA1*, *CTNNA2*, and *CTNNA3* gene regions: a few genes (including *EGR1-4* and *REEP1-4*) that are located near the *CTNNA* genes have four paralogs in the human genome (Fig. S5). This is consistent with the idea that the three nested *lrrtm/ctnna* gene structures may have originated from two rounds of whole genome

	Human	Chicken	Clawed frog	Coelacanth	Zebrafish	Lamprey
Lrrtm1	C2:80.5M<	C4:90.76M>	s1:207.0M>	s126998<	C1:43.1M<	*IrrtmA*
Lrrtm2	C5:138.2M<	C13:2.39M>	**MISSING**	s127522>	C24:35.9M>	s476527:53k>
Lrrtm3	C10:68.7M>	C6:7.97M<	s7:34.2M<	s127465<	**MISSING**	
Lrrtm4	C2:77.0M<	C22:3.33M<	s3:17.8M>	s127611>	C8:x2,C10:x2	*IrrtmB* s476737
Ctnna1	C5:138.0M->	C13:2.34M-<	s3:54.8M->	s127522-<	C24:35.8M-<	
Ctnna2	C2:79.4M->	C4:90.62M-<	s1:206.6M->	s126998-<	C1: 42.6M->	*ctnna*
Ctnna3	C10:67.7M-<	C6:7.78M->	**MISSING**	s127465-<	**MISSING**	s476527:261k-<

Figure 1. List of *lrrtm* and *ctnna* genes and their location in selected vertebrate genomes. The color shading indicates the nested/host gene pairs. Note that the clawed frog genome contains *ctnna1* and *lrrtm3* but lacks the corresponding *lrrtm2* and *ctnna3* orthologs. Both *lrrtm3* and *ctnna3* are absent in zebrafish that has four copies of *lrrtm4* (two adjacent genes in two chromosomes). Lamprey genome has two *lrrtm* genes, one of which (*lrrtmA*) is adjacent to (but not nested in) the single *ctnna* gene. The protein coding region of *lrrtm1* resides within one exon in all vertebrate species analyzed. The other *lrrtm* genes (*lrrtm2, lrrtm3,* and *lrrtm4* and lamprey *lrrtm* genes), have two (or three) protein-coding exons: the first coding exon covers the translation initiation codon and one additional coding nucleotide, while most of the open reading frame is located in the 2nd coding exon. A third coding exon in *lrrtm3* and *lrrtm4* encodes for an alternative C-terminus [35].

duplications in the early vertebrate lineage [23–25]. However, tracing back to these events is difficult. The conserved paralogous genes in human genome (*EGR1-4* and *REEP1-4*) are not immediately adjacent to the *CTNNA* genes and similar regions of conserved synteny (paralogous pairwise clusters) containing *lrrtm/ctnna* were not found in other vertebrate (e.g. chicken or clawed frog) genomes. *Lrrtm4* neighboring genes are not even syntenic between chicken and clawed frog and the *lrrtm4* locus is not assembled in coelacanth genome to allow analysis of synteny.

Lrrtm and ctnna Genes in Lamprey

The genome of sea lamprey (*Petromyzon marinus*), a jawless fish, contains two genes encoding for LRRTMs (annotated in Ensembl as LRRTM3 and LRRTM2, but named here as *lrrtmA* and *lrrtmB*, respectively), of which *lrrtmA* is situated adjacent to, but is not nested in, the single lamprey *ctnna* homolog (Figs. 1 and 3). Both *lrrtmA* and *lrrtmB* possess two protein-coding exons: the first coding exon provides only the first four nucleotides [ATGG] of the open reading frame. The structures of the predicted lamprey LRRTM (PmLRRTM_A and PmLRRTM_B) proteins with 10 LRRs, a single transmembrane domain, and a short cytoplasmic domain (with a C-terminal PDZ binding motif ECEV) are similar to that of mammalian LRRTMs [1,2]. PmLRRTM_A and PmLRRTM_B show higher amino acid sequence identity to LRRTM3 and LRRTM4 (50–55%), than to LRRTM1 and LRRTM2 (40–45%) of other vertebrates (Table 1). In the phylogenic trees (Fig. 2A and Fig. S2) both lamprey LRRTMs branch basal to the LRRTM3-LRRTM4 divergence. Since the *lrrtmA* and *lrrtmB* reside in short scaffolds and many lamprey sequences have unresolved orthologies (possibly due to lineage-specific sequence modifications [26] and independent genome duplications [27]), it is not possible to assign origins to the two lamprey LRRTM sequences by conserved synteny analyses comparing them to other vertebrate genomes. In other words, it remains unclear whether the two sea lamprey LRRTMs originated by an independent duplication after the divergence of lampreys from the vertebrate lineage.

The lamprey *ctnna* gene has a similar structure as other vertebrate α-catenin genes with 17 coding exons, but is much shorter (about 31 kb, compare e.g. to human *CTNNA3* that spans 1.8 Mb). However, the longest intron of lamprey *ctnna* gene (~5.8 kb) is the one between the exons 6 and 7 that hosts *lrrtm* genes in other vertebrates. In the phylogenetic tree, the lamprey α-catenin (PmCTNNA) is basal to the jawed vertebrate branches (Fig. 2B), suggesting that it represents the common ancestor of the tree jawed vertebrate CTNNA subtypes.

Lack of *lrrtm2* and *ctnna3* in Amphibian Genomes

Although the nested *lrrtm/ctnna* gene structure is conserved in most of the analyzed jawed vertebrate species, the genomes of the clawed frogs *Xenopus tropicalis* and *X. laevis* have two notable exceptions. First, the *X. tropicalis* genome [28] lacks an ortholog of *lrrtm2* (Fig. 1). The *X. tropicalis ctnna1*, otherwise similar in structure to α-catenin genes of other jawed vertebrates, is very compact (its length is about 16.5 kb). In particular, the intron between exons 6 and 7 of *X. tropicalis ctnna1* (that would be expected to host *lrrtm2*) is unusually short (434 bp) compared to the corresponding intron of *X. tropicalis ctnna2* (~469 kb) hosting *lrrtm1*, or to the corresponding intron of other jawed vertebrates. Second, an ortholog for *ctnna3* is absent in the *X. tropicalis* genome, although an apparent *lrrtm3* ortholog is present (Fig. 1). Analysis of synteny confirmed that the clawed frog *lrrtm3* is indeed an ortholog of human *LRRTM3* (Fig. 3). Similar to *X. tropicalis*, the draft *X. laevis* genome (xenopus.lab.nig.ac.jp/assembly v7.1 at www.xenbase.org) lacks orthologs of *lrrtm2* and *ctnna3* but contains orthologs for all the other LRRTM and α-catenin genes. We also searched for transcripts corresponding to α-catenins in the extensive *X. tropicalis* and *X. laevis* mRNA databases (www.xenbase.org). While multiple hits are present for *ctnna1* (XB-GENEPAGE-479598) and *ctnna2* (XB-GENEPAGE-5955200), no *ctnna3* mRNAs were found by reciprocal BLAST searches. Similarly, the recently available salamander (axolotl, *Ambystoma mexicanum*) transcriptome (www.ambystoma.org) lacks orthologs of both *lrrtm2* and *ctnna3*, while clear transcripts of all the other LRRTM and α-catenin genes are present.

Experimental Support that the *X. tropicalis* Genome Lacks *ctnna3*

The apparent lack of *ctnna3* in the current amphibian genomes and transcriptomes suggests loss of the *ctnna3* gene in the amphibian lineage during evolution. To obtain further support for this, we carried out polymerase chain reaction (PCR) with degenerate *ctnna3* primers (a3-F and a3-R) designed to amplify *ctnna* sequences from various species (Fig. S3). As a positive control for our strategy, we first used these degenerate primers to amplify corresponding *ctnna* fragments from the chicken genome (Fig. 4A). The primers have no mismatches with chicken *ctnna3*, but have one mismatch with a corresponding region of *ctnna1* and two mismatches with *ctnna2* (Fig. S3). Thus, the primers are expected to primarily amplify *ctnna3*, but may also amplify *ctnna1* though with a lower efficiency. Consistent with this, most of the amplified

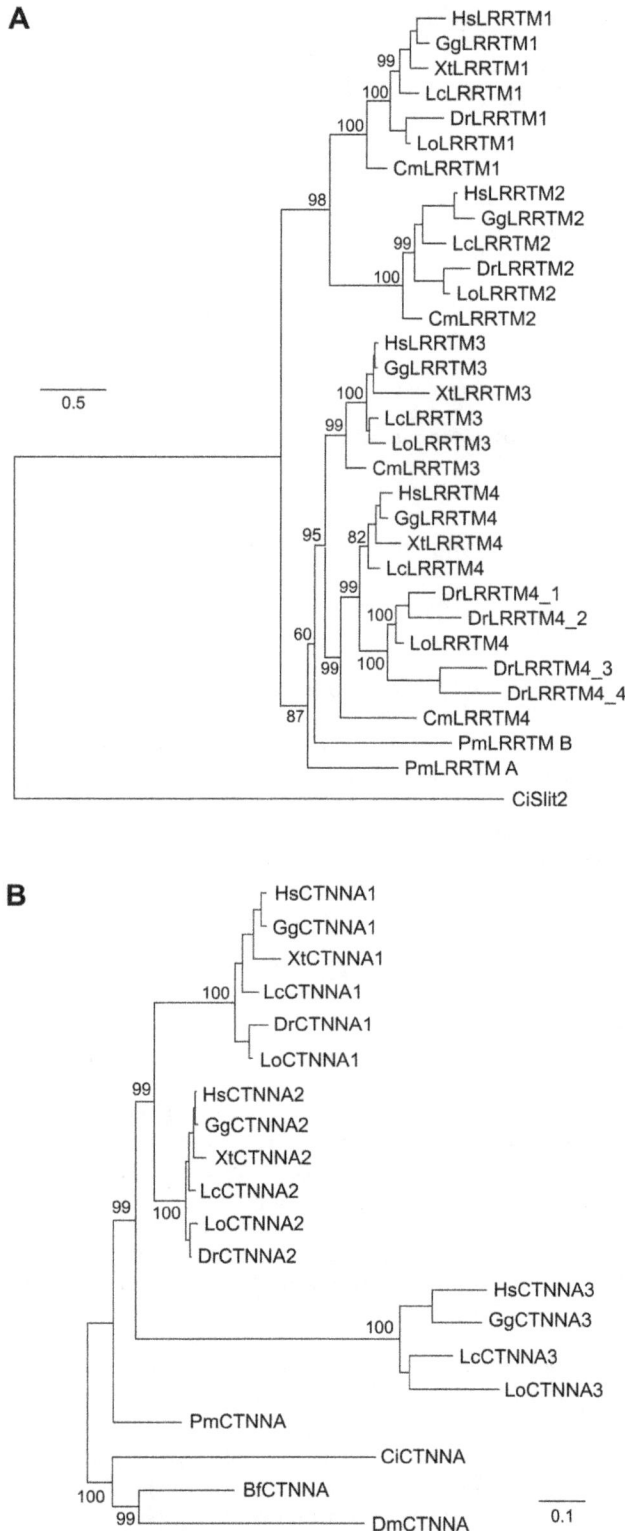

Figure 2. Phylogenetic trees of LRRTM and α-catenin proteins in selected vertebrates. (A) A LRRTM family tree was generated using the alignment shown in Fig. S1 and the maximum likelihood method. A tunicate LRR-domain protein CiSlit2 (one of the best BLAST hits shown in Table S4) is used to root the tree. Notice the absence of XtLRRTM2 and the divergence of XtLRRTM3 from the other vertebrate LRRTM3s. Numbers at each branch point represent bootstrap support for that branch. Bootstrap values of short terminal branches (all >90%) are omitted for clarity. The branch lengths are proportional to the expected

proportion of amino acid sequence divergence (= number of residue substitution) between groups. (B) A maximum likelihood phylogenetic tree of α-catenin proteins. Invertebrate (fruit fly, sea squirt, and lancelet) CTNNAs are included as outgroup. Note that the CTNNA3 clade has diverged more rapidly than the CTNNA1 and CTNNA2 clades during vertebrate evolution. Bf = *Branchiostoma floridae* (lancelet), Ci = *Ciona intestinalis* (sea squirt), Cm = *Callorhinchus milii* (elephant shark), Dm = *Drosophilia melanogaster* (fruit fly), Dr = *Danio rerio* (zebrafish), Gg = *Gallus gallus* (chicken), Hs = *Homo sapiens*, Lc = *Latimeria chalumnae* (coelacanth), Lo = *Lepisosteus oculatus* (spotted gar), Pm = *Petromyzon marinus* (sea lamprey), Xt = *Xenopus tropicalis* (African clawed frog).

product from chicken genomic DNA corresponded to *ctnna3*, but a minor part corresponded to *ctnna1* (Fig. 4A).

The same PCR protocol was then applied to the *X. tropicalis* genomic DNA. Sequencing of the PCR fragment revealed spectra matching only to *X. tropicalis ctnna1* and *ctnna2* (Fig. 4B). Digestion with diagnostic *Hind*III and *Stu*I restriction enzymes confirmed that no other PCR products except for the frog *ctnna1* and *ctnna2* were amplified (Fig. 4C, D).

Analysis of Selected Different-strand Nested Genes in Vertebrates

A previous study of nested genes did not report any phylogenetically old protein-coding nested gene structure that would have been lost in vertebrates [13]. To reassess whether protein-coding nested gene structures are conserved in vertebrates, we identified orthologs for the previously identified human different strand nested genes [12] in zebrafish, coelacanth, clawed frog, and chicken genomes. We included in our analysis only those gene pairs that are conserved in human and mouse, and in which a protein-coding nested gene is flanked by protein-coding exons of the host gene (see Table S3). Most of the analyzed mammalian protein-coding nested gene structures (63/91) have orthologs in the zebrafish and/or coelacanth genomes, but several of these (19/63) cannot be found in the clawed frog or chicken genomes (Table 1 and Table S3). Notably, in majority of these (12/19), the nested gene encodes for an LRR-superfamily protein.

Evolution of Neurexin Alternative Splice Segment

The alternatively spliced segment (AS4) of neurexin protein comprises a loop structure in the binding domain, and deletion of this loop structure (by exon skipping) is required for LRRTM binding in mammals [3–6]. We used BLASTP search to investigate whether an exon homologous to the AS4 exon was present in *nrxn* genes of other species than the jawed vertebrates. Amino acid alignment shows that the fruit fly and sea urchin neurexin proteins lack exactly the region that is homologous to the AS4 amino acid sequence of vertebrate neurexins (Fig. S6). Moreover, in fruit fly and sea urchin *nrxn* genes, the intron between the exons that are homologous to vertebrate AS4-flanking exons is very short. In contrast, an exon homologous to the AS4-exon of mouse *Nrxn1* gene is present in the sea squirt and lamprey *nrxn* genes (Fig. S6). This suggests that the *nrxn* AS4 exon sequence appeared early in chordate evolution.

To further analyze the expression of *nrxn* isoforms lacking AS4 during evolution, we searched the available vertebrate EST databases for *nrxn* transcripts with deletion of the AS4 sequence (Table S6). As in mammals, zebrafish *nrxn* pre-mRNAs are known to undergo alternative splicing, including exon AS4 skipping [15]. We found 9 out of 23 (39%) *nrxn* ESTs that lack AS4 in human, 5 out of 27 (19%) in mouse, 1 out of 9 (11%) in clawed frog, and 1 out of 6 (17%) hits in zebrafish, confirming that this splice variant is expressed throughout the jawed vertebrate class. However, no

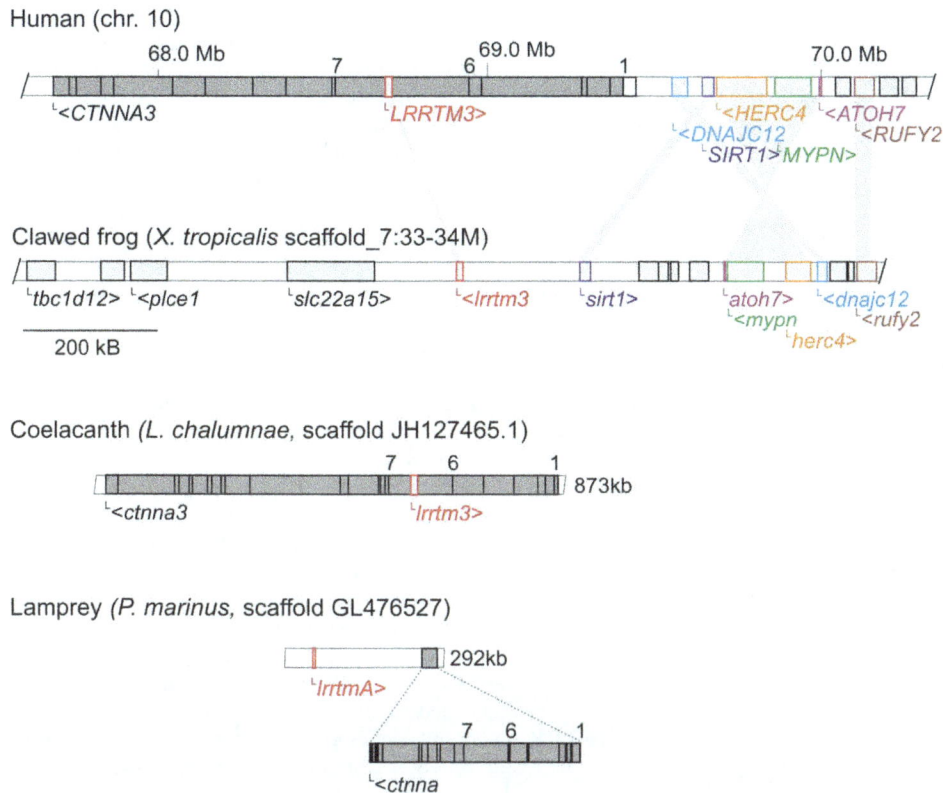

Figure 3. Synteny analysis of the *lrrtm3/ctnna3* locus between the human and the clawed frog genomes. Data is retrieved from Ensembl release 72, Jun 2013 and from Xentr. 7.1. Several genes in *X. tropicalis* scaffold_7 (colored) on one side of *lrrtm3* are orthologs of genes near the human *LRRTM3/CTNNA3* locus. Another cluster of genes (including *tbc1d12* and *plce1*) on the other side of *lrrtm3* is syntenic to another region of human chromosome 10 (~96.1 Mb). However, no sequences orthogous to human *CTNNA3* exons were found in this scaffold. Genomic fragments from coelacanth, which contains a nested *lrrtm3/ctnna3* gene structure, and from lamprey, which contains adjacent *ctnna* and *lrrtmA* genes, are shown for comparison.

nrxn EST transcripts (with or without the AS4 sequence) were found in lamprey, and therefore it remains unclear whether the alternative splicing of *nrxn* was present in jawless vertebrates. In the sea squirt (*C. intestinalis*), none of the few *nrxn* ESTs revealed the AS4 exon deletion according to our selection criteria (Table S6).

Recently it has been reported that alternative splicing of *nrxn* AS4 exon is regulated by RNA binding proteins of KHDBRS family (T-STAR and SAM68) [29,30], which are known to recognize specific tandem repeats of UAAA/UUAA sequences in the introns adjacent to the AS4 exon [29,31]. We therefore analyzed intronic sequences surrounding the *nrxn* AS4 exon in sea squirt (*C. intestinalis*) and sea lamprey for the presence of these repeats. In sea squirt *nrxn*, the introns surrounding the AS4-like exon are short (441 and 430 bp compared to 13620 and 1598 bp in lamprey) and contain only one UWAA (W = U/A) repeat in the upstream and three of them in downstream introns. In contrast, markedly more of the UWAA repeats can be found in both upstream and downstream proximal regions of the large introns surrounding the "AS4-exon" in two out of three lamprey *nrxn* genes (Fig. S7).

Discussion

We show here that the nested *lrrtm/ctnna* gene structure was established in early jawed vertebrates and that a conserved structure of three nested *lrrtm/ctnna* pairs is present in lobe-finned fish (and presumably also in cartilaginous fish) as in amniotes

(Fig. 5A). Based on the available data, we propose a hypothetical sequence of events to explain the evolution of the nested *lrrtm/ctnna* genes (Fig. 5B).

The available genomes of invertebrates (including uro- and cephalochordates) lack clear homologs of *lrrtm*, whereas the jawless vertebrate lamprey has two copies of *lrrtm*. Thus, the first *lrrtm* presumably emerged (by exon shuffling of pre-existing genes containing extracellular LRR, transmembrane, and intracellular domains) in the early vertebrate ancestor. The α-catenin (*ctnna*) gene emerged early in metazoan evolution, presumably from a vinculin-like gene [11]. The first nested *lrrtm/ctnna* gene structure arose presumably by retrotransposition of *lrrtm* into the large intron of the nearby *ctnna* gene.

Since the nested *lrrtm1/ctnna2*, *lrrtm2/ctnna1*, and *lrrtm3/ctnna3* gene structures are similar in all jawed vertebrates (except for amphibians), they share a single ancestral nested *lrrtm/ctnna* gene structure that was duplicated twice presumably as a part of the two-round whole genome duplications (2R-WGD) at the origin of vertebrates [23–25,27]. The two duplications resulted in four nested *lrrtm/ctnna* pairs followed by a loss of *ctnna* around *lrrtm4*. Presumably, one of the two *lrrtm/ctnna*-ancestral gene structures that emerged from the first *lrrtm/ctnna* duplication diverged to become *lrrtm[1/2]/ctnna[1/2]*-ancestral gene structure and was duplicated again, which resulted in the nested *lrrtm1/ctnna2* and *lrrtm2/ctnna1* gene structures. The other nested gene pair diverged to become *lrrtm[3/4]/ctnna[3/4]*-ancestral gene structure that was duplicated to become *lrrtm3/ctnna3* and *lrrtm4/ctnna4*. The putative

Table 1. Nested gene structures lost in clawed frog or chicken genomes.

Host gene	Nested gene	Explanation
ASTN2	TRIM32	nested absent in frog
CACNA2D3	LRTM1	both absent in frog
CASK	GPR82	nested absent in frog
CENPP	ECM2	host gene absent in frog
	ASPN	host gene absent in frog
	OMD	both absent in frog
	OGN	host gene absent in frog
CTNNA1	LRRTM2	nested absent in frog
CTNNA3	LRRTM3	host gene absent in frog
FBXL13	LRRC17	nested absent in frog
FYCO1	CXCR6	host absent in frog, nested absent in chicken
IMMP2L	LRRN3	host gene absent in frog
MED12L	P2RY13	host gene absent in frog
	P2RY12	host gene absent in frog
PC	LRFN4	both absent in chicken
RNF123	AMIGO3	nested absent in frog
SND1	LRRC4	host absent in chicken
SYN1	TIMP1	nested absent in frog
TFB1M	CLDN20	nested absent in frog

The table lists human protein-coding different strand nested gene structures that are also found in coelacanth and/or zebrafish but are absent in clawed frog (*X. tropicalis*) or chicken genomes. The genes were selected (as described in the Methods and Table S3) from a previously published list of human nested genes [12]. The missing host or nested genes are marked in bold. Nested genes that encode LRR-superfamily proteins are underlined.

ctnna4 gene was then lost (Fig. 5B). Consistent with this model, LRRTM1 and LRRTM2 amino acid sequences are more closely related to each other than to LRRTM3 or LRRTM4, LRRTM3 shows the highest amino acid sequence identity to LRRTM4, and CTNNA1 and CTNNA2 amino acid sequences are more close to each other than to CTNNA3.

Based on the lack of *ctnna3* orthologs in most teleost fish and clawed frog genomes, previous studies have suggested that the α3-catenin would be amniote-specific [11,32]. However, clear *ctnna3* (and *lrtm3*) orthologs are present in the elephant shark, spotted gar, and coelacanth genomes. Thus, the *lrtm3/ctnna3* locus was lost in the teleost fish lineage after the divergence of the spotted gar but before the teleost-specific whole genome duplication [33,34]. In a separate event, the *ctnna3* (but not *lrtm3*) gene was also lost in the early amphibian lineage. The assumption that the amphibians lack *ctnna3* is supported by the lack of sequences corresponding to *ctnna3* in the *X. tropicalis*, *X. laevis*, and *A. mexicanum* databases, our synteny analysis (Fig. 3), and PCR data (Fig. 4). Most likely *ctnna3* became non-functional by accumulating mutations and therefore unrecognizable, while *lrtm3* remained intact. Compared to other α-catenins, the amino acid sequence of α3-catenin has diverged rapidly during vertebrate evolution (as is evident from the branch lengths in the phylogenetic tree). The expression of α3-catenin was probably initially widespread but became more restricted during subsequent vertebrate evolution. Consistent with this, mouse α3-catenin regulates the hybrid adhering junctions in the intercalated disks of the heart, which are unique to amniote vertebrates [32].

The loss of *lrtm2* in the amphibian lineage may have occurred concomitant with (or before) the deletion of most parts of the large *ctnna1* intron. Loss of one LRRTM family member is not critical for survival in mice under laboratory conditions [2,10,35,36]. We suggest that LRRTM1, which has an overlapping expression and synaptic function with LRRTM2 [1–6], was likely able to compensate, at least partially, for the lack of LRRTM2 in amphibians.

The precursor of *lrtm4*, which is not nested in *ctnna* of any vertebrates, may have been initially nested in a *ctnna* that became inactive and was lost in the early vertebrates (Fig. 5B). In the teleost fish lineage, a local (probably a head-to-tail tandem) duplication followed presumably by the teleost-specific whole genome duplication [33,34] resulted in four *lrtm4* orthologs. As LRRTM3 and LRRTM4 proteins show highest amino acid similarity to each other, the extra copies of *lrtm4* gene may have taken over the lack of *lrtm3* in teleosts.

No conserved protein-coding nested gene structures were reported in a previous study to have been lost in vertebrates [13]. Therefore, it was rather unexpected that our bioinformatic analysis of 91 protein-coding different strand nested genes conserved between mouse and human [12] (see Table S3) identified 19 protein-coding nested gene structures present also in zebrafish and/or coelacanth but lost in the clawed frog or chicken genomes (Table 1). Interestingly, in 12 out of 19 cases the nested genes encoded LRR superfamily proteins. Therefore, nested LRR superfamily genes may have remained mobile during vertebrate evolution consistent with the idea that many of them have presumably derived *via* retrotransposons [13].

In mice, LRRTMs have been reported to bind specifically those neurexin isoforms that lack the alternatively spliced segment 4 (AS4) [5,6]. *Nrxn* gene structure, including the AS4 exon, is conserved in jawed vertebrates, and *nrxn1-3* transcripts lacking this segment are expressed in zebrafish [14,15]. Hence, the alternative splicing mechanism to skip *nrxn* AS4 exon had apparently evolved already prior to the *nrxn* gene duplications in early vertebrates. The corresponding AS4 exon is also present in the lamprey and sea squirt *nrxn* gene orthologs. However, *nrxn* gene orthologs in the fruit fly and nematode, as well as in the urochordate sea urchin, lack the sequence corresponding to the AS4 exon. This indicates that the *nrxn* AS4 exon emerged *de novo* in evolution of the chordate lineage. Recently, cerebellin (Cbln) family proteins were identified as novel neurexin ligands that may directly bind the AS4 loop [37,38]. Interestingly, putative cerebellin gene orthologs are annotated in vertebrates, as well as in the sea squirt (*C. intestinalis*) but not in the fruit fly or nematode genomes (www.ensembl.org/ Homo_sapiens/Gene/Compara_Tree?db = core;g = ENSG 00000102924). We speculate that the *nrxn* AS4 exon appeared *de novo* at the same time as the gene for its new binding partner cerebellin emerged (by duplication of a related C1q/TNF-superfamily gene) in early chordates.

Recent studies have identified RNA binding proteins of KHDBRS family as key regulators of *neurexin* AS4 exon splicing in mice [29,30]. Multiple AU-rich sequence elements in introns preceding and following AS4 exon act as the response elements including UWAA-rich regions closely downstream of AS4 that are conserved in jawed vertebrates [29,30]. Similar UWAA-rich regions are conserved also in lamprey *neurexin* genes (Fig. S7), and the lamprey genome is known to contain KHDBRS protein orthologs [29]. In contrast, the adjacent short introns in sea squirt *neurexin* contain few UWAA motifs, and all identified *neurexin* EST transcripts from sea squirt retain the AS4 exon sequence, suggesting that the AS4 exon is not skipped in this species. Although additional studies are needed to confirm that the *neurexin*

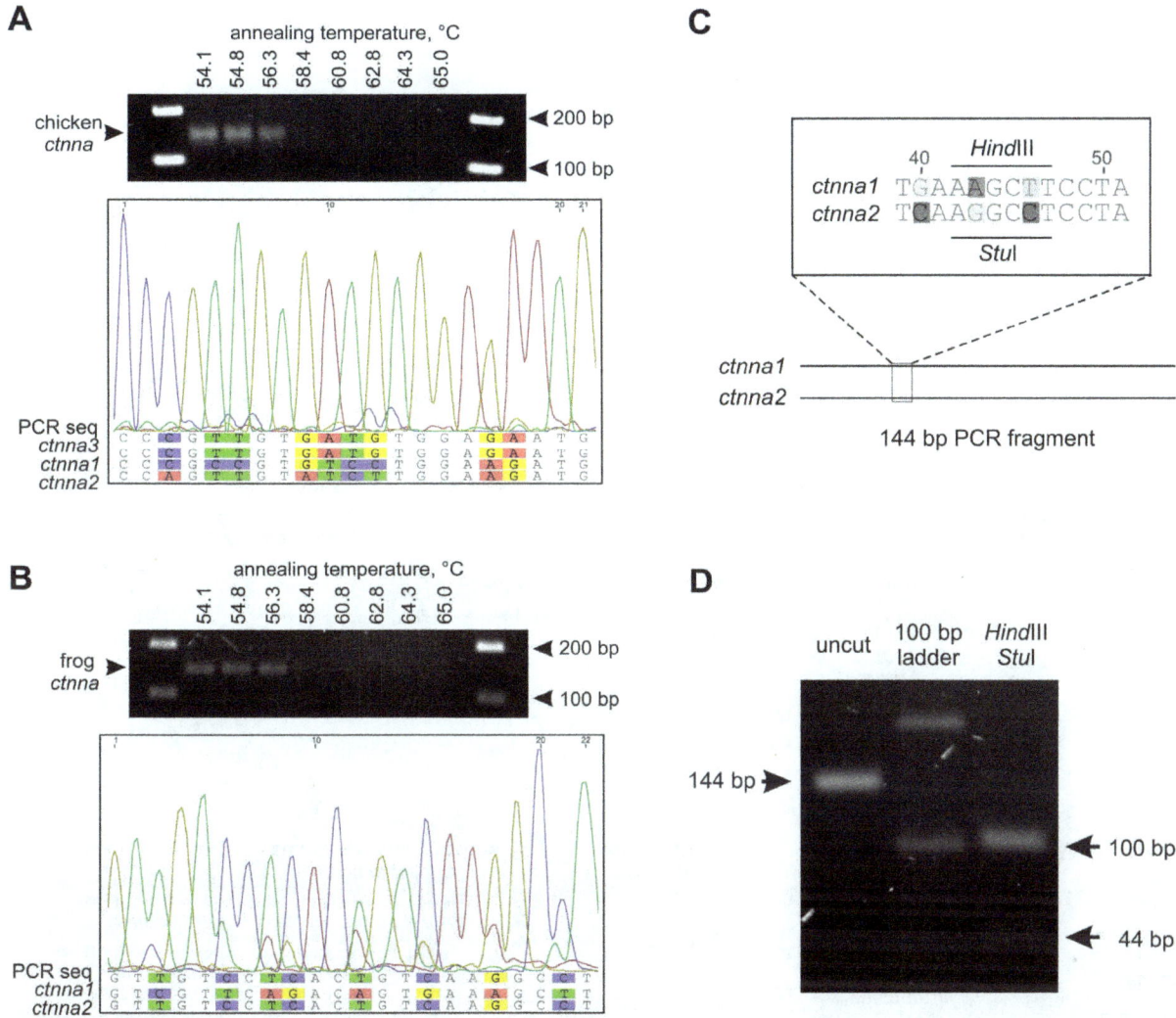

Figure 4. Experimental evidence that the clawed frog genome does not contain a *ctnna*3 ortholog. (A) The degenerate primers a3-F and a3-R (corresponding to the conserved last coding exon of *ctnna3* in vertebrates, see Fig. S3) were used to amplify corresponding fragments of the *ctnna* genes from the chicken genomic DNA. PCR product of the predicted size (about 150 bp) was observed using annealing temperatures from 54.1°C to 56.3°C. Sequencing of the PCR fragment (with the same primers) revealed spectra corresponding mainly to chicken ctnna3. Minor peaks corresponding to chicken ctnna1 PCR product are slightly shifted to the right. (B) The same primers were used to amplify corresponding fragments of the *ctnna* genes from the clawed frog genomic DNA. See also Fig. S4. Sequencing of the PCR fragment revealed spectra corresponding to the frog *ctnna1* and *ctnna2* genes only. Shown is a part of the sequence spectrum obtained with a3-F primer. (C) Schematic drawing of the experimental strategy. A PCR of *X. tropicalis* genomic DNA using degenerate *ctnna3* primers is expected to amplify 144 bp fragments of frog *ctnna1* and *ctnna2* that contain *Hind*III and *Stu*I restriction enzyme sites, respectively. (D) Arrow on the left points at the 144 bp PCR product obtained from the frog genomic DNA. Sequencing of this band is shown in B. Arrows on the right indicate the diagnostic *Hind*III/*Stu*I fragments of the PCR product verifying that the product is solely composed of the predicted *ctnna1* and *ctnna2* fragments.

transcripts lacking AS4 are expressed in lamprey, the present evidence suggests that the LRRTMs and the mechanism of alternative splicing that enabled LRRTM binding to neurexins probably both emerged in early vertebrate evolution, before the divergence of jawed vertebrates.

Several synaptic adhesion molecules, such as neurexins, can be found in less complex metazoan organisms with a simple nervous system. However, the number of genes encoding synaptic adhesion proteins, along with other synaptic components, increased dramatically during the evolution of vertebrates [39]. LRRTMs represent an example of such adhesion proteins that are required to fine tune the formation and maintenance of synapses in the vertebrate brain, while simultaneous diversification of neurexin splice variants contributed towards the same task [7].

Recently it has been found that LRRTM4 and possibly LRRTM3 (but not LRRTM1 or LRRTM2) bind heparan sulphate proteoglycan (HSPG) glypican as a presynaptic ligand [9,10]. We looked at the properties of the vertebrate LRRTM proteins to see if there would be any clues to how the proteins might differ, and when this function might have appeared. We noticed a correlation in total positive charge and the reported HSPG binding function in the LRRTM family: The calculated pI values are higher for the LRR-domains of mouse LRRTM3 and LRRTM4 (pI values of 9.3 and 9.4) than for mouse LRRTM1 and LRRTM2 (pI values of 6.9 and 8.1), resulting in substantial positive charge of LRRTM3 and LRRTM4 that is typical for heparin binding proteins. Similar situation is observed in case of the frog and zebrafish LRRTMs (Table S1). Interestingly, both of

A

B

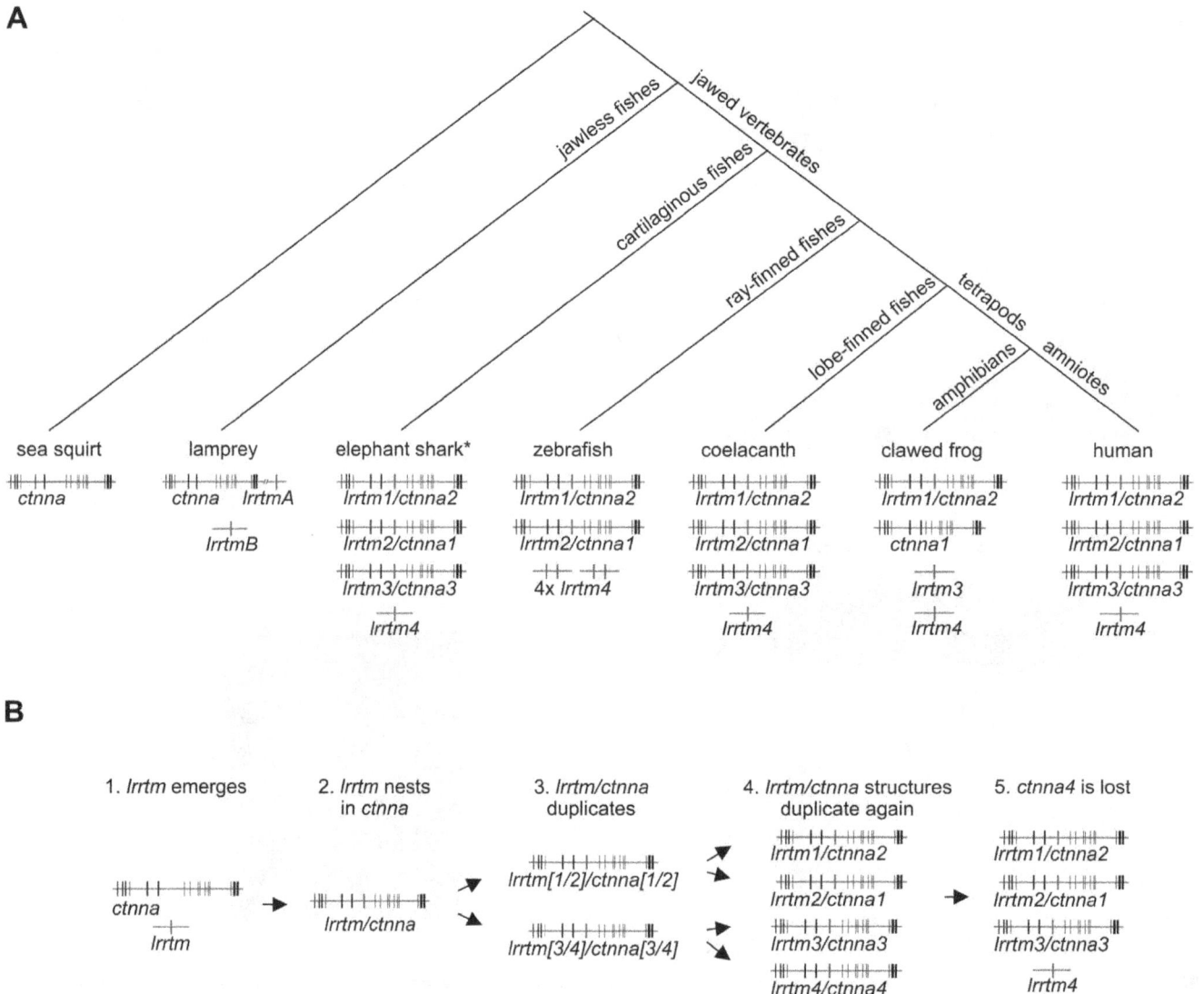

Figure 5. Evolution of LRRTM and α-catenin genes in vertebrates. (A) Structures of the identified *lrrtm* (red) and *ctnna* (black) genes in representative species are shown schematically below a tree of major vertebrate lineages. Intron sizes of individual *ctnna* genes are not in scale. Note that the clawed frog lacks both *lrrtm2* and *ctnna3* orthologs but has the corresponding host (*ctnna2*) and nested (*lrrtm3*) genes. Ray-finned fishes lack *lrrtm3/ctnna3* but have four copies of *lrrtm4*. (B) Hypothetical model of the nested *lrrtm/ctnna* gene structure evolution. (1) The first *lrrtm* gene emerged in the early jawless vertebrates, possibly by exon shuffling. (2) The *lrrtm* gene was translocated into a *ctnna* intron, presumably by retrotransposition (and thereby became intronless). (3–4) The nested *lrrtm/ctnna* gene structure was duplicated twice presumably as part of the two rounds of whole genome duplications that occurred at the base of vertebrates. (5) Loss of one *ctnna* host resulted in *lrrtm4* that is not nested and three nested *lrrtm/ctnna* genes present in the majority of extant jawed vertebrates.

the LRRTMs present in lamprey are highly positively charged (pI values of 9.55 and 9.3). Thus, it seems possible that the HSPG-binding function of LRRTMs might have been present prior to their neurexin binding and then later the HSPG-binding might have been lost in the evolution of LRRTM1 and LRRTM2, which specialized to bind only neurexins. Additional experiments comparing the binding of lamprey LRRTMs to neurexins versus HSPGs are necessary to test this hypothesis.

Conclusions

Our study provides a plausible scenario on how the LRRTMs emerged as new binding partners of neurexins. We show that *lrrtm* became nested in α-catenin gene in the early jawed vertebrates

followed by gene duplications that resulted in three nested *lrrtm/ctnna* gene structures in most vertebrates. The clawed frog genome contains a clear *lrrtm3* ortholog but lacks the corresponding host (*ctnna3*) gene. We identified several other protein-coding nested gene structures that are conserved in jawed vertebrates but either the host or the nested gene is missing in the frog or chicken lineages. Interestingly, majority of these nested genes comprise LRR domains.

Supporting Information

Figure S1   Alignment of LRRTM family protein sequences from selected vertebrates.

Figure S2 Bayesian phylogenetic trees of LRRTM and α-catenin proteins.

Figure S3 Alignment shows that the degenerate *ctnna3* forward (a3-F) and reverse (a3-R) primers have no mismatches with chicken *ctnna3* genomic sequence.

Figure S4 Sequence alignment of the predicted 144 bp *ctnna1* and *ctnna2* PCR fragments.

Figure S5 Paralogous clusters containing α-catenin genes in human genome.

Figure S6 Presence of *neurexin* AS4 exon in selected animal species.

Figure S7 Comparison of UWAA motifs (arrows) within 200 nucleotides (A) upstream and (B) downstream of exon AS4 in human, lamprey and sea squirt *neurexin* genes.

Table S1 Accession numbers, curated LRRTM amino acid sequences and pI values.

Table S2 Accession numbers for the α-catenin sequences used.

Table S3 Conservation of selected human different strand nested gene structures in vertebrate genomes.

Table S4 BLAST analysis of tunicate and lancelet genomes using lamprey LRRTM sequences as query and reciprocal BLAST analysis against vertebrate genomes.

Table S5 Amino acid sequence identity (%) between vertebrate LRRTMs.

Table S6 Alternative splicing of *neurexin* AS4 exon in selected species.

Author Contributions

Conceived and designed the experiments: PU TK MSA. Performed the experiments: PU TK MSA. Analyzed the data: PU TK MSA. Contributed reagents/materials/analysis tools: PU TK MSA. Wrote the paper: PU TK MSA.

References

1. Lauren J, Airaksinen MS, Saarma M, Timmusk TT (2003) A novel gene family encoding leucine-rich repeat transmembrane proteins differentially expressed in the nervous system. Genomics 81: 411–421.
2. Linhoff MW, Lauren J, Cassidy RM, Dobie FA, Takahashi H, et al. (2009) An unbiased expression screen for synaptogenic proteins identifies the LRRTM protein family as synaptic organizers. Neuron 61: 734–749.
3. Ko J, Fuccillo MV, Malenka RC, Sudhof TC (2009) LRRTM2 functions as a neurexin ligand in promoting excitatory synapse formation. Neuron 64: 791–798. 10.1016/j.neuron.2009.12.012.
4. Soler-Llavina GJ, Fuccillo MV, Ko J, Sudhof TC, Malenka RC (2011) The neurexin ligands, neuroligins and leucine-rich repeat transmembrane proteins, perform convergent and divergent synaptic functions in vivo. Proc Natl Acad Sci U S A 108: 16502–16509. 10.1073/pnas.1114028108.
5. Siddiqui TJ, Pancaroglu R, Kang Y, Rooyakkers A, Craig AM (2010) LRRTMs and neuroligins bind neurexins with a differential code to cooperate in glutamate synapse development. J Neurosci 30: 7495–7506. 10.1523/JNEUROSCI.0470-10.2010.
6. de Wit J, Sylwestrak E, O'Sullivan ML, Otto S, Tiglio K, et al. (2009) LRRTM2 interacts with Neurexin1 and regulates excitatory synapse formation. Neuron 64: 799–806. 10.1016/j.neuron.2009.12.019.
7. Krueger DD, Tuffy LP, Papadopoulos T, Brose N (2012) The role of neurexins and neuroligins in the formation, maturation, and function of vertebrate synapses. Curr Opin Neurobiol 22: 412–422. 10.1016/j.conb.2012.02.012.
8. Francks C, Maegawa S, Lauren J, Abrahams BS, Velayos-Baeza A, et al. (2007) LRRTM1 on chromosome 2p12 is a maternally suppressed gene that is associated paternally with handedness and schizophrenia. Mol Psychiatry 12: 1129–39, 1057.
9. de Wit J, O'Sullivan ML, Savas JN, Condomitti G, Caccese MC, et al. (2013) Unbiased discovery of glypican as a receptor for LRRTM4 in regulating excitatory synapse development. Neuron. 10.1016/j.neuron.2013.06.049.
10. Siddiqui TJ, Tari PK, Connor SA, Zhang P, Dobie FA, et al. (2013) An LRRTM4-HSPG complex mediates excitatory synapse development on dentate gyrus granule cells. Neuron. 10.1016/j.neuron.2013.06.029.
11. Zhao ZM, Reynolds AB, Gaucher EA. (2011) The evolutionary history of the catenin gene family during metazoan evolution. BMC Evol Biol 11: 198–2148-11–198. 10.1186/1471-2148-11-198.
12. Ho MR, Tsai KW, Lin WC (2012) A unified framework of overlapping genes: Towards the origination and endogenic regulation. Genomics 100: 231–239. 10.1016/j.ygeno.2012.06.011.
13. Assis R, Kondrashov AS, Koonin EV, Kondrashov FA (2008) Nested genes and increasing organizational complexity of metazoan genomes. Trends Genet 24: 475–478. 10.1016/j.tig.2008.08.003.
14. Tabuchi K, Sudhof TC (2002) Structure and evolution of neurexin genes: Insight into the mechanism of alternative splicing. Genomics 79: 849–859. 10.1006/geno.2002.6780.
15. Rissone A, Monopoli M, Beltrame M, Bussolino F, Cotelli F, et al. (2007) Comparative genome analysis of the neurexin gene family during danio rerio: Insights into their functions and evolution. Mol Biol Evol 24: 236–252. 10.1093/molbev/msl147.
16. Catchen JM, Conery JS, Postlethwait JH (2009) Automated identification of conserved synteny after whole-genome duplication. Genome Res 19: 1497–1505. 10.1101/gr.090480.108.
17. Katoh K, Standley DM (2013) MAFFT multiple sequence alignment software version 7: improvements in performance and usability. Mol Biol Evol 30: 772–780. 10.1093/molbev/mst010.
18. Guindon S, Dufayard JF, Lefort V, Anisimova M, Hordijk W, et al. (2010) New algorithms and methods to estimate maximum-likelihood phylogenies: Assessing the performance of PhyML 3.0. Syst Biol 59: 307–321. 10.1093/sysbio/syq010.
19. Tamura K, Peterson D, Peterson N, Stecher G, Nei M, et al. (2011) MEGA5: Molecular evolutionary genetics analysis using maximum likelihood, evolutionary distance, and maximum parsimony methods. Mol Biol Evol 28: 2731–2739. 10.1093/molbev/msr121.
20. Boyce R, Chilana P, Rose TM (2009) iCODEHOP: A new interactive program for designing COnsensus-DEgenerate hybrid oligonucleotide primers from multiply aligned protein sequences. Nucleic Acids Res 37: W222–8. 10.1093/nar/gkp379.
21. Venkatesh B, Kirkness EF, Loh YH, Halpern AL, Lee AP, et al. (2007) Survey sequencing and comparative analysis of the elephant shark (*Callorhinchus milii*) genome. PLoS Biol 5: e101. 10.1371/journal.pbio.0050101.
22. Amores A, Catchen J, Ferrara A, Fontenot Q, Postlethwait JH (2011) Genome evolution and meiotic maps by massively parallel DNA sequencing: Spotted gar, an outgroup for the teleost genome duplication. Genetics 188: 799–808. 10.1534/genetics.111.127324.
23. Smith JJ, Kuraku S, Holt C, Sauka-Spengler T, Jiang N, et al. (2013) Sequencing of the sea lamprey (*Petromyzon marinus*) genome provides insights into vertebrate evolution. Nat Genet 45: 415–21, 421e1-2. 10.1038/ng.2568.
24. Dehal P, Boore JL (2005) Two rounds of whole genome duplication in the ancestral vertebrate. PLoS Biol 3: e314. 10.1371/journal.pbio.0030314.
25. Nakatani Y, Takeda H, Kohara Y, Morishita S (2007) Reconstruction of the vertebrate ancestral genome reveals dynamic genome reorganization in early vertebrates. Genome Res 17: 1254–1265. 10.1101/gr.6316407.
26. Qiu H, Hildebrand F, Kuraku S, Meyer A (2011) Unresolved orthology and peculiar coding sequence properties of lamprey genes: The KCNA gene family as test case. BMC Genomics 12: 325–2164-12–325. 10.1186/1471-2164-12-325.
27. Mehta TK, Ravi V, Yamasaki S, Lee AP, Lian MM, et al. (2013) Evidence for at least six hox clusters in the japanese lamprey (lethenteron japonicum). Proc Natl Acad Sci U S A 110: 16044–16049. 10.1073/pnas.1315760110.
28. Hellsten U, Harland RM, Gilchrist MJ, Hendrix D, Jurka J, et al. (2010) The genome of the western clawed frog *Xenopus tropicalis*. Science 328: 633–636. 10.1126/science.1183670.
29. Ehrmann I, Dalgliesh C, Liu Y, Danilenko M, Crosier M, et al. (2013) The tissue-specific RNA binding protein T-STAR controls regional splicing patterns of neurexin pre-mRNAs in the brain. PLoS Genet 9: e1003474. 10.1371/journal.pgen.1003474.

30. Iijima T, Wu K, Witte H, Hanno-Iijima Y, Glatter T, et al. (2011) SAM68 regulates neuronal activity-dependent alternative splicing of neurexin-1. Cell 147: 1601–1614. 10.1016/j.cell.2011.11.028.

31. Galarneau A, Richard S (2009) The STAR RNA binding proteins GLD-1, QKI, SAM68 and SLM-2 bind bipartite RNA motifs. BMC Mol Biol 10: 47–2199–10–47. 10.1186/1471-2199-10-47.

32. Li J, Goossens S, van Hengel J, Gao E, Cheng L, et al. (2012) Loss of alphaT-catenin alters the hybrid adhering junctions in the heart and leads to dilated cardiomyopathy and ventricular arrhythmia following acute ischemia. J Cell Sci 125: 1058–1067. 10.1242/jcs.098640.

33. Christoffels A, Koh EG, Chia JM, Brenner S, Aparicio S, et al. (2004) Fugu genome analysis provides evidence for a whole-genome duplication early during the evolution of ray-finned fishes. Mol Biol Evol 21: 1146–1151. 10.1093/molbev/msh114.

34. Meyer A, Schartl M (1999) Gene and genome duplications in vertebrates: The one-to-four (-to-eight in fish) rule and the evolution of novel gene functions. Curr Opin Cell Biol 11: 699–704.

35. Laakso T, Muggalla P, Kysenius K, Lauren J, Paatero A, et al. (2012) LRRTM3 is dispensable for amyloid-beta production in mice. J Alzheimers Dis 31: 759–764. 10.3233/JAD-2012-120193.

36. Voikar V, Kulesskaya N, Laakso T, Lauren J, Strittmatter SM, et al. (2013) LRRTM1-deficient mice show a rare phenotype of avoiding small enclosures - a tentative mouse model for claustrophobia-like behaviour. Behav Brain Res 238: 69–78. 10.1016/j.bbr.2012.10.013.

37. Uemura T, Lee SJ, Yasumura M, Takeuchi T, Yoshida T, et al. (2010) Trans-synaptic interaction of GluRdelta2 and neurexin through Cbln1 mediates synapse formation in the cerebellum. Cell 141: 1068–1079. 10.1016/j.cell.2010.04.035.

38. Matsuda K, Yuzaki M (2011) Cbln family proteins promote synapse formation by regulating distinct neurexin signaling pathways in various brain regions. Eur J Neurosci 33: 1447–1461. 10.1111/j.1460-9568.2011.07638.x.

39. Emes RD, Grant SG (2012) Evolution of synapse complexity and diversity. Annu Rev Neurosci 35: 111–131. 10.1146/annurev-neuro-062111-150433.

15

High-Resolution Melting-Curve Analysis of *obg* Gene to Differentiate the Temperature-Sensitive *Mycoplasma synoviae* Vaccine Strain MS-H from Non-Temperature-Sensitive Strains

Muhammad A. Shahid[1]*, Philip F. Markham[2], Marc S. Marenda[1], Rebecca Agnew-Crumpton[1], Amir H. Noormohammadi[1]

1 Faculty of Veterinary Science, The University of Melbourne, Werribee, Victoria, Australia, **2** Asia-Pacific Centre for Animal Health, Faculty of Veterinary Science, The University of Melbourne, Victoria, Australia

Abstract

Temperature-sensitive (*ts*$^+$) vaccine strain MS-H is the only live attenuated *M. synoviae* vaccine commercially available for use in poultry. With increasing use of this vaccine to control *M. synoviae* infections, differentiation of MS-H from field *M. synoviae* strains and from rarely occurring non-temperature-sensitive (*ts*$^-$) MS-H revertants has become important, especially in countries where local strains are indistinguishable from MS-H by sequence analysis of variable lipoprotein haemagglutinin (*vlhA*) gene. Single nucleotide polymorphisms (SNPs) in the *obg* of MS-H have been found to associate with *ts* phenotype. In this study, four PCRs followed by high-resolution melting (HRM)-curve analysis of the regions encompassing these SNPs were developed and evaluated for their potential to differentiate MS-H from 36 *M. synoviae* strains/isolates. The nested-obg PCR-HRM differentiated *ts*$^+$ MS-H vaccine not only from field *M. synoviae* strains/isolates but also from *ts*$^-$ MS-H revertants. The mean genotype confidence percentages, 96.9$\pm$3.4 and 8.8$\pm$11.2 for *ts*$^+$ and *ts*$^-$ strains, respectively, demonstrated high differentiating power of the nested-obg PCR-HRM. Using a combination of nested-obg and obg-F3R3 PCR-HRM, 97% of the isolates/strains were typed according to their *ts* phenotype with all MS-H isolates typed as MS-H. A set of respiratory swabs from MS-H vaccinated specific pathogen free chickens and *M. synoviae* infected commercial chicken flocks were tested using obg PCR-HRM system and results were consistent with those of *vlhA* genotyping. The PCR-HRM system developed in this study, proved to be a rapid and reliable tool using pure *M. synoviae* cultures as well as direct clinical specimens.

Editor: Mitchell F. Balish, Miami University, United States of America

Funding: Financial support to conduct this study was provided jointly by the University of Melbourne and the Higher Education Commission of Pakistan. The senior author was supported by the Higher Education Commission of Pakistan. The funders had no role in study design, data collection and analysis, decision to publish, or preparation of the manuscript.

Competing Interests: The authors have declared that no competing interests exist.

* E-mail: m.shahid@student.unimelb.edu.au

Introduction

Mycoplasma synoviae causes airsacculitis and infectious synovitis in chickens and turkeys [1]. It causes significant economic losses to the poultry industry due to carcass condemnation, culling of lame birds and deterioration in eggshell quality [2,3]. The temperature-sensitive (*ts*$^+$) strain MS-H (Vaxsafe MS®, Bioproperties Pty. Ltd. Australia) is the only live attenuated vaccine available and is used in several countries to control *M. synoviae* infections in poultry flocks.

Differentiation of MS-H from field strains is an important step to establish whether a flock is free from wild-type *M. synoviae*. It is also important to establish whether the vaccine strain has colonised the respiratory mucosa so as to produce an efficient immune response to protect against wild-type disease. A number of PCR-based techniques have been reported for typing of *M. synoviae* strains, targeting the *vlhA* gene [4-7], 16S rRNA genes [8] or the 16S to 23S rRNA intergenic spacer region [9,10]. Only a small number of these studies included the MS-H vaccine in their

experiments. Jeffery et al. [11] described a combination of PCR and high-resolution melting (HRM) curve analysis of the *vlhA* gene products to discriminate a large number of *M. synoviae* strains, although their system did not differentiate MS-H from several Australian field strains as they shared the same *vlhA* gene sequence. The *vlhA*-based typing system however should be useful in other countries as MS-H like strains are believed to be rare, if not absent, outside Australia. Pulsed-field gel electrophoresis (PFGE) using *Bln*I and *Bam*HI digestions coupled with *vlhA* gene sequencing was useful in differentiating the MS-H from Japanese *M. synoviae* strains/isolates [12]. Also a PCR based cycling probe technology (CPT) developed by Ogino et al. [13], targeting an A→G substitution at 365th nucleotide from the 5′ conserved region of *vlhA* gene, has been claimed useful for MS-H differentiation from Japanese *M. synoviae* strains/isolates. However techniques reported in both of these reports are time consuming and may be difficult to perform on a routine basis in diagnostic laboratories. More importantly, none of the techniques reported

Table 1. *Mycoplasma* strains/isolates used in this study.

Mycoplasma species	ID (*ts* phenotype)[a]	Classification	Origin	Specimen type	Reference
M. synoviae	86079/7NS (−)	V,G2	Australia, NSW, parent strain of MS-H vaccine, palatine cleft	Pure culture	[15]
	MS-H (+)	V,G1	Australia, vaccine strain derived from 86079/7NS	Pure culture	[15]
	MS-H P5 (+)	V,G1	MS-H reisolate, Australia (with 5 consecutive in-vivo passages)	Pure culture	This study
	MS-H[4] (−)	V,G2	MS-H reisolate, Australia	Pure culture	[14]
	MS-H[5] (−)	V,G2	MS-H reisolate, Australia	Pure culture	[14]
	MS-H[3] (−)	V,G2	MS-H reisolate, Australia	Pure culture	[14]
	94036/10-5a (−)	V,G2	MS-H reisolate, Australia	Pure culture	[14]
	93198/5-10a (+)	V,G1	MS-H reisolate, Australia	Pure culture	[14]
	93205/1-2a (+)	V,G1	MS-H reisolate, Australia	Pure culture	[14]
	93205/2-9a (−)	V,G2	MS-H reisolate, Australia	Pure culture	[14]
	93198/6-5b (+)	V,G1	MS-H reisolate, Australia	Pure culture	[14]
	93198/4-19a (+)	V,G1	MS-H reisolate, Australia	Pure culture	[14]
	94036/2-2a (+)	V,G1	MS-H reisolate, Australia	Pure culture	[14]
	93198/3-13b (+)	V,G1	MS-H reisolate, Australia	Pure culture	[14]
	93198/3-15a (+)	V,G1	MS-H reisolate, Australia	Pure culture	[14]
	93205/2-13a (−)	V,G2	MS-H reisolate, Australia	Pure culture	[14]
	93205/9-3a (+)	V,G1	MS-H reisolate, Australia	Pure culture	[14]
	93205/8-9c (−)	V,G2	MS-H reisolate, Australia	Pure culture	[14]
	93205/10-13a (−)	V,G2	MS-H reisolate, Australia	Pure culture	[14]
	94036/9-2a (−)	V,G3	MS-H reisolate, Australia	Pure culture	[14]
	93198/6-1a (−)	V,G3	MS-H reisolate, Australia	Pure culture	[14]
	93198/1-24b (−)	V,G3	MS-H reisolate, Australia	Pure culture	[14]
	94036/2-1a (+)	V,G3	MS-H reisolate, Australia	Pure culture	[14]
	94036/5-5a (−)	V,G2	MS-H reisolate, Australia	Pure culture	[14]
	94036/6-3a (−)	V,G2	MS-H reisolate, Australia	Pure culture	[14]
	94036/8-3a (−)	F,G2	Field strain, Australia	Pure culture	[14,16]
	94041/12a	F,G2	Australia, NSW, field isolate, palatine cleft	Pure culture	[11]
	4GPH3	F,G2	Australia, field isolate, hock joint	Pure culture	[11,34]
	F10-2AS	F,G2	USA, NC, field strain, airsac	Pure culture	[11]
	K1938	F,G2	USA, AR, field strain	Pure culture	[11]
	K870	F,G2	USA, ME, field strain	Pure culture	[11]
	K1858	F,G2	USA, field strain, trachea	Pure culture	[11]
	YA	F,G2	Source unknown	DNA stock	[11,35]
	K1968	F,G2	USA, CO, field strain, turkey, joints	Pure culture	[11]
	K1723	F,G2	USA, AR, field strain, trachea	Pure culture	[11]
	WVU-1853 (−)	F,G2	USA, type strain, joints	Pure culture	[11,36]
	100940-1, -2, -3, -4, -5, -6 and -7 (NA)	S,G2	Australia	Swab samples from non-vaccinated commercial flocks	This study
	100752-A-5T, -B-5T, -C-5T, and -D-5T, 100744-3B	S,G2	Australia	Swab samples from non-vaccinated commercial flocks	This study
	100958-1 to −10 (NA)	S,G2	Australia	Swabs from non-vaccinated commercial flocks	This study
	2774, 2775, 2778, 2781, 2782, 2784 (NA)	S,G1	Australia	Swabs from MS-H vaccinated SPF chickens	This study
M. gallisepticum	ts-11 (+)	NA	Vaccine strain	Pure culture	[37]

[a]*ts* phenotype was determined in previous studies [16,33] except MS-H P5, WVU-1853, 94036/5-5a and 94036/6-3a which was determined in this study.
NA, not applicable; V, MS-H vaccine-related strain; F, field strain; S, swab sample; G, genotype, determined by either nucleotide sequencing of *obg* or nested-obg and obg-F3R3 HRMs, based on *obg* SNPs at position 367 and 629.

above have the capacity to distinguish between MS-H and its non-temperature sensitive isolates rarely isolated from vaccinated flocks [14].

Microtitration followed by incubation at two different temperatures has been used to determine the temperature-sensitive (*ts*) phenotypes of *M. synoviae* strains/isolates [15]. We have recently developed a technique using a combination of differential growth at two different temperatures with a quantitative real-time PCR (*vlhA* Q-PCR) to determine *ts* phenotype of *M. synoviae* strains [16] however this technique still requires culture of the organism and therefore access to live cloned organism.

We have recently compared partial genome sequences of MS-H, its parent strain 86079/7NS and two *ts*⁻ MS-H reisolates (MS-H⁴ and MS-H⁵) and found an SNP (G→A) at nucleotide position 367 in MS-H *obg*, an essential gene encoding highly conserved GTP binding protein Obg found in organisms ranging from human to bacteria. Obg is involved in essential cellular processes such as signal transduction, protein synthesis, ribosome biogenesis, DNA replication initiation, chromosomal segregation and progression through cell cycle [17]. A nucleotide change (G→A) at position 367 in MS-H *obg*, causing an alteration of glycine to arginine at position 123 in Obg fold, was predicted to play a role in temperature sensitivity phenotype of MS-H [18]. Analysis of complete *obg* nucleotide sequences from further 19 MS-H reisolates revealed another SNP (C→T) at position 629, causing amino acid change from alanine to valine at position 210 in GTP binding domain, in 4 MS-H reisolates [18]. These SNPs were used in this study to develop a rapid and reliable test, using HRM-curve analysis, to differentiate MS-H from *ts*⁻ MS-H reisolates and/or *M. synoviae* field strains.

Materials and Methods

Ethics statement

Clinical swab samples were taken from palatine cleft, trachea or sinus of specific pathogen free (SPF) chickens, vaccinated with *M. synoviae* vaccine strain MS-H, after euthanasia using intravenous injection of phenobarbitone as per approval of the Melbourne University Animal Ethics Committee (Approval number

0911472.1). Swabs from field commercial chicken flocks were submitted as diagnostic specimens.

Mycoplasma strains and growth media

All mycoplasma strains used in this study are listed in Table 1. A total of 36 *M. synoviae* strains/isolates and 28 clinical swab samples were used in this study. The collection of MS-H reisolates examined in this study (Table 1) is a unique collection prepared in our laboratory through extensive monitoring of the MS-H vaccinated flocks. For 23 out of 36 *M. synoviae* strains/isolates, the *ts* phenotype was determined in a previous study [16]. For MS-H P5, WVU-1853, 94036/5-5a and 94036/6-3a, the *ts* phenotype was determined in this study using microtitration as described before [16]. Other 9 field strains, characterised as different from MS-H based on the analysis of PCR products from *vlhA* gene single-copy conserved region by single strand conformation polymorphism and HRM genotyping [11] (Table 1), were presumed *ts*⁻. Especially since natural *ts*⁺ *M. synoviae* strains have not been reported so far, examination of temperature sensitivity of the field strains was not considered in this study. The *ts*⁺ phenotype is a property of the MS-H vaccine produced by chemical mutagenesis [15,19-21]. Where necessary, *M. synoviae* strains/isolates were grown in mycoplasma broth (MB) containing 10% swine serum and 0.01% (w/v) of nicotinamide adenine dinucleotide (NAD) [22]. *M. gallisepticum* vaccine strain ts-11 was used as control to test the specificity of *obg* PCR primers.

DNA extraction

Swabs taken from MS-H vaccinated SPF chickens and non-vaccinated commercial chicken flocks were subjected to DNA extraction. Cell pellet of 500 µl MB culture of *M. synoviae* strains/isolates, harvested by centrifugation at 14,000×g for 1 min, or respiratory swabs were placed in 500 µl Qiagen RLT lysis buffer containing 1% of 2-β-mercaptoethanol and incubated at 4°C overnight. After a brief vortex, the swab (where applicable) was removed and 15 µl of Qiaex II matrix (Qiagen, Chadstone, Victoria, Australia) and 300 µl of 70% ethanol added to the lysis buffer. The suspension was mixed and loaded onto a multispin MSK-11 column (Axygen, Union City, California, USA) and

Table 2. Primers used in this study.

Primers	Sequence (5′ to 3′)	Position[a]	PCR product size (bp)	Application
obg-F	GTTGATAAAGGTGGACCAG	88–106	841	Sequencing
obg-R	TTAGTGCAGATATCTCAATG	928–909		
obg-F1	CTTTATTTAGTTGCTAAAGGC	337–357	101	obg-F1R1 HRM
obg-R1	CCGGGCATTCCATTTTCG	437–420		
obg-F1	CTTTATTTAGTTGCTAAAGGC	337–357	60	Nested-obg HRM
obg-Ri2	AGAGGTTTTAAATTTATTATTTCC	396–373		
obg-F1	CTTTATTTAGTTGCTAAAGGC	337–357	335	obg-F1R3 HRM
obg-R3	CCTTTACCTAGTGATGCG	671–654		
obg-F3	TACCTTAGTTCCTCAGTTAGG	573–593	99	obg-F3R3 HRM
obg-R3	CCTTTACCTAGTGATGCG	671–654		

[a]Nucleotide positions of primers in relation to the DNA sequence of *obg* of MS53, GenBank accession number AE017245.

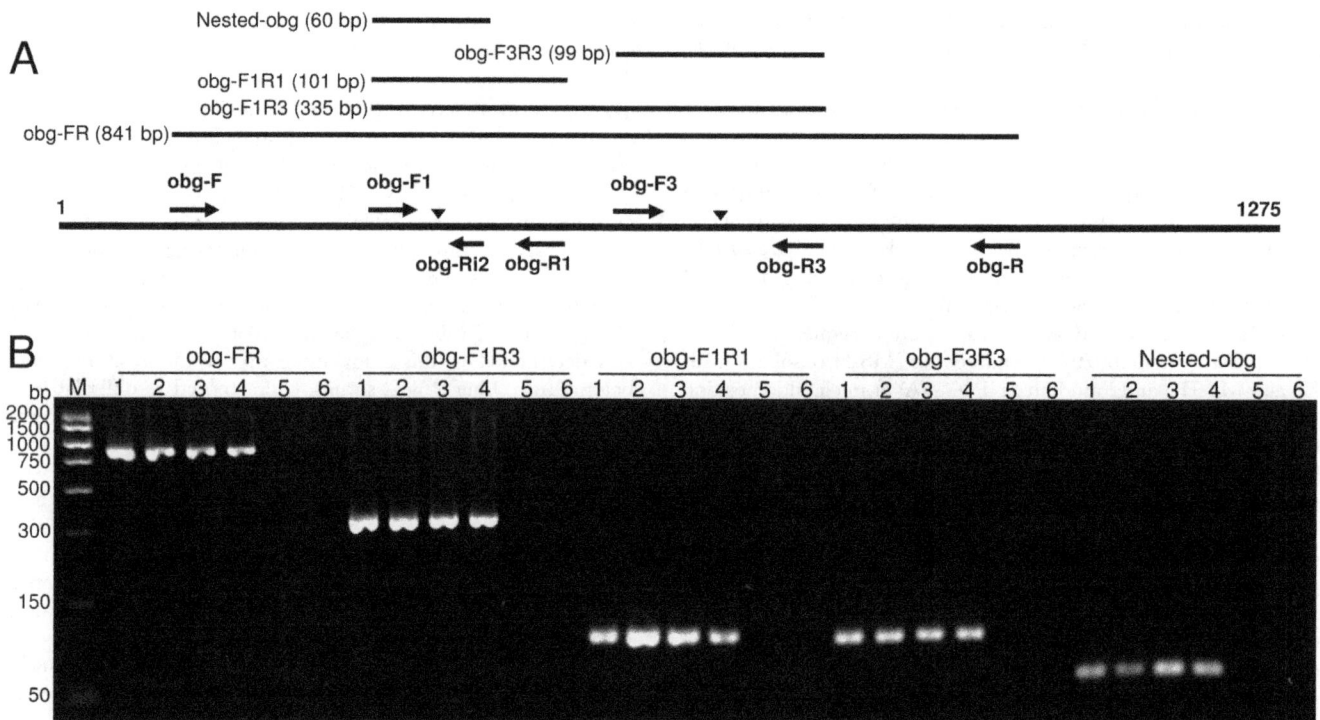

Figure 1. Analysis of different *obg* PCR products by agarose gel electrophoresis. (A) Schematic presentation of *obg* PCRs and the location of primers. Thick and thin lines indicate the extent of full length *obg* and different *obg* PCRs, respectively. Arrows represent primer locations and vertical arrowheads indicate location of *obg* SNPs at positions 367 and 629. (B) Agarose gel electrophoresis of products from *obg* PCRs amplified from MS-H (lane 1), 86079/7NS (lane 2), 94036/1-24b (lane 3), WVU-1853 (lane 4), M. gallisepticum ts-11 (lane 5), and non template control (lane 6). *M, molecular* weight marker (PCR Marker; Sigma, Missouri, USA).

placed in a 1.5 ml microfuge tube and centrifuged for 30 sec at 10,000×g with the flow-through discarded. Columns were washed once with 600 μl of RW1 buffer (Qiagen) and twice with 500 μl of RPE buffer (Qiagen) followed by centrifugation for 30 sec at 10,000×g after each wash. The spin column was dried by centrifugation for 90 sec at 14,000×g. Finally, 50 μl of nuclease free water was added to the columns and DNA was eluted after incubation at room temperature for 5 min and centrifugation at 10,000×g for 60 sec. Similar amount of DNA (~50 ng/μl) was used in all experiments although this was less controllable for clinical specimens submitted as swabs. Extracted DNA was used immediately in PCR or stored at −20°C for future use.

Oligonucleotide primers

The nucleotide primers used in this study, and their sequences, are listed in Table 2 while their location are shown on Figure 1A. All primers were designed using AmplifX version 1.5.4 and PerlPrimer version 1.1.20 [23]. The primers obg-F1 and obg-R1 were designed to flank *obg* SNP 367. Primers obg-F3 and obg-R3 were designed to flank the *obg* SNP 629. Primers obg-F and obg-R were designed for partial sequencing of the *obg* from *M. synoviae* strains/isolates. Specificity of primers was evaluated using BLAST search against non-redundant nucleotide databases.

obg PCRs

Three regions of the *obg*, encompassing both or one of the SNPs detected in MS-H (G→A and C→T at positions 367 and 629, respectively), were targeted by PCR for HRM-curve analysis (Figure 1A). The obg-F1R3 PCR spanned over both SNPs while the obg-F1R1 and obg-F3R3 PCRs spanned over SNP G→A or

C→T, respectively. PCR reactions were carried out in iCycler thermal cycler (Bio-Rad, Gladesville, New South Wales, Australia). A 25 μl PCR reaction mixture contained 1 μl each of 25 μM forward and reverse oligonucleotides (0.1 μl each of 25 μM oligonucleotides for obg-F3R3 PCR), 2 μl of 25 mM MgCl$_2$, 4 μl of 1.25 mM dNTP mixture, 1 U of GoTaq® DNA polymerase (Promega, Alexandria, New South Wales, Australia), 5 μl of 5×GoTaq® flexi green buffer (Promega), 2 μl of 100 μM SYTO 9 green fluorescent nucleic acid stain (Invitrogen, Mount Waverley, Victoria, Australia), 1 μl of *M. synoviae* genomic DNA (~50 ng/μl) and 10.8 μl of nuclease free water. PCR reaction conditions for F1R1 PCR included an initial denaturation at 95°C for 2 min, and then 35 cycles of 95°C for 10 sec, 50°C for 20 sec and 72°C for 25 sec. PCR reaction conditions for obg-F1R3 PCR included an initial denaturation at 95°C for 2 min, and then 35 cycles of 95°C for 30 sec, 49°C for 30 sec and 72°C for 30 sec. PCR reaction conditions for obg-F3R3 PCR included an initial denaturation at 95°C for 2 min, and then 35 cycles of 95°C for 30 sec, 58°C for 30 sec and 72°C for 15 sec. In each set of reaction, nuclease free water was used as negative control. All specimens were tested in triplicates.

To amplify a 841-bp region of the *obg* for sequencing purposes, the obg-FR PCR was conducted using oligonucleotide primers obg-F and obg-R. A 50 μl reaction contained 1 μl each of 25 μM oligonucleotide primers, 4 μl of 25 mM MgCl$_2$, 8 μl of 1.25 mM dNTP mixture (Promega), 0.3 μl of GoTaq® DNA polymerase (Promega), 10 μl of 5×GoTaq® flexi green buffer (Promega), 22.7 μl of nuclease free water and 3 μl of *M. synoviae* genomic DNA (~50 ng/μl). PCR conditions included an initial denaturation at 95°C for 2 min then 45 cycles of 95°C for 10 sec, 48°C for 10 sec and 72°C for 60 sec.

All PCR products were analysed by electrophoresis through 1% agarose gels stained with GelRed™ (Biotium, Hayward, California, USA) and visualised by UV transillumination.

High-resolution melting-curve analysis

High-resolution melting-curve analysis was conducted in a Rotor-Gene 6000 thermal cycler (Corbett Life Science, Mortlake, New South Wales, Australia) and signal detected using an excitation wavelength at 470 nm and detection at 510 nm. Melting-curves were generated by increasing the temperature from 60 to 90°C for obg-F1R1, obg-F3R3 and obg-F1R3 PCR products and recording the fluorescence. To optimise melting conditions for maximum differentiation of sequence differences, PCR products were subjected to different ramp speeds of 0.05, 0.1, 0.2, 0.3 and 0.5°C per sec. The HRM-curve analysis was performed using the software Rotor-Gene 1.7.27 and HRM algorithm provided. Conventional melt-curves were generated automatically. To generate normalised HRM-curves, following normalisation regions were applied: 72.5 to 73.0 and 77.5 to 78.0 for obg-F1R1; 70.9 to 71.9 and 79.2 to 80.2 for obg-F3R3 and 74.5 to 76.0 and 80.5 to 82.0 for obg-F1R3. The MS-H profile was set as 'genotype' and the average HRM genotype confidence percentages (C%) (value attributed to each strain being compared to the genotype, with a value of 100 indicating an exact match) for replicates were automatically calculated by Rotor-Gene 1.7.27. The C% value attributed to all other strains/isolates indicated similarity of the given strain/isolate to the ts^+ MSH. The mean C% of specimen replicates and standard deviations were calculated using Microsoft™ Office Excel 2003.

Nested-obg PCR-HRM

Oligonucleotide primers obg-F1 and obg-Ri2 were used to amplify a 60-bp internal region of obg (harbouring the SNP 367) from products generated in obg-F1R1 PCR (Figure 1A). PCR was performed in 25 μl reaction volumes containing 5 μl of 5×Go-Taq® flexi green buffer (Promega), 0.1 μl each of 25 μM oligonucleotide primers, 2 μl of 25 mM MgCl₂, 2 μl of 1.25 mM dNTP mixture (Promega), 0.2 μl of GoTaq® DNA polymerase (Promega), 2 μl of 100 μM SYTO 9 green fluorescent nucleic acid stain (Invitrogen), 2 μl of 0.01× diluted obg-F1R1 PCR product as template and 10.8 μl of nuclease free water. PCR conditions consisted of denaturation at 95°C for 2 min followed by 35 cycles of 95°C for 10 sec, 52°C for 20 sec and 72°C for 10 sec. All reactions were carried out in triplicate. In each experiment, water instead of template was used as negative control and MS-H and parent strain 86079/7NS genomic DNA were used as ts^+ and ts^- controls, respectively. Following PCR, HRM-curve analysis was carried out in Rotor-Gene 6000 thermal cycler (Corbett Life Science Pty Ltd) as described above. Melting curves were generated by increasing the temperature from 66 to 78°C at ramp speeds of 0.1, 0.2, 0.3 and 0.5°C per sec. Normalisation regions of 66.5 to 67.0 and 76.5 to 77.0 and genotype confidence threshold of 85% were applied to characterise unknown M. synoviae strains/isolates using MS-H as genotype/reference strain.

Nucleotide sequencing and sequence analysis

PCR products generated in obg-FR PCR (841-bp) were separated through 1% agarose gels, bands of expected size were excised, purified using Wizard® SV Gel and PCR clean-Up System (Promega) and cloned into pGEM®-T Easy vector

(Promega) using instructions provided by the manufacturer. The resultant constructs were propagated in α-select competent cells, silver efficiency (Bioline, Alexandria, New South Wales, Australia) and extracted using PureYield™ Plasmid Miniprep (Promega). All purified PCR products or plasmid extracts were subjected to automated sequencing (BigDye Terminator v3.1; Applied Biosystems, Foster City, California, USA) in both directions using primers obg-F and obg-R, or M13 forward and reverse sequencing primers for purified PCR products or cloned PCR products, respectively. Nucleotide sequences were edited using SeqMan™ II and EditSeq programs in DNASTAR. Multiple sequences were aligned using computer program ClustalW2. Nucleotide sequence of complete obg of 25 strains including MS-H and its related strains, belonging to four different genotypes, has been described in our previous study [18]. Nucleotide sequence of partial obg of additional 10 M. synoviae strains, including 94041/12a, 4GPH3, F10-2AS, K1938, K870, K1858, YA, K1968, K1723 and WVU-1853, has been submitted to GenBank under accession numbers KF875990 to KF875999.

Results

PCR amplification of selected regions of the obg from different M. synoviae strains/isolates

In order to evaluate the capacity of obg PCRs for HRM analysis to differentiate MS-H from M. synoviae strains, five sets of oligonucleotide primers, as detailed in Table 2 and Figure 1A, were used to amplify 5 regions of 841, 335, 101, 99 and 60-bp of obg from four M. synoviae strains/isolates. All strains/isolates generated PCR products of the expected size in all PCRs as confirmed by agarose gel electrophoresis (Figure 1B). No PCR product was detected from M. gallisepticum strain ts-11 DNA and no template negative control (Figure 1B) indicating specificity of the obg PCRs.

The obg-F1R3 PCR-HRM curve analysis could not reliably differentiate MS-H from M. synoviae strains/isolates tested

The 335-bp obg-F1R3 PCR products from 16 M. synoviae strains/isolates including MS-H and its related isolates, and a number of field strains from Australia and the USA were subjected to HRM-curve analysis (Figure S1). Conventional melt-curve analysis of the PCR products using a ramp of 0.3°C/sec showed that all strains generated a single peak at 79.7±0.1°C which were also visually very similar in pattern making it difficult to differentiate MS-H from other strains (Figure S1 and Table S1). Visual examination of the normalised HRM-curves also showed very minor differences between curve profiles of MS-H and other M. synoviae strains/isolates (Figure S1). When genotyping was applied to the normalised HRM-curves using MS-H as reference genotype, the C% ± SD for the strains/isolates 93198/1-24b, 94036/5-5a, 4GPH3, K870, WVU-1853 and YA were 78.8±10.0, 83.4±3.8, 71.8±1.6, 87.9±8.8, 89.4±0.5 and 83.2±12.2, respectively. All these strains could be auto-called as 'variation' from MS-H when a genotype confidence threshold of 90% was applied. For other strains/isolates, the C% was above 90% (93.7±3.6) and the normalised melt curves were mostly similar to that of MS-H on visual examination. Therefore the obg-F1R3 HRM-curve analysis was not considered as a reliable tool and was not pursued any further in this study for differentiation of MS-H from other strains/isolates.

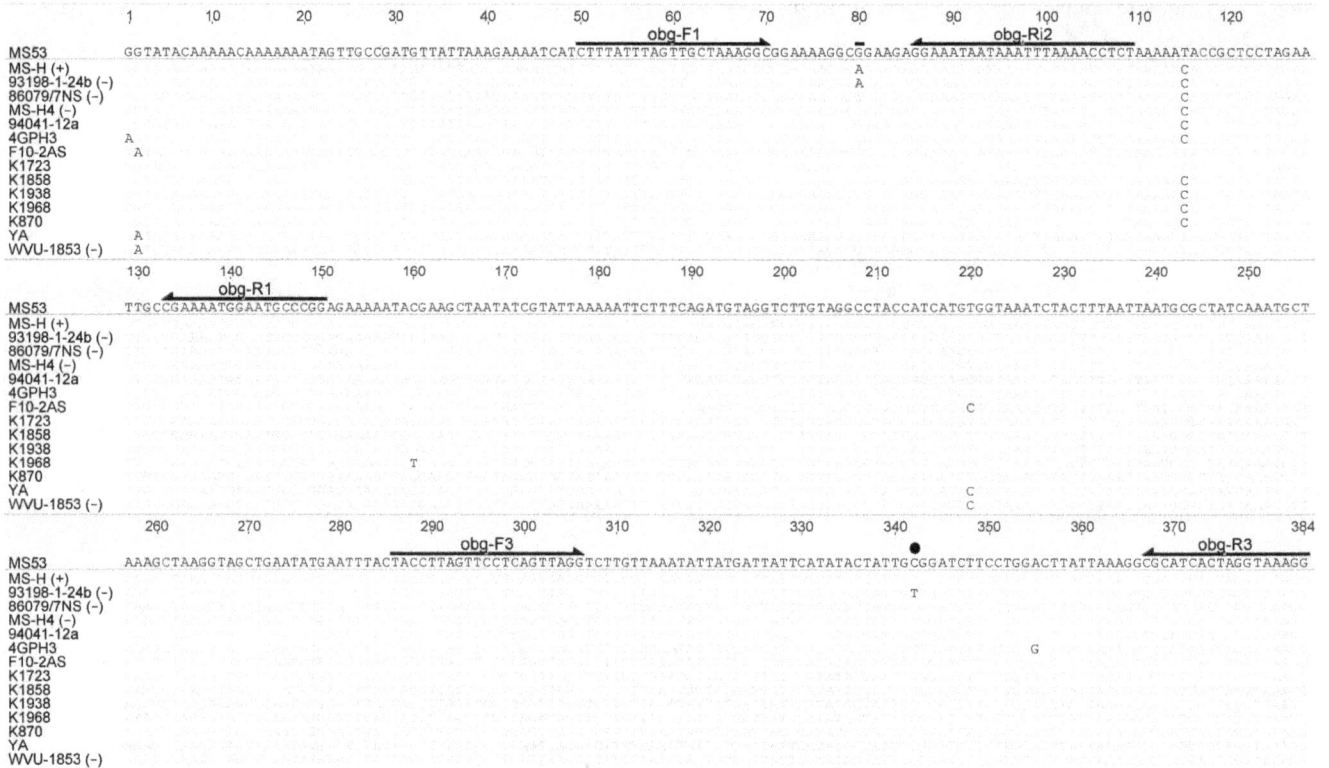

Figure 2. Comparison of partial *obg* nucleotide sequences (corresponding to nt 288-671 of MS53 *obg*; GenBank accession number no. AE017245) from selected *M. synoviae* strains/isolates. Nucleotide differences are highlighted keeping MS53 as reference. Location of primers used in *obg* PCRs as well as SNP G367A discovered in MS-H genome are highlighted with arrows and bar above the sequence, respectively. Location of SNP C629T, observed only in 93198/6-1a, 93198/1-24b, 94036/9-2a and 94036/2-1a [18], is also highlighted with a dot.

The obg-F1R1 PCR-HRM curve analysis differentiated MS-H from all *M. synoviae* field strains/isolates but not from WVU-1853

The 101-bp obg-F1R1 PCR products, spanning over SNP G→A at position 367, from various *M. synoviae* strains/isolates were subjected to HRM-curve analysis. Only a small number of strains/isolates were used in this assay to provide a preliminary evaluation of the assay. Visual examination of conventional melt curves at different ramps revealed that a ramp of 0.3°C/sec generated the most distinct curves and therefore used in the further HRM analysis. The conventional melting-curve analysis showed a single peak for all strains examined. The melting peaks for MS-H vaccine, its ts^+ reisolates, and the US strain WVU-1853 occurred at 75.8±0.0°C while those for all other strains including 86079/7NS and ts^- MS-H reisolates occurred at 76.3±0.1 (Figure S1). Normalised HRM-curve analysis distinctly separated strains into two groups, one for the known ts^+ and the other for the ts^- strains with the exception of the rarely occurring ts^- MS-H reisolates with mutation at position 629 (Figure S1). When genotyping with a C% threshold of 90 was applied, two distinct genotypes were auto-called: one included MS-H and ts^+ MS-H reisolates (mean C%, 97.2±2.6) and the other included ts^- strains (mean C%, 6.1±4.7). The MS strain WVU-1853 and the rarely occurring ts^- MS-H reisolates had normalised HRM-curves identical to that of MS-H (Figure S1). HRM data from different experiments is shown in Table S2.

Alignment of partial *obg* nucleotide sequences revealed further SNPs

Alignment of partial nucleotide sequence of *obg* from MS-H, MS-H reisolates (both ts^+ and ts^-) and field *M. synoviae* strains/isolates revealed further nucleotide variations in *obg* (Figure 2), especially in the region targeted in obg-F1R1 PCR. *M. synoviae* strains F10-2AS, K1723, YA and WVU-1853 had C→T variation at position 402 while 94041/12a had C→A variation at position 434. Therefore, a further oligonucleotide primer (obg-Ri2) was designed to allow targeting of the region spanning over the SNP G→A at position 367 and avoiding other polymorphic sites found in *obg*. Nucleotide sequence alignment of *obg* regions targeted in nested-obg and obg-F3R3 PCR-HRM, for all *M. synoviae* strains/isolates used in this study except field isolate 94036/8-3a, is shown in Figure S2 and S3, respectively. For reasons unknown to the authors, several attempts at sequencing the obg-FR PCR product for 94036/8-3a were failed. Similarly, attempts at sequencing the *vlhA* region of 94036/8-3a were unsuccessful in our previous study [16].

Nested-obg PCR-HRM curve analysis differentiated MS-H from most of the *M. synoviae* field strains/isolates and MS-H reisolates

HRM-curve analysis of the nested-obg PCR product (60-bp in size) at a ramp rate of 0.3°C/sec revealed a single peak of 72.3±0.1°C for MS-H and of 73±0.0°C for 86079/7NS. Visual examination of the conventional and normalised HRM-curves

Table 3. Melting points and genotype confidence percentages (C%) generated in nested-obg HRM from different *M. synoviae* strains/isolates.

Strains/isolates	Nested-obg		
	Melting points (°C) (Mean ± SD)[a]	HRM-curve genotype	C% (Mean ± SD)[a]
MS-H	72.3±0.1	MS-H	99.3±0.5
86079/7NS	73.0±0.0	Variation	5.1±2.2
94036/10-5a	73.0±0.0	Variation	2.9±0.3
93198/5-10a	72.3±0.0	MS-H	97.3±2.6
[b]93198/6-1a	72.3±0.1	MS-H	97.5±3.9
[b]93198/1-24b	72.3±0.1	MS-H	97.9±2.4
[b]94036/9-2a	72.3±0.1	MS-H	98.8±1.7
[b]94036/2-1a	72.3±0.1	MS-H	98.7±0.8
MS-H[4]	73.0±0.1	Variation	3.9±1.9
93205/1-2a	72.3±0.1	MS-H	97.5±2.5
93205/2-9a	73.0±0.0	Variation	4.7±1.7
93198/6-5b	72.4±0.1	MS-H	97.2±3.3
94036/5-5a	73.0±0.1	Variation	28.6±10.4
94036/6-3a	73.0±0.0	Variation	7.3±2.4
93198/4-19a	72.3±0.0	MS-H	99.0±1.2
94036/2-2a	72.4±0.1	MS-H	87.1±3.4
93198/3-13b	72.4±0.1	MS-H	94.4±3.9
93198/3-15a	72.4±0.0	MS-H	98.7±0.8
94041/12a	73.1±0.0	Variation	6.5±0.9
94036/8-3a	72.5±0.0	Variation	45.8±0.9
4GPH3	73.1±0.1	Variation	6.9±3.8
F10-2AS	73.1±0.1	Variation	6.4±1.6
K1938	73.1±0.0	Variation	4.1±0.7
K870	73.1±0.1	Variation	4.9±2.7
K1858	73.1±0.1	Variation	4.3 ± 2.5
YA	73.1±0.0	Variation	4.6±1.8
K1968	73.0±0.1	Variation	5.9±2.7
K1723	73.1±0.0	Variation	4.9±0.7
WVU-1853	73.0±0.0	Variation	3.5±0.3

[a]Melting points and C% values are from one HRM experiment for each strain using each sample DNA tested in triplicate. HRM data for all strains/isolates and swab samples, from different experiments, is shown in Table S3.
[b]These strains are rarely occurring *ts*⁻ MS-H reisolates [18] and can be discriminated from MS-H using obg-F3R3 HRM.

revealed that all known *ts*⁺ MS-H reisolates generated HRM-curves similar to those for MS-H while the *ts*⁻ *M. synoviae* strains/isolates and *ts*⁻ MS-H reisolates (except 93198/1-24b, 94036/9-2a and 93198/6-1a) had HRM-curves similar to those for 86079/7NS. After applying genotyping to the normalised curves using a C% threshold of 84, two distinct genotypes were auto-called: the MS-H type with a mean C% of 96.9±3.4, and variants with a mean C% of 8.8±11.2. MS-H reisolates (with a *ts*⁻ phenotype) that previously could not be differentiated from MS-H, due to identical *vlhA* region, were distinguishable from MS-H in the nested-obg PCR-HRM (Table 3).

Nested-obg HRM melting points for MS-H, 86079/7NS and other strains grouped with either of them exhibited minor variation in melting temperature on different days using different DNA extractions as templates but the melting point differences between MS-H and 86079/7NS remained ≥0.7°C (Table S3). Normalised HRM-curves, in all instances, correctly genotyped all

strains/isolates either with MS-H or 86079/7NS. Furthermore, all tested US strains (F10-2AS, K1723, K1858, K1938, K1968, K870, YA and WVU-1853) were autocalled as variant from MS-H genotype and produced melting-curves (73.1±0.0°C) and C% (5.1±1.8) identical to 86079/7NS, and therefore, characterised as *ts*⁻ (Figure 3A and B and Table 3). HRM data for all (36) *M. synoviae* strains/isolates, used in this study, is shown in Table S3.

Nested-obg PCR-HRM curve analysis successfully applied for direct examination of clinical specimens

The nested-obg PCR-HRM was first optimised using DNA extracted from pure cultures of *M. synoviae* strains/isolates as described above, and then extended to clinical swab specimens taken from sinus, palatine cleft or trachea of SPF and field chickens inoculated intra-ocularly with MS-H. Swabs from palatine cleft and trachea of non-vaccinated commercial chicken flocks were used as negative control. All swabs from MS-H

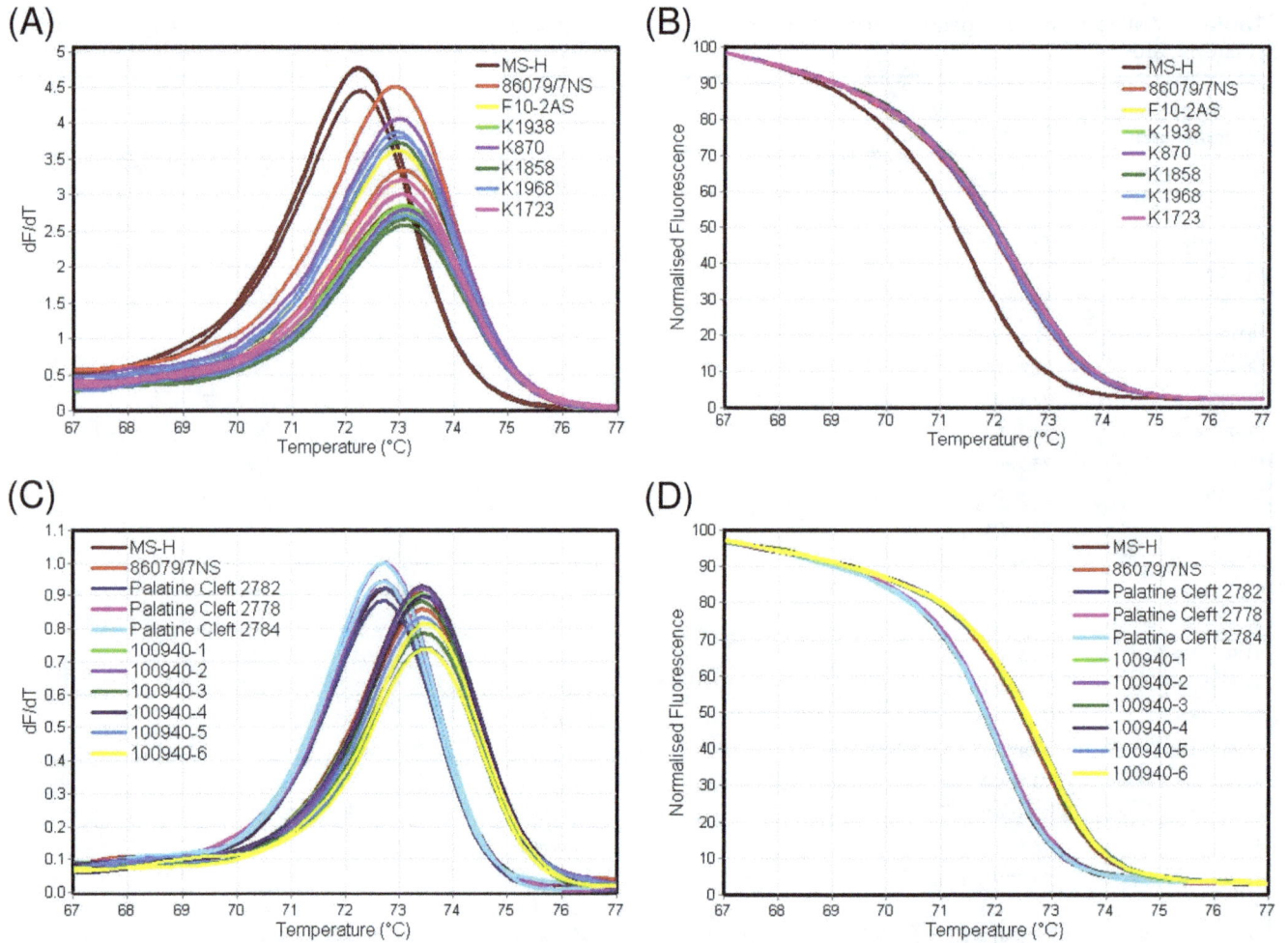

Figure 3. High-resolution melting-curve analysis of *M. synoviae* **strains/isolates using nested-obg PCR products.** (A) Conventional and (B) normalised melt-curves of DNA extracted from pure cultures of *M. synoviae* field strains/isolates indicated 86079/7NS-like genotype, and therefore characterised as *ts⁻*. (C) Conventional and (D) normalised melt-curves of DNA extracted from swabs taken from MS-H vaccinated SPF chickens (palatine cleft 2782, 2778 and 2784) and non-vaccinated commercial chicken flocks (100940-1, -2, -3, -4, -5 and -6). Samples from vaccinated chickens were genotyped as MS-H-like while from non-vaccinated as 86079/7NS-like.

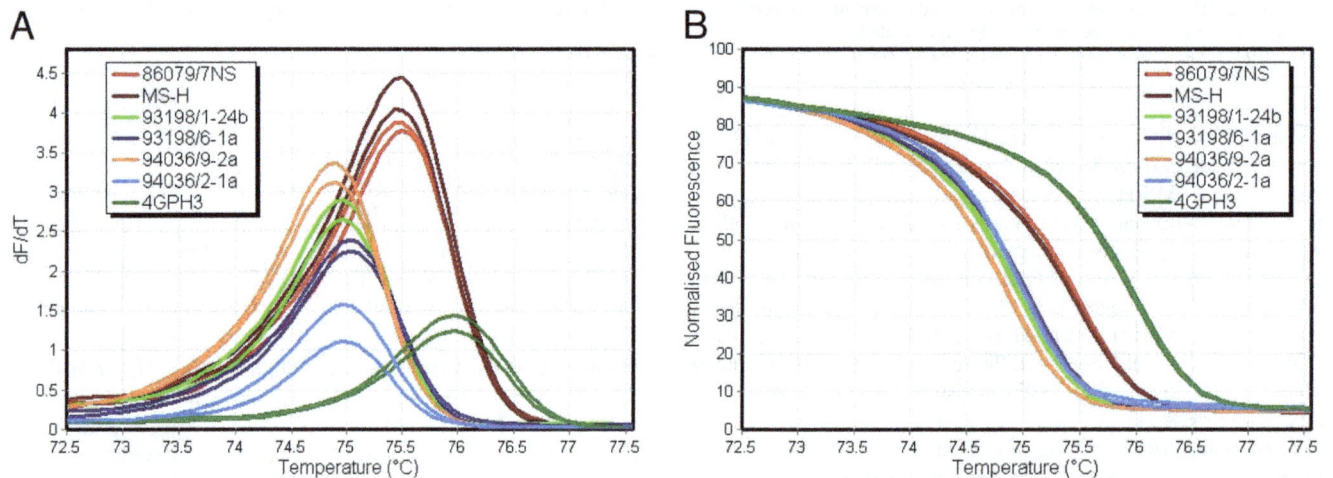

Figure 4. High-resolution melting-curve analysis of *M. synoviae* **strains/isolates using obg-F3R3 PCR products.** (A) Conventional and (B) normalised melt-curves distinguished MS-H from rarely occurring *ts⁻* (93198/1-24b, 93198/6-1a, 94036/9-2a) and *ts⁺* (94036/2-1a) MS-H reisolates and field strains (e.g., 4GPH3).

Table 4. Melting points and genotype confidence percentages (C%) generated in obg-F3R3 HRM from different *M. synoviae* strains/isolates.

Strains/isolates	obg-F3R3		
	Melting points (°C) (Mean ± SD)[a]	HRM-curve genotype	C% (Mean ± SD)[a]
MS-H	75.5±0.0	MS-H	99.9±0.1
86079/7NS	75.5±0.0	MS-H	98.8±1.7
[b]93198/6-1a	75.0±0.0	Variation	12.9±0.9
[b]93198/1-24b	75.0±0.0	Variation	7.6±3.1
[b]94036/9-2a	74.9±0.0	Variation	1.6±0.2
[b]94036/2-1a	75.0±0.0	Variation	10.7±1.6
4GPH3	76.0±0.0	Variation	1.4±0.2

[a]Melting points and C% values are from one HRM experiment for each strain using each sample DNA tested in triplicate.
[b]These strains are rarely occurring *ts*⁻ MS-H reisolates [18] and can only be discriminated from MS-H using obg-F3R3 HRM.

vaccinated SPF chickens produced single melting peak at 72.0±0.0°C while swab samples from non-vaccinated commercial field chicken flocks generated peak at 72.8±0.0°C. Melting peaks for MS-H and 86079/7NS as *ts*⁺ and *ts*⁻ controls were 72.0±0.0°C and 72.8±0.0°C, respectively, thus visual examination of the melting-curves could clearly differentiate the two different melting profiles (Figure 3C). Application of genotyping on normalised curves (using MS-H as reference), distinctively classified the specimens either as MS-H (mean C% of 99.3±0.6) or variation (4.1±1.3) (Figure 3D). Therefore there was an approximate gap of 95% in C% for these two groups. HRM data for all swab samples tested in this study is shown in Table S3.

The obg-F3R3 PCR-HRM curve analysis differentiated MS-H from rare variants of MS-H reisolates and field strains but not from its parent strain 86079/7NS

HRM-curve analysis of 99-bp obg-F3R3 PCR products (encompassing C→T SNP at position 629 in 4 MS-H variants examined), at a ramp of 0.2°C/sec, revealed a single peak at 75.5±0.0°C for MS-H and 86079/7NS and at 75.0±0.1°C for MS-H reisolates 93198/1-24b, 94036/9-2a, 93198/6-1a and 94036/2-1a. The Australian strain 4GPH3 with one base replacement at position 642 of the *obg*, generated a distinguishable (from that of MS-H) single peak at 76.0±0.0°C. Normalised HRM-curves discriminated MS-H from strains/isolates 93198/1-24b, 94036/9-2a, 93198/6-1a and 94036/2-1a, and the field strain 4GPH3, but not from 86079/7NS. When MS-H was selected as genotype, the mean C% of 93198/1-24b, 94036/9-2a,

93198/6-1a and 94036/2-1a was calculated as 8.0±4.8 (Table 4 and Figure 4). For field strain 4GPH3 the mean C% was 1.4±0.2.

Thus, coupling of nested-obg with obg-F3R3 PCR-HRM had 100% accuracy in differentiation of MS-H from all field strains and *ts*⁻ MS-H reisolates. Also a high accuracy (97.2%) was achieved in predicting the *ts* phenotype of *M. synoviae* strains/isolates (Table 5). Irrespective of the (unknown) prevalence of strains with identical nucleotide at position 629 to that of MS-H (e.g., 86079/7NS), the presence of isolates with SNP at position 629 reflects that obg-F3R3 PCR is more useful when combined with the nested obg-PCR.

Discussion

SNPs are the most common type of genetic variation and have been used for species/strain identification of various bacterial pathogens [24–29]. A large number of methods have been utilised for rapid identification of SNPs. These include hybridisation-based methods (molecular beacons, SNP microarrays), enzyme-based methods (restriction fragment length polymorphism, flap endonuclease, primer extension, 5'-nuclease, oligonucleotide ligase assay) and post-amplification methods based on physical properties of the DNA (single strand conformation polymorphism, temperature gradient gel electrophoresis, high resolution melting analysis) [30]. Among these, HRM is thought to be rapid and at the same time most economical where a small number of specimens are to be genotyped [31,32].

In this study, a set of SNPs detected by comparative genomic sequence analysis of *M. synoviae* strains of MS-H lineage were

Table 5. *M. synoviae obg* SNPs-based genotyping scheme and its association with ts phenotype.

Genotypes	SNPs in *obg*		No. of isolates/*ts* phenotype[a] (n = 36)	PCR-HRM for MS-H genotyping
	367	629		
G1 (MS-H)	A	C	10/+	Nested-obg
G2	G	C	22/−	Nested-obg
G3	A	T	3/−	Nested-obg combined with obg-F3R3
G3	A	T	1/+	Nested-obg combined with obg-F3R3

[a]*ts* phenotype was determined by conventional microtitration method [16].

analysed. A selection of MS-H strain specific SNPs was confirmed by Sanger sequencing. SNPs in *obg* associated with change in temperature sensitivity [18] were primarily targeted to enable differentiation of MS-H strain from *ts*⁻ MS-H revertants as well as from *ts*⁻ field strains. Initially a 335-bp obg-F1R3 PCR-HRM was developed but melt curves generated did not provide clear differentiation of MS-H from other strains/isolates. This failure may be related to the small number of sequence variations over a relatively large amplicon and/or balancing effect of the mutations at positions 367 and 629. Previous studies in our group have shown that one nucleotide difference in approximately 400 bp target sequence might be sufficient for differentiation of highly similar sequences [11] but differentiation power of a HRM system may also be influenced by the location of the SNP as well as structure of the DNA surrounding it. In order to develop a more reliable assay for the detection of *obg* SNPs, an alternative PCR (obg-F1R1 PCR) was developed that targeted a smaller (101-bp) region of *obg*, encompassing the *obg* SNP G367A. The obg-F1R1 PCR could differentiate MS-H from *ts*⁻ MS-H reisolates and *ts*⁻ field strains/isolates. However, four field strains (F10-2AS, K1723, WVU-1853 and YA) generated HRM-curves identical to that of MS-H. Despite this limitation, obg-F1R1 PCR-HRM was still considered useful when combined with other strain identification techniques such as *vlhA* gene sequence analysis. To further differentiate MS-H from its closely related isolates, a nested-obg PCR-HRM, targeting a 60-bp region of the *obg* encompassing the SNP G367A was developed. With the exception of 93198/1-24b, 93198/6-1a and 94036/9-2a, the nested-obg PCR-HRM was able to differentiate MS-H from all other field strains/isolates and *ts*⁻ MS-H reisolates.

The potential of nested-obg PCR-HRM was initially evaluated using pure *M. synoviae* cultures available in our laboratory. In order to evaluate the potential of the nested-obg PCR-HRM directly on clinical specimens, two sets of clinical swabs, one from MS-H vaccinated SPF birds and the other from commercial chicken flocks suspected (by serological monitoring) of *M. synoviae* infection were tested. The swabs from MS-H vaccinated SPF birds generated HRM-curves identical to that of MS-H culture while swab samples from infected field birds produced clearly different pattern, identical to that generated by 86079/7NS, indicating infection by a field strain at the time of sampling. This demonstrated the potential of the nested-obg PCR-HRM to rapidly determine whether a flock is infected with a field *M. synoviae* strain or it harbours MS-H vaccine strain. Evaluation of the full potential of this assay for direct examination of clinical specimens should include determination of its sensitivity and specificity although it should be noted that the primary use of this assay in our laboratory has been confined to examination of pure (cloned) *M. synoviae* cultures.

The nested-obg HRM also discriminated *ts*⁺ MS-H from *ts*⁻ field strains and *ts*⁻ MS-H reisolates. Out of total 36 *M. synoviae* strains/isolates, 33 were typed according to their *ts* phenotype. The *ts* phenotype of three MS-H reisolates (93198/1-24b, 93198/6-1a and 94036/9-2a) was not determined in accordance with their temperature sensitivity phenotype (genotyped as 'MS-H' by nested-obg PCR-HRM but exhibited *ts*⁻ phenotype). Complete *obg* sequence of these three strains and one MS-H reisolate 94036/2-1a (*ts*⁺) revealed a secondary mutation (C→T) at position 629. Therefore, an alternative PCR-HRM (obg-F3R3 PCR-HRM), encompassing SNP at position 629 was developed to discriminate these rare variants from all other *M. synoviae* strains/isolates. It is recommended that where an unknown *M. synoviae* strain/isolate

genotyped as MS-H by the nested-obg HRM, the obg-F3R3 HRM should be used to determine its *ts* phenotype and identity. Thus, a combination of nested-obg and obg-F3R3 HRM-curve analysis not only differentiates MS-H from field *M. synoviae* strains and from *ts*⁻ MS-H reisolates, but also exhibited high accuracy (97.2%, 35/36) in predicting the *ts* phenotype of any unknown *M. synoviae* strain/isolate. The exception was the rare isolate 94036/2-1a (*ts*⁺) with *obg* mutations at position 367 and 629. The influence of this second mutation on *ts* phenotype of *M. synoviae* has been discussed in our previous report [18].

Temperature-sensitive bacterial mutants have been produced only in laboratories, mostly by N-methyl-N-nitro-nitrosoguanidine (NTG) [15,19-21]. Such *ts*⁺ mutants are expected to sustain more than one *ts* mutation contributing to overall *ts* phenotype and accounting for the genetic stability of temperature-sensitive mutants [20]. Therefore *obg* SNPs based assays developed in this study may not be ideal for *ts* phenotyping of other organisms although due to their consistency for the MS-H and its reisolates, were found highly useful for *M. synoviae* genotyping purposes.

Wild strain mutation causing *ts*⁺ (MS-H-like phenotype) has never been reported to the best of our knowledge. However reversion from a *ts*⁺ to *ts*⁻ phenotype is expectable under favourable selective pressure [15,33]. The back mutation rate of the MS-H vaccine strain from a *ts*⁺ phenotype to *ts*⁻ phenotype was found in the order of 10⁻⁴. No study on the virulence and transmissibility of *ts*⁺ MS-H reisolates have been conducted although a previous study in our laboratory demonstrated that *ts*⁻ MS-H reisolates did not have the characteristics, including virulence potential, of the vaccine parent strain [33] and that factors other than *ts* phenotype may be involved in loss of virulence of MS-H. Nevertheless the recovery of *ts*⁻ strain/isolate from healthy vaccinated flocks prompted the current study to establish the true identity of the isolate.

The combination of nested-obg and obg-F3R3 PCR-HRM is relatively rapid and can be completed in one day after DNA extraction. High discriminating power of this genotyping system, with an added advantage of predicting the *ts* phenotype, makes it an ideal assay that can be routinely used in veterinary diagnostic laboratories involved in *M. synoviae* genotyping especially in countries where MS-H is routinely used in commercial poultry.

Supporting Information

Figure S1 High-resolution melting-curve analysis of *M. synoviae* strains/isolates using obg-F1R3 and obg-F1R1 PCR products. Using obg-F1R3 HRM, conventional (A) and normalised melt-curves (B) of *M. synoviae* strains were almost identical and thus could not differentiate MS-H from field strains or *ts*– MS-H reisolates. Using obg-F1R1 HRM, conventional (C) and normalised melt-curves (D) of MS-H were distinguishable from all other strains except *M. synoviae* reference strain WVU-1853 and rarely occurring *ts*– MS-H reisolates with *obg* mutation at position 629.

Figure S2 Partial *obg* nucleotide sequence alignment for 35 *M. synoviae* strains/isolates encompassing region harbouring SNP 367.

Figure S3 Partial *obg* nucleotide sequence alignment for 35 *M. synoviae* strains/isolates encompassing region harbouring SNP 629.

Table S1 Melting points and genotype confidence percentages (C%) in obg-F1R3 HRM for different *M. synoviae* strains/isolates.

Table S2 Melting points and genotype confidence percentages (C%) in obg-F1R1 HRM for *M. synoviae* strains/isolates from different experiments.

Table S3 Details of melting points and genotype confidence percentages (C%) in nested-obg HRM for *M. synoviae* strains/ isolates from different experiments.

Acknowledgments

Authors acknowledge the assistance and cooperation from staff of Asia Pacific Centre for Animal Health (APCAH), Faculty of Veterinary Sciences, The University of Melbourne, Australia.

Author Contributions

Conceived and designed the experiments: MAS AHN. Performed the experiments: MAS RAC. Analyzed the data: MAS AHN. Wrote the paper: MAS AHN. Revised the manuscript: MAS PFM MSM AHN.

References

1. Kleven SH, Ferguson-Noel N (2008) *Mycoplasma synoviae* Infection. In: Saif YM, editor.Diseases of Poultry. 12th ed.Ames, Iowa, USA: Blackwell publishing professional. pp. 845–857.
2. Feberwee A, de Wit JJ, Landman WJM (2009) Induction of eggshell apex abnormalities by *Mycoplasma synoviae*: field and experimental studies. Avian Pathol 38: 77–85.
3. Feberwee A, Landman WJM (2010) Induction of eggshell apex abnormalities in broiler breeder hens. Avian Pathol 39: 133–137.
4. Benčina D, Drobnič-Valič M, Horvat S, Narat M, Kleven SH, et al. (2001) Molecular basis of the length variation in the N-terminal part of *Mycoplasma synoviae* hemagglutinin. FEMS Microbiol Lett 203: 115–123.
5. Hammond PP, Ramírez AS, Morrow CJ, Bradbury JM (2009) Development and evaluation of an improved diagnostic PCR for *Mycoplasma synoviae* using primers located in the haemagglutinin encoding gene *vlhA* and its value for strain typing. Vet Microbiol 136: 61–68.
6. Hong Y, García M, Leiting V, Benčina D, Dufour-Zavala L, et al. (2004) Specific detection and typing of *Mycoplasma synoviae* strains in poultry with PCR and DNA sequence analysis targeting the hemagglutinin encoding gene *vlhA*. Avian Dis 48: 606–616.
7. Wetzel AN, Lefevre KM, Raviv Z (2010) Revised *Mycoplasma synoviae vlhA* PCRs. Avian Dis 54: 1292–1297.
8. Buim MR, Buzinhani M, Yamaguti M, Oliveira RC, Mettifogo E, et al. (2010) Intraspecific variation in 16S rRNA gene of *Mycoplasma synoviae* determined by DNA sequencing. Comp Immunol Microbiol Infect Dis 33: 15–23.
9. Ramírez AS, Naylor CJ, Yavari CA, Dare CM, Bradbury JM (2011) Analysis of the 16S to 23S rRNA intergenic spacer region of *Mycoplasma synoviae* field strains. Avian Pathol 40: 79–86.
10. Raviv Z, Kleven SH (2009) The development of diagnostic real-time TaqMan PCRs for the four pathogenic avian mycoplasmas. Avian Dis 53: 103–107.
11. Jeffery N, Gasser RB, Steer PA, Noormohammadi AH (2007) Classification of *Mycoplasma synoviae* strains using single-strand conformation polymorphism and high-resolution melting-curve analysis of the *vlhA* gene single-copy region. Microbiology 153: 2679–2688.
12. Harada K, Kijima-Tanaka M, Uchiyama M, Yamamoto T, Oishi K, et al. (2009) Molecular typing of Japanese field isolates and live commercial vaccine strain of *Mycoplasma synoviae* using improved pulsed-field gel electrophoresis and *vlhA* gene sequencing. Avian Dis 53: 538–543.
13. Ogino S, Munakata Y, Ohashi S, Fukui M, Sakamoto H, et al. (2011) Genotyping of Japanese field isolates of *Mycoplasma synoviae* and rapid molecular differentiation from the MS-H vaccine strain. Avian Dis 55: 187–194.
14. Markham JF, Scott PC, Whithear KG (1998) Field evaluation of the safety and efficacy of a temperature-sensitive *Mycoplasma synoviae* live vaccine. Avian Dis 42: 682–689.
15. Morrow CJ, Markham JF, Whithear KG (1998) Production of temperature-sensitive clones of *Mycoplasma synoviae* for evaluation as live vaccines. Avian Dis 42: 667–670.
16. Shahid MA, Ghorashi SA, Agnew-Crumpton R, Markham PF, Marenda MS, et al. (2013) Combination of differential growth at two different temperatures with a quantitative real time PCR to determine temperature-sensitive phenotype of *Mycoplasma synoviae*. Avian Pathol 42: 185–191.
17. Verstraeten N, Fauvart M, Versées W, Michiels J (2011) The universally conserved prokaryotic GTPases. Microbiol Mol Biol Rev 75: 507–542.
18. Shahid MA, Markham PF, Markham JF, Marenda MS, Noormohammadi AH (2013) Mutations in GTP binding protein Obg of *Mycoplasma synoviae* vaccine strain MS-H: implications in temperature-sensitivity phenotype. PLoS ONE 8: e73954.
19. Brunner H, Greenberg H, James WD, Horswood RL, Chanock RM (1973) Decreased virulence and protective effect of genetically stable temperature-sensitive mutants of *Mycoplasma pneumoniae*. Ann N Y Acad Sci 225: 436–452.
20. Greenberg H, Helms CM, Brunner H, Chanock RM (1974) Asymptomatic infection of adult volunteers with a temperature-sensitive mutant of *Mycoplasma pneumoniae*. Proc Natl Acad Sci U S A 71: 4015–4019.
21. Lopes VC, Back A, Shin HJ, Halvorson DA, Nagaraja KV (2002) Development, characterization, and preliminary evaluation of a temperature-sensitive mutant of *Ornithobacterium rhinotracheale* for potential use as a live vaccine in turkeys. Avian Dis 46: 162–168.
22. Whithear KG (1993) Avian mycoplasmosis. In: Corner LA, Bagust TJ, editors. Australian standard diagnositc techniques for animal diseases. East Melbourne, Australia: CSIRO for the standing committee on agriculutre and resource management. pp. 1–12.
23. Marshall OJ (2004) PerlPrimer: cross-platform, graphical primer design for standard, bisulphite and real-time PCR. Bioinformatics 20: 2471–2472.
24. Easterday WR, Van Ert MN, Simonson TS, Wagner DM, Kenefic LJ, et al. (2005) Use of single nucleotide polymorphisms in the *plcR* gene for specific identification of *Bacillus anthracis*. J Clin Microbiol 43: 1995–1997.
25. Foster JT, Okinaka RT, Svensson R, Shaw K, De BK, et al. (2008) Real-time PCR assays of single-nucleotide polymorphisms defining the major Brucella clades. J Clin Microbiol 46: 296–301.
26. Lilliebridge RA, Tong SYC, Giffard PM, Holt DC (2011) The utility of high-resolution melting analysis of SNP nucleated PCR amplicons—An MLST based *Staphylococcus aureus* typing scheme. PLoS ONE 6: e19749.
27. Stephens AJ, Huygens F, Inman-Bamber J, Price EP, Nimmo GR, et al. (2006) Methicillin-resistant *Staphylococcus aureus* genotyping using a small set of polymorphisms. J Med Microbiol 55: 43–51.
28. U'Ren JM, Van Ert MN, Schupp JM, Easterday WR, Simonson TS, et al. (2005) Use of a real-time PCR TaqMan assay for rapid identification and differentiation of *Burkholderia pseudomallei* and *Burkholderia mallei*. J Clin Microbiol 43: 5771–5774.
29. Van Ert MN, Easterday WR, Simonson TS, U'Ren JM, Pearson T, et al. (2007) Strain-specific single-nucleotide polymorphism assays for the *Bacillus anthracis* Ames strain. J Clin Microbiol 45: 47–53.
30. Twyman RM (2005) Single nucleotide polymorphism (SNP) genotyping techniques—an overview. In: Fuchs J, Podda M, editors.Encyclopedia of Diagnostic Genomics and Proteomics.New York, USA: Marcel Dekker, Inc. pp. 1202–1207.
31. Akey JM, Sosnoski D, Parra E, Dios S, Hiester K, et al. (2001) Melting curve analysis of SNPs (McSNP): a gel-free and inexpensive approach for SNP genotyping. Biotechniques 30: 358–367.
32. Edenberg HJ, Liu Y (2009) Laboratory methods for high-throughput genotyping. Cold Spring Harb Protoc 2009: 1–9.
33. Noormohammadi AH, Jones JF, Harrigan KE, Whithear KG (2003) Evaluation of the non-temperature-sensitive field clonal isolates of the *Mycoplasma synoviae* vaccine strain MS-H. Avian Dis 47: 355–360.
34. Morrow CJ, Bell IG, Walker SB, Markham PF, Thorp BH, et al. (1990) Isolation of *Mycoplasma synoviae* from infectious synovitis of chickens. Aust Vet J 67: 121–124.
35. Noormohammadi AH, Markham PF, Whithear KG, Walker ID, Gurevich VA, et al. (1997) *Mycoplasma synoviae* has two distinct phase variable major membrane antigens, one of which is a putative hemagglutinin. Infect Immun 65: 2542–2547.
36. Olson NO (1956) Studies of infectious synovitis in chickens. Am J Vet Res 17: 747–754.
37. Whithear KG, Soeripto, Harringan KE, Ghiocas E (1990) Safety of temperature sensitive mutant *Mycoplasma gallisepticum* vaccine. Aust Vet J 67: 159–165.

Thermal Manipulation during Embryogenesis Has Long-Term Effects on Muscle and Liver Metabolism in Fast-Growing Chickens

Thomas Loyau[1], Sonia Métayer-Coustard[1], Cécile Berri[1], Sabine Crochet[1], Estelle Cailleau-Audouin[1], Mélanie Sannier[1], Pascal Chartrin[1], Christophe Praud[1], Christelle Hennequet-Antier[1], Nicole Rideau[1], Nathalie Couroussé[1], Sandrine Mignon-Grasteau[1], Nadia Everaert[2,3], Michel Jacques Duclos[1], Shlomo Yahav[4], Sophie Tesseraud[1], Anne Collin[1]*

1 INRA, UR83 Recherches Avicoles, Nouzilly, France, 2 KU Leuven, Department of Biosystems, Leuven, Belgium, 3 University of Liège, Gembloux Agro-Bio Tech, Animal Science Unit, Gembloux, Belgium, 4 Institute of Animal Science, The Volcani Center, Bet Dagan, Israel

Abstract

Fast-growing chickens have a limited ability to tolerate high temperatures. Thermal manipulation during embryogenesis (TM) has previously been shown to lower chicken body temperature (T_b) at hatching and to improve thermotolerance until market age, possibly resulting from changes in metabolic regulation. The aim of this study was to evaluate the long-term effects of TM (12 h/d, 39.5°C, 65% RH from d 7 to 16 of embryogenesis vs. 37.8°C, 56% RH continuously) and of a subsequent heat challenge (32°C for 5 h at 34 d) on the mRNA expression of metabolic genes and cell signaling in the *Pectoralis major* muscle and the liver. Gene expression was analyzed by RT-qPCR in 8 chickens per treatment, characterized by low T_b in the TM groups and high T_b in the control groups. Data were analyzed using the general linear model of SAS considering TM and heat challenge within TM as main effects. TM had significant long-term effects on thyroid hormone metabolism by decreasing the muscle mRNA expression of deiodinase DIO3. Under standard rearing conditions, the expression of several genes involved in the regulation of energy metabolism, such as transcription factor PGC-1α, was affected by TM in the muscle, whereas for other genes regulating mitochondrial function and muscle growth, TM seemed to mitigate the decrease induced by the heat challenge. TM increased DIO2 mRNA expression in the liver (only at 21°C) and reduced the citrate synthase activity involved in the Krebs cycle. The phosphorylation level of p38 Mitogen-activated-protein kinase regulating the cell stress response was higher in the muscle of TM groups compared to controls. In conclusion, markers of energy utilization and growth were either changed by TM in the *Pectoralis major* muscle and the liver by thermal manipulation during incubation as a possible long-term adaptation limiting energy metabolism, or mitigated during heat challenge.

Editor: Shu-Biao Wu, University of New England, Australia

Funding: This study was performed with the financial support of the French Agence Nationale de la Recherche (http://www.agence-nationale-recherche.fr/), Project «Jeunes Chercheuses et Jeunes Chercheurs», ANR-09-JCJC-0015-01, THERMOCHICK. T Loyau is a PhD student supported by a grant from the Ministère de l'Enseignement Supérieur et de la Recherche (http://www.enseignementsup-recherche.gouv.fr/) and Agence Nationale de la Recherche (http://www.agence-nationale-recherche.fr/), Project «Jeunes Chercheuses et Jeunes Chercheurs», ANR-09-JCJC-0015-01, THERMOCHICK. The funders had no role in study design, data collection and analysis, decision to publish, or preparation of the manuscript.

Competing Interests: The authors have declared that no competing interests exist.

* Email: anne.collin@tours.inra.fr

Introduction

Increased ambient temperature is one of the major constraints for poultry production causing lower productivity, morbidity and mortality and thus leading to economic loss [1]. European and American fast-growing strains of chickens are the main genotypes used in meat-type poultry production worldwide and they exhibit limited ability to tolerate high environmental temperatures, probably because of poorer development of cardio-vascular and respiratory organs compared to muscle [2].

Exposure of fast-growing chickens to heat induces several physiological and behavioral readjustments aimed at restoring homeostasis by reducing their resting metabolic rate [3]. Acute heat exposure induces hyperthermia [4] and causes changes in respiratory physiology and plasma ion concentrations [5], affects the thyroid axis, and increases stress markers [6] and oxidative stress in mitochondria [7]. These changes can cause metabolic disorders and may lead to a cascade of irreversible thermoregulatory events and finally death.

Different strategies have been established in order to reduce the potentially negative impact of heat exposure, including thermal manipulation during embryogenesis (TM). The technique of Piestun et al. [6] consisted of increasing the incubation temperature and relative humidity (RH) from 37.8°C and 56% RH to 39.5°C and 65% RH, 12 h/d from embryonic day (E)7 to E16.

This treatment had no effect on hatching parameters but had long-term consequences for chicken physiology. It improved their acquisition of thermotolerance by reducing body temperature in the long term and the mortality of male chickens during heat challenge [6,8–9]. A lower T_b has been associated with a better ability to adapt to heat exposure [10].

At d 34 post-hatching, the glycaemia level of TM broilers submitted to a heat challenge was increased, and TM had significant effects on respiratory parameters, increasing O_2 saturation percentage and decreasing CO_2 partial pressure in venous blood [9]. The mechanisms underlying this acquisition of thermotolerance have hardly been explored. It has been shown that heat manipulation during embryogenesis reduces O_2 consumption and heart rate in embryos, suggesting a lower resting metabolic rate and changes in the vasomotor response [11–12]. It has also been shown to enhance breast meat yield and decrease abdominal fat pad content [9,13]. However, the molecular mechanisms underlying such changes in metabolic rate and body composition of TM animals have not been identified to date. Muscle, a major tissue for metabolic heat production in endotherms [14–15], and liver, a major site of lipogenesis in avian species [16], warrant specific attention. We therefore investigated the effects of TM during embryogenesis on the expression of genes and activity of enzymes regulating energy (cell signaling, mitochondrial functions...) and protein (proteolysis and protein synthesis...) metabolisms in the *Pectoralis major* (PM) muscle and livers of chickens reared in standard conditions or exposed to a heat challenge at marketing age. Tissue samples were obtained from the TM broilers with lowest body temperature and control broilers with highest body temperature.

Materials and Methods

Chemicals

Nitrocellulose membrane, polyacrylamide solution and protein standards were purchased from Bio-Rad Laboratories (Hercules, CA, USA). Antibodies against phospho-extracellular signal-regulated protein kinase p-ERK [T202/Y204], phospho-AMP-activated protein kinase p-AMPK [T172], phospho-p38 mitogen-activated protein (MAP) kinase p-p38 [T180/Y182], phospho-ribosomal protein S6 p-S6 [S235/S236] and S6 were obtained from Cell Signaling Technology (Beverly, MA, USA). Anti-p70S6 kinase or 70 kDa ribosomal protein S6 kinase S6K1 [T389], anti-p38α and anti-ERKα antibodies were from Santa Cruz Biotechnology (Santa Cruz, CA, USA), anti-vinculin from Sigma Chemical Company (St Louis, MO, USA) and anti-AMPKα from Millipore (Paris, France). Alexa Fluor secondary antibodies were purchased from Molecular Probes (Invitrogen, Carlsbad, CA, USA).

Experimental design

All experiments were carried out in accordance with the legislation governing the ethical treatment of animals and approved by the Ethics Committee ("Comité d'Ethique en Expérimentation Animale Val de Loire", Tours, France, N° 2011-9).

One thousand Cobb 500 broiler breeder eggs were incubated in semi-commercial incubators (type 360 E, SMA Coudelou, Rochecorbon, France). Control eggs (C) were maintained at 37.8°C and 56% relative humidity (RH) during the whole incubation period [17]. Thermal manipulation treatment (TM) was applied at 39.5°C and 65% RH for 12 h/d from embryonic day (E)7 to E16 [6]. All eggs were turned through 90° every hour. At hatching, chicks were distributed in floor pens at 33°C and the

temperature was gradually decreased to 21°C at d 25 and were maintained at 21°C until d 34. Water and standard feeds were supplied *ad libitum*.

At d 32, 375 C and 363 TM animals were divided into heat-challenged and non-challenged sub-treatments. Chickens of the heat-challenged group (CCh and TMCh, respectively) were exposed at d 34 to 32°C for 5 hours, whereas non-challenged chickens remained under standard conditions (C and TM groups, respectively). The average T_b of the groups was 40.9±0.1°C (n = 10), 40.7±0.1°C (n = 11), 42.6±0.2°C (n = 10) and 42.6±0.2°C (n = 9), for C, TM, CCh and TMCh, respectively [9]. In earlier studies, Piestun et al. [6,8] reported lower T_b in TM animals than in controls reared in standard conditions and chicken with the lowest T_b were shown to have a better ability to adapt to high ambient temperature [10]. The regulation of energy and protein metabolism was therefore investigated in subsets of 8 birds among 9 to 11 per group per treatment exhibiting the lowest body temperatures in the TM groups (potentially the best acclimated) and the highest temperatures in the control groups to highlight thermoregulatory differences (C: 41.1±0.1°C, TM: 40.6±0.2°C (incubation effect: $P<0.05$), CCh: 42.8±0.1°C; TMCh: 42.5±0.2°C (incubation effect: non-significant) [9]. Chickens were slaughtered by cervical dislocation at d 34.

Tissue sampling

To characterize the pattern of expression of different candidate genes, glycolytic breast PM muscle and livers were removed and snap frozen at d 34 of age in 8 males per treatment. Breast muscle (representing 21% of body mass [9]) was studied in view of the importance of muscle mass in generating metabolic heat. The liver was chosen as a major organ regulating metabolism in birds and mammals, with an additional role in lipogenesis.

RNA extraction, reverse transcription and qPCR

RNA was simultaneously treated with DNAse and Proteinase K, and extracted from both types of tissue using the Qiagen RNAeasy mini kit (Qiagen, The Netherlands) according to the manufacturer's instructions. The amounts and purity of RNA samples were quantified using a NanoDrop ND-1000 UV-Vis Spectrophotometer (Palaiseau, France) and the integrity was checked by electrophoresis.

Five micrograms of total RNA samples were reverse-transcribed using the superscript II kit (Invitrogen, Cergy Pontoise, France) and random hexamers (GE Healthcare, Uppsala, Sweden). Real-time PCR was performed using primers reported in Table S1 that also describes the target gene functions. These genes were chosen on the basis of their involvement in the regulation of thyroid hormone and mitochondrial metabolism [18–19], their response to heat exposure [19] and their role in the regulation of nutrient utilization and cell defense against oxidative stress [20–21]. cDNA samples were subsequently amplified in real time using Sybr Green I Master kit (Roche, Mannheim, Germany) with the LightCycler 480 apparatus (Roche Diagnostics, Meylan, France). A melting curve program was applied from 65 to 95°C in 1 min for each individual sample. Each run included ultrapure water as negative control, samples in triplicate, and control cDNA corresponding to a pool of cDNA from all samples per tissue in duplicate, in addition to the real-time PCR mix.

The relative expression of each target gene was calculated according to the delta-Ct method: Ct (threshold cycle) values for a target gene were normalized to the specimen with the highest expression (minimum Ct value) for that gene, calculated according to the formula: $Q = E \times (min_{Ct} - sample_{Ct})$, where Q is the relative Ct value for a given gene, E the PCR efficiency (ranging from 1 to

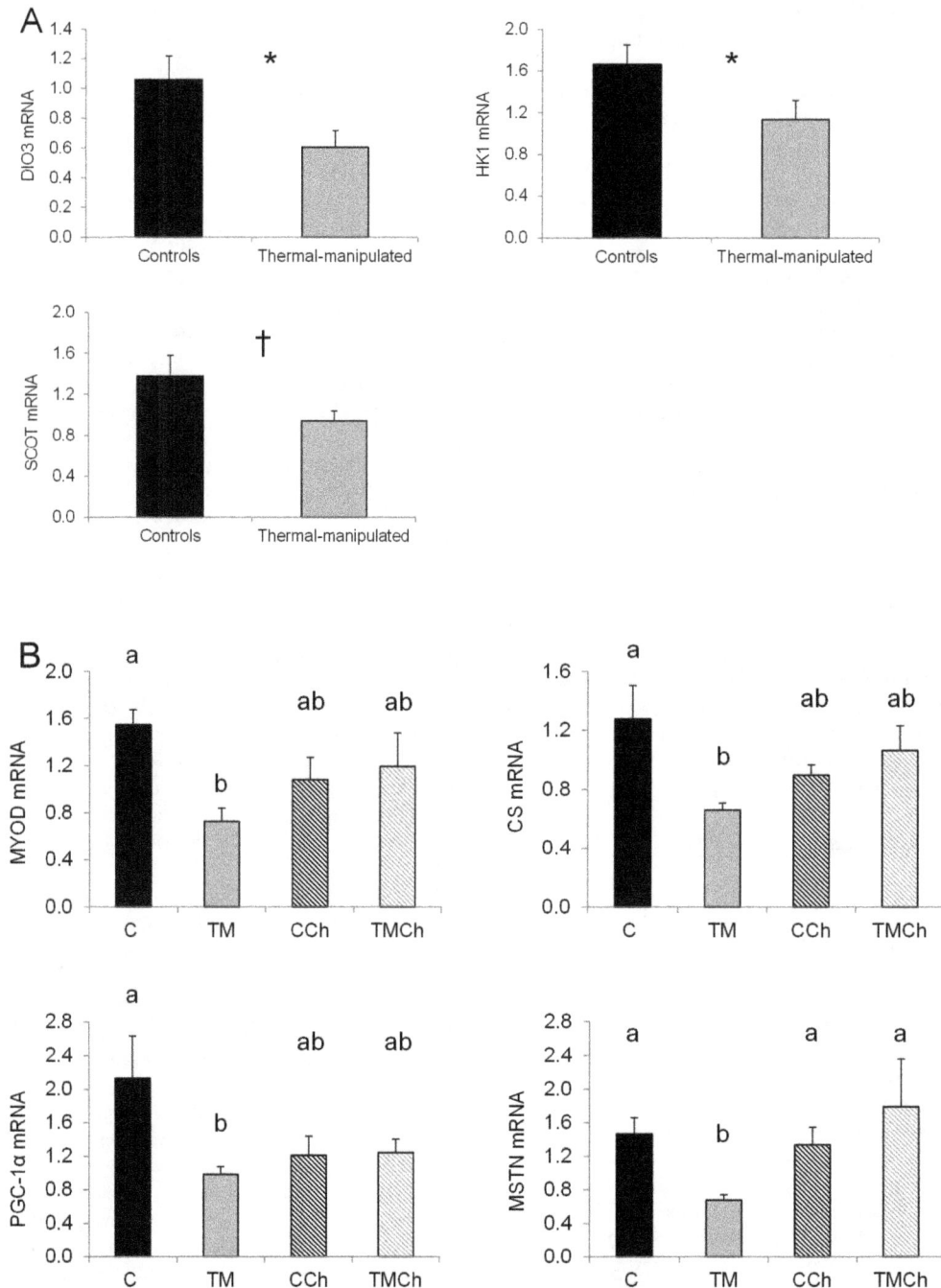

Figure 1. Levels of mRNA of genes affected by thermal manipulation during embryogenesis in the *Pectoralis major* muscle of broiler chickens at d 34. Values were standardized using geNorm factor calculated from the expression of 18S ribosomal RNA, Cytochrome b and β-actin. A) DIO3: deiodinase 3; HK1: hexokinase 1; SCOT: succinyl-CoA: 3-ketoacid CoA transferase; *: $P<0.05$; †: $P<0.10$. B) Chickens were incubated and reared in standard conditions (Controls C), thermally manipulated during embryogenesis and reared in standard conditions (TM), or incubated in standard conditions and exposed to heat challenge at d 34 (CCh) or thermally manipulated during embryogenesis and exposed to heat challenge at d 34 (TMCh). MYOD: myoblast determination protein; GLUT 8: glucose transporter 8; PGC-1α: peroxisome-proliferator-activated receptor (PPAR) γ coactivator 1α; CS: citrate synthase; DIO2: deiodinase 2. Different letters indicate significant differences between treatments (a–b, $P<0.05$) or only a tendency (A–B, $P<0.10$) when both incubation and challenge (incubation) effects or challenge(incubation) effect alone were significant (n = 8 per treatment).

2 with $100\% = 2$) calculated from the standard curve, $\min_{Ct}$ the minimum Ct value for the gene among all specimens, and sample_{Ct} the Ct value of the gene for the current specimen.

To determine a normalization factor (NF), β-actin, Cytochrome b and 18S were checked for expression stability as non-differentially expressed genes using the geNorm software [22].

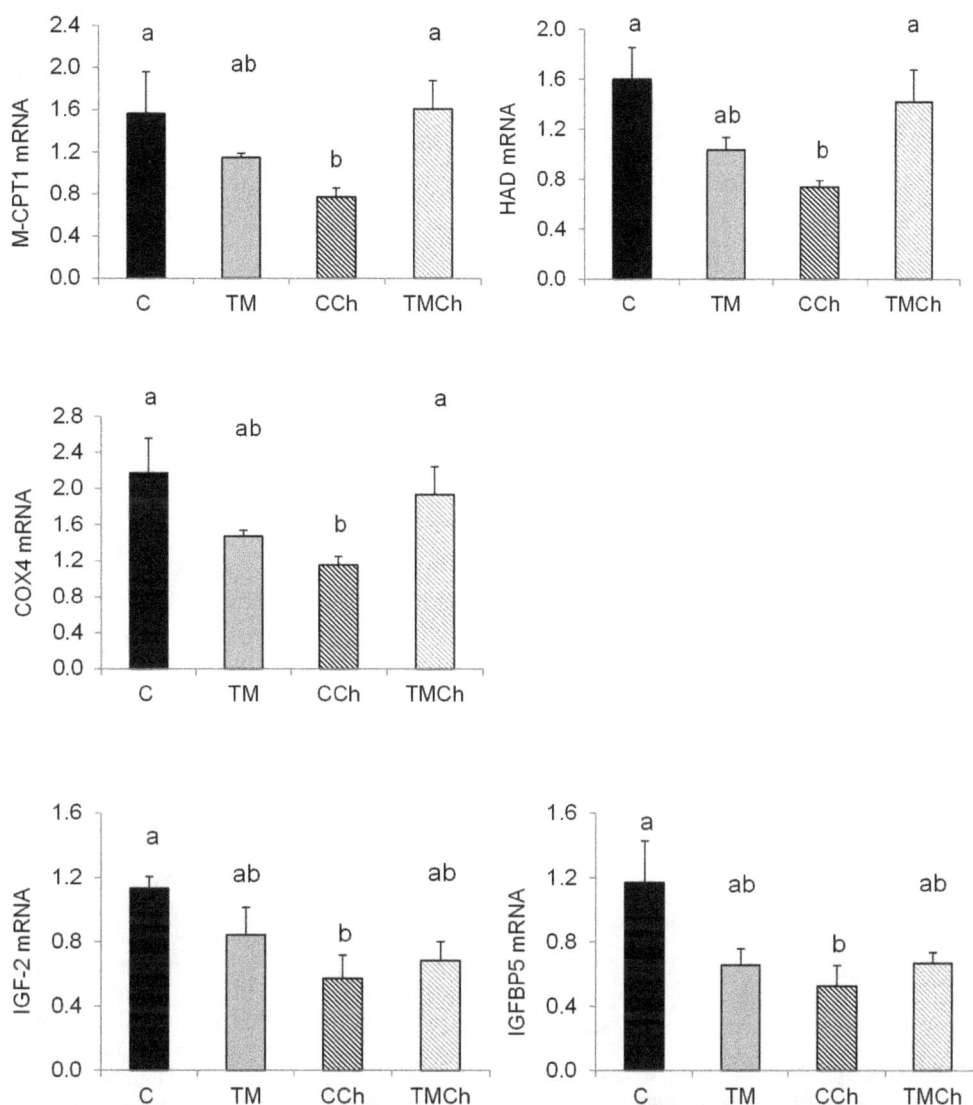

Figure 2. Levels of mRNA of genes affected by heat challenge in the *Pectoralis major* muscle of broiler control chickens at d 34. Chickens were incubated and reared in standard conditions (Controls C), thermally manipulated during embryogenesis and reared in standard conditions (TM), incubated in standard conditions and exposed to heat challenge at d 34 (CCh) or thermally manipulated during embryogenesis and exposed to heat challenge at d 34 (TMCh). Values were standardized using geNorm factor calculated from the levels of expression of 18S ribosomal RNA, Cytochrome b and β-actin. HAD: β -hydroxyacyl-CoA dehydrogenase; COX4: unit 4 of cytochrome c oxidase; IGF-2: insulin growth factor 2; IGFBP5: insulin growth factor binding protein 5. Different letters indicate significant differences between treatments (a–b, $P<0.05$) when both incubation and challenge(incubation) effects or challenge(incubation) effect alone were significant (n = 8 per treatment).

Normalized expression (NE) was calculated as the ratio of Q to NF.

Measurement of levels of β-hydroxyacyl CoA dehydrogenase (HAD), citrate synthase (CS) and lactate dehydrogenase (LDH) activity in PM muscle and liver

HAD, CS and LDH are key enzymes involved in mitochondrial β-oxidation, Krebs cycle and anaerobic glycolysis, respectively. For activity measurement, samples were thawed and homogenized in ice-cold phosphate buffer using an ultra-turrax homogenizer (Ultraturrax, Ilka-Verke, Staufen, Germany). The homogenates were sonicated for 2 minutes at 8.5–8.8 Watts and centrifuged (1,500 g, 10 minutes at 4°C) before collecting the supernatant.

The activity levels of HAD, CS and LDH were determined at 30°C using the spectrophotometric method of Bass et al. [23].

Western blotting

Western blotting (WB) was performed on muscle and liver lysates from 34-day-old chickens to analyze the effects of thermal treatments on signaling pathways involved in the regulation of protein translation and of cell stress response. Muscle and liver lysates were prepared as previously described [20]. Protein concentrations were determined using the Bio-Rad protein assay kit (Bio-Rad, USA). Tissue lysates (60 μg protein) were subjected to SDS-PAGE gel electrophoresis and Western blotting using the appropriate antibody. Membranes were also probed with an anti-vinculin antibody to monitor gel loading and to normalize data.

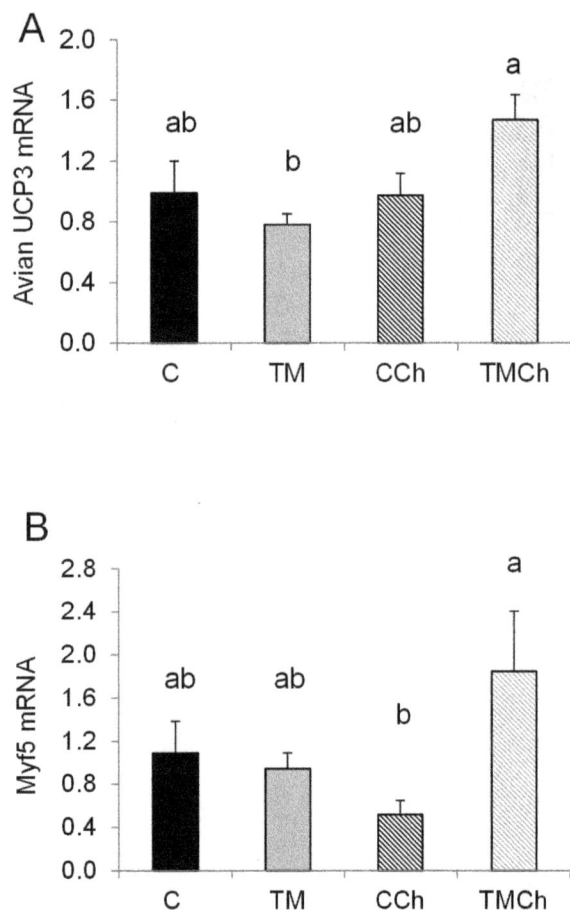

Figure 3. Other patterns of mRNA expression in the _Pectoralis major_ muscle of broiler chickens at d 34. Chickens were incubated and reared in standard conditions (Controls C), or thermally manipulated during embryogenesis and reared in standard conditions (TM), incubated in standard conditions and exposed to heat challenge at d 34 (CCh) or thermally manipulated during embryogenesis and exposed to heat challenge at d 34 (TMCh). Values were standardized using geNorm factor calculated from the expression of 18S ribosomal RNA, Cytochrome b and β-actin. A) avian UCP3: avian uncoupling protein; AdMyHC: adult isoform of myosin heavy chain; Atrogin-1. B) Myf5: myogenic factor 5. Different letters indicate significant differences between treatments (a–b, P<0.05) or only a tendency (A–B, P<0.10) when both incubation and challenge(incubation) effects or challenge(incubation) effect alone were significant (n = 8 per treatment).

After washing, membranes were incubated with an Alexa Fluor secondary antibody (Molecular Probes, Interchim, Montluçon, France). Bands were visualized by Infrared Fluorescence using the Odyssey Imaging System (LI-COR Inc. Biotechnology, Lincoln, NE, USA) and quantified by Odyssey infrared imaging system software (Application software, version 1.2).

Statistical analysis

All data of mRNA expression, protein expression and enzyme activity were analyzed using the GLM procedure of SAS (SAS Inst. Inc., Cary, NC) with the following model: $y_{ijk} = \mu + IT_i + R(IT)_{ij} + e_{ijk}$, where y_{ijk} is the parameter considered for animal k at d 34, μ the general mean, IT_i the fixed effect of incubation treatment (i = Control, TM), $R(IT)_{ij}$ the fixed effect of temperature at d 34 during the challenge j nested within incubation treatment i

(j = non-challenged or heat-challenged) and e_{ijk} the residual pertaining to animal k. The results are presented as least square means (lsmeans) of main effects: incubation treatment (TM during embryogenesis) and the heat challenge within incubation (nested effect). The gene expression values of PGC-1α, LDHA, avian UCP3, HAD, M-CPT1, COX, IGF-1, MSTN in the muscle and of DIO3 and DIO2 in the liver were log-transformed before being analyzed due to heterogeneity of variance between groups measured with the Levene's test using SAS (SAS Inst. Inc., Cary, NC).

Expression profile

The MeV software (MultiExperiment Viewer, http://www.tm4.org/mev/) was used to describe the global expression pattern of genes regulating energy metabolism that were differentially expressed as measured by qPCR for at least one of the factors (incubation effect and/or challenge intra incubation effect).

Results

Messenger RNA expression of metabolic genes in the PM muscle

Levels of messenger RNA expression in the PM muscle that were significantly affected by TM and/or heat challenge within incubation condition are presented in Figures 1 to 3. The expression of other genes studied that were not significantly affected by treatment is reported in Table S2.

Different expression profiles were observed according to the genes. Expression of the first group of genes (Figure 1A) was lower in TM than in control birds. Indeed, TM during incubation decreased the expression of three genes encoding metabolic enzymes, i.e., deiodinase 3 (DIO3) controlling the local availability of T_3, hexokinase 1 (HK1) regulating entry into the glycolytic pathway (P<0.05), and succinyl-CoA-3-ketoacid CoA transferase (SCOT), involved in the production of ketone bodies from fatty acids (P = 0.09, Figure 1A). In the other groups of genes (Figures 1B, 2 and 3) there was an interaction between TM and heat stress at 34 d. In the second group, TM decreased (P<0.05) the expression of four genes but only in unchallenged birds (Figure 1B). These genes were myogenic differentiation factor 1 protein (MYOD), citrate synthase (CS), transcription factor peroxisome-proliferator-activated-receptor gamma coactivator 1 alpha (PGC-1α) and myostatin (MSTN). In the third group mRNA expression was decreased following heat challenge but only in the control chickens (C>CCh; Figure 2). They encoded mitochondrial proteins such as muscle isoform of carnitine palmitoyl transferase (M-CPT1) involved in the entry of fatty acids into mitochondria, HAD and the cytochrome oxidase subunit 4 (COX4), but also insulin-like growth factor 2 (IGF-2) and IGF binding protein 5 (IGFBP5) regulating muscle growth. By contrast, the expression of avian uncoupling protein 3 (avian UCP3) involved in the regulation of oxidative stress was increased (P<0.05) by heat challenge but only in TM birds (TMCh>TM; Figure 3A). Expression of myogenic transcription factor 5 (Myf5) was significantly higher (P<0.05) in the TMCh group than in the CCh group (Figure 3B).

Messenger RNA expression of metabolic genes in the liver

No effect of the incubation treatment alone was observed on the expression of the candidate genes studied in liver tissue. However, the mRNA expression of some genes was affected by heat challenge within incubation conditions, as already observed in PM muscle. The hepatic mRNA expression of 5 genes involved in lipid

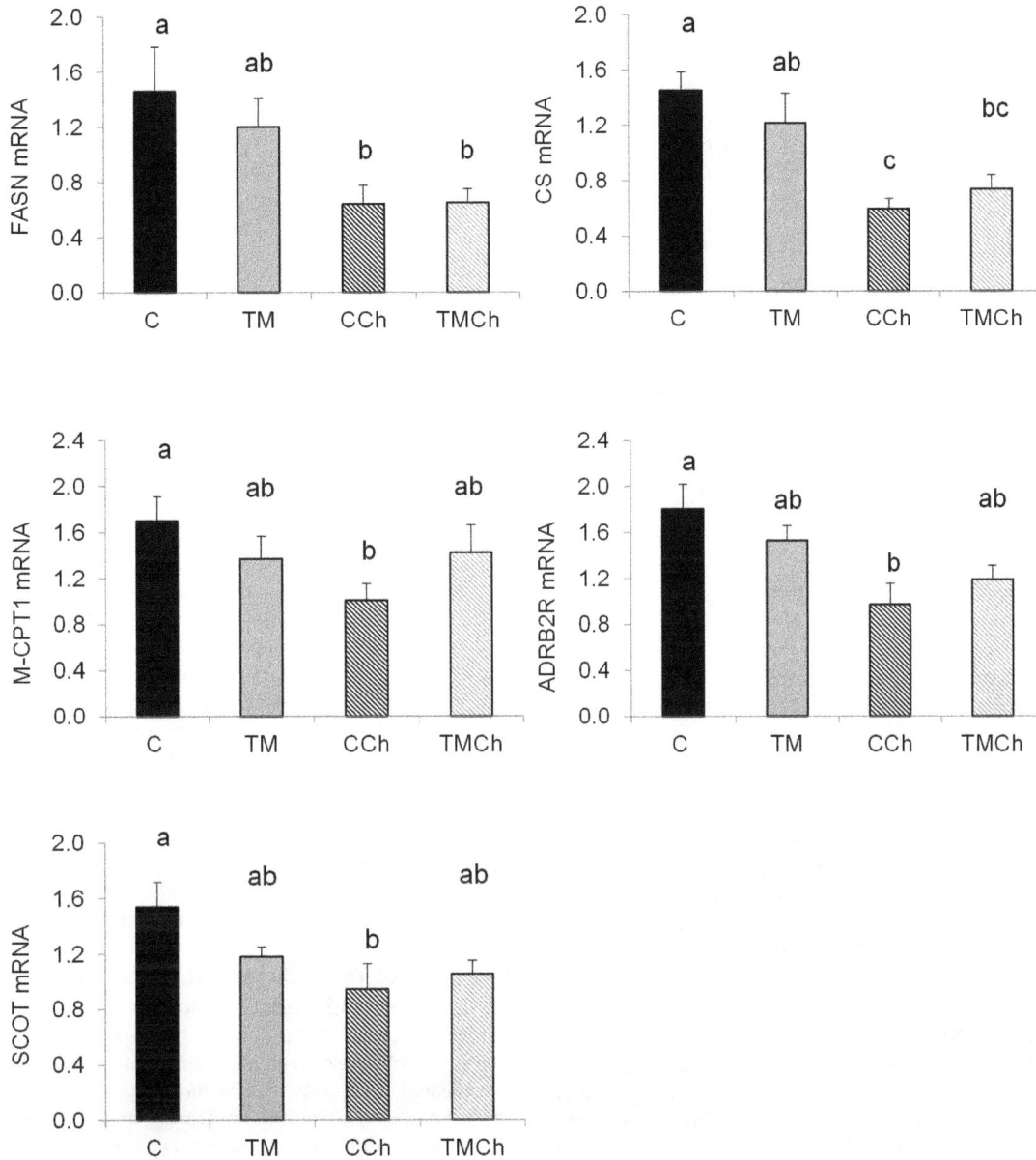

Figure 4. Levels of mRNA of genes affected by heat challenge in the livers of d34 control broiler chickens. Chickens were incubated and reared in standard conditions (Controls C), thermally manipulated during embryogenesis and reared in standard conditions (TM), incubated in standard conditions and exposed to heat challenge at d 34 (CCh), or thermally manipulated during embryogenesis and exposed to heat challenge at d 34 (TMCh). Values were standardized using geNorm factor calculated from the expression of 18S ribosomal RNA, Cytochrome b and β-actin. FASN: fatty acid synthase; CS: citrate synthase; M-CPT1: muscle isoform of carnitine palmitoyltransferase 1; SCOT: succinyl-CoA: 3-ketoacid CoA transferase; ADRB2R: beta-adrenergic receptor 2; HK2: hexokinase 2. Different letters indicate significant differences between treatments (a–b, $P<0.05$) or only a tendency (A–B, $P<0.10$) when both incubation and challenge(incubation) effects or challenge(incubation) effect alone were significant (n = 8 per treatment).

metabolism was lower ($P<0.05$) in CCh compared to the C group: i.e. fatty acid synthase (FASN) involved in lipogenesis, and CS, M-CPT1, SCOT and β-adrenergic receptor 2 (ADRB2R), all of which are involved in the regulation of fatty acid utilization (Figure 4). However, the expression of these genes was not different between TM and TMCh chickens. The mRNA expression of DIO3 was 25-fold higher ($P<0.05$) in heat-challenged (CCh and TMCh) than in non-challenged (C and TM) birds (Figure 5). Deiodinase DIO2 mRNA expression was increased 10 to 21 fold ($P<0.05$) in TM chickens compared to all other groups (Figure 5). The expression of other candidate genes

was not significantly changed in the liver. These are presented in Table S3.

Enzyme activity

Levels of HAD, LDH and CS activity in the *Pectoralis major* muscle were not affected by incubation or by heat challenge within incubation treatment (Table 1). The levels of activity of HAD and LDH in the liver were not affected by treatment, whereas CS activity was decreased ($P<0.05$) by the incubation treatment in liver tissue (Table 1).

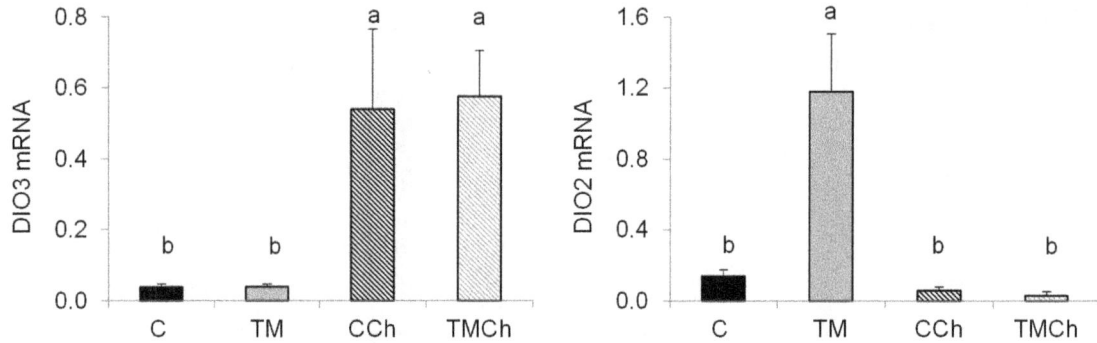

Figure 5. Levels of mRNA of genes altered by heat challenge within incubation treatment in the liver of d 34 broiler chickens. Chickens were incubated and reared in standard conditions (Controls C), thermally manipulated during embryogenesis and reared in standard conditions (TM), incubated in standard conditions and exposed to heat challenge at d 34 (CCh), thermally manipulated during embryogenesis and exposed to heat challenge at d 34 (TMCh). Values were standardized using geNorm factor calculated from the expression of 18S ribosomal RNA, Cytochrome b and β-actin. DIO3: deiodinase 3; DIO2: deiodinase 2; SREBP1: sterol regulatory element binding protein 1. Different letters indicate a tendency to differences between treatments (A–B, $P<0.10$) when both incubation and challenge(incubation) effects or challenge(incubation) effect alone were significant (n = 8 per treatment).

Kinase phosphorylation in the muscle

In the PM muscle, phosphorylation of the extracellular signal-regulated protein kinase (ERK) that is involved in cell survival, proliferation, differentiation and in insulin signaling (Figure 6) tended to be greater ($P = 0.06$) in TM than in control birds whether challenged or not (Figure 7A). Similarly, the phosphorylation level of p38 MAP (mitogen-activated protein) kinase, that is involved in apoptosis and cellular stress, was higher ($P<0.05$) in TM than in control birds (Figure 7A). The phosphorylation level of the energy sensor AMP-activated protein kinase (AMPK) was increased ($P<0.01$) in the TMCh group compared to all other groups (Figure 7B). In the signaling pathway regulating protein translation, lower phosphorylation levels of p70S6 kinase (i.e. S6K1) and of ribosomal protein S6 ($P<0.01$) were found after heat challenge for both control and TM birds (Figure 7C).

Kinase phosphorylation in the liver

The p-S6/S6 ratio in the liver was significantly higher in controls than in all other groups (Figure 8; $P_{incubation}<0.01$; $P_{challenge(incubation)}<0.05$). This was due to both a lower phosphorylation level of p-S6/vinculin in TM than in C ($P<0.10$) and a lower S6/vinculin ratio in C and TM than in TMCh chickens. The phosphorylation levels of other kinases studied in the liver (AMPK, ERK, p38) were not significantly affected, regardless to the nature of the incubation treatment or the challenge within incubation treatment (Table S4).

Discussion

The aim of this study was to investigate the effects of TM during embryogenesis, combined or not with a subsequent post-hatching heat challenge on candidate genes and pathways controlling body composition, muscle biology, protein and energy metabolism. Previous studies using the same experimental model have provided evidence that the chicken physiology and nutritional partition can be affected by such treatments [6,9]. Chickens with lowest T_b for TM and with highest T_b for C were chosen in larger groups in order to better highlight differences between the potentially most and less thermotolerant chickens, respectively. One striking result was that four major patterns of response to thermal treatments were observed among the genes and pathways investigated. First, for a large number of genes or pathways tested there were no

significant differences in the responses of chickens to either treatment, or the response to the heat challenge was similar in both TM and control birds. Secondly, mRNA expression or activity for a group of genes or enzymes was clearly reduced by the TM of the embryo, most differences being observed at 21°C, and more rarely at 32°C. This pattern of expression was mainly found in the muscle tissue. A third group comprised genes for which expression was affected by heat challenge in control but not in TM birds, suggesting a specific impact of TM of the embryo on bird's response to heat stress. This pattern of expression occurred in both tissue types. Finally, some genes were affected by the heat challenge only in TM chickens, also suggesting a role of TM in defining the chicken's response to heat challenge.

Response of TM birds related to the control of body composition and muscle biology

Previous results [8–9,11] have shown positive effects of TM on breast muscle yield and abdominal fat content. We therefore investigated the effects of TM on key genes involved in muscle growth and properties, and in the regulation of body fatness in the present study. Interestingly, under standard conditions at 21°C the muscle mRNA expression of MYOD, a transcription factor involved in the regulation of muscle differentiation [24], decreased in TM chickens. This result suggests that the TM treatment could interfere with the transition between proliferation and differentiation of myoblasts in the muscle. Previous results have shown that TM during late embryogenesis [25] or early post-hatching [26] enhances muscle cell proliferation compared to control conditions and, in the case of posthatch treatment, breast muscle IGF-1 mRNA expression. In our study, Myf 5, that regulates early muscle development events, was not different between TM and control groups when studied under standard rearing conditions (21°C), but was significantly increased following heat challenge in TM birds. The expression of MSTN, a negative regulator of muscle growth, was decreased in the TM group at 21°C, while the mRNA expression of IGF-1, a positive regulator of growth, also tended to decrease in this condition (Table S2). In this study, the heat challenge also affected certain regulators of muscle growth and development, lowering both IGF-2 and IGFBP5 mRNA expression involved in controlling cell survival, differentiation and apoptosis [27], but only in control broilers. These results together suggest that the thermal treatment of embryos may durably affect

Table 1. Effects of thermal manipulation during embryogenesis (TM) and d34 heat challenge on levels of enzyme activity (IU/g protein) in the *Pectoralis major* muscle (PM) and livers of 34-day-old broiler chickens.

		C	TM	CCh	TMCh	P-value Incubation effect	P-value Challenge (Incubation) effect
PM	HAD	12.30±0.91	13.2±0.83	12.19±0.83	13.53±0.91	0.21	0.96
	LDH	5420±669	5235±605	4932±635	4497±669	0.63	0.63
	CS	21.43±1.03	21.87±0.93	21.92±0.98	20.68±1.03	0.69	0.66
Liver	HAD	60.21±7.28	66.85±6.58	78.08±6.90	68.26±7.28	0.82	0.22
	LDH	72.43±7.83	75.58±7.08	89.34±7.42	78.10±7.83	0.60	0.30
	CS	18.05±0.91	15.96±0.82	19.53±0.86	17.04±0.91	0.01	0.35

HAD: β-hydroxyacyl CoA dehydrogenase, CS: citrate synthase and LDH: lactate dehydrogenase activity in PM and Liver. Chickens were incubated and reared in standard conditions (Controls C, n = 9), thermally manipulated during embryogenesis and reared in standard conditions (TM, n = 11), incubated in standard conditions and exposed to heat challenge at 34 d (CCh, n =10) or thermally manipulated during embryogenesis and exposed to heat challenge at 34d (TMCh, n = 9).

the regulation of muscle development, which could contribute to the differential breast yields observed in previous studies between TM and control chickens [9,25]. However, they also indicate possible interference of incubation conditions with the response to subsequent heat challenge, as shown by results concerning the IGF system.

We further studied the regulation of muscle growth and metabolism by measuring the mRNA expression of some key genes involved in proteolysis and by investigating the phosphorylation levels of several kinases regulating protein translation. Indeed, previous studies in our laboratory [19,28] had provided evidence of the regulation of both protein synthesis and degradation by chronic heat exposure. In our experimental conditions, the thermal manipulation of embryos did not significantly affect regulators of protein degradation and synthesis. However, acute heat exposure at d 34 affected some of the genes and proteins involved in protein turnover. The phosphorylation levels of S6K1 and S6, two kinases regulating protein synthesis [29], were dramatically decreased following heat challenge, suggesting lower stimulation of protein synthesis in challenged birds, as already described in the skeletal muscle of heat-exposed chickens by Temim et al. [28]. A tendency was also observed for atrogin-1, involved in protein degradation, to be upregulated during heat challenge. Incubation conditions thus do not seem to affect protein turnover in the breast muscle of 34-day-old chickens directly nor their response to acute heat stress that did not have different effects on regulators of protein turnover in both the control and TM birds.

As reported by Ain Baziz et al. [30] and Lu et al. [31], carcass adiposity is either unchanged or higher in heat-exposed chickens than in control fast-growing chickens. On the other hand, abdominal fatness is lower in TM than in control broilers [9,13]. We therefore investigated the regulation of lipogenic and lipolytic pathways in the liver, the main site of lipid synthesis in birds [16]. Surprisingly, there was no effect of the incubation conditions on the hepatic mRNA expression of FASN, a key enzyme controlling hepatic lipogenesis in chickens, or on SREBP-1, M-CPT1, SCOT, HAD or PPARδ that are all involved in fatty acid utilization. This suggests that post-transcriptional regulation of lipogenic enzyme content or activity and blood lipid transfer may explain the decrease in adiposity previously observed in TM compared to control chickens. Regulation of lipid metabolism by heat exposure was observed. Indeed, a 50% decrease in mRNA expression of FASN was induced by heat challenge, suggesting a considerable negative effect on liver lipogenic activity. Concomitantly, the expression of M-CPT1 and SCOT genes, that regulate fatty acid oxidation and ketone body production, respectively, was also decreased by heat challenge, especially in the control group. Therefore, the regulation of both lipogenic and lipolytic pathways appeared to be affected by acute heat stress, although no change in body composition due to heat challenge was observed in our previous study [9].

Control of heat production and energy metabolism

Thermal manipulation during embryogenesis was reported to decrease body temperature and the plasma triiodothyronine (T_3) concentration that controls heat production and metabolism [32–33], until 28 d in fast-growing chickens [9]. In the present study, a subsample of TM chickens originating from the same study but specifically chosen for low body temperatures did not show lower plasma T_3 concentrations than controls. Deiodinase DIO2 is involved in the conversion of inactive thyroid hormones T_4 and reverse T_3 (rT_3) into active thyroid hormones T_3 and diiodothyronine T_2. Deiodinase DIO3 converts T_3 into T_2 and T_4 into rT_3

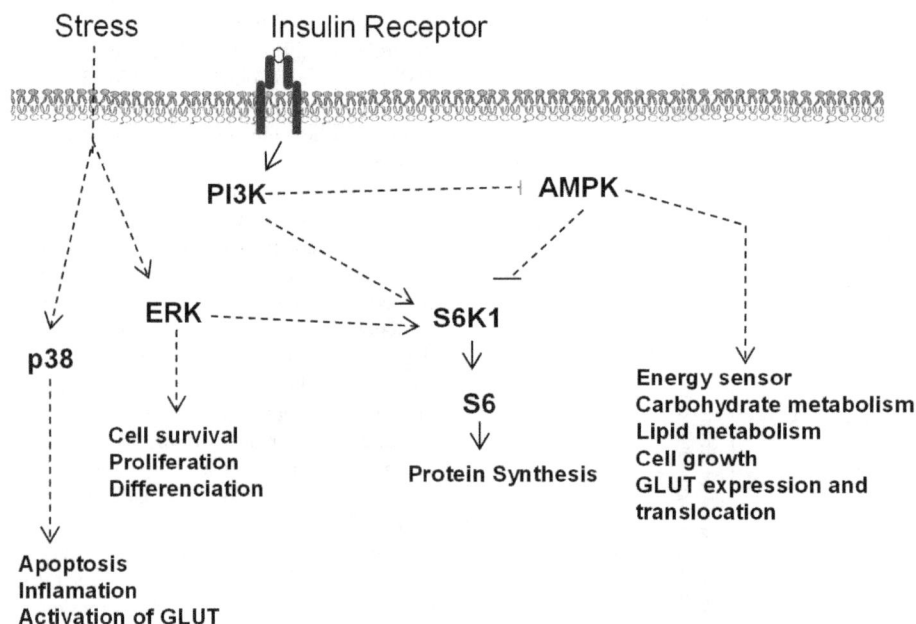

Figure 6. Stress and insulin signaling pathways. ERK: extracellular signal-regulated protein kinase; p38: p38 mitogen-activated protein kinase (p38 MAPK); AMPK: AMP-activated protein kinase; S6K1: p70 S6 kinase or 70 kDA ribosomal protein S6 kinase; S6: ribosomal protein S6. B) Phosphorylation levels of ERK and p38 MAPK (n = 8 per treatment).

[34]. Our results showed higher DIO3 expression in the liver, the major organ converting plasma T_4 into T_3 [34], after heat challenge. This is consistent with the decrease in plasma T_3 previously reported during heat exposure in birds and mammals [6,35]. However, the lack of difference between C and TM chickens for DIO3 expression and the specific increase in DIO2 expression in TM birds at 21°C might explain why plasma T_3 concentrations of TM birds were not lower than those of control birds in the present conditions.

Nevertheless, TM has been shown to decrease O_2 consumption of embryos, indicating a potentially lower metabolic rate in these animals [11–12]. It has also been suggested that TM could affect heat production *via* active local thyroid hormone concentrations. Our study showed that the DIO3 mRNA expression in the breast muscle was significantly decreased by TM in challenged and unchallenged groups, while DIO2 tended to be decreased by TM only in non-challenged chickens (Table S2). These findings suggest that TM has a long-term overall negative effect on thyroid hormone metabolism in the breast muscle. However, the physiological impact of such changes remains to be determined, since the activity of deiodinase enzymes is not necessarily correlated with their mRNA expression.

As a transcriptional gene coactivator, T_3 binds to its receptor (TR) to interact with thyroid hormone receptor response elements on DNA. A candidate gene for regulating thyroid-stimulated metabolic pathways and mitochondrial biogenesis is transcription factor Peroxisome-Proliferator-Activated-Receptor γ Coactivator 1 α (PGC-1α) [36–37]. Indeed, the increase in plasma T_3 levels and functional maturation of thyroid hormones in chickens coincides with an upregulation of PGC-1α during embryogenesis [38]. It has also been reported that cold exposure upregulates PGC-1α expression in skeletal muscle during establishment of endothermy in birds (around E15; [39]), while one week of chronic heat exposure reduced its mRNA expression in 4-wk-old broiler chickens [19]. In our conditions, PGC-1α was significantly lower in the PM muscle of TM animals than in C, suggesting subsequent

modifications in the regulation of genes involved in mitochondrial function. Energy production pathways might thus be modified in the long term by incubation conditions in chickens reared at 21°C, as indicated by decreased levels of gene expression of HK1 (key enzyme in glycolysis), and CS (a key enzyme in the Krebs cycle) in TM, as compared to C chicken muscles. SCOT, involved in the production of ketone bodies, also tended to be affected in muscle by the TM treatment. The heat challenge had no additional effect on the expression of these genes in the muscle of TM animals. These results suggest that the regulation of both mitochondrial metabolism and glycolysis may have been affected in the long term by TM, probably contributing to an overall decrease in energy metabolism in the muscle tissue of TM animals characterized by lower body temperature.

Levels of expression of M-CPT1, HAD and COX4 mRNA were decreased by the heat challenge in the PM muscle of control but not of TM chickens. This indicates that the TM treatment may limit the impact of the heat challenge on the expression of these genes involved in the regulation of the β-oxidation of lipids and of the respiratory chain, respectively [18]. Although Azad et al. [40] had previously reported decreased HAD activity in the muscle of chickens exposed to 34°C for 15 d we did not observe any effect of TM or heat challenge on the activity of enzymes involved in muscle energy metabolism in the present study, probably due to the lower intensity and shorter duration of the thermal exposure applied.

In order to obtain an overall picture of the expression pattern of target genes regulating energy metabolism in the PM muscle (controlling mitochondrial metabolism, fatty acid utilization or glycolytic metabolism), we represented differentially expressed genes for at least one factor (incubation and/or challenge intra incubation condition) on the same Figure. Our results showed that expression of genes regulating energy metabolism tended overall to be lower in TM, TMCh and CCh chickens than in controls (Figure S1). Metabolic heat production may thus be as reduced in TM broilers as in heat-challenged animals. This suggests that

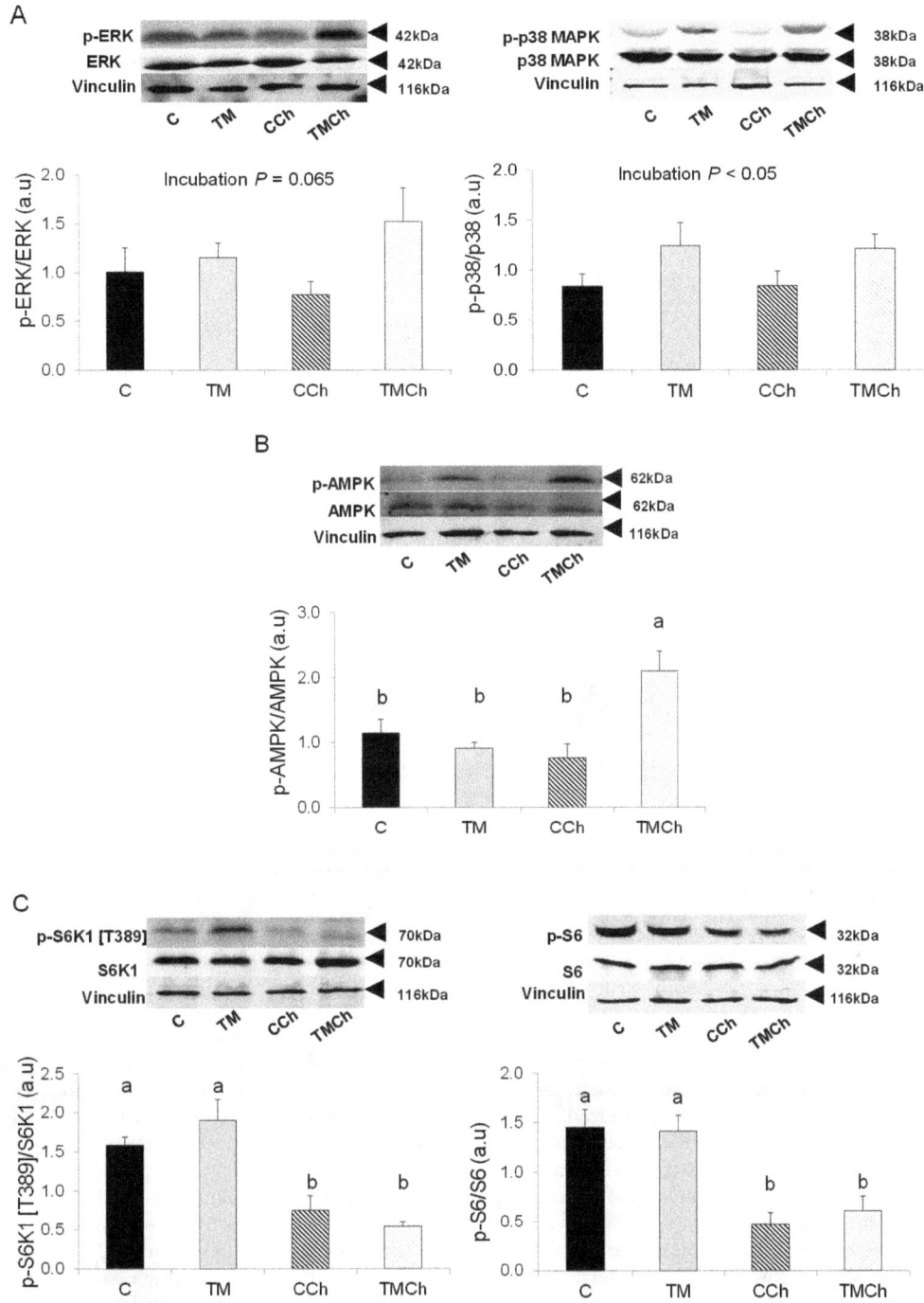

Figure 7. Phosphorylation levels of kinases involved in the regulation of protein and energy metabolism and cellular stress in the _Pectoralis major_ muscle. Chickens were incubated and reared in standard conditions (Controls C), thermally manipulated during embryogenesis and reared in standard conditions (TM), incubated in standard conditions and exposed to heat challenge at d 34 (CCh), or thermally manipulated during embryogenesis and exposed to heat challenge at d 34 (TMCh). All western-blots were performed using anti-vinculin antibody as protein loading control. Results are presented as phosphorylated (p-) to total protein ratios. ERK: extracellular signal-regulated protein kinase; p38: p38 mitogen-activated protein kinase (p38 MAPK); AMPK: AMP-activated protein kinase; S6K1: p70 S6 kinase or 70 kDA ribosomal protein S6 kinase; S6: ribosomal protein S6. A) Phosphorylation levels of ERK and p38 MAPK. B) Phosphorylation level of AMPK. C) Phosphorylation levels of S6K1 and S6. Different letters indicate significant differences ($P<0.05$) between treatments (a–b) when both incubation and challenge(incubation) effects or challenge(incubation) effect alone were significant (n = 8 per treatment).

effective thermal manipulation inducing low T_b may already prepare animals to tolerate high ambient temperatures by downregulating key genes involved in energy production pathways, meaning that subsequent acute heat exposure induces no or only slight further modification of the expression of these genes.

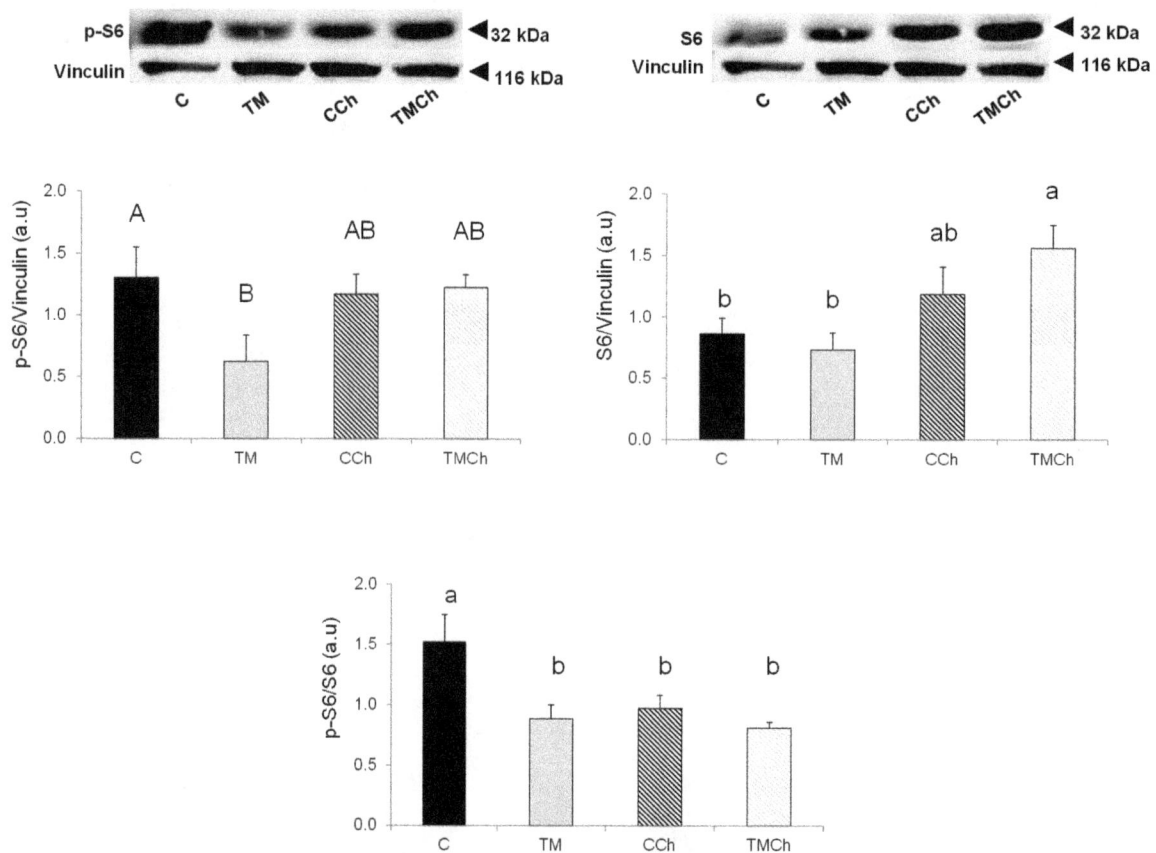

Figure 8. Phosphorylation levels of kinases involved in the regulation of protein metabolism in the liver. Chickens were incubated and reared in standard conditions (Controls C), thermally manipulated during embryogenesis and reared in standard conditions (TM), incubated in standard conditions and exposed to heat challenge at d 34 (CCh), or thermally manipulated during embryogenesis and exposed to heat challenge at d 34 (TMCh). All western-blots were performed using anti-vinculin antibody as protein loading control. Results are presented as phosphorylated (p-) to total protein ratios. S6: ribosomal protein S6. Different letters indicate significant differences between treatments (a–b, $P<0.05$) or only a tendency (A–B, $P<0.10$) when both incubation and challenge(incubation) effects or challenge(incubation) effect alone were significant (n = 8 per treatment).

The changes in mRNA expression observed in the muscle in response to TM were concomitant with modifications of the activation of AMPK involved in energy sensing. Indeed, in the present study, we found an increase in AMPK phosphorylation in TMCh chickens as compared to all other groups. AMPK has been shown to trigger skeletal muscle glucose utilization in chicken embryos [41]. In mammals activation of AMPK induces the membrane translocation of the GLUT transporter in skeletal muscle [42]. We previously demonstrated that the regulation of glucose utilization was modulated by combined embryo and postnatal thermal treatment, with higher plasma glucose concentrations in TMCh chickens than in all other groups, despite unchanged plasma insulin concentrations in TMCh and CCh chickens [9]. The higher phosphorylation level of AMPK in TMCh chickens might thus represent a signal inducing the transport of glucose to the skeletal muscle *via* translocation of the glucose transporter in response to high glycaemia. This might also be the result of an increased need for energy production pathways in response to the acute heat challenge in TM chickens characterized in standard conditions by down-regulated ATP-generating pathways as suggested by lower levels of PGC-1α mRNA expression.

In order to characterize the metabolic changes induced by our treatments in major organs regulating body composition and animal metabolism, we also focused on target pathways controlling hepatic lipogenesis and energy utilization. In the liver, citrate synthase activity was lower in TM than in C animals, possibly reflecting lower intensity of energy transfer in mitochondria affected by the embryo treatment, and consistent with an overall decrease in metabolic intensity in TM birds. This effect was however not the same as that observed at the mRNA level, where the expression of CS was lower at 32°C than at 21°C only in control chickens. Moreover, gene expression in the muscle and in the liver was affected differently by treatments, and only DIO2 mRNA expression in the latter was changed by TM. In accordance with previous results [43–44] showing hepatic metabolic modifications during heat exposure, lower levels of mRNA expression of SCOT, M-CPT1, the β-adrenergic receptor ADRB2R and of HK2, were found in the liver following heat challenge. However, these lower expression levels were mainly observed in control birds, and were intermediate in TM chickens. This might reflect a possible limitation of the heat challenge effect on fatty acid mitochondrial utilization, production of ketone bodies, response to β-adrenergic pathway and glycolysis, in the livers of TM chickens compared to control chickens.

Mechanisms involved in the stress responses of chickens

Thermal manipulation during incubation induced specific effects on mechanisms controlling stress responses and apoptosis in the PM muscle, but not in the liver. One of these was the MAP kinase signaling pathway. It was recently shown that the activation of p38 MAPK is induced by oxidative stress and that its upregulation is responsible for the downregulation of generation of free radicals and for *in vitro* survival of mammalian cell lines [45]. The upregulation of p38 MAPK observed in TM chickens and, to a lesser extent, of the MAP kinase ERK also involved in cell stress response, may represent an adaptive mechanism for regulating oxidative stress and cell survival in both thermoneutral and heat challenge conditions. UCP is also known to control oxidative stress. Expression of avian UCP3 mRNA was upregulated in the muscle of TMCh compared to TM birds. Avian UCP3 has previously been shown to be affected by nutrition [46], genotype [18,47], and ambient temperature [48–50], and its expression is regulated by the thyroid axis and the beta-adrenergic pathway [20–21,51]. It is thought to protect muscle tissue from oxidative injury by reducing oxidative stress that is particularly increased during acute heat exposure [52]. The overexpression of avian UCP3 in the muscle during heat challenge in TM animals may thus contribute to protection against oxidative stress, whereas this pathway seemed not to be affected in heat-challenged control chickens. In addition to these different responses to heat-induced oxidative stress, we have previously reported potentially lower stress responses in TM animals during heat challenge, as indicated by modified plasma corticosterone concentration and blood heterophil to lymphocyte ratio [9], a well-known marker of the stress response in avian species [53].

To conclude, chickens submitted to TM during embryogenesis and characterized by low T_b exhibited long-term modifications of their metabolism. TM may contribute to a decrease in the intensity of energy metabolism in the liver and breast muscle, potentially resulting in lower heat production, or may mitigate the effects of heat stress later in life. We also report modifications of pathways regulating muscle cell stress responses and development that may contribute to the greater tolerance of thermal-manipulated chickens subsequently exposed to heat stress. These potentially "programing" effects of thermal manipulation of the embryo may be partly due to epigenetic regulation that has already been suggested to be involved in the modification of gene expression in the case of early post-hatch thermal exposure. Whether such mechanisms are involved in the regulation observed in the present study remains to be elucidated.

Supporting Information

Figure S1   Expression profiles of target genes involved in energy metabolism and differentially expressed in at least one condition (whether incubation treatment and/or heat challenge (intra incubation)). Genes included were peroxisome proliferator activated receptor coactivator 1 alpha, citrate synthase, glucose transporter 8, hexokinase 1, succinyl-CoA: 3-ketoacid CoA transferase, cytochrome oxidase subunit 4, β-hydroxyl-acyl CoA dehydrogenase, muscle isoform of carnitine palmitoyl transferase 1 with the average expression of these genes in red. Blue dashes correspond to the highest or lowest points of the average line (n = 8 per treatment).

Table S1   Primers used for qRT-PCR.

Table S2   Levels of m-RNA expression in the *Pectoralis major* muscle of 34-day-old broiler chickens.

Table S3   Levels of m-RNA expression in the livers of 34-day-old broiler chickens.

Table S4   Levels of phosphorylation of kinases in the livers of 34-day-old broiler chickens.

File S1   Individual data of mRNA expressions, phosphorylation levels of kinases and metabolic enzyme activities in the *Pectoralis major* muscle and in the livers of 34-day-old broiler chickens.

Acknowledgments

The authors wish to thank T. Bordeau, E. Godet, E. Baeza, M. Couty, L. Bouyer, V. Coustham (INRA, UR83 Recherches Avicoles, F-37380 Nouzilly, France) and F. Mercerand, J. Delaveau, C. Rat, H. Rigoreau, J.M. Brigant, O. Callut, N. Sellier (INRA, UE1295 Pôle d'Expérimentation Avicole de Tours, F-37380 Nouzilly).

Author Contributions

Conceived and designed the experiments: AC SMC SMG CHA MJD CB ST CP SY. Performed the experiments: TL SMC SC ECA MS PC NC AC NR. Analyzed the data: TL AC SMG CHA MS SMC SC ECA. Contributed reagents/materials/analysis tools: AC SMC CB SC ECA PC CHA NR SMG MJD. Contributed to the writing of the manuscript: TL AC SMC CB SC ECA MS PC CP CHA NR NC SMG NE MJD SY ST.

References

1. St-Pierre NR, Cobanov B, Schnitkey G (2003) Economic Losses from Heat Stress by US Livestock Industries. J Dairy Sci 86, Suppl: E52–E77.

2. Yahav S, Collin A, Shinder D, Picard M (2004) Thermal manipulations during broiler chick embryogenesis: effects of timing and temperature. Poult Sci 83: 1959–1963.

3. Geraert PA, Padilha JC, Guillaumin S (1996) Metabolic and endocrine changes induced by chronic heat exposure in broiler chickens: biological and endocrinological variables. Br J Nutr 75: 205–216.

4. Yahav S (2009) Alleviating heat stress in domestic fowl – different strategies. World Poult Sci J 65: 719–732.

5. Arad Z, Marder J (1983) Acid-base regulation during thermal panting in the fowl (Gallus domesticus): Comparison between breeds. Comp Biochem Physiol A. 74: 125–130.

6. Piestun Y, Shinder D, Ruzal M, Halevy O, Brake J, et al. (2008) Thermal manipulations during broiler embryogenesis: effect on the acquisition of thermotolerance. Poult Sci 87: 1516–1525.

7. Mujahid A, Akiba Y, Warden CH, Toyomizu M (2007) Sequential changes in superoxide production, anion carriers and substrate oxidation in skeletal muscle mitochondria of heat-stressed chickens. FEBS Lett 581: 3461–3467.

8. Piestun Y, Druyan S, Brake J, Yahav S (2013) Thermal manipulations during broiler incubation alters performance of broilers to 70 days of age. Poult Sci 92: 1155–1163.

9. Loyau T, Berri C, Bedrani L, Métayer-Coustard S, Praud C, et al. (2013) Embryo thermal manipulations modifies the physiology and body compositions of broiler chickens reared in floor pens without altering breast meat processing quality. J Anim Sci 91: 3674–3685.

10. De Basilio V, Requena F, León A, Vilariño M and Picard M (2003) Early age thermal conditioning immediately reduces body temperature of broiler chicks in a tropical environment. Poult Sci 82: 1235–1241.

11. Tona K, Onagbesan O, Bruggeman V, Collin A, Berri C, et al. (2008) Effects of heat conditioning at d 16 to 18 of incubation or during early broiler rearing on embryo physiology, post-hatch growth performance and heat tolerance. Arch Geflugel 72: S75–S83.

12. Piestun Y, Halevy O, Yahav S (2009) Thermal manipulations of broiler embryos, the effect on thermoregulation and development during embryogenesis. Poult Sci 88: 2677–2688.

13. Piestun Y, Halevy O, Shinder D, Ruzal M, Druyan S, et al. (2011) Thermal manipulations during broiler embryogenesis improves post-hatch performance under hot conditions. J Therm Biol 36: 469–474.

14. Rolfe DF, Brand MD (1996) Contribution of mitochondrial proton leak to skeletal muscle respiration and to standard metabolic rate. Am J Physiol Cell Physiol 271: C1380–C1389.

15. Ueda M, Watanabe K, Sato K, Akiba Y, Toyomizu M (2005) Possible role for avPGC-1[alpha] in the control of expression of fiber type, along with avUCP and avANT mRNAs in the skeletal muscles of cold-exposed chickens. FEBS Lett 579: 11–17.

16. Leveille GA, O'Hea EK, Chakrabarty K (1968) In vivo lipogenesis in the domestic chicken. Proc Soc Exp Biol Med 128: 398–401.

17. Bruzual JJ, Peak SD, Brake J, Peebles ED (2000) Effects of relative humidity during the last five days of incubation and brooding temperature on performance of broiler chicks from young broiler breeders. Poult Sci 79: 1385–1391.Skiba-Cassy S, Collin A, Chartrin P, Médale F, Simon J, et al. (2007) Chicken liver and muscle carnitine palmitoyltransferase 1: nutritional regulation of messengers. Comp Biochem Physiol B Biochem Mol Biol 147: 278–287.

18. Collin A, Swennen Q, Skiba-Cassy S, Buyse J, Chartrin P, et al. (2009) Regulation of fatty acid oxidation in chicken (*Gallus gallus*): interactions between genotype and diet composition. Comp Biochem Physiol B Biochem Mol Biol 153: 171–177.

19. Boussaid-Om Ezzine S, Everaert N, Metayer-Coustard S, Rideau N, Berri C, et al. (2010) Effects of heat exposure on Akt/S6K1 signaling and expression of genes related to protein and energy metabolism in chicken (Gallus gallus) pectoralis major muscle. Comp Biochem Physiol B Biochem Mol Biol 157: 281–287.

20. Joubert R, Métayer Coustard S, Swennen Q, Sibut V, Crochet S, et al. (2010) Beta-adrenergic system is involved in the regulation of the expression of the avian uncoupling protein in chicken. Domest Anim Endocrinol 38: 115–125.

21. Joubert R, Métayer-Coustard S, Crochet S, Cailleau-Audouin E, Dupont J, et al. (2011) Regulation of the expression of the avian uncoupling protein 3 by isoproterenol and fatty acids in chick myoblasts: possible involvement of AMPK and PPARalpha? Am J Physiol Regul Integr Comp Physiol 301: R201–R208.

22. Vandesompele J, De Preter K, Pattyn F, Poppe B, Van Roy N, et al. (2002) Accurate normalization of real-time quantitative RT-PCR data by geometric averaging of multiple internal control genes. Genome Biology 3(7): RESEARCH0034. Epub 2002 Jun 18.

23. Bass A, Brdiczka D, Eyer P, Hofer S, Pette D (1969) Metabolic differentiation of distinct muscle types at the level of enzymatic organization. Eur J Biochem 10(2): 198–206.

24. Weintraub H (1993) The MyoD family and myogenesis: redundancy, networks and threshold. Cell 75: 1241–1244.

25. Piestun Y, Harel M, Barak M, Yahav S, Halevy O (2009) Thermal manipulations in late-term chick embryos have immediate and longer term effects on myoblast proliferation and skeletal muscle hypertrophy. J Appl Physiol 106: 233–240.

26. Halevy O, Krispin A, Leshem Y, McMurtry JP, Yahav S (2001) Early-age heat exposure affects skeletal muscle satellite cell proliferation and differentiation in chicks. Am J Physiol Regul Integr Comp Physiol 281: R302–R309.

27. Beattie J, Allan GJ, Lochrie JD, Flint DJ (2006) Insulin-like growth factor-binding protein-5 (IGFBP-5): a critical member of the IGF axis. Biochem J 395: 1–19.

28. Temim S, Chagneau AM, Peresson R, Tesseraud S (2000) Chronic heat exposure alters protein turnover of three different skeletal muscles in finishing broiler chickens fed 20 or 25% protein diets. J Nutr 130: 813–819.

29. Duchêne S, Audouin E, Crochet S, Duclos MJ, Dupont J, et al. (2008) Involvement of the ERK1/2 MAPK pathway in insulin-induced S6K1 activation in avian cells. Domest Anim Endocrinol 34: 63–73.

30. Ain Baziz H, Geraert PA, Padilha JC, Guillaumin S (1996) Chronic heat exposure enhances fat deposition and modifies muscle and fat partition in broiler carcasses. Poult Sci 75: 505–513.

31. Lu Q, Wen J, Zhang H (2007) Effect of chronic heat exposure on fat deposition and meat quality in two genetic types of chicken. Poult Sci 86: 1059–1064.

32. Klandorf H, Sharp PJ, Macleod MG (1981) The relationship between heat production and concentrations of plasma thyroid hormones in the domestic hen. Gen Comp Endocrinol 45: 513–520.

33. Collin A, Joubert R, Swennen Q, Damon M, Métayer Coustard S, et al. (2009) Involvement of thyroid hormones in the regulation of mitochondrial oxidations in mammals and birds. In Novascience, editor, In: Thyroid hormones: Functions, related diseases and uses. Francis S. Kuehn and Mauris P. Lauzad, New York.

34. Darras VM, Verhoelst CH, Reyns GE, Kühn ER, Van der Geyten S (2006) Thyroid hormone deiodination in birds. Thyroid 16: 25–35.

35. Collin A, Vaz MJ, Le Dividich J (2002) Effects of high temperature on body temperature and hormonal adjustments in piglets. Reprod Nutr Dev 42: 45–53.

36. Seebacher F, Schwartz TS, Thompson MB (2006) Transition from ectothermy to endothermy: the development of metabolic capacity in a bird (*Gallus gallus*). Proc Roy Soc Lond B Biol Sci 273: 565–570.

37. Wulf A, Harneit A, Kröger M, Kebenko M, Wetzel, et al. (2008) T3-mediated expression of PGC-1[alpha] via a far upstream located thyroid hormone response element. Mol Cell Endocrinol 287: 90–95.

38. Walter I, Seebacher F (2009) Endothermy in birds: underlying molecular mechanisms. J Exp Biol 212: 2328–2336.

39. Walter I, Seebacher F (2007) Molecular mechanisms underlying the development of endothermy in birds (*Gallus gallus*): a new role of PGC-1{alpha}? Am J Physiol Regul Integr Comp Physiol 293: R2315–R2322.

40. Azad MA, Kikusato M, Maekawa T, Shirakawa H, Toyomizu M (2010) Metabolic characteristics and oxidative damage to skeletal muscle in broiler chickens exposed to chronic heat stress. Comp Biochem Physiol A Mol Integr Physiol 155: 401–406.

41. Walter I, Hegarty B, Seebacher F (2010) AMP-activated protein kinase controls metabolism and heat production during embryonic development in birds. J Exp Biol 213: 3167–3176.

42. Holmes BF, Kurth-Kraczek EJ, Winder WW (1999) Chronic activation 5'-AMP- activated protein kinase increases Glut-4, hexokinase, and glycogen in muscle. J Appl Physiol 87: 1990–1995.

43. Dridi S, Temim S, Derouet M, Tesseraud S, Taouis M (2008) Acute cold- and chronic heat-exposure upregulate hepatic leptin and muscle uncoupling protein (UCP) gene expression in broiler chickens. J Exp Zool A Ecol Genet Physiol 309: 381–388.

44. Tan GY, Yang L, Fu YQ, Feng JH, Zhang MH (2010) Effects of different acute high ambient temperatures on function of hepatic mitochondrial respiration, antioxidative enzymes, and oxidative injury in broiler chickens. Poult Sci 89: 115–22.

45. Gutiérrez-Uzquiza A, Arechederra M, Bragado P, Aguirre-Ghiso J, Porras A (2012) p38α Mediates cell survival in response to oxidative stress. J Biol Chem 287: 2632–2642.

46. Collin A, Malheiros RD, Moraes VM, Van As P, Darras VM, et al. (2003) Effects of dietary macronutrient content on energy metabolism and uncoupling protein mRNA expression in broiler chickens. Br J Nutr 90: 261–269.

47. Sibut V, Hennequet-Antier C, Le Bihan-Duval E, Marthey S, Duclos MJ, et al. (2011) Identification of differentially expressed genes in chickens differing in muscle glycogen content and meat quality. BMC Genomics. 12: 112. doi: 10.1186/1471-2164-12-112.

48. Taouis M, De Basilio V, Mignon-Grasteau S, Crochet S, Bouchot C, et al. (2002) Early-age thermal conditioning reduces uncoupling protein messenger RNA expression in pectoral muscle of broiler chicks at seven days of age. Poult Sci 81: 1640–1643.

49. Collin A, Buyse J, van As P, Darras VM, Malheiros RD, et al. (2003) Cold-induced enhancement of avian uncoupling protein expression, heat production, and triiodothyronine concentrations in broiler chicks. Gen Comp Endocrinol 130: 70–77.

50. Mujahid A, Sato K, Akiba Y, Toyomizu M (2006) Acute heat stress stimulates mitochondrial superoxide production in broiler skeletal muscle, possibly via downregulation of uncoupling protein content. Poult Sci 85: 1259–1265.

51. Collin A, Taouis M, Buyse J, Ifuta NB, Darras VM, et al. (2003) Thyroid status, but not insulin status, affects expression of avian uncoupling protein mRNA in chicken. Am J Physiol Endocrinol Metab 284: E771–E777.

52. Mujahid A, Akiba Y, Toyomizu M (2009) Olive oil-supplemented diet alleviates acute heat stress-induced mitochondrial ROS production in chicken skeletal muscle. Am J Physiol Regul Integr Comp Physiol 297: R690-R698. Erratum in: Am J Physiol Regul Integr Comp Physiol (2010) 299: R386.

53. Gross WB, Siegel HS (1983) Evaluation of the heterophil lymphocyte ratio as a measure of stress in chickens. Avian Dis 27: 972–997.

17

A New Method to Monitor the Contribution of Fast Food Restaurants to the Diets of US Children

Colin D. Rehm[1]*, Adam Drewnowski[1,2]

1 Center for Public Health Nutrition, University of Washington, Seattle, Washington, United States of America, **2** Institute for Cardiometabolism and Nutrition, Groupe Hospitalier Pitié-Salpêtrière, Paris, France

Abstract

Background: American adults consume 11.3% of total daily calories from foods and beverages from fast food restaurants. The contribution of different types of fast food restaurants to the diets of US children is unknown.

Objective: To estimate the consumption of energy, sodium, added sugars, and solid fats among US children ages 4–19 y by fast food restaurant type.

Methods: Analyses used the first 24-h recall for 12,378 children in the 2003–2010 cycles of the nationally representative National Health and Nutrition Examination Survey (NHANES 2003–2010). NHANES data identify foods by location of origin, including stores and fast food restaurants (FFR). A novel custom algorithm divided FFRs into 8 segments and assigned meals and snacks to each. These included burger, pizza, sandwich, Mexican, Asian, fish, and coffee/snack restaurants. The contribution of each restaurant type to intakes of energy and other dietary constituents was then assessed by age group (4–11 y and 12–19 y) and by race/ethnicity.

Results: Store-bought foods and beverages provided 64.8% of energy, 61.9% of sodium, 68.9% of added sugars, and 60.1% of solid fats. FFRs provided 14.1% of energy, 15.9% of sodium, 10.4% of added sugars and 17.9% of solid fats. Among FFR segments, burger restaurants provided 6.2% of total energy, 5.8% of sodium, 6.2% of added sugars, and 7.6% of solid fats. Less energy was provided by pizza (3.3%), sandwich (1.4%), Mexican (1.3%), and chicken restaurants (1.2%). Non-Hispanic black children obtained a greater proportion of their total energy (7.4%), sodium (7.1%), and solid fats (9.5%) from burger restaurants as compared to non-Hispanic white children (6.0% of energy, 5.5% of sodium, and 7.3% of solid fat).

Conclusions: These novel analyses, based on consumption data by fast food market segment, allow public health stakeholders to better monitor the effectiveness of industry efforts to promote healthier menu options.

Editor: Harry Zhang, Old Dominion University, United States of America

Funding: This study was funded by a research grant from McDonald's Corp. to the University of Washington. The funders had no role in study design, data collection and analysis, decision to publish, or preparation of the manuscript.

Competing Interests: CR has no conflicts to report. AD advises McDonald's on nutrition issues and has also received grants, honoraria, and consulting fees from numerous food and beverage companies and other commercial and nonprofit entities with interests in nutrition. The University of Washington has received grants, donations, and contracts from both the public and the private sector.

* Email: crehm@uw.edu

Background

Consumption of foods away from home (FAFH) is thought to contribute to poor diet quality among children and adults [1–6]. FAFH represent a broad category that can include meals and snacks eaten at restaurants, schools, or at entertainment or sports events. The impact of fast foods on diet quality of children has received much research attention [7–11]. Nutritional analyses of fast food menu offerings have focused on the amount of energy, sodium, added sugars, and fats for fast food meals and menu items [12–14].

Since 2003, the National Health and Nutrition Examination Survey (NHANES), the primary source of dietary surveillance data in the US, has coded all foods and beverages consumed by their location of origin, which is reported by survey respondents. Among such locations were supermarkets and grocery stores, fast food restaurants/pizza (FFR), full-service restaurants (FSR), school cafeterias, and vending machines, among many others. The fast food industry generally segments fast food restaurants into 8 different types: burger, pizza, sandwich, chicken, Mexican, Asian, and fish restaurants, as well as coffee/snack shops [15]. However, the NHANES dataset does not distinguish among different FFRs by restaurant type [10,16].

The present study provides the first analysis of children's consumption patterns by the type of FFR. Custom algorithms, developed by the authors, subdivided the NHANES FFR/pizza category into 8 different types. Meals and snacks consumed at each type of restaurant were then analyzed for energy, sodium, added sugars, and solid fats. The present focus was on the relative contribution of different types of FFR to US children's diets.

Analysis of actual consumption data, instead of menu offerings, can provide an alternate picture of the contribution of the fast food

Table 1. Mean and % of total for energy and selected dietary factors (and standard errors) by location of origin.

	Energy (kcal)		Sodium (mg)		Added sugars (kcal)		Solid fats (kcal)	
	Mean	% of total	Mean	% of total	Mean	% of total	Mean	% of total
Overall (age 4–19 y)								
Store	1345 (13)	64.8 (0.6)	2005 (25)	61.9 (0.7)	238 (4)	68.9 (0.6)	252 (3)	60.1 (0.7)
Fast food restaurant	293 (9)	14.1 (0.4)	515 (16)	15.9 (0.5)	36 (1)	10.4 (0.4)	75 (2)	17.9 (0.5)
Full-service restaurant	116 (7)	5.6 (0.4)	226 (16)	7.0 (0.5)	16 (1)	4.5 (0.3)	26 (2)	6.2 (0.4)
School	150 (9)	7.2 (0.4)	247 (17)	7.6 (0.5)	17 (1)	4.9 (0.4)	34 (2)	8.1 (0.5)
Other	171 (5)	8.2 (0.3)	248 (9)	7.6 (0.3)	39 (2)	11.4 (0.5)	32 (1)	7.7 (0.4)
Age 4–11 y								
Store	1246 (16)	65.6 (0.8)	1854 (26)	63.9 (0.9)	204 (4)	66.9 (1.0)	245 (5)	62.3 (1.0)
Fast food restaurant	212 (11)	11.2 (0.6)	356 (20)	12.3 (0.7)	27 (2)	8.8 (0.5)	54 (3)	13.8 (0.8)
Full-service restaurant	84 (7)	4.5 (0.4)	158 (14)	5.5 (0.5)	12 (1)	4.1 (0.4)	18 (2)	4.7 (0.5)
School	173 (11)	9.1 (0.6)	277 (19)	9.5 (0.6)	19 (1)	6.3 (0.5)	39 (3)	10.0 (0.7)
Other	184 (8)	9.7 (0.4)	258 (12)	8.9 (3.9)	42 (3)	13.9 (0.8)	36 (2)	9.2 (0.5)
Age 12–19 y								
Store	1441 (19)	64.2 (0.7)	2150 (40)	60.3 (9.5)	271 (6)	70.3 (0.7)	258 (5)	58.2 (0.9)
Fast food restaurant	371 (12)	16.5 (0.5)	668 (24)	18.7 (0.7)	45 (2)	11.6 (0.4)	95 (3)	21.4 (0.7)
Full-service restaurant	147 (10)	6.5 (0.5)	291 (26)	8.2 (0.7)	19 (2)	4.9 (0.4)	33 (3)	7.5 (0.6)
School	127 (11)	5.6 (0.5)	219 (20)	6.1 (0.5)	15 (2)	3.8 (0.4)	29 (3)	6.5 (0.6)
Other	159 (7)	7.1 (0.4)	237 (14)	6.7 (0.4)	36 (2)	9.4 (0.5)	29 (2)	6.5 (0.4)

industry to the diets of US children and youth. Detailed analyses by FFR type will also allow for better monitoring of food industry trends by interested public health stakeholders. The present study is the first-ever analysis of children's diets by specific type of fast food restaurant. Rather than assess the caloric contribution of specific food items (e.g. burgers or pizza) [16,17], the present analyses provide the first estimate of calories, sodium, added sugars and solid fats from burger or pizza restaurants.

Methods

Dietary intake data sources

Data analyses were based on the first 24-h recall from 4 cycles of the nationally representative National Health and Nutrition Examination Survey (NHANES) for the years 2003–2010. The NHANES 24-h dietary recall utilizes a multi-pass method, where all foods and beverages consumed in the preceding 24-h, from midnight to midnight were reported. The name, time and place of consumption for each eating occasion were also measured. For children 4–5 y, a parent or guardian completed the dietary recall. For children 6–11 y, the child was the primary respondent, but the parent was present and able to assist. For children 12–19 y, the child was the primary respondent, but could be assisted by an adult [18].

Data from the MyPyramid Equivalents Database (MPED) were used to assess intakes of added sugars and solid fats [19]. The MPED 2.0 database was updated for use with more recent NHANES cycles by imputing the MPED equivalents for a limited number of foods (n = 291). Energy from solid fats was estimated as 9 kcal/g of solid fat. Since the MPED database provided added sugars in teaspoon equivalents, this value was converted to energy (1 tsp = 16 kcal) [20].

The 5 principal locations of origin for all foods and beverages were obtained from NHANES data. These were: grocery stores, fast food restaurants (FFR), full-service restaurants (FSR), school cafeterias, and other [21]. The "other" category included foods or beverages from someone else or as a gift, child-care centers, sports/recreational facilities, vending machines, and grown or caught among other categories [21]. Since no further data about the FFR category are provided in NHANES data, a multi-step algorithm was developed by the authors to assign FFR meals and

snacks into one of 8 market segments as defined by the restaurant industry [15].

The FFR segmentation

The 8 FFR segments were burger restaurants (e.g., McDonald's or Burger King), pizza restaurants (e.g., Pizza Hut or Domino's), sandwich restaurants (e.g., Subway or Quiznos), chicken restaurants (e.g., KFC or Chick-fil-A), Mexican restaurants (e.g., Taco Bell), Asian restaurants (e.g., Panda Express), fish restaurants (e.g., Long John Silver's) and coffee/snack shops (e.g., Starbucks or Baskin-Robbins). The NHANES data included meals and snacks from national chains as well as regional or local restaurants: the specific brand names provided above are for reference only.

FFR meals were assigned to one of the eight pre-defined segments based on the foods and beverages consumed at that meal. First, unique meals were defined as eating occasions that occurred at the same time and at the same location (i.e., at-home or away from home). Both fast foods consumed at-home and away-from-home were included in the coding algorithm; the place of eating was simply used to define unique eating occasions. Second, an iterative algorithm scanned the 24-h recall individual foods file for "sentinel" foods that could be used to identify a specific market segment. Among the 26 sentinel foods were burgers, pizza, Mexican dishes (e.g., tacos or burritos), chicken strips/nuggets, fried chicken, submarine/deli sandwiches, and hot breakfast items as well as pretzels, hot dogs, side-dishes often served with fried chicken (e.g., baked beans, mashed potatoes and gravy, biscuits, potato salad, where no fried chicken was served), French fries alone, or soda alone.

Foods and food combinations (e.g., hamburger, fried chicken, or pizza) indicated the type of fast food restaurant. If a given meal occasion included only one of these sentinel main dishes, it was coded as such. Examples of meal components that were readily coded and assigned to a given FFR segment were burgers, pizza, Mexican dishes, fried chicken (but not chicken strips/nuggets), or Asian dishes (e.g., fried rice or General Tso's chicken). Meals that contained multiple potential main dishes were flagged for further refinement (described below). Most of the meal components could be unambiguously assigned to a single and identifiable FFR segment. The initial scan coded 77.3% of single items, while the remainder was assessed manually.

Table 2. Mean energy and % of total energy (and standard errors) by FFR segment and age group.

	Total (n = 12,378)		Age 4–11 y (n = 5,681)		Age 12–19 y (n = 6,697)	
	Mean	% of total	Mean	% of total	Mean	% of total
Burger	128 (6.1)	6.2 (0.3)	102 (8.1)[†]	5.4 (0.4)[‡]	154 (6.7)	6.8 (0.3)
Pizza	69 (4.8)	3.3 (0.2)	51 (5.9)[†]	2.7 (0.3)[‡]	87 (7.3)	3.9 (0.3)
Sandwich	29 (2.2)	1.4 (0.1)	15 (2)[†]	0.8 (0.1)[†]	42 (3.9)	1.9 (0.2)
Chicken	26 (2.3)	1.2 (0.1)	23 (3.3)	1.2 (0.2)	28 (3.6)	1.2 (0.2)
Mexican	27 (2.4)	1.3 (0.1)	12 (2.7)[†]	0.6 (0.1)[†]	41 (4)	1.8 (0.2)
Asian	9 (1.4)	0.4 (0.1)	4 (1.2)[‡]	0.2 (0.1)[‡]	13 (2.4)	0.6 (0.1)
Coffee/snack	3 (0.5)	0.2 (0.0)	3 (0.6)	0.2 (0.0)	4 (0.8)	0.2 (0.0)
Fish	3 (0.9)	0.1 (0.0)	1 (0.7)	0.1 (0.0)	4 (1.6)	0.2 (0.1)
Other sources	1782 (14.2)	85.9 (0.4)	1688 (13.8)[‡]	88.8 (0.6)[†]	1873 (21.7)	83.5 (0.5)
Total energy	2076 (13.7)	–	1900 (15.4)[†]	–	2244 (21.0)	–

[†]p<0.001; [‡]0.001<p<0.01 for difference between two age groups. Significance testing conducted separately for mean and % of total energy. Value in parentheses is the standard error.

Table 3. Mean sodium and % of total sodium (and standard errors) by FFR segment and age group.

	Total (n = 12,378)		Age 4–11 y (n = 5,681)		Age 12–19 y (n = 6,697)	
	Mean	% of total	Mean	% of total	Mean	% of total
Burger	189 (9.8)	5.8 (0.3)	142 (12.1)[†]	4.9 (0.4)[†]	233 (12.3)	6.5 (0.4)
Pizza	142 (9.7)	4.4 (0.3)	102 (11.2)[†]	3.5 (0.4)[‡]	180 (15.4)	5.0 (0.4)
Sandwich	62 (4.8)	1.9 (0.2)	34 (4.4)[†]	1.2 (0.2)[†]	89 (8.8)	2.5 (0.2)
Chicken	42 (4.0)	1.3 (0.1)	37 (4.6)	1.3 (0.2)	48 (7.0)	1.3 (0.2)
Mexican	49 (4.9)	1.5 (0.1)	21 (5.1)[†]	0.7 (0.2)[†]	76 (8.0)	2.1 (0.2)
Asian	22 (3.5)	0.7 (0.1)	13 (3.5)[‡]	0.5 (0.1)[¶]	31 (5.7)	0.9 (0.2)
Coffee/snack	5 (1.1)	0.1 (0.0)	4 (1.5)	0.1 (0.1)	5 (1.5)	0.1 (0.0)
Fish	4 (1.4)	0.1 (0.0)	2 (0.8)	0.1 (0.0)	6 (2.5)	0.2 (0.1)
Other sources	2726 (33.9)	84.1 (0.5)	2548 (29.3)[†]	87.7 (0.7)[†]	2897 (51.7)	81.3 (0.7)
Total sodium	3242 (30.9)	–	2903 (26.7)[†]	–	3566 (49.8)	–

[†]p<0.001; [‡]0.001<p<0.01; [¶]0.01<p<0.05 for difference between two age groups. Significance testing conducted separately for mean and % of total sodium. Value in parentheses is the standard error.

For some meals, the assignment to a FFR segment was more complex. For example, chicken nuggets/strips are typically sold at both burger and chicken restaurants. Other examples include soda consumed alone, hot breakfast dishes, French fries alone, or ice cream. These meals were randomly assigned to each segment according to a deterministic probability based on weights from sales data for the 50 largest chains. For example, all chicken and burger establishments selling a chicken nugget/strip product were identified and assigned a weight to each segment that randomly divided such meals according to weighted sales. This approach assumes that the relative revenues from each product were similar across different brand segments. For the chicken strips/nugget meal component, 84.3% were assigned to the burger segment while 15.7% were applied to the chicken segment [15].

The reliability of the algorithm was evaluated by an independent coder who assigned 138 random meals (412 individual food items) into the 8 FFR segments. The chance-corrected concordance (Kappa) was estimated at 0.89, indicating a high level of agreement. Some refinements were made to the algorithm following the reliability sub-study. Specifically, sweet and sour sauce was originally coded as an indicator food for Asian-type dishes, but it became clear it was most often associated with eating chicken nuggets/strips.

Analytical approach

Analyses were conducted for all children (ages 4–19 y) and by age group (ages 4–11 and 12–19 y). Three race/ethnicity groups were examined: non-Hispanic whites, Mexican-Americans and non-Hispanic blacks. The race/ethnicity analysis excluded a subsample of the population for which presentation of race/ethnicity specific estimates is not recommended [22]. The three race/ethnicity groups examined constituted 89% of the population.

Dependent measures were dietary energy (kcal/d), sodium (mg/d), added sugars (kcal/d) and solid fats (kcal/d). These dietary constituents were most frequently cited in past analyses of foods away from home, and our important components of summary

Table 4. Mean energy from added sugars and % of total energy from added sugars (and standard errors) by FFR segment and age group.

	Total (n = 12,378)		Age 4–11 y (n = 5,681)		Age 12–19 y (n = 6,697)	
	Mean	% of total	Mean	% of total	Mean	% of total
Burger	21 (1.1)	6.2 (0.3)	17 (1.3)[†]	5.5 (0.4)[¶]	26 (1.3)	6.7 (0.3)
Pizza	3.4 (0.4)	1.0 (0.1)	3.0 (0.5)	1.0 (0.2)	3.9 (0.6)	1.0 (0.1)
Sandwich	3.3 (0.4)	1.0 (0.1)	1.7 (0.3)[†]	0.5 (0.1)[†]	4.9 (0.6)	1.3 (0.2)
Chicken	3.2 (0.4)	0.9 (0.1)	2.9 (0.6)	1.0 (0.2)	3.5 (0.6)	0.9 (0.2)
Mexican	3.0 (0.3)	0.9 (0.1)	1.3 (0.2)[†]	0.4 (0.1)[†]	4.6 (0.6)	1.2 (0.1)
Asian	0.5 (0.1)	0.2 (0.0)	0.2 (0.1)[¶]	0.1 (0.0)[¶]	0.8 (0.2)	0.2 (0.1)
Coffee/snack	0.8 (0.2)	0.2 (0.0)	0.8 (0.2)	0.3 (0.1)	0.8 (0.3)	0.2 (0.1)
Fish	0.3 (0.1)	0.1 (0.0)	0.2 (0.1)	0.1 (0.0)	0.4 (0.2)	0.1 (0.1)
Other sources	310 (4.0)	89.6 (0.4)	278 (4.8)[†]	91.2 (0.5)[†]	340 (5.8)	88.4 (0.4)
Total added sugars	346 (4.5)	–	305 (5.1)[†]	–	385 (6.4)	

[†]p<0.001; [¶]0.01<p<0.05 for difference between two age groups. Significance testing conducted separately for mean and % of total added sugars. Value in parentheses is the standard error.

Table 5. Mean energy from solid fats and % of total energy from solid fats (and standard errors) by FFR segment and age group.

	Total (n = 12,378)		Age 4–11 y (n = 5,681)		Age 12–19 y (n = 6,697)	
	Mean	% of total	Mean	% of total	Mean	% of total
Burger	32 (1.7)	7.6 (0.4)	25 (2.1)[†]	6.4 (0.5)[†]	38 (1.9)	8.6 (0.5)
Pizza	20 (1.6)	4.8 (0.4)	15 (2.1)[†]	3.8 (0.5)[‡]	25 (2.3)	5.7 (0.5)
Sandwich	6.5 (0.5)	1.6 (0.1)	3.8 (0.5)[†]	1.0 (0.1)[†]	9.2 (0.8)	2.1 (0.2)
Chicken	6.1 (0.6)	1.5 (0.1)	5.7 (0.7)	1.4 (0.2)	6.6 (0.9)	1.5 (0.2)
Mexican	7.2 (0.7)	1.7 (0.2)	3.2 (0.8)[†]	0.8 (0.2)[†]	11.1 (1.2)	2.5 (0.3)
Asian	1.3 (0.3)	0.3 (0.1)	0.5 (0.1)[‡]	0.1 (0.0)[¶]	2.1 (0.6)	0.5 (0.1)
Coffee/snack	0.9 (0.2)	0.2 (0.0)	0.8 (0.2)	0.2 (0.1)	1.0 (0.2)	0.2 (0.1)
Fish	0.7 (0.3)	0.2 (0.1)	0.3 (0.1)	0.1 (0.0)	1.2 (0.5)	0.3 (0.1)
Other sources	344 (4.4)	82.1 (0.5)	339 (5.3)	86.2 (0.8)[†]	349 (6.5)	78.6 (0.7)
Total energy from solid fats	419 (4.0)	–	393 (5.3)[†]	–	443 (6.5)	–

[†]p<0.001; [‡]0.001<p<0.01; [¶]0.01<p<0.05 for difference between two age groups. Significance testing conducted separately for mean and % of total solid fats. Value in parentheses is the standard error.

measures of diet quality [12–14,23,24]. Two summary measures were assessed: the survey-weighted mean and the survey-weighted population proportion. The population proportion is the percent of each dietary constituent provided by FFR segment. This measure, interpreted as a ratio of the means, rather than a mean of the ratios, is best suited for examinations of population-level habits [25].

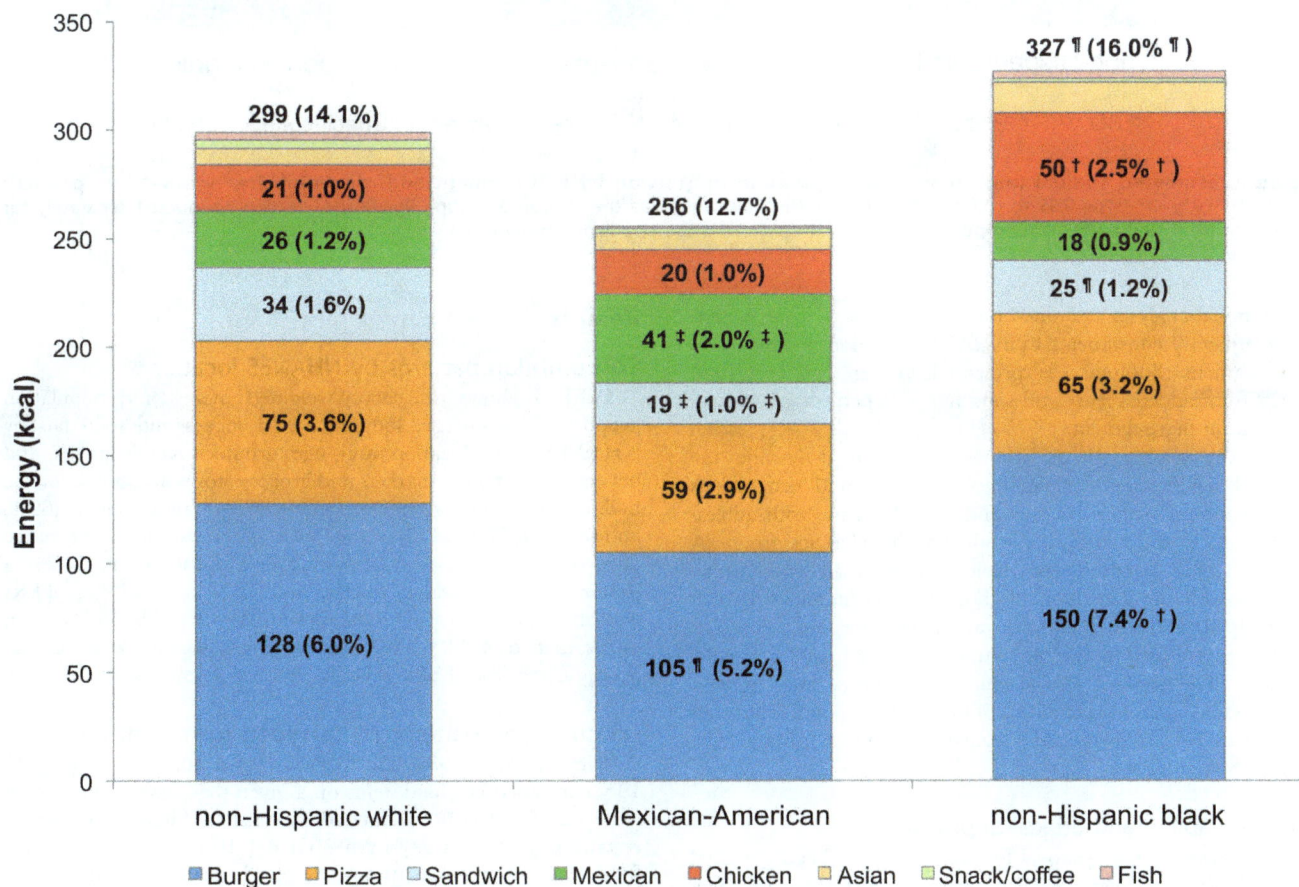

Figure 1. Estimated energy (kcal) intake and population proportion by FFR segment and race/ethnicity, age 4–19 y. [†]p<0.001; [‡]0.001<p<0.01; [¶]0.01<p<0.05 for difference, with non-Hispanic whites as the reference group. Significance testing conducted separately for population mean and population proportion (value in parentheses is the population proportion).

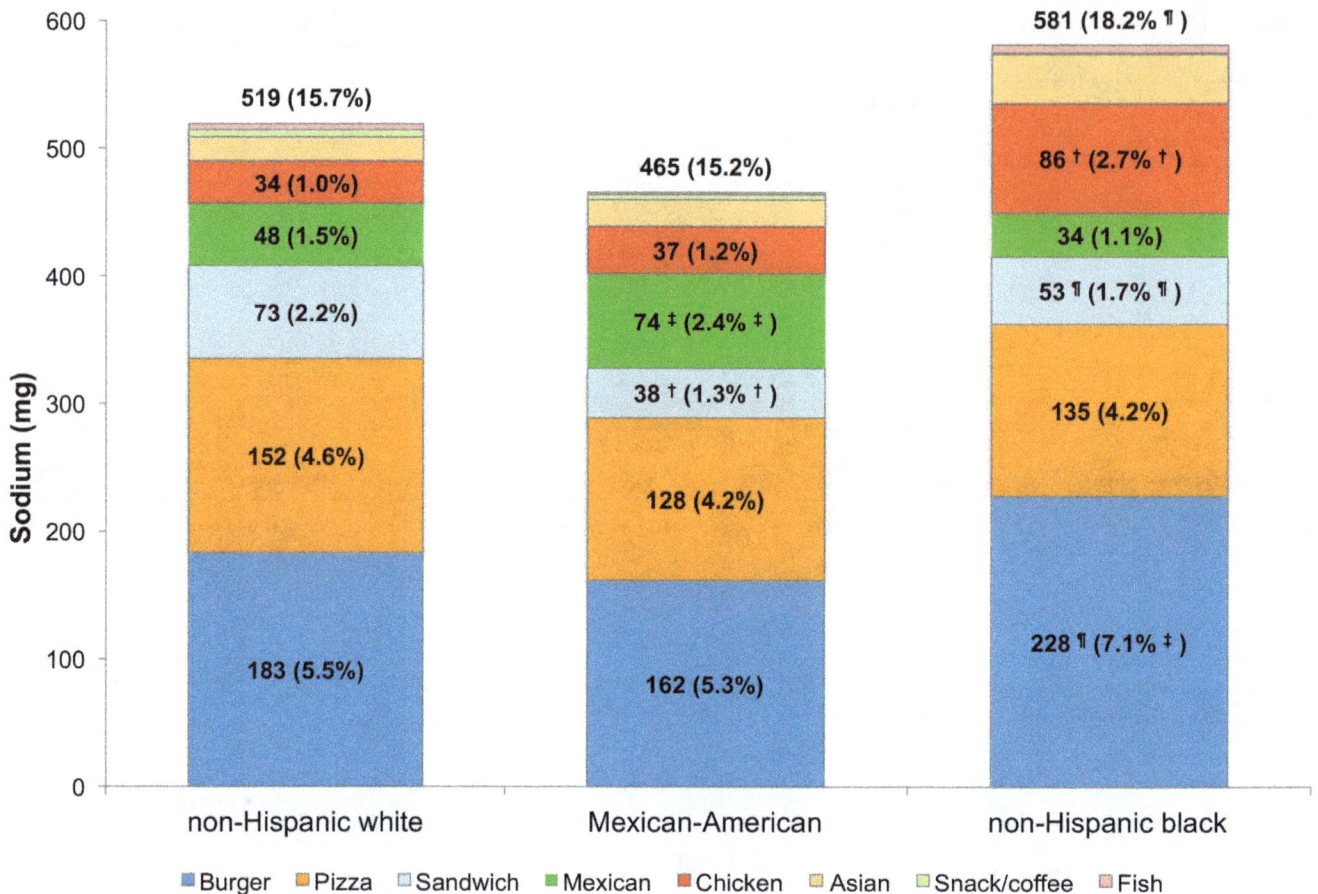

Figure 2. Estimated sodium (mg) intake and population proportion by FFR segment and race/ethnicity, age 4–19 y. †p<0.001; ‡0.001<p<0.01; ¶0.01<p<0.05 for difference, with non-Hispanic whites as the reference group. Significance testing conducted separately for population mean and population proportion (value in parentheses is the population proportion).

Additional analyses examined the energy-adjusted contribution of these dietary constituents by NHANES location of origin and FFR segment. Sodium was presented in mg/1000 kcal, and energy from added sugars and solid fats as a percentage of total energy from that segment.

A survey-weighted Wald test was used to determine the statistical significance of differences in means and proportions between sub-populations by age and race/ethnicity, with adolescents (12–19 y) and non-Hispanic adults as the reference groups. A total of 12,378 children and adolescents were included in all analyses, with the exception of those by race/ethnicity, which included 10,847. All analyses accounted for the complex NHANES stratified multistage sampling design and for over-sampling. The analysis also accounted for survey non-response and all results were representative of the US child population from 2003–2010. Analyses were conducted using Stata 12.1 (StataCorp, College Station, TX).

Data availability and ethical approval

The necessary IRB approval for NHANES had been obtained by the National Center for Health Statistics (NCHS) [26]. The study was exempt from human subjects review per University of Washington policies. All data used here are publicly available on the NCHS website [27].

Results

Consumption patterns by NHANES location

Table 1 shows the survey-weighted means and population proportion for energy, sodium, added sugars and solid fats by NHANES food location and age group. Overall, foods and beverages from supermarkets and grocery stores accounted for the bulk of dietary energy (65%), sodium (62%), added sugars (69%) and solid fats (60%). FFRs were the second most important source of these dietary factors, contributing 14% of total energy, 16% of sodium, 10% of added sugars, and 18% of solid fats. FFRs contributed more energy to the diets of 12–19 y olds (16.5%) than to the diets of 4–11 y olds (11.2%). Approximately 21% of total energy came from FSRs, schools, and other sources.

Consumption patterns by fast food restaurant type

Overall, 35.7% (95% CI 33.9–37.7%) of all children ages 4–19 y consumed fast food items on a given day. Seventeen percent (95% CI 16.1–18.8%) of all children consumed items from burger restaurants, whereas 9.0% (95% CI 8.0–10.1%) consumed items from pizza restaurants. About 4% of all children consumed any items from sandwich, chicken, and Mexican FFRs.

Table 2 provides the mean and population proportion of energy by fast food restaurant type and age group. Within the FFR category, burger restaurants provided the most energy (6.2% overall, 5.4% for 4–11 y olds and 6.8% for 12–19 y olds). Among

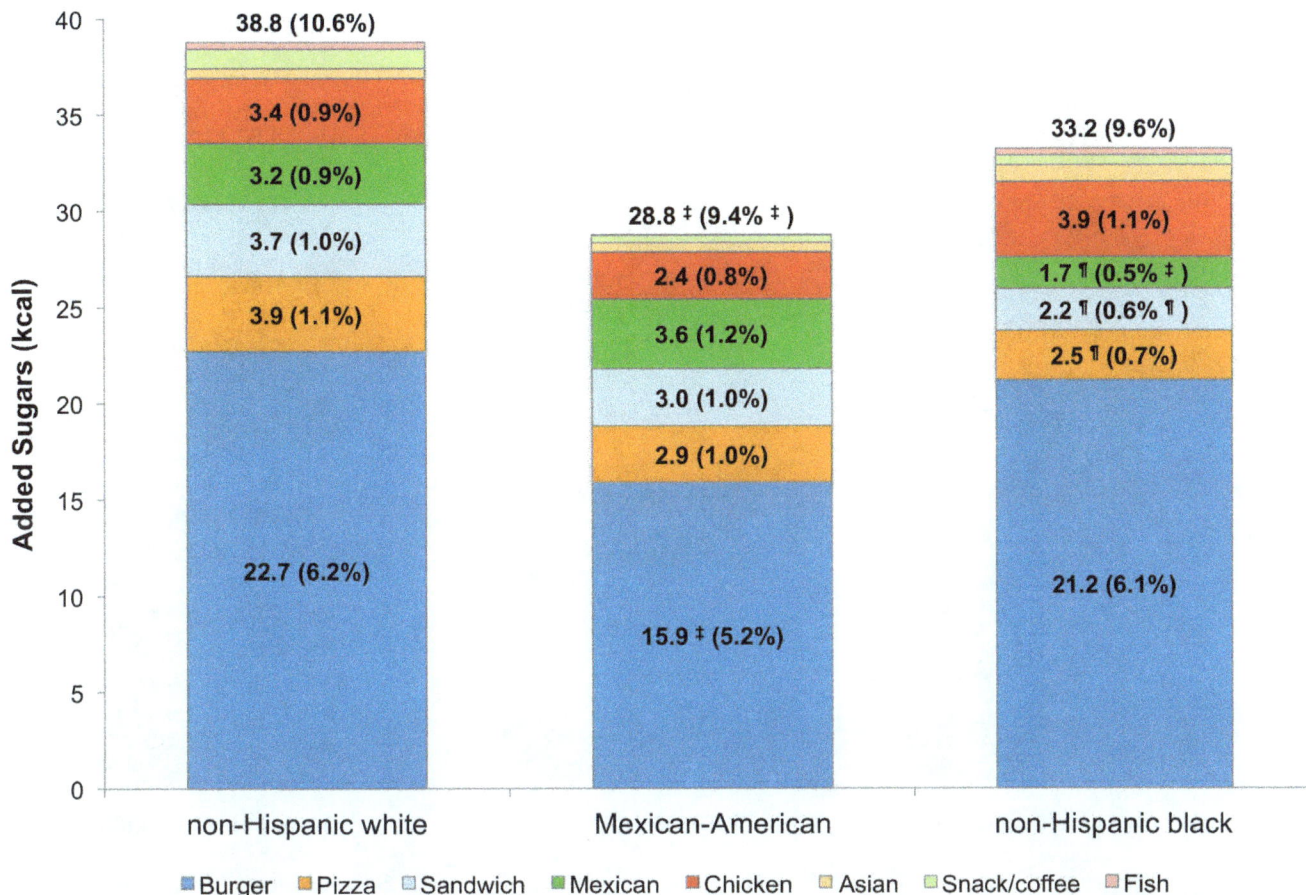

Figure 3. Estimated intake of energy from added sugars (kcal) and population proportion by FFR segment and race/ethnicity, age 4–19 y. [†]p<0.001; [‡]0.001<p<0.01; [¶]0.01<p<0.05 for difference, with non-Hispanic whites as the reference group. Significance testing conducted separately for population mean and population proportion (value in parentheses is the population proportion).

the FFR types, pizza restaurants provided the second most energy (3.3% overall, 2.7% for 4–11 y olds, and 3.9% for 12–19 y olds). Overall, sandwich, chicken and Mexican restaurants each provided less than 1.5% of total energy. The Asian, coffee/snack and fish segments each provided less than 0.5% of total energy. Given the strong correlation between energy and other dietary components, the contribution of each FFR segment to sodium (**Table 3**), added sugars (**Table 4**) and solid fats (**Table 5**) tracked the results for energy. Among the different FFR segments, burger restaurants provided the most sodium, added sugars and solid fats.

Figure 1 shows total energy intakes analyzed by fast food restaurant type and race/ethnicity. The distribution of energy intakes by FFR segment varied by race/ethnicity. Non-Hispanic black children derived more energy from fast food restaurants (327 kcal and 16.0% of total) than did non-Hispanic white children (299 kcal and 14.1% of total). Compared to non-Hispanic white children (6.0%), Mexican-American children derived significantly less energy from burger restaurants, whereas non-Hispanic black children derived significantly more (7.4%). By contrast, non-Hispanic white children derived significantly more energy from sandwich restaurants compared to the other race/ethnicity groups, while Mexican-American children derived more calories from Mexican restaurants when compared to non-Hispanic white children. Non-Hispanic black children derived significantly more energy from chicken restaurants than did non-

Hispanic white children. From sandwich, Mexican, and chicken FFRs combined, there were no differences in energy by race/ethnicity (81 kcal for non-Hispanic white children, 80 kcal for Mexican-American children and 93 kcal for non-Hispanic black children). Comparable findings were obtained for the other dietary components (see **Figures 2–4** for sodium, energy from added sugars and energy from solid fats, respectively).

Energy-adjusted consumption of sodium, added sugars and solid fats by fast food restaurant type

Table 6 shows the energy-adjusted amounts of sodium and percent of total energy from added sugars and solid fats by FFR type. The data were expressed as a percent of total energy, for solid fats and added sugars, or as mg/1,000 kcal for sodium. Overall, fast food restaurants contributed significantly more sodium per 1,000 kcal and percent of energy from solid fats than stores, while stores contributed significantly more energy from added sugars than FFRs. Pizza, sandwich, Mexican and Asian FFRs contributed significantly more sodium per 1,000 kcal than the burger segment. By contrast, pizza, sandwich, chicken, Mexican and Asian FFRs accounted for a lower proportion of their total energy from added sugars, when compared to burger restaurants. For percent of energy from solid fats, pizza FFRs contributed a greater proportion of total energy from solid fats

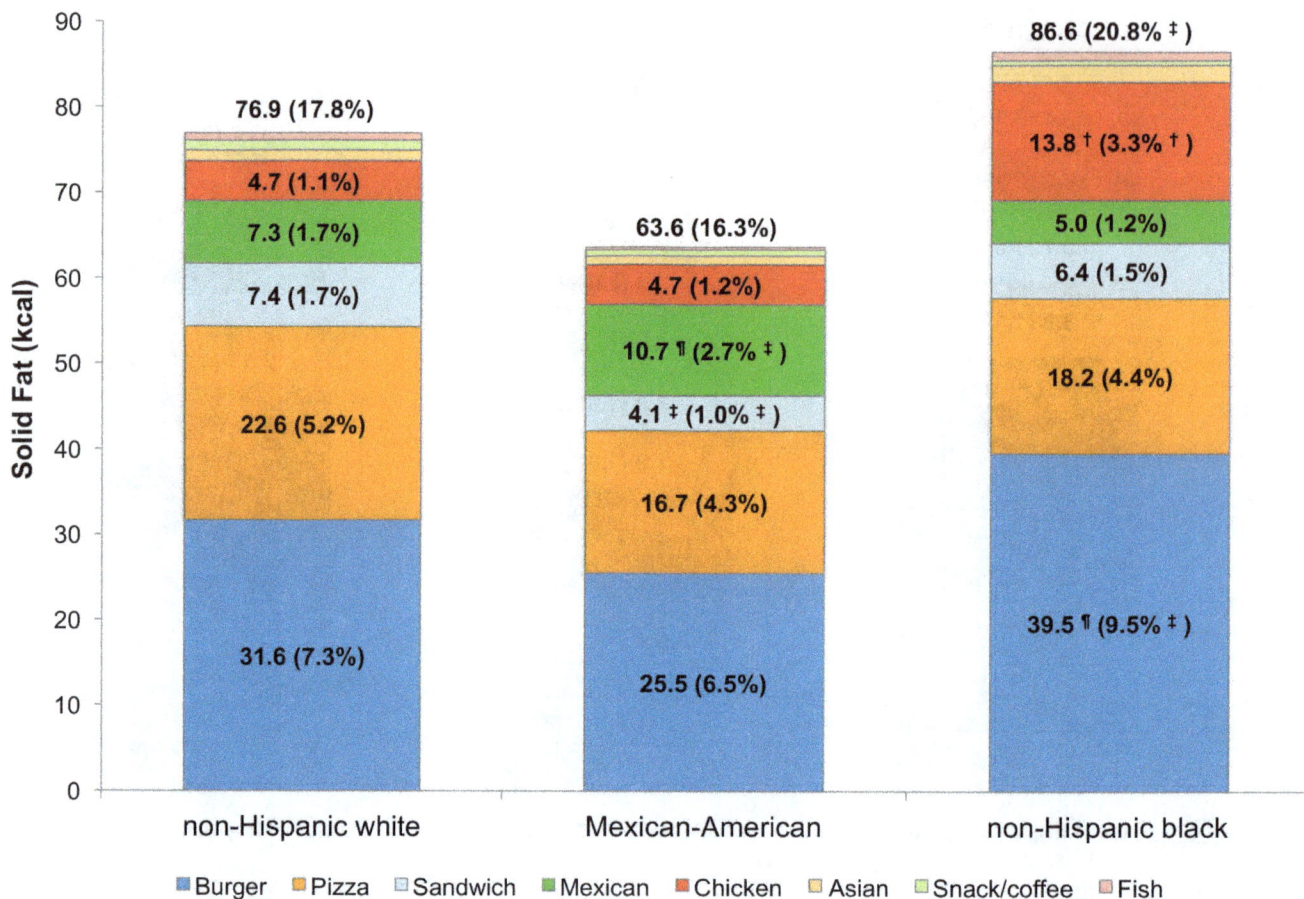

Figure 4. Estimated intake of energy from solid fats (kcal) and population proportion by FFR segment and race/ethnicity, age 4–19 y. [†]p<0.001; [‡]0.001<p<0.01; [¶]0.01<p<0.05 for difference, with non-Hispanic whites as the reference group. Significance testing conducted separately for population mean and population proportion (value in parentheses is the population proportion).

than did burger FFRs, while Asian FFRs contributed the lowest proportion of total energy from solid fats.

Discussion

The present analyses are the first example of stratifying consumption data for a nationally representative sample of US children by fast food restaurant type. First, analyses of consumption data are a useful supplement to the numerous prior analyses of fast food menu offerings [12–14]. Second, separating the FFR category into specific market segments as defined by the restaurant industry allows interested stakeholders to monitor the impact of public health policies and programs aimed at improving the quality of children's diets.

The present analyses complement published studies of US diets by NHANES location of origin, where foods were aggregated based on a National Cancer Institute coding scheme [16,17]. The present analyses are unique in that they accounted for all foods and beverages eaten at different types of fast food restaurants. Items consumed at burger restaurants may include burgers but may also include fries, chicken, salad, beverages, milkshakes, desserts, fruit and coffee. Thus, burgers alone contributed 1.2% of energy to the diets of the 6–11 y age group and 1.9% of energy to the 12–19 y age group, as reported previously [16]. By contrast, the total contribution of calories from burger restaurants was

much higher: 6.2% overall, 5.4% for 4–11 y olds, and 6.8% for 12–19 y olds.

The present estimates, based on nationally representative federal datasets, were generally consistent with FFR segment sales, as published by industry sources [15]. Among FFRs, burger restaurants are the leading segment, as reflected in the consumption data. In 2009, total sales for burgers among leading national chains were $66 billion (or 47.7% of all FFR sales). The present estimate was that 43.5% (95% CI 40.2–47.4%) of FFR calories consumed by US children came from the burger segment. In this age group, pizza was far and away the second most important source of energy. Nationally, the sandwich, snack, chicken, pizza and Mexican segments were the 2nd–6th leading segments in terms of sales, respectively. The infrequent consumption of coffee among children/adolescents explains why the snack/coffee segment was not observed as an important source of energy in this study. The present approach allows publicly available federal data to be used for identifying the relative importance of different FFR segments for specific demographic groups of interest, a definite advantage over the use of menu offerings or sales data. This approach can also be used to monitor population-based trends in dietary intakes from different types of FFRs.

The present consumption-based analyses complement past studies on the nutrient density of foods/meals from FFRs that are generally based on menu offerings [12–14,28–31]. No nationally representative data on food consumption by FFR

Table 6. Mean sodium per 1,000 kcal and percent of energy from added sugars and solid fats (and standard errors) by FFR segment.

	Sodium		Added sugars		Solid fats	
	Sodium (mg) per 1,000 kcal	% of total sodium	% of total energy from added sugars[1]	% of total added sugars	% of total energy from solid fats[1]	% of total solid fats
FFR segment						
Burger (ref)	1458 (24)	5.8 (0.3)	16.5 (0.5)	6.2 (0.3)	25.3 (0.5)	7.6 (0.4)
Pizza	2028 (33)¶	4.4 (0.3)	4.5 (0.6)¶	1.0 (0.1)	30.6 (1.3)¶	4.8 (0.4)
Sandwich	2005 (76)¶	1.9 (0.2)	12.1 (0.9)¶	1.0 (0.1)	22.6 (0.9)	1.6 (0.1)
Chicken	1577 (67)	1.3 (0.1)	13.0 (1.1)¶	0.9 (0.1)	23.9 (0.8)	1.5 (0.1)
Mexican	1841 (36)¶	1.5 (0.1)	11.5 (0.9)¶	0.9 (0.1)	27.0 (0.8)	1.7 (0.2)
Asian	2590 (204)¶	0.7 (0.1)	6.2 (1.2)¶	0.2 (0.0)	16.6 (2.5)¶	0.3 (0.1)
Coffee/snack	1579 (267)	0.1 (0.0)	22.2 (4.4)	0.2 (0.0)	26.2 (1.8)	0.2 (0.0)
Fish	1420 (97)	0.1 (0.0)	14.3 (4.2)	0.1 (0.0)	28.8 (1.5)	0.2 (0.1)
Location of origin						
All FFR (ref)	1756 (20)	15.9 (0.5)	12.0 (0.4)	10.4 (0.4)	26.3 (0.4)	17.9 (0.5)
Store	1466 (11)¶	61.9 (0.7)	18.3 (0.3)¶	68.9 (0.6)	18.2 (0.1)¶	60.1 (0.7)
FSR	1892 (61)	7.0 (0.5)	13.8 (0.8)	4.5 (0.3)	23.1 (0.7)¶	6.2 (0.4)
School	1635 (78)	7.6 (0.5)	11.5 (0.5)	4.9 (0.4)	23.1 (0.5)¶	8.1 (0.5)
Other	1499 (40)¶	7.6 (0.3)	20.6 (0.9)¶	11.4 (0.5)	19.9 (0.7)¶	7.7 (0.4)

[1]Value can be interpreted as the proportion of energy from each segment that is from added sugars or solid fats. For example, of all energy from burger FFRs, 16.5% came from added sugars.
¶Indicates p-value <0.05. Reference group for segment analysis is burger. Reference group for location of origin is all FFR. Hypothesis testing only conducted for sodium (mg) per 1,000 kcal, and percent of total energy from added sugars and solid fats. Value in parentheses is the standard error.

segment has previously been published. Merging market segmentation and public health surveillance approaches, the present method has the potential to transform ways in which the contribution of different food sources to the total diet can be evaluated and monitored over time. In particular, given efforts by the food and restaurant industry to improve the composition of menus and food products, a surveillance system needs to be established. The approach used here can also be adapted for a large number of dietary outcomes, including fruit, vegetable, low-fat dairy, fiber, potassium or any other dietary constituent or food group of interest.

The present study had several limitations. First, the present analyses were based on a 24-hour recall, which may result in under-reporting of foods perceived to be less healthful [32,33]. This may result in a falsely minimized estimation of energy from both full-service and fast food restaurants or from food groups such as desserts, salty snacks, pizza or soda. However, such underreporting should not affect the relative rankings of consumption patterns by FFR segment. For younger children, dietary reporting by proxy respondent may result in under-reporting of foods consumed while the parent is not present. However, such a reporting error is less likely for restaurant foods, where parents are likely to be present, as opposed to school/childcare settings.

The NHANES food location of origin may also have some error. In particular, stores include supermarkets, grocery stores, convenience stores and pharmacies where people buy food. The FFR-segment coding algorithm also had some limitations; a small proportion of eating occasions could not be unambiguously assigned to a specific FFR category and were coded instead based on the relative market share of each segment.

Despite these limitations, the present work represents the only assessment of the contribution of different types of fast food restaurants to the diets of US children. The present analyses advance the field in two important ways. First, consumption data can supplement the ongoing analyses of menu offerings, while offering better insight into what is eaten. Second, intakes of energy and dietary constituents of public health concern can now be assigned to different types of fast food restaurants: burger, pizza, sandwich, chicken, Mexican, Asian, fish, and coffee/snack. The present approach can be usefully applied to monitor the effectiveness of industry and public health policies aimed at improving the dietary habits of American children.

Author Contributions

Conceived and designed the experiments: CDR AD. Analyzed the data: CDR. Wrote the paper: CDR AD.

References

1. Bauer KW, Larson NI, Nelson MC, Story M, Neumark-Sztainer D (2009) Fast food intake among adolescents: secular and longitudinal trends from 1999 to 2004. Prev Med 48: 284–287.
2. Mancino L, Todd J, Lin B-H (2009) Separating what we eat from where: Measuring the effect of food away from home on diet quality. Food Policy 34: 557–562.
3. U.S. Food and Drug Administration: Backgrounder - Keystone Forum on Away-From-Home Foods: Opportunities for Preventing Weight Gain and Obesity. Available at: http://www.fda.gov/Food/LabelingNutrition/ReportsResearch/ucm082064.htm.
4. Guthrie JF, Lin BH, Frazao E (2002) Role of food prepared away from home in the American diet, 1977–78 versus 1994–96: changes and consequences. J Nutr Educ Behav 34: 140–150.
5. Todd J, Mancino L, Lin BH (2010) The impact on food away from home on adult diet quality. Economic Research Report 2010.

6. Jeffery RW, French SA (1998) Epidemic obesity in the United States: are fast foods and television viewing contributing? Am J Public Health 88: 277–280.

7. Bowman SA, Gortmaker SL, Ebbeling CB, Pereira MA, Ludwig DS (2004) Effects of fast-food consumption on energy intake and diet quality among children in a national household survey. Pediatrics 113: 112–118.

8. Paeratakul S, Ferdinand DP, Champagne CM, Ryan DH, Bray GA (2003) Fast-food consumption among US adults and children: dietary and nutrient intake profile. J Am Diet Assoc 103: 1332–1338.

9. Taveras EM, Berkey CS, Rifas-Shiman SL, Ludwig DS, Rockett HR, et al. (2005) Association of consumption of fried food away from home with body mass index and diet quality in older children and adolescents. Pediatrics 116: e518–524.

10. Powell LM, Nguyen BT (2013) Fast-food and full-service restaurant consumption among children and adolescents: effect on energy, beverage, and nutrient intake. JAMA Pediatr 167: 14–20.

11. Schmidt M, Affenito SG, Striegel-Moore R, Khoury PR, Barton B, et al. (2005) Fast-food intake and diet quality in black and white girls: the National Heart, Lung, and Blood Institute Growth and Health Study. Arch Pediatr Adolesc Med 159: 626–631.

12. Hearst MO, Harnack LJ, Bauer KW, Earnest AA, French SA, et al. (2013) Nutritional quality at eight U.S. fast-food chains: 14-year trends. Am J Prev Med 44: 589–594.

13. Wu HW, Sturm R (2014) Changes in the Energy and Sodium Content of Main Entrees in US Chain Restaurants from 2010 to 2011. J Acad Nutr Diet 114: 209–219.

14. Wu HW, Sturm R (2013) What's on the menu? A review of the energy and nutritional content of US chain restaurant menus. Public Health Nutr 16: 87–96.

15. Oches S (2011) Top 50 breakdown by market segments. QSR Magazine online. Available at: http://www.qsrmagazine.com/reports/top-50-breakdown-market-segments.

16. Drewnowski A, Rehm CD (2013) Energy intakes of US children and adults by food purchase location and by specific food source. Nutr J 12: 59.

17. Drewnowski A, Rehm CD (2013) Sodium intakes of US children and adults from foods and beverages by location of origin and by specific food source. Nutrients 5: 1840–1855.

18. National Health and Nutrition Examination Survey: MEC In-Person Dietary Interviewers Procedure Manual. Available at: http://www.cdc.gov/nchs/data/nhanes/nhanes_09_10/DietaryInterviewers_Inperson.pdf.

19. Bowman SA, Friday JE, Moshfegh A (2008). MyPyramid Equivalents Database, 2.0 for USDA Survey Foods, 2003–2004 [Online] Food Surveys Research Group. Beltsville Human Nutrition Research Center, Agricultural Research Service, U.S. Department of Agriculture, Beltsville, MD.

20. Ervin RB, Ogden CL (2013) Consumption of added sugars among U.S. adults, 2005–2010. NCHS Data Brief: 1–8.

21. National Health and Nutrition Examination Survey. Dietary Interview: Individual Foods – First Day (DR1IFF_F) Codebook. Available at: http://www.cdc.gov/nchs/nhanes/nhanes2009-2010/DR1IFF_F.htm.

22. (2011) National Health and Nutrition Examination Survey. Analytic Note Regarding 2007–2010 Survey Design Changes and Combining Data Across other Survey Cycles. Available at: http://www.cdc.gov/nchs/data/nhanes/analyticnote_2007-2010.pdf.

23. Guenther PM, Casavale KO, Reedy J, Kirkpatrick SI, Hiza HA, et al. (2013) Update of the Healthy Eating Index: HEI-2010. J Acad Nutr Diet 113: 569–580.

24. Guenther PM, Reedy J, Krebs-Smith SM (2008) Development of the Healthy Eating Index-2005. J Am Diet Assoc 108: 1896–1901.

25. Krebs-Smith SM, Kott PS, Guenther PM (1989) Mean proportion and population proportion: two answers to the same question? J Am Diet Assoc 89: 671–676.

26. National Health and Nutrition Examination Survey. NCHS Research Ethics Review Board (ERB) Approval. Available at: http://www.cdc.gov/nchs/nhanes/irba98.htm.

27. (2014) National Center for Health Statistics, Centers for Disease Control and Prevention. National Health and Nutrition Examination Survey Questionnaires, Datasets and Related Documentation. Available at: http://www.cdc.gov/nchs/nhanes/nhanes_questionnaires.htm.

28. Kirkpatrick SI, Reedy J, Kahle LL, Harris JL, Ohri-Vachaspati P, et al. (2013) Fast-food menu offerings vary in dietary quality, but are consistently poor. Public Health Nutr: 1–8.

29. Namba A, Auchincloss A, Leonberg BL, Wootan MG (2013) Exploratory analysis of fast-food chain restaurant menus before and after implementation of local calorie-labeling policies, 2005–2011. Prev Chronic Dis 10: E101.

30. Bauer KW, Hearst MO, Earnest AA, French SA, Oakes JM, et al. (2012) Energy content of U.S. fast-food restaurant offerings: 14-year trends. Am J Prev Med 43: 490–497.

31. Bruemmer B, Krieger J, Saelens BE, Chan N (2012) Energy, saturated fat, and sodium were lower in entrees at chain restaurants at 18 months compared with 6 months following the implementation of mandatory menu labeling regulation in King County, Washington. J Acad Nutr Diet 112: 1169–1176.

32. Rasmussen LB, Matthiessen J, Biltoft-Jensen A, Tetens I (2007) Characteristics of misreporters of dietary intake and physical activity. Public Health Nutr 10: 230–237.

33. Lafay L, Mennen L, Basdevant A, Charles MA, Borys JM, et al. (2000) Does energy intake underreporting involve all kinds of food or only specific food items? Results from the Fleurbaix Laventie Ville Sante (FLVS) study. Int J Obes Relat Metab Disord 24: 1500–1506.

18

Ecological, Social and Biological Risk Factors for Continued *Trypanosoma cruzi* Transmission by *Triatoma dimidiata* in Guatemala

Dulce M. Bustamante[1], Sandra M. De Urioste-Stone[2], José G. Juárez[3], Pamela M. Pennington[3]*

1 Department of Biology, San Carlos University, Guatemala City, Guatemala, 2 School of Forest Resources, University of Maine, Orono, Maine, United States of America, 3 Health Studies Center, Universidad del Valle de Guatemala, Guatemala City, Guatemala

Abstract

Background: Chagas disease transmission by *Triatoma dimidiata* persists in Guatemala and elsewhere in Central America under undefined ecological, biological and social (eco-bio-social) conditions.

Methodology: Eco-bio-social risk factors associated with persistent domiciliary infestation were identified by a cross-sectional survey and qualitative participatory methods. Quantitative and qualitative data were generated regarding *Trypanosoma cruzi* reservoirs and triatomine hosts. Blood meal analysis and infection of insects, dogs and rodents were determined. Based on these data, multimodel inference was used to identify risk factors for domestic infestation with the greatest relative importance (>0.75).

Principal Findings: Blood meal analysis showed that 64% of 36 bugs fed on chickens, 50% on humans, 17% on dogs; 24% of 34 bugs fed on *Rattus rattus* and 21% on *Mus musculus*. Seroprevalence among 80 dogs was 37%. Eight (17%) of 46 *M. musculus* and three (43%) of seven *R. rattus* from households with infected triatomines were infected with *T. cruzi* Distinct Typing Unit I. Results from interviews and participatory meetings indicated that vector control personnel and some householders perceived chickens roosting and laying eggs in the house as bug infestation risk factors. House construction practices were seen as a risk factor for bug and rodent infestation, with rodents being perceived as a pest by study participants. Multimodel inference showed that house infestation risk factors of high relative importance are dog density, mouse presence, interior wall plaster condition, dirt floor, tile roofing and coffee tree presence.

Conclusions/Significance: Persistent house infestation is closely related to eco-bio-social factors that maintain productive *T. dimidiata* habitats associated with dogs, chickens and rodents. Triatomine, dog and rodent infections indicate active *T. cruzi* transmission. Integrated vector control methods should include actions that consider the role of peridomestic animals in transmission and community members level of knowledge, attitudes and practices associated with the disease and transmission process.

Editor: Claudio R. Lazzari, University of Tours, France

Funding: This investigation received financial support from the UNICEF/UNDP/World Bank/WHO Special Programme for Research and Training in Tropical Diseases (TDR), grant Number A90299 to PMP, in the context of the TDR/IDRC research initiative Towards Improved Chagas and Dengue Disease Control through Innovative Ecosystem Management and Community-Directed Interventions: An Eco-Bio-Social Research Programme on Chagas and Dengue Disease Control in Latin America and the Caribbean", supported by Canada's International Development Research Centre (IDRC). The funders had no role in study design, data collection and analysis, decision to publish, or preparation of the manuscript.

Competing Interests: The authors have declared that no competing interests exist.

* Email: pamelap@uvg.edu.gt

Introduction

Chagas disease is a vector-borne neglected tropical disease that continues to affect the most vulnerable populations across Latin America. Integrated vector management (IVM) was proposed by the World Health Organization (WHO) as part of the 2008–2015 strategy to control neglected tropical diseases [1]. Tropical Disease Research/WHO implemented an initiative to study the ecological, biological and social (eco-bio-social) factors that lead to the presence of dengue and Chagas disease vectors under different eco-epidemiological settings at the local level [2,3]. The local ecology of vector-borne diseases must be considered by epidemi-ologists, public health officials and policy makers in the development of novel IVM and disease control strategies.

Despite advances in the control of Chagas disease vectors, insecticide-based control is limited by local conditions that lead to persistent infestation foci, sometimes derived from peridomestic habitats [4–8]. The transmission cycle includes mammalian reservoirs of the parasite and triatomine species, such as *Triatoma dimidiata* and *Triatoma infestans*, which colonize domestic and peridomestic environments [6,9]. Indoor residual insecticide spraying and changes in house construction methods are the only public health vector control tools currently available. Meanwhile, unidentified peridomestic habitats persist as a vector control challenge.

In Central America, a Chagas disease control initiative was launched in 2001 based largely on indoor residual spraying with insecticides [5]. Entomological surveys at the Municipality geopolitical level were performed in Guatemala to determine vector distribution. These surveys were followed by the implementation of a National Chagas Disease Control Program that applied residual pyrethroid insecticides in all affected Municipalities. Evaluations of the initiative showed that *T. dimidiata* infestation levels were reduced up to nine-fold in many Municipalities [5]. However, despite multiple insecticide applications, infestation levels remained clustered in some communities [8,10,11].

Chagas disease eco-epidemiology is related to local environmental conditions combined with socio-economic and cultural factors that lead to the presence of animal nests harboring triatomines near or within households [3,9,12,13]. Dogs, cats and rodents are synanthropic reservoirs of *T. cruzi* and common blood meal sources [14–16]. Chickens are a blood source for various triatomines and are thought to play a role in house colonization, even though they are refractory to *T. cruzi* [17,18]. Thus, the definition of risk factors related to vector infestation and parasite transmission should consider the eco-bio-social factors that lead to the presence of blood meal sources and reservoir hosts in and around the household.

This study used a multidisciplinary mixed methods approach to identify eco-bio-social factors of persistent triatomine intradomiciliary infestation in a region of Guatemala where community-wide insecticide applications were less effective compared to other areas of the country.

Methods

Ethics statement

The study protocol was approved by the Universidad del Valle de Guatemala (IRB 00002049, FWA 00001902) and World Health Organization Institutional Review Boards. Individual written informed consent was obtained from study participants before household surveys and group written consent was obtained before each group meeting. Consents included permission to take photographs and make video recordings of activities. This study was performed in strict accordance with the recommendations in the Guide for the Care and Use of Laboratory Animals of the National Institutes of Health. The protocol was approved by the Institutional Animal Care and Use Committee of the Universidad del Valle de Guatemala (AWLAW No. A5847-01).

Study area and population

This study was conducted in the municipalities of Comapa and Zapotitlán, department of Jutiapa, in eastern Guatemala (Figure 1). Comapa is located at: −89°54′46.8″ and 14°6′38.6748″, and Zapotitlán at −89° 49′ 33.1314″ and 14° 8′ 15.1866″. More than 80% of the populations of Comapa and Zapotitlán live in rural areas, more than 80% live in poverty and more than 30% live in extreme poverty [19].

Study design

A multilevel triangulation mixed methods design was used to converge and validate quantitative results and qualitative information [20] from multiple disciplines. This design is used to address different levels within an eco-bio-social system; each method is treated as one level or data layer. Quantitative and qualitative methods from diverse fields are utilized. Findings from all levels are integrated for an overall interpretation. A combination of biophysical and social science research methods were used including: household and entomological surveys, rodent survey, canine serological survey, qualitative data collection methods (semi-structured interviews and participatory community meetings).

A cross-sectional study was performed between January and February 2011 among randomly selected villages, to produce a baseline survey for entomological, ecological and social factors. A follow-up animal/entomological survey was undertaken in a subset of the houses from March to July 2011, including canine serological and rodent surveys at the household level. Interviews were conducted between January and August 2011. Participatory group meetings were held with selected communities from March to May 2011.

Community selection and sample size

Baseline survey communities were selected based on location, >800 meters above sea level, >20% baseline infestation prevalence for *T. dimidiata* (i.e. before the Ministry of Health control program in 2004–2010) and a history of multiple insecticide applications with persistent infestation by this vector [11]. A total of 40 communities were identified, with a total of 3,944 households (population). Thirty-two communities (30 from Comapa and two from Zapotitlán) were randomly selected with two-stage cluster sampling that considered probability proportional to size of clusters and systematic probability sampling of households [21–23].

The first stage used a probability proportional to size selection of clusters since the number of elements in each community was not the same [21–23]; clusters were selected based on a geo-referenced sampling frame of communities with ArcMap 9.3.1 (ESRI, Redlands, CA, USA). Fifty-nine clusters of eight households each were selected. To reduce selection/interviewer error in the field, the second stage included systematic probability sampling [21] to randomly select households from the geo-referenced sampling frame using ArcMap 9.3.1. For each selected community an interval was estimated using the number of households to be sampled (eight, 16, 24, 32, or 40) and the total population of households for that community; in addition a random start was chosen for each community [21]. Once the interval and random start were selected, a systematic random sampling of households was performed for each community, utilizing ArcMap 9.3.1. Maps with randomly selected households were printed for each community. In the field, selected houses where identified and visited to conduct the household demographic and Knowledge Attitude and Practices (KAP) questionnaire and entomological surveys. In the case a household previously selected was not found, interviewers searched the next house to the right on the direction previously established.

A follow-up animal survey included a subset of 23 randomly selected communities (all were from Comapa due to random selection) and 248 systematically selected households surveyed at baseline. In two cases, the originally sampled house was not available for the follow-up survey and a new household was systematically recruited. Households without an adult at the time of the survey were not eligible for the study. Each household was georeferenced with a geographical positioning device (Garmin, Schaffhausen, Switzerland).

The sample size was 472 households for the baseline survey and 248 households for the follow-up survey. The baseline survey sample assumed a 20% infestation with 7% precision, and a 0% false-negative error rate, for a design effect of 2. The sample allowed detection of risk factors with a prevalence of 20% to 80%

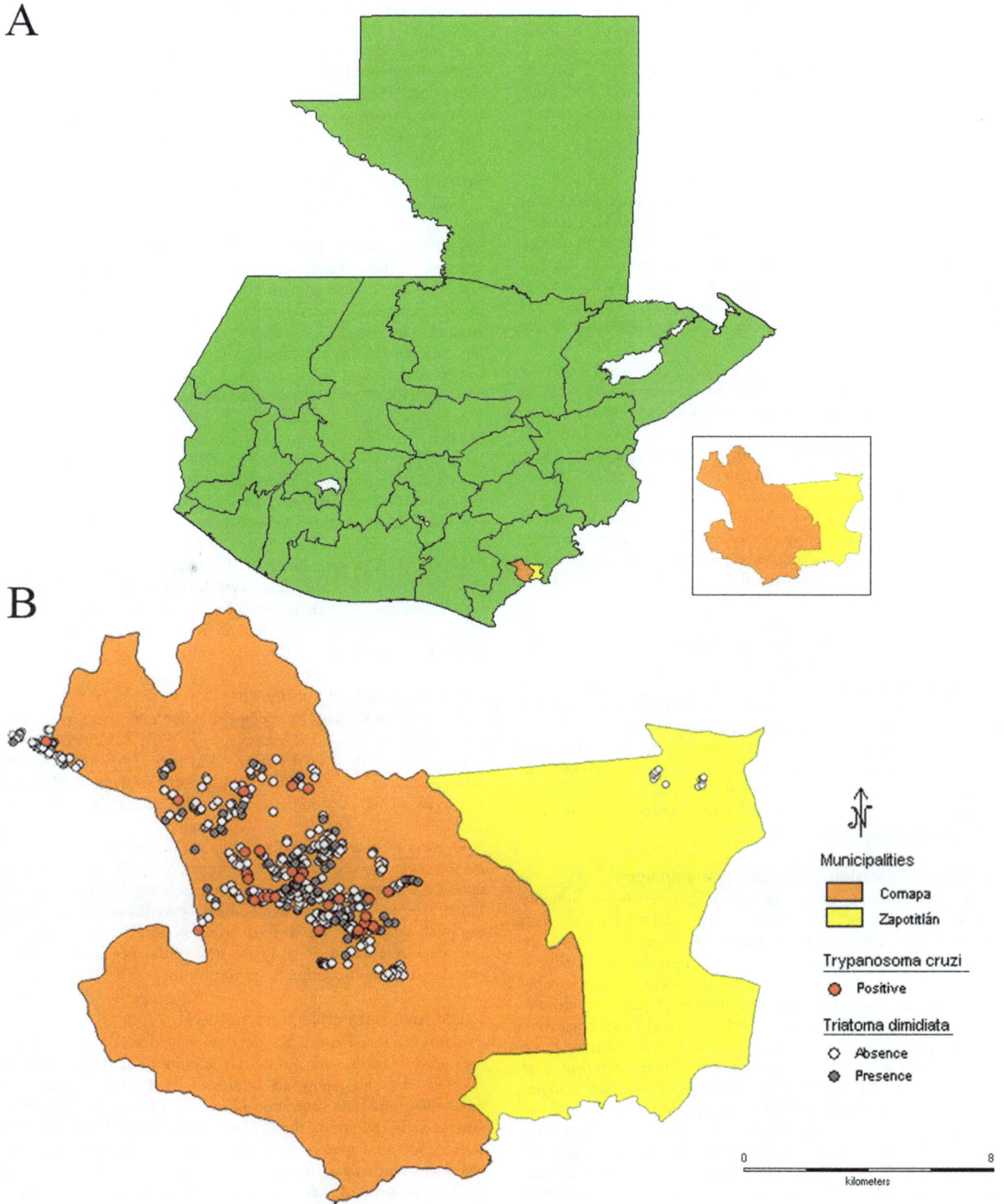

Figure 1. Study site location and household infestation status in 2011. A. Map of Guatemala showing the location of Comapa and Zapotitlán, Jutiapa. **B**. Map of Comapa and Zapotitlán showing the presence (grey circles) and absence (white circles) of household infestation of *Triatoma dimidiata*, and triatomines infected with *Trypanosoma cruzi* (red circle). No triatomines were collected in the municipality of Zapotitlán. *Because of conflicts between municipalities three communities considered part of Comapa are located outside the map boundary.

and a minimum detectable odds ratio (OR) of 2 for logistic regression with a single binary covariate, according to the Wald test [24,25]. Standard logistic regressions are likely underestimating both effect-sizes (ORs) and their variances. The follow-up survey sample size was based on the expected 30% dog seroprevalence and number of households reporting dog ownership at baseline.

Household and entomological surveys

To characterize households, a structured questionnaire was used for a face-to-face household demographic and KAP survey [26], according to Dillman's Tailored Design Method [27]. The questionnaire consisted of closed-ended, semiclosed-ended and ranking items related to Chagas disease, triatomines, animal management, rodents, household construction materials and peridomestic structures. The questionnaire was validated in the field previous to its implementation.

For the entomological survey during the baseline and follow-up surveys, each house was searched by two members of the team in the domestic and peridomestic environments (15 min each ecotope) using the person-hour method [28]. Triatomine presence, abundance and location were recorded. Specimens were transported to the laboratory for processing.

During the follow-up animal survey, householders were asked to search for triatomines during three consecutive nights to gain an understanding of community participation in triatomine surveillance. Given that the collection ecotope was not noted, the results of this participant entomological survey were not used in univariate analyses nor model generation.

Descriptive statistics and normality tests were conducted on all variables from the KAP surveys. According to the results from the descriptive tests, level of measurement and normality tests, contingency tables, chi-squared, phi, Cramer's V and odds ratios were used to compare all variables from the KAP surveys against the entomological variable of triatomine presence in domestic environments as determined by the person-hour method (triatomines collected in peridomestic environments were not included in these analyses). The p value was set at 0.05. Phi values of 0.001 to 0.10 indicated weak association, and values between 0.11 and 0.30 indicated moderate associations [29].

Triatomine infection and blood meal analysis

Bias in triatomine processing was introduced by a mortality during transport of 477 (71%) of 669 bugs, due to high environmental temperatures. For each household, live bugs were randomly selected for dissection. A rectal puncture was analyzed by screening 20 fields at 40× by phase contrast microscopy. Individual live bugs from each household were screened until the first positive specimen was identified. Live and dead bugs were preserved in 95% ethanol for molecular analysis. Midgut DNA was extracted from all triatomines (n = 192) that arrived alive at the laboratory and from 9 dead specimens. Triatomine midguts were dissected from ethanol-preserved specimens and air dried; then 25–50 mg was homogenized with a metal bead in 600 µL DNAzol (Invitrogen, Carlsbad, CA, USA) on a Mix Mill MM 400 homogenizer (Retsch, Haan, Germany). After overnight incubation at 4°C, 400 µL DNAzol was added, followed by incubation for 1 h at room temperature and centrifugation at 13,000 g for 15 min at room temperature. The DNA was precipitated with 500 µL absolute ethanol and washed twice with 70% ethanol. All extracted samples (n = 201) were screened with a universal vertebrate mitochondrial cytochrome b (cyt b) PCR (Table S1 in File S1). Samples were run on 2% agarose/1×Tris acetic acid EDTA buffer gels containing ethidium bromide (Promega,

Madison, WI,USA) and visualized under ultraviolet light (UVP, Upland, CA, USA). Thirty six samples positive for universal cyt b were subjected to additional individual PCR with specific primers for Homo sapiens (β-globin), Gallus gallus, Canis familiaris, Rattus rattus and Mus musculus. Polymerase chain reaction and fragment size analysis for T. cruzi minicircle was performed on β-globin positive triatomines (Table S1 in File S1).

Rodent survey and tissue processing

During the follow-up survey, rodent infestation was estimated for the subset of houses by placing five 7.6×8.9×22.9 cm Sherman traps inside each house for three consecutive nights with daily checking. Captured animals were sedated with pentobarbital and euthanized by cardiac puncture followed by cervical dislocation. A blood smear was prepared and heart tissue was collected in 95% ethanol. The carcass was preserved in ethanol and transported to the laboratory for taxonomic characterization. All preserved rodent hearts were analyzed for T. cruzi DNA from 15 of 17 households with at least one microscopically positive triatomine. Polymerase chain reaction for T. cruzi characterization was performed on rodent heart DNA extracted as described above. Selected samples were typed at the Distinct Typing Unit level by glucose-6-phosphate isomerase (GPI) sequence analysis. The PCR product of GPI was obtained according to conditions in Table S1 in File S1, purified with a Wizard PCR purification kit (Promega, Winsconsin, USA) and sequenced at Macrogen, Inc. (Rockville, MD, USA). The remainder of the samples were screened for T. cruzi DNA with maxicircle cytochrome c oxidase subunit II (COII), confirming positive samples by minicircle hypervariable DNA amplification and fragment analysis according to conditions in Table S1 in File S1 [30].

Canine serological survey

A sample of 80 dogs (one dog per household) was screened to detect an estimated 30% seroprevalence (90% Confidence level, 0.1 precision) [31], based on the rapid tests (Trypanosoma cruzi Detect-Canine; Inbios, Seattle, WA, USA) sensitivity of 91% and specificity of 98% in dogs experimentally infected with T. cruzi [32].

Owners were asked to select an animal in the household based on the inclusion criteria: age (>6 months), health (no emaciated animals) and reproductive status (non-lactating females). Canine blood sample (2–5 mL) was obtained by venipuncture from the braquial vein and rapid tests were performed according to manufacturers instructions. Proportion confidence intervals were calculated without continuity correction [33].

Qualitative data collection methods

Semi-structured interviews were conducted with key stakeholders in the region and community members: participants were selected through snowball strategies [34]. Interviews were conducted until data saturation [20,35] was reached-no new data emerged. In addition, three communities from Comapa were selected for participatory group meetings [36] based on infestation level, social organization, socioeconomic and political dynamics. A cyclical process of joint learning, reflection, and exploration occurred during the five sessions with each community [36]. Guided discussions included food production systems, management of peridomestic animals, socioeconomic and land tenure systems, triatomine and Chagas disease knowledge and awareness, problem identification associated with Chagas disease, and community-based generation of future goals and strategies to reduce triatomine infestation.

Qualitative data were transcribed, encoded and analyzed with NVivo 9 (QSR int, Massachusetts, USA). Free nodes from patterns were created by open coding [37] and tree nodes were based on the research objectives. Pattern coding aided in the generation of categories and themes [38,39], according to the PRECEDE-PROCEED model [40]. In addition to the emerging patterns, advanced search routines with selected keywords and nodes were run to analyze participants' perceptions regarding risk factors detected by multimodel inference.

Identification of risk factors for house infestation with multimodel inference

Information generated in the household and entomological surveys was used to determine the relative importance (RI) of a set of variables in explaining the presence of *T. dimidiata* in domestic environments in Comapa and Zapotitlán households. The eco-bio-social variables included in the analysis were selected *a priori* based on evidence of association with infestation from published studies, known infection reservoirs, blood meal analysis and animal infection data gathered throughout the present study. Not all selected variables had significance in the univariate analysis from the KAP and entomological surveys. These variables constituted our hypothesis to explain the presence of *T. dimidiata* inside houses. Multimodel inference was used to determine the RI for each variable in the hypothesis and to explore model uncertainty for reduced models. Records (houses) with missing values were removed [41]. The resulting dataset included 449 records (houses) and 25 variables (Table S2 in File S1). Rodent survey information (*M. musculus* and *R. rattus* presence) was available for a subset of 220 houses.

To model the variable "bug presence in domestic environments" logistic regression was used with the glm function in R [42]. The Hosmer and Lemeshow test was used to evaluate the goodness of fit of the model to the data. Multicollinearity was evaluated by calculating the variance inflation factors using the vif function from the car library in R. Collinear variables were identified and dropped from the model. Multimodel inference was used to determine the relative importance of variables that best explained the presence or absence of *T. dimidiata* in the households [43]. A subset of reduced models with different combinations of variables was compared with the Akaike information criterion with a correction for finite samples (AICc) to assess model uncertainty using Akaike weights. Due to model uncertainty, weighted parameter estimates were calculated. Variable RI was estimated by adding the weights of the models with the lowest AICc in which the variables appeared.

The subset of reduced models with different variable combinations was explored using the glmulti package for R [44,45]. Ten genetic algorithm runs (GAR) were conducted with the parameter value combinations (Methods in File S1) for the dataset that included all the houses. For each GAR the RI, weighted parameter estimates, and weighted parameter variances of all variables were obtained using the coef function in glmulti, and then averaged across all runs. Odds ratios and their confidence intervals were calculated from these average values; we interpret these intervals with caution given that they are derived from the average of weighted parameters. After the most important variables were identified with the larger dataset, 10 more GAR were conducted using these variables and the rodent information with the smaller subset of houses. The RI, weighted parameter estimates, and variance for these analyses are also reported.

Triangulation

Methodological triangulation [34,35] of biophysical, quantitative and qualitative social science data collection methods was used to enhance the strengths of interpretations and conclusion about risk factors. Information from interviews, document reviews, reflections, and group meeting was triangulated with information identified through surveys, multimodel inference, rodent infection and blood meal analysis. Analysis for this paper utilized a triangulation approach for converging and validating results using multiple methods. In addition, triangulation across sources [46] was utilized to understand perspectives among different interviewees.

Results

Biological variables associated with infestation

The baseline entomological survey revealed domestic and/or peridomestic infestation in 120 (25%) of 472 houses. Domestic infestation was detected in 101 (21%) of 472 houses. Peridomestic infestation, defined as infestation in structures not sharing a common roof with sleeping quarters, was detected in 33 (7%) of 472 houses. Infestation in both domestic and peridomestic environments was detected in 14 (3%) of 472 houses. The results of household participant searches during the follow-up survey showed that 144 (30%) of 477 houses were infested. Considering both the baseline and follow-up surveys together, colonization (i.e. nymph presence) was observed in 104 (72%) of 144 houses. Of all adults collected by the person-hour method, 38% were collected in bedrooms, 10% in chicken coops and 8% in kitchens. Interestingly, similar proportions were observed for nymphs collected in these locations (data not shown).

The self-reported presence of domestic and peridomestic animals was evaluated as a risk factor for house infestation (presence of adults, nymphs or both). The reported ownership of at least one dog showed a significant association with triatomine infestation (Table 1). Rodent surveys showed 61% infestation, with *M. musculus* and *R. rattus* as the primary rodent species. *Peromyscus mexicanus*, *Liomys salvini*, and *Oryzomys* spp were collected sporadically inside homes. The presence of at least one rodent per house (any species) showed a significant association with triatomine infestation (Table 1). The presence of chickens roosting or laying eggs inside the house showed a weak to moderate association (Table 1).

The biological relevance of the association between peridomestic animals and triatomine infestation was confirmed by blood meal analyses to detect DNA from humans, dogs, chickens and the two major rodent species, *R. rattus* and *M. musculus* (Table 2). Vertebrate blood was detected in 37 (18%) of 201 specimens. Most insects had fed on chickens (65%), followed by humans (51%), *R. rattus* (22%), dogs (19%) and *M. musculus* (19%). Only 12 (36%) bugs had single blood meals (6 from chickens, 5 from humans and 1 from *M. musculus*). More than half of the blood meals had two to four different blood sources. All blood meal types were identified in triatomines collected in both domestic and peridomestic environments. Three nymphs with human-mouse or human-rat blood meals were collected in sleeping quarters. A human-mouse blood meal was detected in a third instar collected in a corral located 4-8 m from domestic premises. Infection was detected in 11 (31%) of 35 triatomines with a recent vertebrate blood meal, including 33% of insects with chicken blood meals. No detectable association was identified between blood meal source and infection status (Fishers exact test, p>0.5).

Table 1. Association of triatomine infestation according to person-hour survey and animal ownership as reported by householders, and rodent presence as recorded by rodent surveys.

Animal	No. present/Total no. (%)	Range of numbers of animals per house	Yates chi-square	p	df	OR (95% CI)
Horse	99/467 (21)	0–4	1.20	0.273	1	
Donkey	11/467 (2)	0–2	0.73	0.395	1	
Rabbit	4/467 (1)	0–4	0.62	0.430	1	
Cow	57/467 (12)	0–12	0.63	0.429	1	
Cat	147/467 (31)	0–6	0.96	0.328	1	
Pig	62/467 (13)	0–10	0.01	0.938	1	
Dog[a]	333/466 (71)	0–9	7.61	0.006	1	2.3 (1.3–4.0)
Hen	329/467 (70)	0–100	0.26	0.600	1	
Fowl	131/467 (28)	0–50	0.80	0.373	1	
Chicks	16/467 (3)	0–41	0.33	0.566	1	
Hens laying eggs inside[b]	22/466 (5)		3.27	0.071	1	2.2 (0.9–5.5)
Hens roosting inside[c]	90/457 (20)		6.55	0.010	1	2.0 (1.2–3.3)
Rodents[d]	149/249 (61)	0–13	5.44	0.002	1	4.0 (1.1–14.1)

[a]Significant associations between reported ownership of animal and presence of triatomines.
[b]Weak association.
[c]Moderate association.
[d]Significant association between presence of any rodent species and presence of triatomines.

Trypanosoma cruzi transmission indicators

Trypanosoma cruzi transmission was evaluated by analysis of infection in vectors, dogs and rodents (Table 3). Microscopic screening showed that 40 (31%) of 130 of the infested households had at least one infected triatomine. Infections were detected in females, males, second, third and fourth instars. Twentyeight (35%) of 80 adult dogs were seroreactive. The true prevalence adjusted for test specificity and sensitivity is estimated at 37% (25%-50%, Blakers 95% Confidence Limits) (Table 3). Eleven (21%) of 53 rodents captured in homes with microscopically infected triatomines had *T. cruzi* DNA in heart tissue confirmed by COII amplification and minicircle fragment size or GPI sequence analysis. Eight (17%) of 46 *M. musculus* and three (43%) of seven *R. rattus* had *T. cruzi* infection confirmed by GPI sequence or minicircle fragment size analysis. The *T. cruzi* strain in two mice and two rats from the same household was confirmed as DTU I by sequencing GPI (Accession Nos. KJ682643-6). Eight (44%) of 18 triatomines with human blood meals had *T. cruzi* DNA as determined by minicircle fragment size.

Ecological variables associated with infestation

Reports of triatomines or bats entering or approaching the house at night were associated with infestation (Table 4). The presence in the patio of fruit-bearing trees such as avocado, coffee and jocote (*Spondias purpurea*) was found to be positively associated with triatomine presence (Table 4). Figure 2 shows a storage area in a cinder block house where nymphs were detected. As the area was investigated, rodent feces and jocote fruit pits were found on the dirt floor, suggesting a link between rodents and fruit-bearing trees in the area.

Socioeconomic variables associated with infestation

Table 5 and Table S3 in File S1 present all variables from the KAP survey and observations showing an association between socioeconomic variables and household construction with triatomine infestation. As an indication of previous vector control activity in the region, the presence of a Ministry of Health Household Vector Card in the house was found in 187 (41%) of 454 surveyed households (Table S3 in File S1). This card is placed by the vector control personnel inside houses that are visited for

Table 2. Triatomine blood meal sources of *T. dimidiata* and *T. cruzi* infection status according to blood meal source.

Identified blood meal	No. blood meals/No. examined (%)	No. infected by PCR/No. examined (%)
Chicken	23/36 (64)	8/21 (38)
Dog	6/36 (17)	2/6 (33)
Human	18/36 (50)	6/16 (38)
Mus musculus	7/34 (21)	0/6 (0)
Rattus rattus	8/34 (24)	2/6 (33)
Vertebrate	3/36 (8)	0/3 (0)

Mixed blood meals include human, chicken, dog and mouse (2), human-chicken-dog (1), human-chicken-rat (4), chicken-dog-rat (2), human-chicken-mouse (1), human-mouse-rat (1), chicken-dog (1), chicken-mouse (1), chicken-rat (2), human-chicken (4), human-dog (1), human-rat (1) and human-mouse (1).

Ecological, Social and Biological Risk Factors for Continued Trypanosoma cruzi Transmission...

207

Table 3. Indicators of *T. cruzi* transmission in Comapa and Zapotitlán, Jutiapa, Guatemala 2011.

Transmission indicators	No. present/Total no. (%)	95% CI
Triatomine infested households with at least one microscopically positive triatomine	40/130 (31)	23–39
Seropositive adult dogs*	28/80 (37)	25–50
PCR-positive rodents in households with infected triatomines	14/54 (26)	16–39
PCR-positive triatomines with detectable vertebrate blood meals	11/33 (33)	20–50

* True prevalence adjusted for test specificity and sensitivity (Blakers 95% Confidence Limits) [61].

surveillance or insecticide applications. Households having this card had 1.7 (1.1–2.7, 95% CI) higher odds of infestation than those without it. Only 241 (53%) of 454 surveyed had heard about Chagas disease, with 52 (22%) of these individuals having received information from the vector control personnel and 94 (39%) from the Health Center. The odds of infestation were 0.4 (0.2–0.8) lower in households where someone had received information from the Health Center, and 2.5 (1.2–5.0) higher when they received information from vector control personnel. Of those who

had heard about the disease, 132 (54%) of 245 reported using insecticides to control it, showing a weak positive association with infestation (phi = 0.106).

Only 26% of houses were constructed in part from cinder blocks; most houses were built of adobe (45%) and bajareque (mud with sticks and sometimes wood) or mud (46%). Sixteen percent of households still had a tile roof. In addition, 66% of houses had dirt floors. For ventilation purposes, eaves (open space between the roof and the wall), another house construction feature in the area,

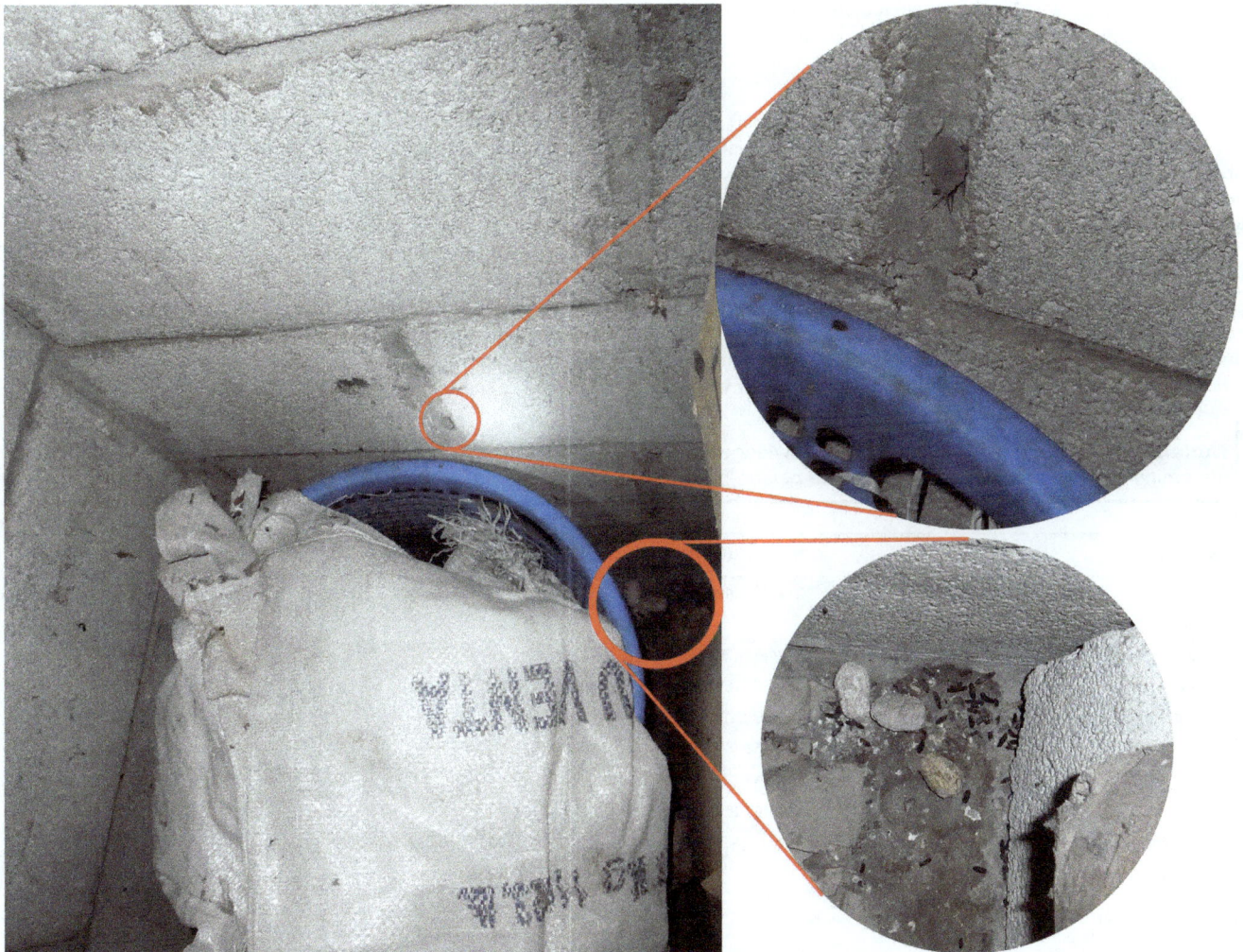

Figure 2. Storage facility associated with triatomine and rodent infestation. *Triatoma dimidiata* nymph found in association with a storage area with rodent feces and jocote fruit pits. Upper inset: Nymph. Lower inset: Fruit pits surrounded by rodent feces.

Table 4. Ecologic factors from the Knowledge-Attitude-Practices and entomological survey showing association with triatomine infestation, as determined by the person-hour method (baseline survey n = 472).

Variable	No. positive/Total no. (%)	Phi (∅)	Odds ratio (95% CI)
Bugs enter or approach the house at night	243/469 (52)	0.215[b]	3.0 (1.9–4.9)
Bats enter or approach the house at night	307/469 (66)	0.097[a]	1.7 (1.0–2.8)
Avocado trees in the patio	230/469 (49)	0.098[a]	1.6 (1.0–2.5)
Coffee trees in the patio	307/469 (65)	0.119[b]	1.9 (1.2–3.2)
Jocote trees in the patio	346/469 (74)	0.100[a]	1.8 (1.1–3.2)

[a]Weak association.
[b]Moderate association.

were present in 85% of houses. Household partial wall plastering was positively associated with infestation. Cement or tile floors were protective (OR = 0.3, 95% CI 0.2–0.7) whereas tile roofs increased infestation risk (OR = 1.9, 95% CI 1.1–3.3). Cinder block walls were protective. Walls containing bajareque, but not adobe, were a risk factor (OR = 1.9, 95% CI 1.2–3.9). There was a moderate positive association between socioeconomic indices and the presence of triatomines. Firewood was used for cooking by 98% of households. The house was often used as a storage facility: 72% stored grains in metal containers, 41% stored grains in bags, and 20% stored firewood.

Identification of risk factors for house infestation

Logistic regression analysis was used to model the variable "bug presence in domestic environments" (Hosmer-Lemeshow fit test, chi-squared = 4.20, p = 0.84) for the first dataset including all the houses. The variables "total number of humans in the house" and "type of wall plaster (exterior)" were dropped due to collinearity, but the variables "number of persons per room" and "type of wall plaster (interior)" were retained. Implementation of the glmulti genetic algorithm made it possible to fit a subset of reduced logistic models with different variable combinations, which allowed to estimate the relative importance of the variables in the hypothesis

and to assess model uncertainty. The average number of generations per run was 1,653±726 (100 models fitted per generation) before convergence. The exploration of multiple models with the genetic algorithm revealed that there was high uncertainty and that there was no single best model among all the models compared in the 10 GAR. The reduced model with the lowest AICc in the subset of models explored was "bugsInside~Intercept + CoffeeTrees + RoofType + FloorType + InternalPlaster + CellPhone + Dogs", which had an Akaike weight of 0.26 and an AICc of 433.26 compared to an AICc of 468.32 in the full model (24 variables). An Akaike weight under 0.90 indicates that there is not a single reduced model that could be considered the best model (= hypothesis) among those compared. Therefore a model using weighted parameters like the one presented here is recommended. After obtaining the average across all runs, we found that the most important variables (RI≥0.75, Table 6), among those considered to explain the presence of bugs in houses in this region of Jutiapa were the type of internal wall plastering (RI = 0.98), type of floor (RI = 0.94) and number of dogs (RI = 0.98) (Table 6). The presence of coffee trees around the house and the type of roof also had high RI values (RI = 0.85 and 0.83, respectively). All other variables had an RI under less than 0.75 (Table S4 in File S1).

Table 5. Socioeconomic factors from the Knowledge-Attitude-Practices and entomological survey showing associations with triatomine infestation, as determined by the person-hour method (baseline survey, n = 472).

Variable	No. positive/Total no. (%)	Cramer's V	Phi (∅)	Odds ratio (95% CI)
Bajareque (mud and stick) walls present	214/468 (46)		0.134[b]	1.9 (1.2–3.9)
Plastering condition (partial, complete, absent)		0.139[b]		
Cinder block walls	122/468 (26)		−0.146[b]	0.4 (0.2–0.7)
Tile roofs	76/468 (16)		0.107[a]	1.9 (1.1–3.3)
Cement tile floors	94/468 (20)		−0.146[b]	0.3 (0.2–0.7)
Earthen floors	309/468 (66)		0.201[b]	3.4 (1.9–6.0)
Cement slab floors	137/468 (29)		−0.121[b]	0.5 (0.3–0.8)
Eaves	378/447 (85)		0.110[b]	2.5 (1.1–5.3)
Socioeconomic index 1		0.118[b]		
Socioeconomic index 2		0.222[b]		

[a]Weak association.
[b]Moderate association.
Socioeconomic index 1: variables for having electricity at home, owning a cellular phone, using electricity for cooking, and using electricity for lighting the house at night, Socioeconomic index 2: variables for having electricity at home; owning a cellular phone; using electricity for cooking; using electricity for lighting the house at night; and number of chickens, cows, and pigs owned. Table S2 in File S1 shows the variables associated with infestation that were used to derive Socioeconomic index 1 and 2.

Table 6. Model results showing the most important individual factors that increase (or decrease) the relative odds of *Triatoma dimidiata* presence inside households, keeping all other conditions constant.

Factors (odds$_1$/odds$_2$)	Average relative importance (RI)	Average weighted estimate	Average weighted variance	Odds ratio (95% CI)
Intercept	1	−2.48	0.52	
1–2 dogs/0 dogs	0.99	0.74	0.12	2.09 (1.06–4.13)
>2 dogs/0 dogs	0.98	1.08	0.16	2.95 (1.36–6.40)
Partial interior plaster/complete interior plaster	0.98	1.44	0.32	4.20 (1.37–12.83)
No interior plaster/complete interior plaster	0.98	0.34	0.19	1.40 (0.59–3.33)
Non-dirt floor/dirt floor	0.94	−0.82	0.12	0.44 (0.22–0.87)
Coffee trees yes/no	0.85	0.50	0.11	1.64 (0.85–3.16)
Clay tile or plant material roof/metal-sheet roof	0.83	0.51	0.12	1.67 (0.84–3.29)

The table shows only variables with RI>0.80.

Of these variables, the ones that appeared to increase the relative odds of bug presence and detection, keeping all other conditions constant, were (a) partial or no plastering in the house, in comparison to houses with complete plastering; (b) having at least one dog; (c) having coffee trees around the house; and (d) having a roof made of materials other than metal-sheet like clay tiles or plant leaves. On the other hand, cement or tile floors reduced the odds of bug presence inside the house in comparison to dirt floors.

A logistic regression model using only the most important variables identified previously (type of roof, type of floor, type of interior plastering, number of dogs and presence of coffee trees), and the variables altitude, *M. musculus* presence and *R. rattus* presence was fitted to a subset of data with rodent information. This model showed a good fit to explain the infestation data (Hosmer-Lemeshow fit test, chi-squared = 5.32, p = 0.72). The exploration of multiple reduced models with the genetic algorithm (with an average 214±28 generations per run) again revealed high uncertainty and there was no single best model among the models compared in the 10 GAR. The model with the lowest AICc "BugsInside ~ Intercept + RoofType + FloorType+ Dogs + *Mus musculus* presence", with an Akaike weight of 0.05 and an AICc of 212.38 compared to an AICc of 217.92 in the full model. Using multimodel inference it was determined that the most important variables (RI≥0.75) among those considered to explain the presence of bugs in this subset of houses were *M. musculus* presence (RI = 0.89, OR = 2.13, 95% CI 1.93–4.89) and type of roof (RI = 0.77, OR = 2.02, 95% CI 0.71–5.77). All other variables had an RI less than 0.75 for this subset of data.

Perceptions and practices related to chickens, rodents, construction and environment as risk factors

Several study participants mentioned having seen triatomines more often around chickens and chicken coops, especially at night. Qualitative data from participant group meetings indicated that hens and chickens were left inside the house at night to roost and during the day to lay eggs; this seemed to be an established practice in the study area. According to the following quote from a group session, one participant recognized that having chickens inside the house represented a risk for triatomine infestation.

"I have proved that chickens do in fact attract the bugs, because previously we left them all inside, a while ago. Where they slept, there were the bugs. Then my husband decided to take them outside, in a coop outside" (*Group meeting 9.1, 2011*).

In addition, vector control personnel recognize chickens laying eggs under the bed as a risk factor.

"People continue to have chickens lay their eggs underneath the beds and, unfortunately due to extreme poverty, the walls have all the conditions for the bug to keep reproducing inside the houses" (Interview 1, February 25, 2011).

The majority of participants (68%) were concerned about rodents transmitting diseases. Interestingly, 61% of the households reported having few mice and 50% few rats. Nevertheless, 53% perceived that there were many mice in the community and 56% that there were many rats.

Several participants emphasized the influence of house construction materials and styles on the presence of triatomines and rodents in the household. Innovations mentioned by participants to prevent house infestation by triatomines and rodents include refurbishing walls, placing new roof tiles, building a cement floor and closing eaves.

"The interviewee mentioned he had refurbished the walls himself. He also mentioned having sealed the eaves, to prevent bugs and rats, and in fact the wall (mud renovated) reached the ceiling (sheet). He even mentioned that before renovating the house walls and eaves he had problems with animals, but not after that." (Reflective journal EP, October 12, 2012).

"More triatomines could be found inside the houses when we had tile roofs" (Group meeting, November, 2012).

Traditional house construction styles in the area have been considered risk factors for the presence of triatomines by vector control personnel.

"...Usually we find in these families all the factors that contribute to triatomine presence, such as a (dirt) floor, bajareque or mud walls –they make own adobe –, thatched roofs, and they share their habitat with pets: dogs, pigs, hens [...]. They are people with limited resources [...]. They cook with firewood, and store firewood around the house" (Interview I.11, June 17, 2011).

Finally, several study participants believed leaf litter to be associated with bug infestation. "Having leaf litter around the house makes having bugs (triatomines) there more likely" (Reflective journal EP from group meeting 2.3, November 12, 2012).

Discussion

The factors that were found to play a role in persistent intradomiciliary triatomine infestation in this region of Guatemala include the presence of parasite reservoir hosts such as dogs and rodents, the presence of chickens as blood sources, house construction methods, and environmental conditions around the household that are related to cultural practices and socioeconomic conditions. We developed a multilevel triangulation mixed methods design that combines biostatistical, biological and social science methods to improve our understanding of the eco-epidemiology of Chagas disease at the local level, and that provides evidence for IVM.

Our study showed that house construction methods and materials related to local cultural and socioeconomic conditions are associated with infestation. Construction materials such as bajareque (mud and sticks) and tile roofs are associated with infestation, whereas cement and brick floors are protective. The association of construction materials and triatomines is not homogeneous. A study in a nearby municipality identified primarily adobe walls as a risk factor [12] whereas tile roofs were found to be a protective factor at the department level [28]. In agreement with our study, King *et al* found dirt floors to be a risk factor in Jutiapa but not in a northern department of the country [28]. The presence of different *T. dimidiata* phylogenetic groups across the endemic area from Mexico to Ecuador suggests that different vector populations may be associated with different ecotopes, depending on environmental conditions [47]. These local differences indicate that eco-bio-social conditions vary across regions, and generalizations regarding house construction materials as risk factors may not be possible for *T. dimidiata*.

In our study, partial wall plastering was a risk factor when compared to complete wall plastering. On the other hand, the lack of wall plastering was associated with infestation in a different region of Guatemala [48]. Improving the quality of plastering has been suggested as a control method for triatomines in Central America [49]. Our results indicate that house improvement methods will need to take into account local socioeconomic conditions and practices related to construction, which will vary across regions.

The role of vegetation around the household needs more analysis. The type of vegetation (i.e. cropland and grassland) was previously shown to be related to *T. dimidiata* infestation in different regions of Guatemala [28]. In our study, the presence of coffee trees appears to be associated with triatomine infestation. Whether this is related to the availability of host nesting sites close to the house or other environmental conditions remains to be determined. Householders plant coffee, avocado and jocote (local fruit) trees in their patios for personal consumption. These plants produce fruits and leaf litter, providing ideal breeding sites and food for reservoir hosts such as rats in and around the house [50]. Interestingly, *R. rattus* is commonly called the "tile roof rat" because of its arboreal nesting habits. The importance of tile roofs, coffee trees and rodent infestations as shown in the model may be

related to the presence of domestic and peridomestic rodent nests that act as links with the domestic transmission cycle.

Practices regarding animals that pose a risk factor for infestation and transmission may also vary according to local perceptions and needs. Blood meal analyses showed that chickens, rodents and dogs are important blood sources in this region, as in another study in a nearby Municipality [51]. Chickens are an important part of household nutrition and economy, and are allowed to nest inside the house. The practice of having chickens laying eggs in the house showed a weak association with infestation and was identified by participants as a risk factor that could be reduced by moving the animals outside of the house. The importance of chickens nesting inside the house was shown for *T. infestans* [18], whereas keeping chickens in coops was a risk factor for *T. dimidiata* in Mexico [3]. Similar to an earlier study in Mexico, we found that the presence of more than two dogs per household was an important risk factor, suggesting that animal densities are a determinant of infestation [3]. The bug mortality during transport could lead to bias in the relative frequencies of the different blood meals which could change the relative importance of these animals in maintaining triatomine colonies. However, we were able to triangulate the identification of chicken, dog, mouse and rat blood meals with the evidence of infection in dogs, mice and rats and the perceptions regarding rodents and chickens as associated with infestation. These data taken together strengthen the inclusion of these animals in the model. The use of species-specific PCR limits the identification of other potential important blood sources to be included in the model, such as cats. Moreover, several *cyt b* positive samples did not disclose any specific blood source. Future studies to sequence the universal *cyt b* gene should provide additional insight into other potential habitats for triatomines and the role of other mammals in transmission in and around the house.

This study confirms *T. cruzi* transmission risk in this region. The high proportion of infected triatomines having human blood meals, together with the evidence of previous infection in dogs and *T. cruzi* DTU I in rodents, suggests a high risk for infection in domestic environments. The association between rodent and triatomine infestation; the presence of rodent blood meals and the presence of infected rodents in households with infected triatomines indicate that rodents are an important reservoir in the study area.

Although synanthropic rodents were previously found to be infected with *T. cruzi* [52,53], their role in domestic transmission has not been fully documented. The molecular detection of a blood meal indicates blood meals taken within the last 2–3 weeks [53]. The detection of multiple blood meals in nymphs, including human blood meals in outdoor structures, suggests that immature stages actively move in search of blood meals, as found in another Municipality [51] and for *T. infestans* in Peru and Argentina [4,6]. This may be a situation that facilitates house re-infestation, as nymphs may find peridomestic shelters in rodent nesting areas, found within and outside the household, that do not receive insecticide treatment. A participant perceived construction materials as risk factors for triatomines and rodent infestation. Also, leaf litter found outside the household was noted as associated with triatomines. It is tempting to speculate that these factors represent an ecological niche for rodent nests that link peridomestic and domestic transmission cycles. Control methods such as environmental management to reduce rodent habitats could be developed to target nesting sites, as recently suggested for *Triatoma brasiliensis* [54].

The variables included in the multivariate model represent our hypothesis of the factors that are associated with (or that could be the cause of) house infestation by *T. dimidiata* in the region, as a proxy to transmission risk to humans. Our modeling technique (multimodel inference) allowed us to estimate the relative importance of the variables in our hypothesis. We were also able to explore a subset of reduced models with different combinations of the variables in the hypothesis in order to detect model uncertainty: given the high uncertainty found, a model with weighted parameters is recommended for this data. One limitation of our model is that triatomine infestation was detected by the person-hour method. This method is known to have a low sensitivity, especially at low vector densities and in complex peridomestic environments [54,55]. Including reports by householders could improve infestation detection [55]. However, the current study focused on identifying risk factors for intradomiciliary infestation and collections by participants did not include capture site information. Future studies of this species in Guatemala could include repeated sampling to estimate detection error for distinct ecotopes [54].

A second limitation of the model is that it is dependent on the sample size and the data obtained. Some of the factors we hypothesized to be associated with infestation fell below the 20% prevalence level (e.g., the prevalence of chickens entering the house to lay eggs was 5%), and had to be aggregated to create a new factor ("presence of animals inside the house"); as a result they could not be tested individually. Additionally, some of the factors included in the model (i.e., wall plastering) are complex to measure and were secondarily derived from field observations (detailed data for the extent of plastering of each wall), which may have affected the ultimate association with infestation data. Finally, the presence of vector control cards in 41% of the households indicates a previous visit by the Ministry of Health that may include insecticide spraying or other control activities. We were not able to include in the model previous spraying activity because not all households had this information. Given that reinfestation odds vary depending on the evaluation time after spraying, the model could be improved if previous spraying information were gathered for each house [8].

The triangulation of data from multimodel inference, blood meal analysis, biological indicators of transmission and social science analyses supports the notion that rodents and dogs may be involved in triatomine infestation and in maintaining the parasite transmission cycle in the house, as has been observed for dogs, cats [56] and guinea pigs [6] in connection with *T. infestans*. Adult dog seroprevalence was in a range similar to that observed in other studies with domestic dogs in the region, ranging from 10% to 30% [57,58]. Considering the high seroprevalence and the low number of dog blood meals from bugs collected inside the

households, it is tempting to speculate that dogs may become infected orally by ingesting infected rodents or triatomines. In a natural reserve in Brazil, a similar proportion of dogs were found to be infected, showing that transmission in dogs is not circumscribed to synanthropic environments [59].

Future studies should focus on clarifying the role that rodents and dogs play in triatomine infestation and disease transmission, and may provide control strategies to reduce their impact as reservoirs. Given the high rodent infestation and evidence of infection of mice and rats with *T. cruzi* in the study area in addition to their importance in the transmission of other diseases [60], we propose that an IVM program should also include synanthropic rodent control to reduce *T. cruzi* transmission.

Supporting Information

File S1 Supplementary material including Tables S1, S2, S3 and S4.

Acknowledgments

We thank the communities of Comapa and Zapotitlán that participated in the study. We also thank the Ministry of Health vector control personnel, in particular Ranfery Trampe, and the chief epidemiologist of Jutiapa, Dr. Elsa Berganza, for their collaboration in this study. We thank Celia Cordón-Rosales for helpful discussions throughout the study. We thank Paola Cáceres and Gabriela Samayoa for the molecular analyses. We thank Hugo Enríquez and Jorge Paniagua for their contributions in the animal surveys. We thank Edgar Pereira, Elizabeth Pellecer and Teresa Aguilar for their implementation of Participatory Action Research. We thank Martin Llewellyn and the Chagas EpiNet for the GPI nested primer sequences.

This article was drafted and revised during a scientific writing workshop organized by the Communicable Diseases Research Program, Communicable Diseases and Health Analysis Department, Pan American Health Organization to support the dissemination of IDRC-funded research carried out from an eco-bio-social approach. The authors thank Zaida E. Yadón for organizing the workshop, Karen Shashok for editorial guidance, and Ricardo E. Gürtler for helpful comments on prior versions of the manuscript.

Author Contributions

Conceived and designed the experiments: PMP SMDU JGJ. Performed the experiments: JGJ PMP. Analyzed the data: PMP SMDU DMB JGJ. Contributed reagents/materials/analysis tools: PMP SMDU DMB JGJ. Wrote the paper: PMP SMDU DMB JGJ. Final approval of the manuscript: PMP SMDU DMB JGJ.

References

1. WHO (2007) WHO Report of the global partners' meeting on neglected tropical diseases 2007- a turning point. Geneva, Switzerland: World Health Organization. 75 p.

2. Sommerfeld J, Kroeger A (2012) Eco-bio-social research on dengue in Asia: a multicountry study on ecosystem and community-based approaches for the control of dengue vectors in urban and peri-urban Asia. Pathog Glob Health 106: 428–435.

3. Dumonteil E, Nouvellet P, Rosecrans K, Ramirez-Sierra MJ, Gamboa-Leon R, et al. (2013) Eco-Bio-Social determinants for house infestation by non-domiciliated *Triatoma dimidiata* in the Yucatan Peninsula, Mexico. PLoS Negl Trop Dis 7: e2466.

4. Vazquez-Prokopec GM, Spillmann C, Zaidenberg M, Gurtler RE, Kitron U (2012) Spatial heterogeneity and risk maps of community infestation by *Triatoma infestans* in rural northwestern Argentina. PLoS Negl Trop Dis 6: e1788.

5. Hashimoto K, Alvarez H, Nakagawa J, Juarez J, Monroy C, et al. (2012) Vector control intervention towards interruption of transmission of Chagas disease by *Rhodnius prolixus*, main vector in Guatemala. Mem Inst Oswaldo Cruz 107: 877–887.

6. Levy MZ, Quispe-Machaca VR, Ylla-Velasquez JL, Waller LA, Richards JM, et al. (2008) Impregnated netting slows infestation by *Triatoma infestans*. Am J Trop Med Hyg 79: 528–534.

7. Ceballos LA, Piccinali RV, Marcet PL, Vazquez-Prokopec GM, Cardinal MV, et al. (2011) Hidden sylvatic foci of the main vector of Chagas disease *Triatoma infestans*: threats to the vector elimination campaign? PLoS Negl Trop Dis 5: e1365.

8. Manne J, Nakagawa J, Yamagata Y, Goehler A, Brownstein JS, et al. (2012) Triatomine infestation in Guatemala: spatial assessment after two rounds of vector control. Am J Trop Med Hyg 86: 446–454.

9. Zeledon R, Montenegro VM, Zeledon O (2001) Evidence of colonization of man-made ecotopes by *Triatoma dimidiata* (Latreille, 1811) in Costa Rica. Mem Inst Oswaldo Cruz 96: 659–660.

10. Aiga H, Sasagawa E, Hashimoto K, Nakamura J, Zúniga C, et al. (2012) Chagas disease: Assessing the existence of a threshold for bug infestation rate. Am J Trop Med Hyg 86: 972–979.

11. Hashimoto K, Cordon-Rosales C, Trampe R, Kawabata M (2006) Impact of single and multiple residual sprayings of pyrethroid insecticides against *Triatoma*

dimidiata (Reduviiade; Triatominae), the principal vector of Chagas disease in Jutiapa, Guatemala. Am J Trop Med Hyg 75: 226–230.

12. Bustamante DM, Monroy C, Pineda S, Rodas A, Castro X, et al. (2009) Risk factors for intradomiciliary infestation by the Chagas disease vector *Triatoma dimidiata* in Jutiapa, Guatemala. Cad Saude Publica 25 Suppl 1: S83–92.

13. Gurevitz JM, Ceballos LA, Gaspe MS, Alvarado-Otegui JA, Enriquez GF, et al. (2011) Factors affecting infestation by *Triatoma infestans* in a rural area of the humid Chaco in Argentina: a multi-model inference approach. PLoS Negl Trop Dis 5: e1349.

14. Zeledon R, Solano G, Swartzwelder JC (1970) Sources of blood for *Triatoma dimidiata* (Hemiptera: Reduviidae) in an endemic area of Chagas' disease in Costa Rica. J Parasitol 56: 102.

15. Gurtler RE, Ceballos LA, Ordonez-Krasnowski P, Lanati LA, Stariolo R, et al. (2009) Strong host-feeding preferences of the vector *Triatoma infestans* modified by vector density: implications for the epidemiology of Chagas disease. PLoS Negl Trop Dis 3: e447.

16. Grijalva MJ, Palomeque FS, Villacis AG, Black CL, Arcos-Teran L (2010) Absence of domestic triatomine colonies in an area of the coastal region of Ecuador where Chagas disease is endemic. Mem Inst Oswaldo Cruz 105: 677–681.

17. Schwarz A, Helling S, Collin N, Teixeira CR, Medrano-Mercado N, et al. (2009) Immunogenic salivary proteins of *Triatoma infestans*: development of a recombinant antigen for the detection of low-level infestation of triatomines. PLoS Negl Trop Dis 3: e532.

18. Cecere MC, Gurtler RE, Chuit R, Cohen JE (1997) Effects of chickens on the prevalence of infestation and population density of *Triatoma infestans* in rural houses of north-west Argentina. Med Vet Entomol 11: 383–388.

19. Beteta H, Castro M, Rodríguez E, Benedicto Estrada S, Anzueto M, et al. (2006) Mapas de pobreza en Guatemala al 2002. Guatemala: Gobierno de Guatemala, SEGEPLAN. 1–47 p.

20. Creswell J, Plano Clark V (2007) Designing and conducting mixed methods research. Thousand Oaks, CA.: Sage Publications, Inc. 275 p.

21. Scheaffer RL, Mendenhall III W, Ott RL, Gerow KG (2012) Elementary survey sampling. Boston, MA: Brooks/Cole. 436 p.

22. Bennett S, Woods T, Liyanage WM, Smith DL (1991) Simplified general method for cluster-sample surveys of health in developing countries. Rapport trimestriel de statistiques sanitaires mondiales 44: 98–106.

23. World Health Organization (2005) Immunization coverage cluster survey: Reference manual. Geneva, Switzerland: WHO Document Production Services. 115 p.

24. Demidenko E (2007) Sample size determination for logistic regression revisited. Stat Med 26: 3385–3397.

25. Demidenko E (2008) Sample size and optimal design for logistic regression with binary interaction. Stat Med 27: 36–46.

26. WHO (2008) Advocacy, communication and social mobilization for TB control: A guide to developing knowledge, attitude and practice surveys. Geneva: World Health Organization. 60 p.

27. Dillman DA, Smyth JD, Melani L (2009) Internet, mail, and mixed-mode surveys: The tailored design method. Hoboken, NJ: John Wiley & Sons, Inc. 512 p.

28. King RJ, Cordon-Rosales C, Cox J, Davies CR, Kitron UD (2011) *Triatoma dimidiata* infestation in Chagas disease endemic regions of Guatemala: comparison of random and targeted cross-sectional surveys. PLoS Negl Trop Dis 5: e1035.

29. Healey J (2012) Statistics: A tool for social research. Belmont, CA: Wadsworth Cengage Learning. 512 p.

30. Messenger LA, Llewellyn MS, Bhattacharyya T, Franzen O, Lewis MD, et al. (2012) Multiple mitochondrial introgression events and heteroplasmy in *Trypanosoma cruzi* revealed by maxicircle MLST and next generation sequencing. PLoS Negl Trop Dis 6: e1584.

31. Malhotra RK, Indrayan A (2010) A simple nomogram for sample size for estimating sensitivity and specificity of medical tests. Indian J Ophthalmol 58: 519–522.

32. Rosypal AC, Hill R, Lewis S, Barr SC, Valadas S, et al. (2011) Evaluation of a rapid immunochromatographic dipstick test for detection of antibodies to *Trypanosoma cruzi* in dogs experimentally infected with isolates obtained from opossums (*Didelphis virginiana*), armadillos (*Dasypus novemcinctus*), and dogs (*Canis familiaris*) from the United States. J Parasitol 97: 140–143.

33. Newcombe RG (1998) Two-sided confidence intervals for the single proportion: Comparison of seven methods. Stat Med 17: 857–872.

34. Patton M (2002) Qualitative research & evaluation methods Thousand Oaks, CA: Sage Publications. A2 p.

35. Padgett DK (2012) Qualitative and mixed methods in public health. Thousand Oaks, CA.: Sage Publications, Inc. 285 p.

36. Chevalier J, Buckles D (2013) Participatory action research: Theory and methods for engaged inquiry. New York, NY: Routledge. pp. 496.

37. Strauss A, Corbin J (1998) Basics of Qualitative Research, techniques and procedures for developing grounded theory. Thousand Oaks, CA: Sage Publications. 312 p.

38. Stake RE (2010) Qualitative research: Studying how things work. New York, NY: The Guilford Press. 243 p.

39. Gibbs GR (2007) Analyzing qualitative data. Thousand Oaks, CA: SAGE Publications. 160 p.

40. Green LW, Kreuter MW (2005) Health program planning: An educational and ecological approach. New York, NY: McGraw-Hill. 621 p.

41. Zuur AF, Ieno E, Walker N, Saveliev A, Smith G (2009) Mixed effects models and extensions in ecology with R. New York: Springer. 574 p.

42. Team R (2013) R: A language and environment for statistical computing. In: Computing RFfS, editor. Vienna, Austria. Available: http://www.R-project.org/ Accessed 2014 Mar 10.

43. Burnham K, Anderson D (2002) Model selection and multimodel inference. New York: Springer. 488 p.

44. Calcagno V, Mazancourt C (2010) glmulti: an R package for easy automated model selection with (generalized) linear models. J Stat Softw 34: 1–29.

45. Calcagno V (2013) glmulti: Model selection and multimodel inference made easy. R package version 1.0.7. Available: http://cran.r-project.org/web/packages/glmulti/index.html Accessed 2014 Mar 10.

46. Erlandson D, Harris E, Skipper B, Allen S (1993) Doing naturalistic inquiry: A guide to methods. Newbury Park, CA.: Sage Publications. 198 p.

47. Monteiro FA, Peretolchina T, Lazoski C, Harris K, Dotson EM, et al. (2013) Phylogeographic pattern and extensive mitochondrial DNA divergence disclose a species complex within the Chagas disease vector *Triatoma dimidiata*. PLoS One 8: e70974.

48. Weeks EN, Cordon-Rosales C, Davies C, Gezan S, Yeo M, et al. (2013) Risk factors for domestic infestation by the Chagas disease vector, *Triatoma dimidiata* in Chiquimula, Guatemala. Bull Entomol Res: 1–10.

49. Lucero DE, Morrissey LA, Rizzo DM, Rodas A, Garnica R, et al. (2013) Ecohealth interventions limit triatomine reinfestation following insecticide spraying in La Brea, Guatemala. Am J Trop Med Hyg 88: 630–637.

50. Clark D (1982) Foraging behavior of a vertebrate omnivore (*Rattus rattus*): Meal structure, sampling, and diet breadth. Ecology 63: 763–772.

51. Pellecer MJ, Dorn PL, Bustamante DM, Rodas A, Monroy MC (2013) Vector blood meals are an early indicator of the effectiveness of the Ecohealth approach in halting Chagas transmission in Guatemala. Am J Trop Med Hyg 88: 638–644.

52. Zeledon R, Solano G, Burstin L, Swartzwelder JC (1975) Epidemiological pattern of Chagas' disease in an endemic area of Costa Rica. Am J Trop Med Hyg 24: 214–225.

53. Pinto J, Roellig DM, Gilman RH, Calderon M, Bartra C, et al. (2012) Temporal differences in blood meal detection from the midguts of *Triatoma infestans*. Rev Inst Med Trop Sao Paulo 54: 83–87.

54. Valenca-Barbosa C, Lima MM, Sarquis O, Bezerra CM, Abad-Franch F (2014) Modeling disease vector occurrence when detection is imperfect II: Drivers of site-occupancy by synanthropic *Triatoma brasiliensis* in the Brazilian northeast. PLoS Negl Trop Dis 8: e2861.

55. Abad-Franch F, Vega M, Rolón M, Santos W, Rojas de Arias A (2011) Community participation in Chagas disease vector surveillance: Systematic review. PLoS Negl Trop Dis 5: e1207.

56. Cardinal MV, Lauricella MA, Marcet PL, Orozco MM, Kitron U, et al. (2007) Impact of community-based vector control on house infestation and *Trypanosoma cruzi* infection in *Triatoma infestans*, dogs and cats in the Argentine Chaco. Acta Trop 103: 201–211.

57. Pineda V, Saldana A, Monfante I, Santamaria A, Gottdenker NL, et al. (2011) Prevalence of trypanosome infections in dogs from Chagas disease endemic regions in Panama, Central America. Vet Parasitol 178: 360–363.

58. Montenegro VM, Jimenez M, Dias JC, Zeledon R (2002) Chagas disease in dogs from endemic areas of Costa Rica. Mem Inst Oswaldo Cruz 97: 491–494.

59. Rocha FL, Roque AL, de Lima JS, Cheida CC, Lemos FG, et al. (2013) *Trypanosoma cruzi* infection in neotropical wild carnivores (Mammalia: Carnivora): at the top of the *T. cruzi* transmission chain. PLoS One 8: e67463.

60. Davis S, Calvet E, Leirs H (2005) Fluctuating rodent populations and risk to humans from rodent-borne zoonoses. Vector Borne Zoonotic Dis 5: 305–314.

61. Reiczigel J, Földi J, Ózsvári L (2010) Exact confidence limits for prevalence of a disease with an imperfect diagnostic test. Epidemiol Infect 138: 1674–1678.

Identification of Genes Related to Beak Deformity of Chickens Using Digital Gene Expression Profiling

Hao Bai⁹, Jing Zhu⁹, Yanyan Sun, Ranran Liu, Nian Liu, Dongli Li, Jie Wen, Jilan Chen*

Key Laboratory of Genetics Resources and Utilization of Livestock, Institute of Animal Science, Chinese Academy of Agricultural Sciences, Beijing, China

Abstract

Frequencies of up to 3% of beak deformity (normally a crossed beak) occur in some indigenous chickens in China, such as and Beijing-You. Chickens with deformed beaks have reduced feed intake, growth rate, and abnormal behaviors. Beak deformity represents an economic as well as an animal welfare problem in the poultry industry. Because the genetic basis of beak deformity remains incompletely understood, the present study sought to identify important genes and metabolic pathways involved in this phenotype. Digital gene expression analysis was performed on deformed and normal beaks collected from Beijing-You chickens to detect global gene expression differences. A total of >11 million cDNA tags were sequenced, and 5,864,499 and 5,648,877 clean tags were obtained in the libraries of deformed and normal beaks, respectively. In total, 1,156 differentially expressed genes (DEG) were identified in the deformed beak with 409 being up-regulated and 747 down-regulated in the deformed beaks. qRT-PCR using eight genes was performed to verify the results of DGE profiling. Gene ontology (GO) analysis highlighted that genes of the keratin family on GGA25 were abundant among the DEGs. Pathway analysis showed that many DEGs were linked to the biosynthesis of unsaturated fatty acids and glycerolipid metabolism. Combining the analyses, 11 genes (*MUC*, *LOC426217*, *BMP4*, *ACAA1*, *LPL*, *ALDH7A1*, *GLA*, *RETSAT*, *SDR16C5*, *WWOX*, and *MOGAT1*) were highlighted as potential candidate genes for beak deformity in chickens. Some of these genes have been identified previously, while others have unknown function with respect to thus phenotype. To the best of our knowledge, this is the first genome-wide study to investigate the transcriptome differences in the deformed and normal beaks of chickens. The DEGs identified here are worthy of further functional characterization.

Editor: Marinus F.W. te Pas, Wageningen UR Livestock Research, Netherlands

Funding: This work was funded by 1. National High Technology Research and Development Program (863 Program)(2011AA100305,2008AA101009), National Natural Science Foundation of China. http://www.nsfc.gov.cn/; and 2. Agricultural Science and Technology Innovation Program (ASTIP-IAS04), Chinese Academy of Agricultural Sciences, China. http://www.caas.cn/en/. The funders had no role in study design, data collection and analysis, decision to publish, or preparation of the manuscript.

Competing Interests: The authors have declared that no competing interests exist.

* Email: chen.jilan@163.com

⑨ These authors contributed equally to this work.

Introduction

The beak is an external structure of birds consisting of the upper and lower mandibles covered with a thin keratinized layer of epidermis [1]. It is used for many important activities such as feeding, drinking, fighting, and preening. In addition to striking morphological differences between species, beak deformities of different forms (noticeably elongated, crossed, bent at right angles and so on) have been documented in many wild birds including Japanese quails [2], Brown-headed Cowbird [3], Black Capped Chickadees [4], Northwestern Crows [5], African Seedcracker [6], and Senegal Parrots [7]. Frequencies of up to 3% beak deformity (Figure 1) have been found in some indigenous breeds, such as Beijing-You chickens, the breed studied here [8] and Qingyuanma chickens. Chickens with deformed beaks have reduced feed intake, growth rate, and impaired normal behaviors like preening and social contact with their mates. Beak deformity, therefore, represents an economic as well as an animal welfare problem in the poultry industry.

Intensive researches have been performed to seek the pinpoint causes of the malformation in birds, especially the wild birds. Diseases, environmental contaminants, and lack of nutriments are some likely contributing factors [9,10]. Further complexity stems from the very early embryonic determination of rostral development likely influencing subsequent growth, which continues long after hatching [11]. The underlying pathology and possible mechanisms of the disorder, however, remain incompletely understood and have baffled scientists for a number of years.

According to our observations in a pedigreed Beijing-You chickens population, in the absence of known environmental factors contributing to the malformation, birds with deformed beaks appear consistently in each generation and cannot be eliminated from a population simply on the basis of the phenotype. This indicated the genetic effects for the formation of beak deformity. Previously recognized genetic factors associated with beak deformity include the reported candidate genes such as fibroblast growth factor 8 (*FGF8*) [12] and bone morphogenetic protein 4 (*BMP4*) [13]. The over-expression of homeobox A1 (*HOXA1*) and homeobox D3 (*HOXD3*) may result in beak

Figure 1. The deformed (left) and normal (right) beaks of Beijing-You chickens. The chicken with a deformed beak had problem with feeding, drinking, and preening and therefore showed lower body weight and poor mental condition.

deformity in chicks [14]. The approaches used in these existing studies, while mechanistically rigorous, have the limitation of lower-throughput for comprehensive screening and critical underlying genes can be overlooked.

Based on the new generation of high-throughput sequencing and high-performance computing technology, digital gene expression (DGE) can be used to capture the entire transcriptome of a given tissue. The aim of this first study was to identify genes related to the beak deformity phenotype using DGE as a starting point for revealing the molecular genetic mechanisms underlying the condition. The results identified sets of up-regulated and down-regulated genes in the deformed beak compared to normal. Combined with the follow-up gene ontology and pathway enrichment analyses of the differentially expressed genes (DEGs), some candidate genes and pathways related to beak deformity are proposed and discussed.

Materials and Methods

Samples and RNA isolation

The lower mandibles of the beaks were collected from 6 Beijing-You cocks of 56 days of age: 2 with crossed beaks (individuals 1 and 2) and 4 with normal beaks (individuals 3, 4, 5, and 6). Individuals 1, 3, 5, and 6 were full sibs; as were individuals 2 and

4. The chickens had been incubated and housed under the same conditions. Total RNA of the lower mandibles of the beaks was isolated using TRIzol (Invitrogen, USA) according to the manufacturer's instructions followed by RNase-free DNase treatment (TIANGEN, China). The quality and quantity of the total RNA was assessed with a Biophotometer Plus (Eppendorf, Germany). RNA from the beaks of individual 1 (deformed) and 6 (control) was used to determine the DEGs at the genome-wide level using DGE profiling. Four RNA sample pairs (2 and 4, 1 and 3, 1 and 5, and 1 and 6) were used to determine relative abundance of 8 transcripts using quantitative real-time PCR (qRT-PCR).

DGE-tag profiling and sequence annotation

cDNA was prepared from RNA1 and 6, digested and ligated with Illumina adaptors, as recommended, to generate 17 bp tagged fragments for Solexa sequencing, which was performed by BGI (Beijing, China).

Clean-tags were obtained by filtering the adaptor sequences and removing low-quality sequences (those with some ambiguous bases) then mapped to the reference genome and genes of chickens (*Gallus_gallus-4.0*) available at http://www.ncbi.nlm.nih.gov. Only tags with a perfect match or 1 mismatch were further considered and annotated. The expression level of each gene was

Table 1. Gene-specific primers used in qRT-PCR.

Gene[a]	GenBank No.	Primer Sequence	Tm (°C)	Product length (bp)
NPM3	XM_001233763	F: 5′-CCATCGTGCCTGCCAAGAAG-3′	60	132
		R: 5′-AAAAAAGGCAGCAAAAGTTG-3′		
LPL	NM_205282	F: 5′-GGTAGACCAGCCATTCCTGA-3′	60	192
		R: 5′-CACCAGTCTGACCAGCTGAA-3′		
BMP4	NM_205237	F: 5′-AGTCCGGAGAAGAGGAGGAG-3′	60	164
		R: 5′-GCTGAGGTTGAAGACGAAGC-3′		
SGOL1	NM_001199648	F: 5′-ACAGCGTGAGCAGAAACTCC-3′	60	159
		R: 5′-GGTGCGTTCCCTTTGCTTAG-3′		
KRT19	NM_205009	F: 5′-AAGATCCTGGCCGATATGAG-3′	60	122
		R: 5′-TCGGTATTGACGGCTAACTC-3′		
sKer	XM_428869	F: 5′-GCTATGGAGGATCTCAGGGT-3′	60	138
		R: 5′-AGAGGCCAGAGCTGTAGGAC-3′		
MMP7	NM_001006278	F: 5′-GGAAGAGGTGGCACATTAGC-3′	60	167
		R: 5′-ACATTTGAGTGGGCGAGTCC-3′		
MUC	XM_421033	F: 5′-GCAACGGCATCAATGACTTC-3′	60	168
		R: 5′-TGCTACACTGCTTCTCTGAC-3′		
β-actin	NM_205518	F: 5′-GAGAAATTGTGCGTGACATCA-3′	60	152
		R: 5′-CCTGAACCTCTCATTGCCA-3′		

[a]NPM3 = nucleophosmin/nucleoplasmin 3; LPL = lipoprotein lipase; BMP4 = bone morphogenetic protein 4; SGOL1 = shugoshin-like 1; KRT19 = Keratin 19; sKer = similar to Scale keratin; MMP7 = matrix metalloproteinase 7; MUC = mucin protein.

estimated by the frequency of clean tags and then normalized as TPM (number of transcripts per million clean tags) [15,16].

Identification of DEGs

The differential expression of genes between the deformed and normal beaks was determined with a rigorous algorithm method, which has been developed to identify DEGs between two samples by the BGI (Beijing, China), referring to the methods published by Audic and Claverie [17]. Denote the number of unambiguous clean tag from gene A as x, as every gene's expression occupies only a small part of the library, the p(x) is in the Poisson distribution:

$$p(x) = \frac{e^{-\lambda}\lambda^x}{x!} \quad (\lambda \text{ is the real transcripts of the gene})$$

The total clean tag number of the sample 1 is N1, and total clean tag number of sample 2 is N2; gene A holds x tags in sample1 and y tags in sample 2. The probability of gene A expressed equally between two samples can be calculated with:

$$p(y|x) = (\frac{N_2}{N_1})^y \frac{(x+y)!}{x!y!(1+\frac{N_2}{N_1})^{(x+y+1)}}$$

P-Value corresponds to differential gene expression test. Genes were deemed as being significantly differentially expressed with a P-value <0.005, FDR (false discovery rate) <0.001, and a 2-fold relative change threshold (|log2-Ratio (deformed beak/normal beak)| >1) [18] in the sequence counts across the 2 libraries. All

Table 2. Statistics of tag mapping against reference gene and genome sequence of the chicken.

	Deformed beak		Normal beak	
	Total number	Distinct tag number	Total number	Distinct tag number
Raw data	6,027,921	262,179	5,807,305	251,861
Clean tags	5,864,499	106,180	5,648,877	100,893
All tags mapping to genes	2,935,689	41,530	2,925,791	42,794
Unambiguous tag mapping to genes	2,566,707	38,420	2,452,094	39,666
All tag-mapped genes	10,425	10,425	10,725	10,725
Unambiguous tag-mapped genes	9,096	9,096	9,328	9,328
Mapping to genome	1,444,399	38,761	1,275,093	36,268
Unknown tags	1,484,411	25,889	1,447,993	21,831

Table 3. The number of total reference genes and tags.

	Number	Percentage
Reference gene	19,131	
Genes with CATG site	17,655	92.28%
Total reference tags	153,121	
Unambiguous Tags	145,689	95.15%
Ambiguous tags	7,432	4.85%

the genes in the present study were referenced from the gene dataset of GenBank.

Gene ontology (GO) and pathway enrichment analysis of DEGs

Based on the Gene Ontology Database (http://www.geneontology.org/) [19] and the KEGG pathway (http://www.genome.jp/kegg/) [20], GO functional enrichment analysis and pathway enrichment analysis were used to identify significantly enriched functional classification and metabolic pathways in the DEGs, respectively [21]. The formula was [22]:

$$P = 1 - \sum_{i=0}^{m-1} \frac{\binom{M}{i}\binom{N-M}{n-i}}{\binom{N}{n}},$$

where N is the total number of genes with GO/KEGG functional annotations, and n is the number of DEGs in N, M is number of the genes with specific GO/KEGG annotations, and m is the number of DEGs in M. In this study, P-value ≤ 0.05 indicated significantly enriched GO terms and pathways.

qRT-PCR analysis for validating the DEGs

Eight genes (Table 1), selected from the DEGs, were analyzed by qRT-PCR with an ABI 7500 Real-time Detection System (Applied Biosystems, USA) to validate the DGE results. The total RNA samples (1 μg) of four comparison pairs were used. The housekeeping gene β-actin was used as the endogenous control. Gene-specific primers were designed from the reference Unigene sequences available in GenBank using Primer Premier 5.0 (Table 1). qRT-PCR was performed in triplicate according to manufacturer specifications (TaKaRa SYBR PrimeScript RT-PCR Kit, Dalian, China). After denaturing at 95°C for 30 s, 40 cycles of 95°C for 5 s and 60°C for 32 s were used, followed by thermal denaturing to generate melting curves to verify amplification specificity. The comparative CT method ($2^{-\Delta\Delta CT}$) was used to determine the fold-changes in transcript abundance. Student's t-tests were used to evaluate the differences between the RNA samples of deformed and normal beaks.

Ethics statement

The Institutional Animal Care and Use Committee at Institute of Animal Science, Chinese Academy of Agricultural Sciences approved all procedures involving the use of animals. All efforts were made to minimize the suffering of animals.

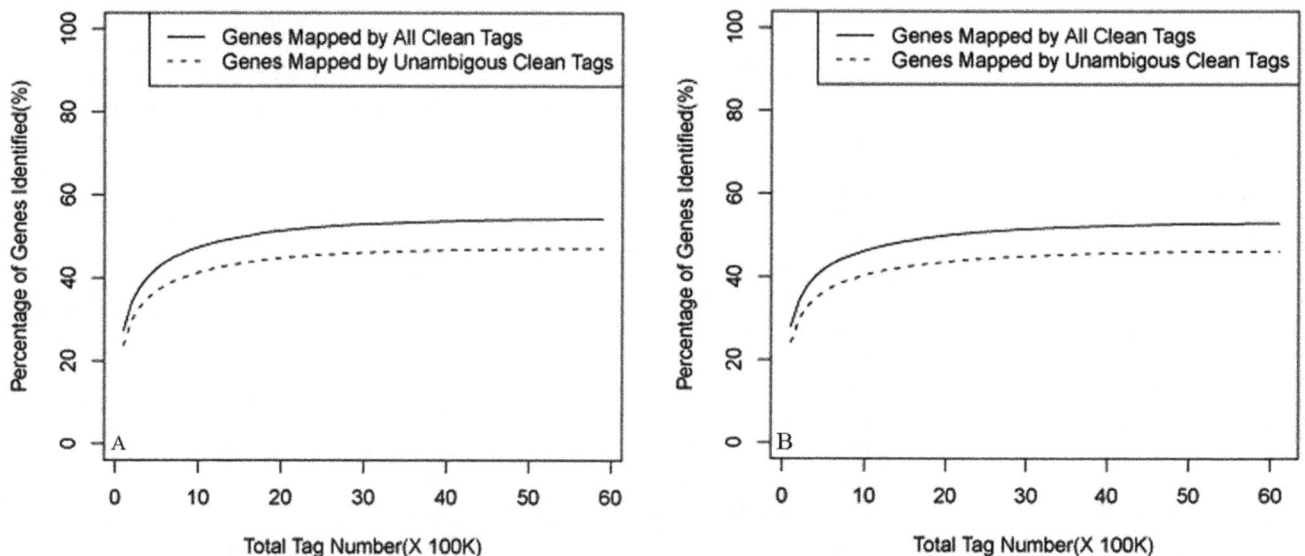

Figure 2. Sequencing saturation analysis for RNA from the deformed (A) and normal beak (B).

Gene Expression Level 11 vs 1

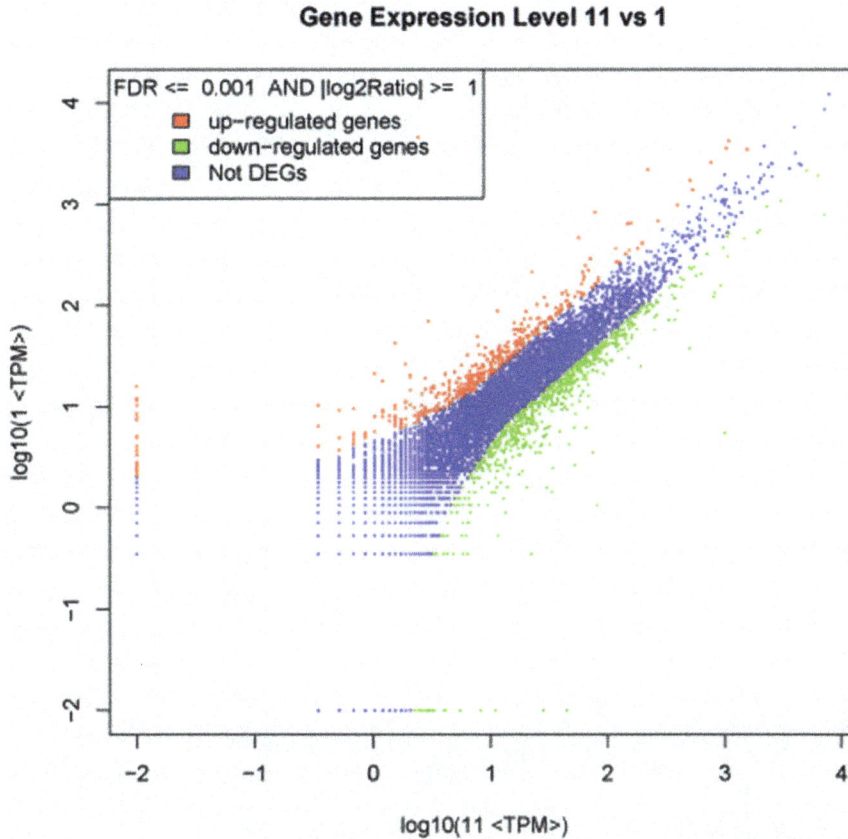

Figure 3. Comparison of gene expression between libraries of deformed and normal beak RNA. Blue dots represent the transcripts with no significant differential expression. Red and green dots represent transcripts more abundant in the test sample and control, respectively. FDR < 0.001 and |log2-Ratio| ≥ 1 were used as the thresholds to judge the significance of gene expression difference.

Results

Analysis of DGE libraries and tag mapping

The total number of sequenced tags (6,027,921 for RNA1 (deformed beak) (Sequence Read Archive accession number: SRR1514179) and 5,807,305 for RNA6 (normal beak) (Sequence Read Archive accession number: SRR1514178)) were obtained from the Solexa sequencing. After filtering the adaptor sequences and removing the low-quality tags, 5,864,499 and 5,648,877 clean tags were retained (Table 2). Considering the robustness of subsequent data analysis, only tags with more than 1 copy were considered for further analysis.

A reference gene database that included 19,131 Unigene sequences was preprocessed for tag mapping. As shown in Table 3, genes containing a CATG motif, accounted for 92.28% of all reference genes. Unambiguous tags accounted for 95.15% of the total reference tags. The data indicated that a high proportion of the entire chicken genome was expressed in the beak samples.

Saturation analysis of the sequencing data showed that the genes that were mapped by all clean tags and unambiguous clean tags were saturated when the tag counts approached 2 million (Figure 2). Therefore, the sequencing depth used here was sufficient for the transcriptome coverage.

Comparison and analysis of DEGs

The transcripts detected with at least 2-fold differences (FDR < 0.001 and |log2-Ratio (deformed beak/normal beak)| ≥ 1) in the deformed-beak library compared to the normal-beak library are shown in Figure 3. Most (99.06%) of the ratios of distinct tag copy number between the 2 libraries were within 5-fold (Figure 4).

Distribution of Ratio of Distinct Tag Copy Number between Two Library

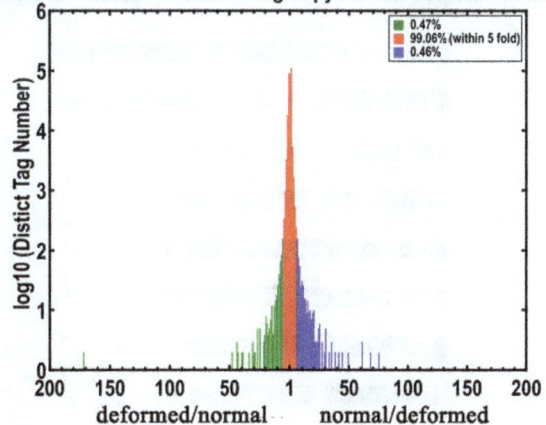

Figure 4. Distribution of ratios of distinct tag copy number between libraries of deformed and normal beak RNA. Red region represents the differentially expressed tags with differential expression less than 5-fold. Blue and green region represent the up- and down-regulated tags of more than 5-fold, respectively.

Table 4. Some extremely differentially expressed genes (|log2-Ratio (deformed beak/normal beak)| ≥ 9).

Species	Gene[a]	log2-Ratio(deformed/normal)
Gallus gallus	LOC426217	10.91
Gallus gallus	NPM3	10.62
Gallus gallus	LPL	10.25
Gallus gallus	RBP7	10.17
Gallus gallus	NUBP2	10.00
Gallus gallus	ARL6IP1	9.73
Gallus gallus	ABF1	9.61
Gallus gallus	RNG213	9.54
Gallus gallus	THOC3	9.00
Gallus gallus	DGCR14	−9.09
Gallus gallus	C3orf38	−9.09
Gallus gallus	KRT19	−9.68
Gallus gallus	SGOL1	−9.68
Gallus gallus	sKer	−10.09
Gallus gallus	NMP7	−11.45
Gallus gallus	MUC	−12.11

[a]LOC426217 = claw keratin-like; NPM3 = nucleophosmin/nucleoplasmin 3; LPL = lipoprotein lipase; RBP7 = retinol binding protein 7 cellular; NUBP2 = nucleotide binding protein 2 (MinD homolog E. coli); ARL6IP1 = ADP-ribosylation factor-like 6 interacting protein 1; ABF1 = activated B-cell factor 1; RNG213 = ring finger protein 213; THOC3 = THO complex 3; DGCR14 = DiGeorge syndrome critical region gene 14; C3orf38 = chromosome 3 open reading frame 38; KRT19 = Keratin 19; SGOL1 = shugoshin-like 1; sKer = similar to Scale keratin; MMP7 = matrix metalloproteinase 7; MUC = mucin protein.

From these 2 libraries, 1,156 significant DEGs were identified, of which 409 (35.4%) were higher and 747 (64.6%) were lower in the deformed beak. A number of transcripts with extreme differences in abundance (|log2-Ratio (deformed beak/normal beak)| ≥ 9) are shown in Table 4. The detailed lists of DEGs are provided in Table S1 (up-regulated) and Table S2 (down-regulated). Based on the differential ratios, claw keratin-like (LOC426217) (|log2-Ratio (deformed beak/normal beak)| = 10.91) and mucin protein (MUC) (|log2-Ratio (deformed beak/normal beak)| = −12.11) are the most up- and down-regulated genes, respectively. They can be denoted as being likely important candidate genes related to beak deformity.

GO enrichment analysis of DEGs

The ontology database covers 3 domains: cellular component, molecular function, and biological process. When GO analysis results of the presently identified DEGs were examined, there were no significantly enriched GO terms on the ontology of molecular function or biological process. The DEGs were significantly

Top 10 entriment in cellular component

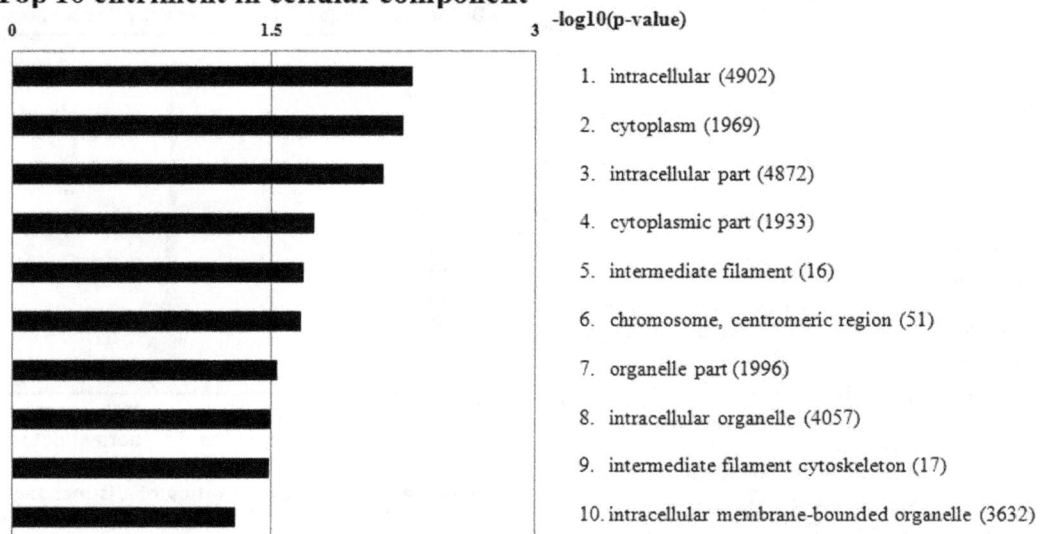

-log10(p-value)

1. intracellular (4902)
2. cytoplasm (1969)
3. intracellular part (4872)
4. cytoplasmic part (1933)
5. intermediate filament (16)
6. chromosome, centromeric region (51)
7. organelle part (1996)
8. intracellular organelle (4057)
9. intermediate filament cytoskeleton (17)
10. intracellular membrane-bounded organelle (3632)

Figure 5. Top 10 enrichment cellular components. The numbers of genes involved in the components are indicated in parentheses.

Top 10 pathway

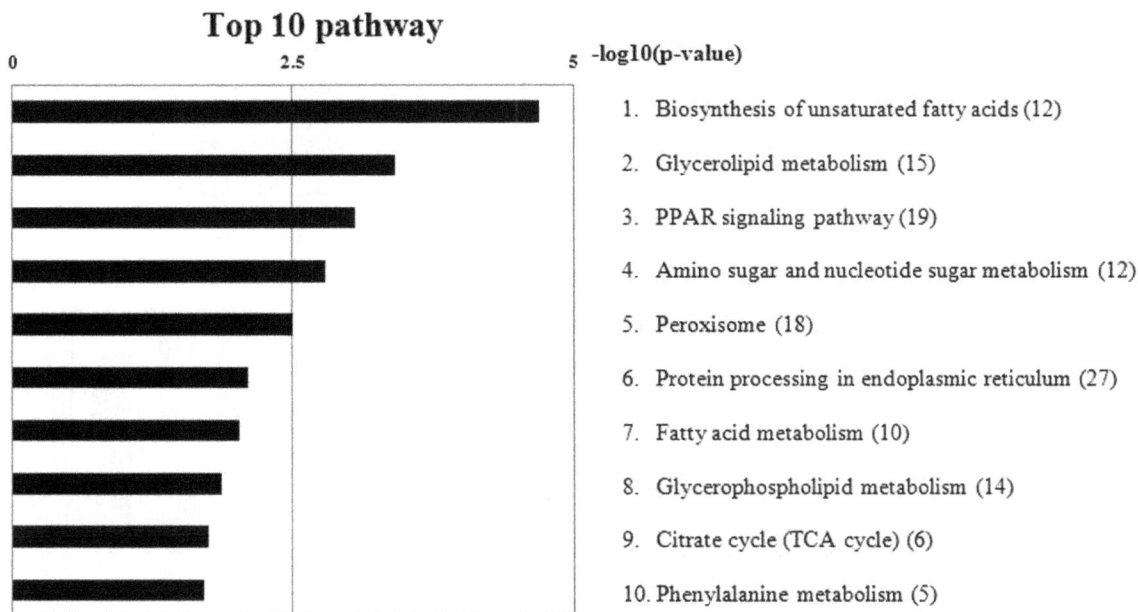

-log10(p-value)

1. Biosynthesis of unsaturated fatty acids (12)

2. Glycerolipid metabolism (15)

3. PPAR signaling pathway (19)

4. Amino sugar and nucleotide sugar metabolism (12)

5. Peroxisome (18)

6. Protein processing in endoplasmic reticulum (27)

7. Fatty acid metabolism (10)

8. Glycerophospholipid metabolism (14)

9. Citrate cycle (TCA cycle) (6)

10. Phenylalanine metabolism (5)

Figure 6. Top 10 enrichment pathways. The numbers of genes involved in the pathways are indicated in parentheses.

enriched, however, on 9 terms of the ontology of cellular component category (Figure 5): GO0005622 intracellular (426 genes), GO0005737 cytoplasm (197 genes), GO0044424 intracellular part (423 genes), GO0044444 cytoplasmic part (191 genes), GO0005882 intermediate filament (7 genes), GO0000775 chromosome, centromeric region (13 genes), GO0044422 organelle part (195 genes), GO0043229 intracellular organelle (359 genes), and GO0045111 intermediate filament cytoskeleton (7 genes). These DEGs, therefore, were mainly associated with intracellular, cell organelle, and cytoskeletal terms, where they play their roles.

Table 5. The two most significantly enriched pathways and the involved differentially expressed genes (DEG).

Biosynthesis of unsaturated fatty acids		Glycerolipid metabolism	
DEGs[a]	log2-Ratio (deformed/normal)	DEGs	log2-Ratio (deformed/normal)
SCD5	1.41	LPL	10.25
GLG1	−1.01	PNPLA2	8.54
ELOVL6	−1.08	PPAP2A	7.96
FAD	−1.19	PPAP2B	2.06
PTPLB	−1.20	AGPAT6	1.46
ACAA1	−1.35	SHROOM3	1.18
PECR	−1.43	ALDH7A1	−1.06
MFSD4	−1.54	AKR1B10	−1.14
DLEC1	−1.87	DGKD	−1.20
PTPLAD1	−1.92	GK5	−1.23
HSDL1	−2.31	GLA	−1.26
LOC423119	−3.37	MOGAT1	−1.28
		AR	−3.01
		LIPG	−3.01

[a]SCD5 = stearoyl-CoA desaturase 5; GLG1 = golgi glycoprotein 1; ELOVL6 = ELOVL family member 6; PTPLB = protein tyrosine phosphatase-like B; ACAA1 = acetyl-CoA acyltransferase 1; PECR = privacy and electronic communications regulations; MFSD4 = major facilitator superfamily domain containing 4; DLEC1 = deleted in lung and esophageal cancer 1; PTPLAD1 = protein tyrosine phosphatase-like A domain containing 1; HSDL1 = hydroxysteroid dehydrogenase like gene; LOC423119 = Gallus gallus fatty acid desaturase 1-like; LPL = lipoprotein lipase; PNPLA2 = patatin-like phospholipase domain containing 2; PPAP2A = phosphatidic acid phosphatase 2A; PPAP2B = phosphatidic acid phosphatase 2B; AGPAT6 = 1-acylglycerol-3-phosphate O-acyltransferase 6; SHROOM3 = shroom family member 3; ALDH7A1 = aldehyde dehydrogenase 7 family, member A1; AKR1B10 = aldo-keto reductase family 1, member B10; DGKD = diacylglycerol kinase; GK5 = glycerol kinase 5; GLA = galactosidase, alpha; MOGAT1 = monoacylglycerol O-acyltransferase 1; AR = androgen receptor; LIPG = endothelial lipase.

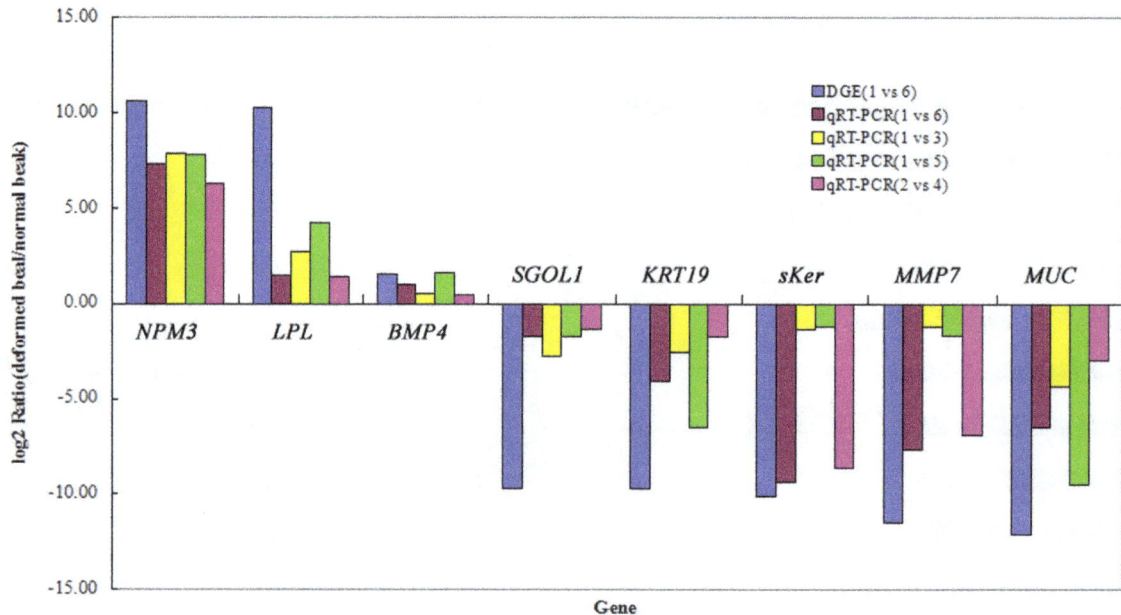

Figure 7. qRT-PCR of 8 transcripts for validating the DGE results. The horizontal axis identifies the 8 transcripts examined by qRT-PCR; the vertical axis shows the relative gene expression level in deformed (individuals 1 and 2) versus normal (individuals 3, 4, 5, and 6) beak tissues. The 4 pairs of individuals were all full sibs. The first bar shows the value obtained from DGE using 1 and 6. *NPM3* = nucleophosmin/nucleoplasmin 3; *LPL* = lipoprotein lipase; *BMP4* = bone morphogenetic protein 4; *SGOL1* = shugoshin-like 1; *KRT19* = Keratin 19; *sKer* = similar to Scale keratin; *MMP7* = matrix metalloproteinase 7; *MUC* = mucin protein.

Pathway enrichment analysis of DEGs

Pathway functional significant enrichment analysis showed that some DEGs were associated with biochemical metabolism and signal transduction pathways. All the DEGs were involved in 210 pathways and the top 10 enrichment pathways of the DEGs are shown in Figure 6. The results revealed that, *LOC426217* was not involved in any pathways. *MUC* was from salivary secretion pathway. *BMP4* was involved in the pathways of cancer, basal cell carcinoma, hedgehog signaling, TGF-beta signaling and cytokine-cytokine receptor interaction. Significantly enriched pathways included biosynthesis of unsaturated fatty acid (KO 01040 with 12 genes) and glycerolipid metabolism (KO 00561 with 14 genes) (Table 5). The details of the two pathways are shown in Figures S1 and S2, respectively. Four DEGs were involved in the two pathways viz. acetyl-CoA acyltransferase 1 (*ACAA1*), Lipoprotein Lipase (*LPL*), Aldehyde Dehydrogenase 7 family member A1 (*ALDH7A1*), and Alpha-galactosidase A (*GLA*). Although not detected as significantly enriched, retinol acid metabolism pathway was highlighted as 4 DEGs viz. Retinol saturase (*RETSAT*), short chain dehydrogenase/reductase family 16C, member 5 (*SDR16C5*), WW domain containing oxidoreductase (*WWOX*), and Monoacylglycerol O-acyltransferase 1 (*MOGAT1*) were identified to be involved.

DGEs results were confirmed by qRT-PCR

To verify the foregoing DGE analysis, 8 genes *NPM3*, *LPL*, *BMP4*, *SGOL1*, *KRT19*, *sKer*, *MMP7* (matrix metalloproteinase 7), and *MUC* from the DEGs were examined by qRT-PCR using RNA from the beaks of 4 full-sib pairs, with and without the deformity phenotype. The results (Figure 7) showed good agreement with the DGE analysis pattern, indicating reliability of the latter.

Discussion

The molecular genetic mechanism underlying beak deformity is likely to be very complex. Based on the DGE and bioinformatics analyses used in the present study, we identified the DEGs in the deformed and normal beaks, some of which were quite extreme, and found the enriched pathways of the DEGs. Some DEGs and pathways were selected as the likely candidate genes related to beak deformity.

DEGs between the deformed and normal beaks

In the present study, 1,156 DEGs with more than 2-fold expression difference were identified. Of these, about a third were up-regulated and the remainder were down-regulated in the anomalous beak. Three genes were selected as promising candidate genes on the basis of their known functions as well as their differential ratios, identified here. The most profoundly up-regulated gene (log2-Ratio (deformed beak/normal beak) = 10.91) was *LOC426217*, a hypothetical gene member of the keratin family [23]. Keratin is an important protein of the cytoskeleton playing a critical role in maintaining cell morphology [24]. It is also an intermediate filament protein that has essential functions in maintaining the structural integrity of epidermis and its appendages [25], presumably including the beak. Although, the key mechanistic role of *LOC426217* is not yet known, our DEGs results suggest its having a role in maintaining beak morphology. At the other extreme, *MUC* was the most down-regulated gene (log2-Ratio (deformed beak/normal beak) = −12.11, i.e. >4000-fold). It is also a gene nearby many gene families [26], the expression product of which is mucin, a cohort of highly glycosylated proteins produced by epithelial tissues and implicated in bone formation [27]. Already shown by others to be a key gene in the development of beak morphology [13,28,29,30], and shown here to be somewhat up-regulated (log2-Ratio (deformed beak/

normal beak) = 1.55), *BMP4* is included in our group of provisional candidate genes associated with beak deformity.

GO and pathway enrichment analysis of DEGs

GO enrichment analysis showed that the DEGs were significantly enriched on the ontology of cellular component where they were mainly associated with intracellular, cell organs, and cytoskeleton categories. This indicated that the enriched DEGs play their roles mainly in these categories. It is worth mentioning that 7 DEGs including the most up-regulated gene *LOC426217* are members of the keratin family, located on GGA 25. As stated earlier, keratin is crucial for maintaining normal cell morphology and change of its structure results in dysmorphic cells [24]. Variation of keratin structure can lead to beak deformity [31]. The cytoskeleton is a complex of intracellular proteins that contribute to shape, support, and movement of cells [32]. The enrichment of keratin family gene DEGs in the cytoskeleton category may indicate their role in cell support. Their abnormal expression in the beak may therefore result in a deformed beak, although further functional study of these genes, especially that of *LOC426217*, is needed.

To shed more lights into the functional roles of DEGs responsible for the deformity beak, biological metabolic pathways were investigated by the enrichment analysis of DEGs. Pathway enrichment analysis showed that those involving biosynthesis of unsaturated fatty acids and glycerolipid metabolism were significantly enriched pathways of the DEGs exposed here. They could be target pathways in the development of beak deformity and the genes selected from the pathways above could be important candidate genes related to beak deformity. Four DEGs from the 2 pathways were selected, consisting of *ACAA1* from unsaturated fatty acid biosynthesis, and *LPL*, *ALDH7A1*, and *GLA* from glycerolipid metabolism. *ACAA1* is known to be related to the formation and development of teeth [33]. *LPL* encodes the extracellular lipase enabling acyltransfer between blood and tissue lipids. *ALDH7A1* functions in lipid peroxidation, and *GLA* is involved in glycerolipid hydrolysis. The relevance of these genes and their involved pathways to beak deformity may relate to the high content (up to 21%) of lipid in beak tissue [8] where it has a critical function as an extracellular matrix for the keratinized cells. Previous studies indicated that the disruption of retinol metabolism can cause abnormal development of the beak [34,35] and influences the first branchial arch cartilages of chicken embryos [36]. Our results, however, did not show enrichment of the DEGs on this pathway but 4 DEGs identified here did relate to retinol acid metabolism, Retinol saturase (*RETSAT*), short chain dehydrogenase/reductase family 16C, member 5 (*SDR16C5*), WW domain containing oxidoreductase (*WWOX*), and Monoacylglycerol O-acyltransferase 1 (*MOGAT1*). They may be interesting candidate genes for beak deformity, worth further study.

The morphology of the mature beak reflects the sum of all proceeding developmental and growth processes, much of which remains poorly understood. There has been extensive investigation of normal chicken craniofacial development and some of the determining factors such as *BMPs*, *Shh*, and *FGFs*, have been

documented [9,29,34]. For these reasons, exposing the anomalous transcriptome at 56 days of age cannot be expected to identify all causative genes, nor can the DEG analysis alone fully expose possible post-transcriptional defects underlying the deformity. This approach does, however, identify genes with atypical expression and potentially establishes a cohort of candidates for seeking gene variants and post-transcriptional modifiers. To increase the scope and power of future planned studies, a breeding population of Beijing-You chickens, from founders with deformed beaks, is being established.

Conclusions

To the best of our knowledge, this is the first time the genes related to beak deformity have been identified using genome-wide gene expression analysis. The results suggested that 11 candidate genes are deserving of further study: *MUC*, *LOC426217*, *BMP4*, *ACAA1*, *LPL*, *ALDH7A1*, *GLA*, *RETSAT*, *SDR16C5*, *WWOX*, and *MOGAT1*. Several of the genes correspond to the cellular component GO category and KEGG analysis showed the importance of pathways of unsaturated fatty acid biosynthesis and glycerolipid metabolism. Subsequent studies including exploring, analysis of network, sequencing analysis, and functional verification will be done in these candidate genes, which could ultimately reveal the pinpoint causes of beak deformity and the underlying mechanisms of the disorder in chickens, as well as wild birds.

Supporting Information

Figure S1 The biosynthesis of unsaturated fatty acid pathway. The up-regulated genes were marked with red color and down-regulated genes with green color.

Figure S2 The glycerolipid metabolism pathway. The up-regulated genes were marked with red color and down-regulated genes with green color.

Table S1 The up-regulated genes with the (log2-Ratio (deformed beak/normal beak) ≥ 2).

Table S2 The down-regulated genes with the (log2-Ratio (deformed beak/normal beak) ≤ -2).

Acknowledgments

W. Bruce Currie, (Emeritus Professor, Cornell University, Ithaca, USA) is acknowledged for his valuable comments to improve the manuscript.

Author Contributions

Conceived and designed the experiments: JC JZ HB. Performed the experiments: JZ HB NL. Analyzed the data: JZ YS RL. Contributed reagents/materials/analysis tools: RL DL JW. Contributed to the writing of the manuscript: HB YS JZ.

References

1. Seki Y, Bodde SG, Meyers MA (2010) Toucan and hornbill beaks: A comparative study (vol 6, pg 331, 2010). Acta Biomaterialia 6: 2363–2363.
2. Tsudzuki M, Nakane Y, Wada A (1998) Brief communication. Short beak: a new autosomal recessive semilethal mutation in Japanese quail. Journal of Heredity 89: 175–178.
3. Rintoul DA (2005) Beak deformity in a brown-headed cowbird, with notes on causes of beak deformities in birds. Kansas Ornithol Soc Bull 56: 29–32.
4. Handel CM, Pajot LM, Matsuoka SM, Van Hemert C, Terenzi J, et al. (2010) Epizootic of Beak Deformities among Wild Birds in Alaska: An Emerging Disease in North America? Auk 127: 882–898.
5. Van Hemert C, Handel CM (2010) Beak Deformities in Northwestern Crows: Evidence of a Multispecies Epizootic. Auk 127: 746–751.

6. Clabaut C, Herrel A, Sanger TJ, Smith TB, Abzhanov A (2009) Development of beak polymorphism in the African seedcracker, Pyrenestes ostrinus. Evolution & Development 11: 636–646.

7. Demery ZP, Chappell J, Martin GR (2011) Vision, touch and object manipulation in Senegal parrots Poicephalus senegalus. Proceedings of the Royal Society B-Biological Sciences 278: 3687–3693.

8. Zhu J, Chen JL, Liu RR, Zheng MQ, Hu J, et al. (2012) Chemical analysis of chicken beak tissue. China Poultry 34: 2.

9. Gartrell BD, Alley MR, Kelly T (2003) Bacterial sinusitis as a cause of beak deformity in an Antipodes Island parakeet (Cyanoramphus unicolor). New Zealand Veterinary Journal 51: 196–198.

10. Jheon AH, Schneider RA (2009) The Cells that Fill the Bill: Neural Crest and the Evolution of Craniofacial Development. Journal of Dental Research 88: 12–21.

11. Chan T, Burggren W (2005) Hypoxic incubation creates differential morphological effects during specific developmental critical windows in the embryo of the chicken (Gallus gallus). Respiratory Physiology & Neurobiology 145: 251–263.

12. MacDonald ME, Abbott UK, Richman JM (2004) Upper beak truncation in chicken embryos with the cleft primary palate mutation is due to an epithelial defect in the frontonasal mass. Developmental dynamics 230: 335–349.

13. Wu P, Jiang TX, Suksaweang S, Widelitz RB, Chuong CM (2004) Molecular shaping of the beak. Science 305: 1465–1466.

14. Jaszczak K, Malewski T, Parada R, Malec H (2006) Expression of Hoxa1 and Hoxd3 genes in chicken embryos with exencephaly. Journal of Animal and Feed Sciences 15: 463.

15. Hoen PAC't, Ariyurek Y, Thygesen HH, Vreugdenhil E, Vossen RHAM, et al. (2008) Deep sequencing-based expression analysis shows major advances in robustness, resolution and inter-lab portability over five microarray platforms. Nucleic Acids Research 36: e141–e141.

16. Feng L, Liu H, Liu Y, Lu Z, Guo G, et al. (2010) Power of deep sequencing and agilent microarray for gene expression profiling study. Mol Biotechnol 45: 101–110.

17. Audic S, Claverie JM (1997) The Significance of Digital Gene Expression Profiles. Genome Research 7: 986–995.

18. Morrissy AS, Morin RD, Delaney A, Zeng T, McDonald H, et al. (2009) Next-generation tag sequencing for cancer gene expression profiling. Genome Research 19: 1825–1835.

19. Consortium TGO (2008) The Gene Ontology project in 2008. Nucleic Acids Research 36: D440–444.

20. Kanehisa M (2002) The KEGG database. In Silico Simulation of Biological Processes 247: 91–103.

21. Shen Y, Jiang Z, Yao X, Zhang Z, Lin H, et al. (2012) Genome expression profile analysis of the immature maize embryo during dedifferentiation. PLoS One 7: e32237.

22. Benjamini Y, Hochberg Y (1995) Controlling the False Discovery Rate - a Practical and Powerful Approach to Multiple Testing. Journal of the Royal Statistical Society Series B-Methodological 57: 289–300.

23. Eckhart L, Dalla Valle L, Jaeger K, Ballaun C, Szabo S, et al. (2008) Identification of reptilian genes encoding hair keratin-like proteins suggests a new scenario for the evolutionary origin of hair. Proceedings of the National Academy of Sciences of the United States of America 105: 18419–18423.

24. Steinert PM (1993) Structure, function, and dynamics of keratin intermediate filaments. J Invest Dermatol 100: 729–734.

25. Torma H (2011) Regulation of keratin expression by retinoids. Dermatoendocrinol 3: 136–140.

26. Golder HM, Geier MS, Forder REA, Hynd PI, Hughes RJ (2011) Effects of necrotic enteritis challenge on intestinal micro-architecture and mucin profile. British Poultry Science 52: 500–506.

27. Van Patter HT, Whittick JW (1955) Heterotopic ossification in intestinal neoplasms. The American journal of pathology 31: 73.

28. Pennisi E (2004) Developmental biology. Bonemaking protein shapes beaks of Darwin's finches. Science 305: 1383.

29. Abzhanov A, Protas M, Grant BR, Grant PR, Tabin CJ (2004) Bmp4 and morphological variation of beaks in Darwin's finches. Science 305: 1462–1465.

30. Neves H, Dupin E, Parreira L, Le Douarin NM (2012) Modulation of Bmp4 signalling in the epithelial-mesenchymal interactions that take place in early thymus and parathyroid development in avian embryos. Developmental Biology 361: 208–219.

31. D'Alba L, Saranathan V, Clarke JA, Vinther JA, Prum RO, et al. (2011) Colour-producing beta-keratin nanofibres in blue penguin (Eudyptula minor) feathers. Biology Letters 7: 543–546.

32. Frixione E (2000) Recurring views on the structure and function of the cytoskeleton: A 300-year epic. Cell Motility and the Cytoskeleton 46: 73–94.

33. Lawrence JW, Li Y, Chen SY, DeLuca JG, Berger JP, et al. (2001) Differential gene regulation in human versus rodent hepatocytes by peroxisome proliferator-activated receptor (PPAR) alpha -PPAR alpha fails to induce peroxisome proliferation-associated genes in human cells independently of the level of receptor expression. Journal of Biological Chemistry 276: 31521–31527.

34. Song Y, Hui JN, Fu KK, Richman JM (2004) Control of retinoic acid synthesis and FGF expression in the nasal pit is required to pattern the craniofacial skeleton. Developmental Biology 276: 313–329.

35. Choi S, Park S, Kim S, Lim C, Kim J, et al. (2012) Recombinant fusion protein of albumin-retinol binding protein inactivates stellate cells. Biochemical and Biophysical Research Communications 418: 191–197.

36. Plant MR, MacDonald ME, Grad LI, Ritchie SJ, Richman JM (2000) Locally released retinoic acid repatterns the first branchial arch cartilages in vivo. Developmental Biology 222: 12–26.

PERMISSIONS

LIST OF CONTRIBUTORS

Philipp Olias, Anne Meyer and Achim D. Gruber
Institute of Veterinary Pathology, Freie Universität Berlin, Berlin, Germany

Iris Adam and Constance Scharff
Institute of Biology, Department of Animal Behavior, Freie Universität Berlin, Berlin, Germany

Na Wang
State Key Laboratory of Pollution Control and Resource Reuse, School of the Environment, Nanjing University, Nanjing, 210093, China
Nanjing Institute of Environmental Science, Ministry of Environmental Protection of China, Nanjing, 210042, China

Shixiang Gao
State Key Laboratory of Pollution Control and Resource Reuse, School of the Environment, Nanjing University, Nanjing, 210093, China

Shaojun Jiao
Nanjing Institute of Environmental Science, Ministry of Environmental Protection of China, Nanjing, 210042, China

Xiaohong Yang, Jun Zhang and Boping Ye
3 School of Life Science and Technology, China Pharmaceutical University, Nanjing, 210009, China

Wei Zhang, Jing Wang and Peilong Wang
Key Laboratory of Agrifood Safety and Quality, Ministry of Agriculture, Beijing, P.R. China
Institute of Quality Standards and Testing Technology for Agriculture Products, China Agricultural Academy of Science, Beijing, P.R. China

Jing Jia and Ruohua Zhu
Department of Chemistry, Capital Normal University, Beijing, China

Yvette A. Halley, Eric Bhattarai, Ian R. Tizard, Donald J. Brightsmith and Christopher M. Seabury
Department of Veterinary Pathobiology, College of Veterinary Medicine, Texas A&M University, College Station, Texas, United States of America

Scot E. Dowd
Molecular Research LP, Shallowater, Texas, United States of America

Jeremy F. Taylor and Jared E. Decker
Division of Animal Sciences, University of Missouri, Columbia, Missouri, United States of America

Paul M. Seabury
ElanTech Inc., Greenbelt, Maryland, United States of America

Charles D. Johnson
Genomics and Bioinformatics Core, Texas A&M AgriLife Research, College Station, Texas, United States of America

Dale Rollins
Rolling Plains Quail Research Ranch, Rotan, Texas, United States of America

Markus J. Peterson
Department of Wildlife and Fisheries Sciences, Texas A&M University, College Station, Texas, United States of America

Elske N. de Haas, J. Elizabeth Bolhuis and Bas Kemp
Adaptation Physiology Group, Department of Animal Science, Wageningen University and Research, Wageningen, The Netherlands

Ton G. G. Groothuis
Behavioural Biology, Centre for Behaviour and Neuroscience, University of Groningen, Groningen, The Netherlands

T. Bas Rodenburg
Behavioural Ecology Group, Department of Animal Sciences, Wageningen University and Research, Wageningen, The Netherlands

Domingo Miranzo-Navarro and Katharine E. Magor
Department of Biological Sciences and the Li Ka Shing Institute of Virology, University of Alberta, Edmonton, Alberta, Canada

Hanne Sørup Tastesen, Alexander Krokedal Rønnevik and Lise Madsen
Department of Biology, University of Copenhagen, Copenhagen, Denmark National Institute of Nutrition and Seafood Research, Bergen, Norway

Karsten Kristiansen and Kamil Borkowski
Department of Biology, University of Copenhagen, Copenhagen, Denmark

Bjørn Liaset
National Institute of Nutrition and Seafood Research, Bergen, Norway

Erick Contreras-Vallejos, Elías Utreras, Daniel A. Bórquez and Christian Gonzá lez-Billault
Laboratory of Cellular and Neuronal Dynamics, Department of Biology, Faculty of Sciences, Universidad de Chile, Santiago, Chile

Michaela Prochazkova, Anita Terse and Ashok B. Kulkarni
Functional Genomics Section, National Institute of Dental and Craniofacial Research, National Institutes of Health, Bethesda MD, USA

Howard Jaffe
Protein and Peptide Facility, National Institute of Neurological Disorders and Stroke, National Institutes of Health, Bethesda MD, USA

Andrea Toledo and Cristina Arruti
Laboratorio de Cultivo de Tejidos, Sección Biología Celular, Departamento de Biología Celulary Molecular, Facultad de Ciencias, Universidad de la República, Montevideo, Uruguay

Harish C. Pant
Laboratory of Neurochemistry, National Institute of Neurological Disorders and
Stroke, National Institutes of Health, Bethesda MD, USA

Gudrun S. Freidl, Janko van Beek and Marion P. G. Koopmans
Department of Viroscience, Erasmus Medical Center, Rotterdam, The Netherlands
Emerging Infectious Diseases, Division of Virology, Centre for Infectious Diseases
Research, Diagnostics and Perinatal Screening, Centre for Infectious Disease Control, National Institute for Public Health and the Environment (RIVM), Bilthoven, The Netherlands

Henk-Jan van den Ham
Department of Viroscience, Erasmus Medical Center, Rotterdam, The Netherlands

Erwin de Bruin and Johan Reimerink
Emerging Infectious Diseases, Division of Virology, Centre for Infectious Diseases
Research, Diagnostics and Perinatal Screening, Centre for Infectious Disease Control, National Institute for Public Health and the Environment (RIVM), Bilthoven, The Netherlands

Sjaak de Wit
GD Animal Health Service (AHS), Deventer, The Netherlands

Guus Koch
Central Veterinary Institute (CVI), Lelystad, The Netherlands

Lonneke Vervelde
Utrecht University, Utrecht, The Netherlands

Hendra Wibawa
Commonwealth Scientific and Industrial Research Organisation (CSIRO), Australian Animal Health Laboratory, Geelong, Australia
School of Veterinary Science, The University of Queensland, Brisbane, Australia
Disease Investigation Centre Regional IV Wates, Yogyakarta, Indonesia

Harimurti Nuradji
Commonwealth Scientific and Industrial Research Organisation (CSIRO), Australian Animal Health Laboratory, Geelong, Australia
School of Veterinary Science, The University of Queensland, Brisbane, Australia
Indonesian Research Center for Veterinary Science, Bogor, West Java, Indonesia

John Bingham, Sue Lowther, Jean Payne, Jenni Harper and Deborah Middleton
Commonwealth Scientific and Industrial Research Organisation (CSIRO), Australian Animal Health Laboratory, Geelong, Australia

Joanne Meers
School of Veterinary Science, The University of Queensland, Brisbane, Australia

Akhmad Junaidi
Disease Investigation Centre Regional IV Wates, Yogyakarta, Indonesia

Yanfang Li, Sujuan Chen, Xiaojian Zhang, Qiang Fu, Zhiye Zhang, Shaohua Shi, Yinbiao Zhu, Min Gu, Daxin Peng and Xiufan Liu
College of Veterinary Medicine, Yangzhou University, Jiangsu Co-Innovation Center for the Prevention and Control of Important Animal Infectious Disease and Zoonoses, Yangzhou University, Yangzhou, Jiangsu, P.R. China

Findley R. Finseth and Richard G. Harrison
Department of Ecology and Evolutionary Biology, Cornell University, Ithaca, New York, United States of America

Pavel Uvarov and Matti S. Airaksinen
Institute of Biomedicine, Anatomy, University of Helsinki, Helsinki, Finland

Tommi Kajander
Institute of Biotechnology, University of Helsinki, Helsinki, Finland

Muhammad A. Shahid, Marc S. Marenda, Rebecca Agnew-Crumpton and Amir H. Noormohammadi
Faculty of Veterinary Science, The University of Melbourne, Werribee, Victoria, Australia

Philip F. Markham
Asia-Pacific Centre for Animal Health, Faculty of Veterinary Science, The University of Melbourne, Victoria, Australia

Thomas Loyau, Sonia Métayer-Coustard, Cécile Berri, Sabine Crochet, Estelle Cailleau-Audouin, Mélanie Sannier, Pascal Chartrin, Christophe Praud, Christelle Hennequet-Antier, Nicole Rideau, Nathalie Couroussé, Sandrine Mignon-Grasteau, Michel Jacques Duclos, Sophie Tesseraud and Anne Collin
INRA, UR83 Recherches Avicoles, Nouzilly, France

Nadia Everaert
KU Leuven, Department of Biosystems, Leuven, Belgium
University of Liége, Gembloux Agro-Bio Tech, Animal Science Unit, Gembloux, Belgium

Shlomo Yahav
Institute of Animal Science, The Volcani Center, Bet Dagan, Israel

Colin D. Rehm
Center for Public Health Nutrition, University of Washington, Seattle, Washington, United States of America

Adam Drewnowski
Center for Public Health Nutrition, University of Washington, Seattle, Washington, United States of America
Institute for Cardiometabolism and Nutrition, Groupe Hospitalier Pitié-Salpê triére, Paris, France

Dulce M. Bustamante
Department of Biology, San Carlos University, Guatemala City, Guatemala

Sandra M. De Urioste-Stone
School of Forest Resources, University of Maine, Orono, Maine, United States of America

José G. Juárez and Pamela M. Pennington
Health Studies Center, Universidad del Valle de Guatemala, Guatemala City, Guatemala

Hao Bai, Jing Zhu, Yanyan Sun, Ranran Liu, Nian Liu, Dongli Li, Jie Wen and Jilan Chen
Key Laboratory of Genetics Resources and Utilization of Livestock, Institute of Animal Science, Chinese Academy of Agricultural Sciences, Beijing, China

Index

A

Adipose Tissue, 11, 76, 78, 81, 83, 86

Anaerobic, 181

Anas Platyrhynchos, 1, 3, 132

Antibiotic Resistance Genes, 13, 22

Antibiotic-resistant Bacteria, 13, 22

Avian Influenza A Viruses, 112, 132

Avian Influenza Viruses, 120-122, 132-133, 143-144

Avian Species, 1-3, 5, 8, 10-11, 37, 40-41, 44-45, 47, 62, 64, 72, 189

B

Bacillus, 13, 15, 20-21, 23, 177

Bacteria, 11, 13-17, 20, 22, 169

Bacterial Strains, 14

Bobwhite, 34-50

C

Calorimetry, 76-78, 80, 82, 84-86

Carbohydrate, 76-78, 86

Casein, 76-83, 85-86

Chicken, 1-5, 7-8, 10-17, 20, 24-26, 30-37, 40-50, 52-54, 58, 62, 65, 67, 69, 71-83, 85, 90, 92, 94, 98-99, 101, 103-104, 106-108, 110, 113-114, 117-121, 123-125, 127-128, 133-135, 141, 143-144, 146-154, 156-160, 162-165, 167, 169, 173-176, 178-179, 184, 186, 188, 190-191, 193-199, 205-206, 209-210, 214-215, 217, 221-222

Chicken Tissue, 24, 26, 32

Cockatiel, 1, 3, 7-8, 10

Colinus Virginianus, 34, 37-39, 42, 45, 48-50

Corticosterone, 52-53, 55-57, 62, 189

Crane, 1, 3, 7-8

Cytoskeletal Dynamics, 88

D

De Novo Genome, 34-36, 41, 44-45, 49

E

Eagle, 1, 3, 7

Environmental Pollutants, 13

Escherichia Coli, 20, 22-23

G

Gallus Gallus Domesticus, 1, 3, 63

Gene, 1-4, 7-15, 17-20, 22, 34, 36, 40-43, 46-47, 49-51, 58, 62-64, 73-75, 79-80, 82, 89-92, 94, 98-99, 112, 124, 133-136, 138, 141, 143-145, 147-162, 164-167, 169, 171, 176-180, 182, 186, 188-190, 210, 213-222

Genomic Dna, 13-15, 17-20, 45, 158, 160, 163, 170-171

Glucose, 76-80, 82-83, 85-86, 180, 188-189, 204

H

Haliaeetus Albicilla, 1, 3

Hemagglutinin, 112, 119-121, 124, 143-144, 177

Heterozygous Polymorphisms, 34, 37

Hippocampal Neurons, 51, 88, 90, 92, 94-95

Humboldt Penguin, 1, 3, 7

I

Incubation, 12, 14, 20, 53, 113, 124, 136, 169-170, 178-190, 204, 222

Indonesian H5n1, 122-123, 131-132

Influenza A, 64, 67, 69, 73, 75, 112, 120-121, 124, 129, 132-133, 143-144

L

Limits Of Detection, 24

Liquid Chromatography, 24, 33

Livestock Farms, 13

M

Magnetic Nanoparticles, 24-27, 33

Mallard, 1, 3, 7-8, 114, 122-123, 129, 131, 133-136, 139-143

Manures, 13

Mass Spectrometry, 24, 33, 64, 69, 72, 74-75, 88-89

Meleagris Gallopavo F. Domestica, 1, 3

Microbiopsies, 1, 3, 5, 8, 10

Mining, 101-102, 104-106, 108-111

Mitochondrial Antiviral Signaling, 64

Mutagenesis, 64, 73, 75, 134, 169

N

Nanoparticles, 24-28, 32-33

Neural Cells, 88, 94, 98

Neuraminidase Stalk, 133, 143-144

Nymphicus Hollandicus, 1, 3

O

Oropharynx, 122, 140

P

Parental Stock, 52

Phenylarsonic Acids, 24-25, 27-30, 32-33

Phosphoproteomics, 88, 92, 94, 98

Phylogenetic Tree, 1, 10, 159-160, 162

Plasmids, 13-14, 17, 20, 22, 65, 71, 73, 134, 138

Pleistocene, 34, 38, 49

Poultry, 13-14, 20, 24-25, 49-50, 52-53, 62-63, 112-113, 120-122, 133, 136-137, 140, 143-144, 167, 176-178, 213, 222

Protein Phosphorylation, 88-89, 99

Pseudomonas, 13, 15, 20-21, 23

Q

Quantitative Real-time Pcr, 1, 11, 135, 169, 214

R

Respiration, 190

Retinoic Acid Inducible Gene I, 64

Rna, 1, 5, 11, 15, 22, 37, 40-41, 64-65, 67, 74-75, 79, 91, 112, 123-127, 129-131, 133, 135, 141, 143-155, 161-162, 165-166, 179-184, 189-190, 214, 216-217, 220

Roxarsone, 33

S

Scarlet Macaw, 34-50

Severe Feather Pecking, 52, 55, 57-59, 63

Shigella, 13, 15, 20-21, 23

Smelting, 101-102, 108, 110-111

Soil, 13-17, 20, 22, 101-111

Soil Pollution, 13, 104

Species, 1-3, 5, 7-11, 14-15, 20-21, 25, 28, 32, 34-35, 37, 40-41, 43-45, 47, 49-50, 52, 62, 64-65, 72, 74-75, 95, 102-103, 112, 117, 122-123, 130, 132-134, 145-146, 150, 153-154, 157-160, 162, 164-165, 168, 175, 179, 189, 201, 205-206, 210-213, 218

Sphenicus Humboldti, 1, 3

Sul-positive Genera, 13

Sulfonamide, 13-15, 20, 22-23

T

Tryptophan, 2, 78

Turkey, 1, 3, 7, 11, 35-37, 40, 46-47, 49-50, 114, 121, 168

Tyrosine, 2, 50, 88, 98, 219

V

Veterinary Antibiotics, 13-14, 22

www.ingramcontent.com/pod-product-compliance
Lightning Source LLC
Chambersburg PA
CBHW061244190326
41458CB00011B/3576